Post-translational Modification of Protein Biopharmaceuticals

Edited by
Gary Walsh

Related Titles

Lutz, S., Bornscheuer, U. T. (eds.)

Protein Engineering Handbook

2008

ISBN: 978-3-527-31850-6

Behme, S. (ed.)

Manufacturing of Pharmaceutical Proteins

From Technology to Economy

2009

ISBN: 978-3-527-32444-6

Dübel, S. (ed.)

Handbook of Therapeutic Antibodies

2007

ISBN: 978-3-527-31453-9

Desiderio, D. M., Nibbering, N. M.

Redox Proteomics

From Protein Modifications to Cellular Dysfunction and Diseases

2006

ISBN: 978-0-471-72345-5

Knäblein, J. (ed.)

Modern Biopharmaceuticals

Recent Success Stories

2009

ISBN: 978-3-527-32283-1

Gruber, A. C.

Biotech Funding Trends

Insights from Entrepreneurs and Investor

2008

ISBN: 978-3-527-32435-4

Borbye, L. (ed.)

Industry Immersion Learning

Real-Life Industry Case-Studies in Biotechnology and Business

2009

ISBN: 978-3-527-32408-8

Post-translational Modification of Protein Biopharmaceuticals

Edited by
Gary Walsh

WILEY-BLACKWELL

WILEY-VCH Verlag GmbH & Co. KGaA

The Editor

Prof. Dr. Gary Walsh
Industrial Biochemistry Program
CES Department
University of Limerick
Castletroy, Limerick City
Ireland

Library of Congress Card No.: applied for

British Library Cataloguing-in-Publication Data
A catalogue record for this book is available from the British Library.

Bibliographic information published by the Deutsche Nationalbibliothek
The Deutsche Nationalbibliothek lists this publication in the Deutsche Nationalbibliografie; detailed bibliographic data are available on the Internet at http://dnb.d-nb.de.

Composition Thomson Digital, Noida, India
Printing Strauss GmbH, Mörlenbach
Bookbinding Litges & Dopf GmbH, Heppenheim
Cover Design Adam Design, Weinheim

Printed in the Federal Republic of Germany
Printed on acid-free paper

ISBN: 978-3-527-32074-5

Contents

Preface *XV*
List of Contributors *XVII*

1 **Post-Translational Modifications in the Context of Therapeutic Proteins: An Introductory Overview** *1*
Gary Walsh
1.1 Introduction *1*
1.2 Biopharmaceuticals and the Biopharmaceutical Sector *1*
1.3 Protein Post-Translational Modification *2*
1.4 PTMs in the Context of Biopharmaceuticals *7*
1.5 Some Specific PTMs *8*
1.5.1 Glycosylation *8*
1.5.2 Disulfide Bond Formation and Proteolytic Cleavage *10*
1.5.3 γ-Carboxylation and β-Hydroxylation *11*
1.5.4 Amidation and Sulfation *12*
1.6 Extending and Engineering PTM Profiles *12*
1.7 Conclusion *13*
References *14*

Part One Glycosylation *15*

2 **Protein Glycosylation: The Basic Science** *17*
Susan A. Brooks
2.1 Introduction – Glycosylated Proteins *17*
2.2 Basic Building Blocks of Glycosylation in Human Cells *18*
2.3 Formation of Complex Glycan Structures *19*
2.3.1 α and β Glycosidic Bonds *19*
2.3.2 Structural Complexity of Glycoprotein Glycans *19*
2.4 Glycan Synthesis is Catalyzed by Enzymes of Glycosylation – the "Glycozymes" *22*

Post-translational Modification of Protein Biopharmaceuticals. Edited by Gary Walsh

ISBN: 978-3-527-32074-5

2.5 Protein Glycosylation – Relationship Between N-linked and O-linked Glycoproteins 23
2.6 N-linked Glycoproteins 24
2.6.1 Where Does N-linked Glycosylation of Proteins Take Place? 24
2.6.2 Why is it Called N-linked? 24
2.6.3 *N*-Glycosylation Step-by-Step 24
2.6.3.1 Polypeptide Enters the RER 24
2.6.3.2 Building and Positioning the Dolichol Oligosaccharide Precursor 24
2.6.3.3 Attachment of the Dolichol Oligosaccharide Precursor to the Polypeptide Chain 25
2.6.3.4 Trimming the Dolichol Oligosaccharide Precursor in the RER 25
2.6.3.5 Processing of N-linked Oligosaccharides in the Golgi Apparatus 25
2.6.4 Three Classes of N-linked Glycoproteins 25
2.6.5 What Determines Whether a Potential Site of *N*-Glycosylation on the Polypeptide Chain is Occupied or Not? 27
2.7 O-linked Glycoproteins 27
2.7.1 Why is it Called O-linked? 28
2.7.2 Different Types of O-linked Glycosylation 28
2.7.3 Overview of O-linked Mucin-Type Protein Glycosylation 28
2.7.4 No Consensus Sequence for Mucin-Type O-linked Glycoprotein Glycosylation 29
2.7.5 Synthesis of O-linked Mucin-Type Glycan Core Structures Step-by-Step 29
2.7.5.1 Initiation – Synthesis of GalNAcα1→ Ser/Thr (Tn Epitope) 29
2.7.5.2 Synthesis of NeuNAc(α2→6)GalNAcα1→Ser/Thr (Sialyl Tn) 29
2.7.5.3 Synthesis of Core 1, Gal(β1→3)GalNAcα-Ser/Thr, or the Thomsen–Friedenreich (T or TF) Antigen 29
2.7.5.4 Synthesis of Core 2, GlcNAc(β1→6)[Galβ(1→3)]GalNAcα1-Ser/Thr 31
2.7.5.5 Synthesis of Core 3, GlcNAc(β1→3)GalNAcα1→Ser/Thr 31
2.7.5.6 Synthesis of Core 4, GlcNAc(β1→ 6)[GlcNAc(β1→ 3)]GalNAcα1-Ser/Thr 31
2.7.5.7 Less Common Core Structures 31
2.7.6 Regulation of *O*-Glycan Synthesis 32
2.8 O- and N-linked Glycan Chain Extension and Commonly Occurring Glycan Motifs 32
2.8.1 Type I Chains or Neolactosamine Units, Gal(β1→3)GlcNAc/GalNAc 32
2.8.2 Type 2 Chains or Lactosamine Units, Gal(β1→4)GlcNAc 33
2.8.3 Termination of Glycan Chains 33
2.8.4 Blood Group Antigens 33
2.9 Analytical Methodologies Developed to Detect and Characterize Glycosylation 33
2.9.1 Glycan Analysis is Complex and Requires a Number of Techniques 33

2.9.2 Detection of Glycans 34
2.9.2.1 Periodic Acid-Schiff (PAS) Reaction 34
2.9.2.2 Recognition Through Lectin Binding 34
2.9.2.3 Pulsed Amperometric Detection (PAD) of Unlabelled Free Glycans 35
2.9.2.4 Labeling Glycans with Radiolabels or Fluorescent Labels 35
2.9.3 Profiling Glycans Using Microarrays 35
2.9.4 One- and Two-Dimensional Gel Electrophoresis – Exploring the Glycome 36
2.9.5 Chemical Release and Analysis of Monosaccharides 36
2.9.6 Chemical Release of Intact Oligosaccharides 37
2.9.7 Enzymatic Release of Intact Oligosaccharides 37
2.9.8 Sequential Exoglycosidase Digestions to Provide Monosaccharide Sequence and Linkage Data 37
2.9.9 Oligosaccharide Separation and Mapping by High-Performance Liquid Chromatography (HPLC) 38
2.9.9.1 Normal Phase HPLC (NP-HPLC) 38
2.9.9.2 Weak Anion Exchange High-Performance Liquid Chromatography (WAX-HPLC) 38
2.9.9.3 High-Performance (or High pH) Anion Exchange Chromatography (HPAEC) 39
2.9.10 Separation and Mapping of Oligosaccharides by Fluorophore-Assisted Carbohydrate Electrophoresis (FACE) 39
2.9.11 Separation and Mapping of Oligosaccharides by Capillary Electrophoresis (CE) 39
2.9.12 Oligosaccharide Analysis by Nuclear Magnetic Resonance (NMR) 40
2.9.13 Determining the Mass of an Oligosaccharide Using Mass Spectrometry (MS) 40
2.9.14 Gas–Liquid Chromatography (GLC)/MS for Determining the Linkage Position of Monosaccharides in Oligosaccharides 41
2.10 Conclusion 42
References 42

3 Mammalian Cell Lines and Glycosylation: A Case Study *51*
Michael Butler
3.1 Introduction 51
3.2 The Choice of Cell Line for Glycoprotein Production 51
3.3 Effect of Growth and Protein Production Rate on Glycosylation 54
3.4 Enzymes Associated with Glycan Heterogeneity 55
3.4.1 *N*-Acetyl Glucosaminyltransferases 55
3.4.2 Fucosylation 57
3.4.3 Sialylation 58
3.5 Immunogenicity of Non-Human Glycans 61
3.6 Culture Parameters that may Affect Glycosylation 61
3.6.1 Nutrient Depletion 61

3.6.2 Fed-Batch Cultures and Supplements *63*
3.6.2.1 Glucosamine as a Supplement *64*
3.6.2.2 Galactose as a Supplement *65*
3.6.3 Ammonia *65*
3.6.4 pH *66*
3.6.5 Oxygen *66*
3.7 Functional Glycomics *67*
3.8 Conclusion *68*
References *69*

4 Antibody Glycosylation *79*
Roy Jefferis
4.1 Introduction *79*
4.2 Antibodies *80*
4.2.1 Basic Structure/Function *80*
4.2.2 Antibody (Immunoglobulin) Isotypes *84*
4.3 Glycosylation *84*
4.3.1 Glycosylation of Normal Human IgG *84*
4.3.2 Impact of Glycosylation on Structure *86*
4.3.3 Impact of Glycosylation on Stability *87*
4.4 IgG-Fc Effector Functions *88*
4.4.1 Inflammatory Cascades *88*
4.4.2 Catabolism, Pharmacokinetics and Placental Transport *90*
4.5 Individual IgG-Fc Glycoforms *91*
4.5.1 Individual IgG-Fc Glycoforms and Effector Activities *91*
4.5.2 Sialylation of IgG-Fc Oligosaccharides *92*
4.5.3 Influence of Galactosylation on IgG-Fc Activities *92*
4.5.4 Influence of Fucose and Bisecting *N*-Acetylglucosamine on IgG-Fc Activities *94*
4.6 IgG-Fab Glycosylation *96*
4.7 Recombinant Monoclonal Antibodies for Therapy *98*
4.8 Conclusions and Future Perspectives *100*
References *100*

5 Gonadotropins and the Importance of Glycosylation *109*
Alfredo Ulloa-Aguirre, James A. Dias, and George R. Bousfield
5.1 Introduction *109*
5.2 Structure of Gonadotropins *111*
5.3 Glycosylation of Gonadotropins and Structural Microheterogeneity *114*
5.4 Role of Glycosylation in the Function of Gonadotropins *121*
5.4.1 Role in Folding, Subunit Assembly and Secretion *121*
5.4.2 Metabolic Clearance Rate *122*
5.4.3 Binding and Signal Transduction *123*
5.4.4 LH and FSH Glycoforms and Gonadotropin Function *124*

5.4.5 Chorionic Gonadotropin Glycoforms and Function *126*
5.5 Regulation of Gonadotropin Glycosylation *127*
5.5.1 Effects of Estrogens *127*
5.5.2 Effects of Androgens *128*
5.5.3 Effects of Gonadotropin-Releasing Hormone *129*
5.6 Therapeutic Applications of Gonadotropins *130*
5.7 Conclusions *132*
References *133*

6 Yeast Glycosylation and Engineering in the Context of Therapeutic Proteins *149*
Terrance A. Stadheim and Natarajan Sethuraman
6.1 Introduction *149*
6.2 *N*-Glycosylation in Fungi *150*
6.3 *O*-Glycosylation in Fungi and Mammals *152*
6.4 Remodeling Yeast Glycosylation for Therapeutic Protein Production *154*
References *160*

7 Insect Cell Glycosylation Patterns in the Context of Biopharmaceuticals *165*
Christoph Geisler and Don Jarvis
7.1 Introduction *165*
7.2 Recombinant *N*-Glycoproteins in the BEVS Product Pipeline *166*
7.2.1 Chimigen™ Vaccines *167*
7.2.2 FluBlok™ *167*
7.2.3 Influenza Virus-Like Particles *167*
7.2.4 Provenge® *167*
7.2.5 Specifid™ *168*
7.3 Insect Glycoprotein *N*-Glycan Structure *168*
7.3.1 Typical *N*-Glycan Structures *168*
7.3.2 Hybrid/Complex *N*-Glycans *170*
7.3.3 Sialylated *N*-Glycans *170*
7.3.4 Summary *170*
7.4 *N*-Glycan Processing Enzymes in the BEVS *171*
7.4.1 Processing β-*N*-Acetylglucosaminidase *171*
7.4.2 Core α1,3 Fucosyltransferase *172*
7.4.3 Lack of Mannose-6-Phosphate *172*
7.5 Lack of Glycosyltransferases *173*
7.5.1 *N*-Acetylglucosaminyltransferases II–IV *173*
7.5.2 *N*-acetylgalactosaminyl-Galactosyl- and Sialyltransferase *173*
7.5.3 Glycosyltransferase Donor Substrates *174*
7.6 Use of Baculoviruses to Extend BEVS *N*-Glycosylation *174*
7.6.1 Promoter Choice *174*
7.6.2 Baculovirus Encoded Glycosyltransferases *175*

7.6.2.1 *N*-Acetylglucosaminyltransferase I *175*
7.6.2.2 Galactosyltransferase *175*
7.6.2.3 Sialyltransferase *175*
7.6.2.4 *N*-acetylglucosaminyltransferase II *176*
7.6.2.5 Trans-Sialidase *176*
7.6.3 Baculoviruses Encoded Sugar Processing Genes *176*
7.6.3.1 UDP-GlcNAc 2-Epimerase/*N*-Acetylmannosamine Kinase *176*
7.6.3.2 Sialic Acid Synthetase *177*
7.6.3.3 CMP-Sialic Acid Synthetase *177*
7.6.4 Summary *177*
7.7 Transgenic Insect Cell Lines *178*
7.7.1 Proof of Concept *178*
7.7.2 Transgenic Glycosyltransferases *179*
7.7.2.1 Galactosyltransferase *179*
7.7.2.2 Sialyltransferase *179*
7.7.2.3 *N*-Acetylglucosaminyltransferase II *180*
7.7.3 Transgenic Sugar Processing Genes *180*
7.7.4 Use of Transposon-Based Systems *181*
7.7.5 Summary *181*
7.8 Sugar Supplementation *182*
7.9 Future Directions *182*
7.9.1 Completing the *N*-Glycosylation Pathway *183*
7.9.2 Reducing Deleterious Activities *183*
References *184*

8 Getting Bacteria to Glycosylate *193*
Michael Kowarik and Mario F. Feldman
8.1 Introduction *193*
8.1.1 Overview and Background *193*
8.1.2 Bacterial Protein Glycosylation *194*
8.2 *N*-Glycosylation *194*
8.2.1 Introduction *194*
8.2.2 The Acceptor Protein *195*
8.2.2.1 Primary Acceptor Consensus *196*
8.2.2.2 Conformational Requirements for *N*-Glycosylation *196*
8.2.2.3 Crystal Structures of Bacterial *N*-Glycoproteins *197*
8.2.3 The LLO Substrate and the N-OTase, PglB *197*
8.3 *O*-Glycosylation *199*
8.3.1 *O*-Glycosylation in *Pseudomonas aeruginosa* *200*
8.3.1.1 Introduction *200*
8.3.1.2 The Acceptor Protein for *O*-Glycosylation *200*
8.3.1.3 Glycan Structures in *P. aeruginosa* *O*-Glycosylation *201*
8.3.2 *O*-Glycosylation in *Neisseria* *201*
8.3.2.1 Introduction *201*
8.3.2.2 PglL, the *O*-OTase of *N. meningitidis* *201*

8.3.2.3 Glycan Substrates for *N. meningitidis O*-Glycosylation *202*
8.4 Exploitation of N- and O-Linked Glycosylation *203*
8.4.1 Therapeutic (Human) Proteins *203*
8.4.2 Bioconjugate Vaccines *204*
8.4.3 Glycoengineering *205*
References *206*

Part Two Other Modifications *209*

9 Biopharmaceuticals: Post-Translational Modification Carboxylation and Hydroxylation *211*
Mark A. Brown and Leisa M. Stenberg
9.1 Introduction *211*
9.2 γ-Carboxylation *211*
9.2.1 Biological Function of γ-Carboxylation *211*
9.2.2 The Gla Domain *214*
9.2.3 Biosynthesis of Gla *217*
9.2.4 γ-Carboxylated Biopharmaceuticals *219*
9.2.4.1 Factor IX *219*
9.2.4.2 Factor VIIa *221*
9.2.4.3 Protein C/Activated Protein C *222*
9.2.4.4 Prothrombin *223*
9.2.4.5 Conotoxins *224*
9.2.5 Enhancement of Cellular Carboxylation Capacity *224*
9.2.5.1 Enhanced Expression of the γ-Carboxylation Machinery *225*
9.2.5.2 Inhibition of Calumenin Expression *226*
9.2.5.3 Propeptide/Propeptidase Engineering *226*
9.2.6 Purification of γ-Carboxylated Proteins *226*
9.2.7 Analytical Characterization of γ-Carboxylated Proteins *227*
9.2.7.1 Methods for Detecting Gla *227*
9.2.7.2 Metal Content *227*
9.2.7.3 Metal Binding-Induced Structural Changes *228*
9.2.7.4 Phospholipid Membrane Binding Assays *228*
9.2.7.5 γ-Carboxylase Enzyme Assays *228*
9.3 Post-Translational Hydroxylation *229*
9.3.1 Biological Function of Hydroxylation *229*
9.3.2 Biosynthesis of Hydroxylated Amino Acids *231*
9.3.2.1 Biosynthesis of Hya/Hyn *231*
9.3.2.2 Biosynthesis of Hydroxyproline *232*
9.3.3 Hydroxylated Biopharmaceuticals *233*
9.3.3.1 Factor IX *233*
9.3.3.2 Protein C/Activated Protein C *233*
9.3.3.3 Conotoxins *234*
9.3.4 Analytical Characterization of β-Hydroxylated Proteins *235*
9.3.4.1 Methods for Detecting Hya/Hyn *235*

9.3.4.2 β-Hydroxylase Enzyme Assays *235*
9.4 Conclusions *235*
References *236*

10 C-Terminal α-Amidation *253*
Nozer M. Mehta, Sarah E. Carpenter, and Angelo P. Consalvo
10.1 Introduction *253*
10.2 Substrate Specificity of PAM *253*
10.3 Activity of PAM *254*
10.3.1 Assays for Measurement of PAM Activity *254*
10.3.2 Mechanism of Action *255*
10.3.3 Species Distribution of α-Amidated Peptides and PAM *257*
10.4 Genomic Structure and Processing of PAM *257*
10.4.1 Organization of the PAM Gene *257*
10.4.2 Tissue-Specific Forms of PAM *258*
10.5 Structure–Activity Relationships (SAR) for Rat PAM Activity *258*
10.6 α-Amidation of Glycine-Extended Peptides *260*
10.6.1 *In Vitro* α-Amidation *260*
10.6.2 Optimization of the PAM Reaction *In Vitro* *261*
10.7 Cloning and Expression of Various Forms of PAM *263*
10.7.1 PHM, PHMcc and PAL *264*
10.7.2 Bifunctional PAM *264*
10.7.3 Co-expression of PAM with Glycine-Extended Peptides *265*
10.8 A Process for Recombinant Production of α-Amidated Peptides *265*
10.8.1 Expression of Glycine-Extended Peptides in *E. coli* by a Direct Expression Process *266*
10.8.2 Purification of the Glycine-Extended Peptides *266*
10.8.3 Expression of PAM in CHO Cells *267*
10.8.4 Purification of Recombinant PAM (rPAM) *267*
10.8.5 Post-Amidation Purification *268*
10.8.6 Expression Levels of Peptides by Direct Expression Technology *269*
10.9 Marketed Peptides *269*
10.9.1 Marketed α-Amidated Peptides *269*
10.10 Conclusions *271*
References *271*

11 Disulfide Bond Formation *277*
Hayat El Hajjaji and Jean-François Collet
11.1 Introduction *277*
11.2 Disulfide Bonds have a Stabilizing Effect *278*
11.3 Disulfide Bond Formation is a Catalyzed Process *278*
11.4 Disulfide Bond Formation in the Bacterial Periplasm *278*
11.4.1 The Oxidation Pathway: DsbA and DsbB *279*
11.4.1.1 DsbA, a very Oxidizing Protein *279*
11.4.1.2 DsbB *280*

11.4.1.3 DsbB is Reoxidized by the Electron Transport Chain 280
11.4.1.4 Engineering of a New Oxidation Pathway 281
11.4.2 Disulfide Isomerization Pathway 282
11.4.2.1 DsbC, a Periplasmic Protein Disulfide Isomerase 282
11.4.2.2 DsbC is a Dimeric Protein 284
11.4.2.3 DsbC is Kept Reduced by DsbD 284
11.4.2.4 DsbC can Function Independently of DsbD 285
11.4.2.5 DsbG, a Controversial Protein Disulfide Isomerase 285
11.5 Disulfide Bond Formation in the Cytoplasm 286
11.6 Formation of Protein Disulfide Bond in Heterologous Proteins Expressed in *E. coli* 288
11.7 Disulfide Bond Formation in the Endoplasmic Reticulum 288
11.7.1 PDI Functions both as an Oxidase and an Isomerase 288
11.7.2 PDI is Reoxidized by Ero1 289
11.7.3 Regulation of Ero1 290
11.8 Conclusions 290
References 290

Part Three Engineering of PTMS 295

12 Glycoengineering of Erythropoietin 297
Steve Elliott
12.1 Introduction 297
12.2 Endogenous Epo, rHuEpo and their Attached Carbohydrates 298
12.2.1 Rules for Attachment of Carbohydrates 300
12.2.2 Carbohydrate Glycoforms and their Effect on Structure and Activity 302
12.2.3 Glycoengineering of New Molecules – Darbepoetin Alfa 303
12.2.4 Glycoengineering of New Molecules – AMG114 307
12.2.5 Glycoengineered ESAs and Biological Activity 308
12.3 Effect of Carbohydrate on Clearance, Mechanism of Clearance 309
References 311

13 Glycoengineering: Cerezyme as a Case Study 319
Scott M. Van Patten and Tim Edmunds
13.1 Introduction 319
13.2 Basis for Glycan-Directed Enzyme Replacement Therapy for LSDs 319
13.2.1 Lysosomal Storage Diseases 319
13.2.2 Gaucher Disease 320
13.2.3 Glucocerebrosidase 321
13.2.4 Enzyme Replacement as a Therapy for LSDs 322
13.3 Use of Placental Glucocerebrosidase for ERT 323
13.3.1 Initial Clinical Studies with Unmodified GCase Isolated from Placenta 323

13.3.2 Identification of Mannose Receptor and its Role in Macrophage Uptake *324*
13.3.3 Glycoengineering via Sequential Removal of Glycans *324*
13.3.4 Development of First ERT for Gaucher Disease – Ceredase *325*
13.4 Development of a Second-Generation ERT using Recombinant Technology *327*
13.4.1 Production of a CHO Cell-Expressed Recombinant Human GCase – Cerezyme *327*
13.4.2 Biochemical Comparison of Cerezyme to Ceredase *327*
13.4.3 Comparison of Ceredase to Cerezyme *In Vivo* *329*
13.5 Alternative Strategies for Glycoengineered GCase *330*
13.5.1 Overview of Possible Strategies for Targeting Mannose Receptor *330*
13.5.2 Use of Mannosidase Inhibitors *330*
13.5.3 Use of Mutant Cell Lines *333*
13.5.4 Use of Alternative Expression Systems *333*
13.6 Summary *334*
References *334*

14 Engineering in a PTM: PEGylation *341*
Gian Maria Bonora and Francesco Maria Veronese
14.1 Protein PEGylation *341*
14.2 General Properties of PEG *343*
14.3 Chemically Activated PEGs and the Process of PEGylation *344*
14.4 Potential Effects of PEGylation that are of Therapeutic Relevance *347*
14.5 Some Specific PEGylated Biopharmaceuticals *349*
14.6 Conclusions *352*
References *353*

Index *359*

Preface

The majority of approved therapeutic proteins, as well as those currently in development, naturally undergo some form of post-translational modification (PTM). An increasing appreciation of the central importance of such PTMs to the application-relevant properties of these proteins continues to emerge.

This book aims to provide a comprehensive overview of protein post-translational modifications specifically in the context of biopharmaceuticals. Chapter 1 introduces the topic, to provide an appropriate context for the remainder of the book. The subsequent 13 chapters focus upon various specific post-translational modifications. Several chapters are devoted to various aspects of glycosylation, as this PTM is by far the most complex and significant associated with therapeutic proteins. Individual chapter authors are drawn from both academia and industry, and from various global regions.

The book will serve as a reference source for those working or wishing to work in the biopharmaceutical sector. Its scope should also render it a useful reference text for third level students undertaking healthcare-related programs of study (e.g., undergraduate or taught postgraduate programs in pharmacy, pharmaceutical science and biotechnology, as well as in biochemistry). Likewise, it should serve as a useful reference for academic and industry researchers whose research interests relate to biopharmaceuticals.

February 2009

G. Walsh
University of Limerick

Post-translational Modification of Protein Biopharmaceuticals. Edited by Gary Walsh

ISBN: 978-3-527-32074-5

List of Contributors

Gian Maria Bonora
University of Trieste
Department of Chemical Science
Via Giorgieri
34126 Trieste
Italy

George R. Bousfield
Wichita State University
Department of Biological Sciences
1845 Fairmount
Wichita, KS 67260
USA

Susan A. Brooks
Oxford Brookes University
School of Life Sciences
Gipsy Lane
Headington
Oxford OX3 0BP
UK

Mark A. Brown
Marine Biological Laboratory
Woods Hole, MA 02543
USA
and
Center for Hemostasis and Thrombosis Research
Beth Israel Deaconess Medical Center
and Harvard Medical School
Cambridge, MA 02139
USA

Michael Butler
University of Manitoba
Faculty of Science
Department of Microbiology
Winnipeg
Manitoba R3T 2N2
Canada

Sarah E. Carpenter
Unigene Laboratorics, Inc.
110 Little Falls Road
Fairfield, NJ 07004
USA

Jean-François Collet
Université catholique de Louvain
de Duve Institute
75-39 Avenue Hippocrate
1200 Brussels
Belgium

Post-translational Modification of Protein Biopharmaceuticals. Edited by Gary Walsh

ISBN: 978-3-527-32074-5

Angelo P. Consalvo
Unigene Laboratories, Inc.
110 Little Falls Road
Fairfield, NJ 07004
USA

James A. Dias
State University of New York at Albany
New York State Department of Health
and Department of Biomedical Sciences
Wadsworth Center
David Axelrod Institute for
Public Health
120 New Scotland Ave
Albany, NY 12208
USA

Tim Edmunds
Therapeutic Protein Research
Genzyme Corporation
One Mountain Road
Framingham, MA 01701
USA

Steve Elliott
Amgen Inc.
One Amgen Center Drive
Thousand Oaks, CA 91320
USA

Hayat El Hajjaji
Université catholique de Louvain
de Duve Institute
75-39 Avenue Hippocrate
1200 Brussels
Belgium

Mario Feldman
University of Alberta
Department of Biological Sciences
Alberta Ingenuity Centre for
Carbohydrate Sciences
Edmonton T6G 2E9
Canada

Christoph Geisler
University of Wyoming
Department of Molecular Biology
1000. E. University Avenue
Laramie, WY 82071
USA

Don Jarvis
University of Wyoming
Department of Molecular Biology
1000. E. University Avenue
Laramie, WY 82071
USA

Roy Jefferis
University of Birmingham
The Division of Immunity & Infection
Birmingham B15 2TT
UK

Michael Kowarik
GlycoVaxyn AG
Grabenstrasse 3
8952 Schlieren
Switzerland

Nozer M. Mehta
Unigene Laboratories, Inc.
110 Little Falls Road
Fairfield, NJ 07004
USA

Scott M. Van Patten
Therapeutic Protein Research
Genzyme Corporation
One Mountain Road
Framingham, MA 01701
USA

Natarajan Sethuraman
GlycoFi, a wholly-owned subsidiary of Merck and Co.
21 Lafayette Street
Suite 200
Lebanon, NH 03766
USA

Terrance A. Stadheim
GlycoFi, a wholly-owned subsidiary of Merck and Co.
21 Lafayette Street
Suite 200
Lebanon, NH 03766
USA

Leisa M. Stenberg
Marine Biological Laboratory
Woods Hole, MA 02543
USA
and
Center for Hemostasis and Thrombosis Research
Beth Israel Deaconess Medical Center, and Harvard Medical School
Cambridge, MA 02139
USA

Francesco Maria Veronese
University of Padova
Department of Pharmaceutical Sciences
Via F. Marzolo
35131 Padova
Italy

Alfredo Ulloa-Aguirre
Hospital de Gineco-Obstetricia "Luis Castelazo Ayala"
Research Unit in Reproductive Medicine
Instituto Mexicano del Seguro Social
Mexico D.F.
Mexico

Gary Walsh
University of Limerick
Industrial Biochemistry Program
Limerick
Ireland

1 Post-Translational Modifications in the Context of Therapeutic Proteins: An Introductory Overview

Gary Walsh

1.1 Introduction

Many proteins, particularly those derived from eukaryotic sources, undergo covalent modification either during their ribosomal synthesis or (more usually) after synthesis is complete, giving rise to the concept of co-translational and post-translational modification. While the stage during protein synthesis/maturation at which the co- or post-translational modification occurs obviously differs, both are often referred to in the literature simply as post-translational modifications (PTMs), a convention also followed in this chapter. These modifications invariably influence some structural aspect or functional role of the affected protein. The main aims of this first chapter is to provide an introductory overview of such modifications, particularly in the context of therapeutic proteins. As such it serves to contextualize the remaining chapters, each of which focuses in detail upon some specific aspect or type of post-translational modification. The initial portion of this chapter serves to introduce biopharmaceuticals and the biopharmaceutical sector. A brief overview of the more common protein PTMs is next provided. The final section then builds upon these two initial sections and overviews issues concerning the PTM complement of therapeutic proteins.

1.2 Biopharmaceuticals and the Biopharmaceutical Sector

It is estimated that there are in excess of 4000 biotech companies currently in existence. These are based mainly within the United States and Europe, while the Asia pacific area represents the third largest region [1]. A significant majority of these companies focus upon healthcare biotechnology and the vast majority are research intensive, spending an average of 28% of revenues on research and development [1].

The biopharmaceutical sector represents the backbone of the global biotech industry.

Post-translational Modification of Protein Biopharmaceuticals. Edited by Gary Walsh

ISBN: 978-3-527-32074-5

This sector can trace its roots back to the late 1970s when the advent of genetic engineering and monoclonal antibody technology underpinned the establishment of hundreds of start-up biopharmaceutical companies. The term "biopharmaceutical" was coined at that time and described therapeutic products produced by modern biotechnological techniques. During the 1980s this equated exclusively with recombinant proteins and monoclonal antibody-based products. During the1990s the concept of nucleic acid-based drugs for use in gene therapy and antisense technology, as well as the concept of engineered stem and other therapeutic cell lines came to the fore. Such products are also considered biopharmaceuticals but, as outlined below, the vast majority of biopharmaceuticals approved and in clinical trials remain protein-based.

Currently almost 200 biopharmaceutical products have gained approval for general medical use within the EU and/or USA and likely in the region of 350 million people worldwide have been treated with these drugs to date. By the mid-2000s the sector was generating revenues in excess of $30 billion, approximately double its global value at the end of the 1990s [2]. Continued strong growth is expected, with the industry projected to reach $70 billion by the end of the decade [3, 4]. Typically 8–10 new biopharmaceuticals are approved by global regulatory authorities each year, representing approximately 1 in 4 of all genuinely new drugs coming onto the market.

Profiles of biopharmaceutical products approved thus far are available in various publications (e.g., References [5, 6]) and the major sub-categories of biopharmaceuticals include:

- Antibody-based products (used for *in vivo* diagnostic and therapeutic purposes).
- Hormone-based products (e.g., recombinant insulins, glucagon, gonadotropins and human growth hormone).
- Subunit vaccines (containing a recombinant antigenic component, e.g., recombinant hepatitis B subunit vaccines).
- Recombinant interferons, interleukins and other cytokines.
- Haematopoietic growth factors (e.g., erythropoietins and colony stimulating factors).
- Recombinant blood factors (factors VIIa, VIII and IX).
- Recombinant thrombolytics (mainly tissue plasminogen activator based products).
- Enzyme-based products (e.g., recombinant DNase and glucocerebrosidase).

Table 1.1 provides a summary of individual products approved within the last 3 years.

1.3
Protein Post-Translational Modification

Proteins can potentially undergo well in excess of 100 different PTMs. Some such modifications (e.g., glycosylation) are common in their occurrence, while others

Table 1.1 Biopharmaceutical products (recombinant proteins, monoclonal antibody and nucleic acid-based products) approved in the USA or European Union (EU) from January 2005 to December 2007[a].

Product	Company	Therapeutic Indication	Approved
Fortical (r salmon calcitonin produced in *E. coli*)	Upsher-Smith laboratories/Unigene	Postmenopausal osteoporosis	2005 (USA)
GEM 21S (growth factor enhanced matrix; contains rhPDGF-BB produced in *S. cerevisiae*, in addition to tricalcium phosphate)	Luitpold pharmaceuticals, BioMimetic pharmaceuticals	Periodontally related defects	2005 (USA)
Hylenex (rh hyaluronidase produced in CHO cells)	Baxter/Halozyme therapeutics	Adjuvant to increase absorption and dispersion of other drugs	2005 (USA)
Increlex (mecasermin, rh IGF-1 produced in *E. coli*)	Tercica/Baxter	Long-term treatment of growth failure in children with severe primary IGF-1 deficiency or with growth hormone gene deletion	2005 (USA) 2007 (EU)
IPLEX (mecasermin rinfabate, a complex of rh IGF-1 and rh IGFBP-3 produced separately in *E. coli*)	Insmed	Long-term treatment of growth failure in children with severe primary IGF-1 deficiency or with growth hormone gene deletion	2005 (USA)
Naglazyme (galsulfase, rh *N*-acetylgalactosamine 4 sulfatase produced in a CHO cell line)	BioMarin	Long-term enzyme replacement therapy in patients suffering from Mucopolysaccharidosis VI	2005 (USA) 2006 (EU)
Orencia (abatacept, soluble fusion protein produced in a mammalian cell line)	Bristol-Myers Squibb	Rheumatoid arthritis	2005 (USA) 2007 (EU)

(Continued)

Table 1.1 (*Continued*)

Product	Company	Therapeutic Indication	Approved
Atryn (rh antithrombin, from milk of transgenic goats)	Genzyme/Leopharma	Hereditary antithrombin deficiency	2006 (EU)
Elaprase (rh Iduronate-2-sulfatase, produced in a human cell line)	Shire human genetic therapies inc.	Mucopolysaccharidosis II (Hunter's syndrome)	2006 (USA) 2007 (EU)
Exubera (rh Insulin, produced in *E. coli*)	Pfizer	Diabetes	2006 (USA and EU)
Gardasil (human papillomavirus vaccine, type 6,11,16,18, recombinant, produced in *S. cerevisiae*). Also marketed as Silgard in EU	Merck (USA) Sanofi Pasteur (EU) Merck Sharp & Dome (EU under trade name Silgard)	Vaccine against cervical cancer and related conditions caused by HPV	2006 (USA and EU)
Lucentis (ranibizumab, a humanized IgG fragment produced in *E. coli*. Binds and inactivates VEGF-A)	Genentech	Neovascular (wet) age-related macular degeneration	2006 (USA) 2007 (EU)
Myozyme (rh acid-α-glucosidase, produced in a CHO cell line)	Genzyme	Pompe's disease (glycogen storage disease type II)	2006 (EU and USA)
Omnitrope (rhGH, produced in *E. coli*)	Sandoz	GH deficiency/growth failure	2006 (EU and USA)
Preotach (rh parathyroid hormone, produced in *E. coli*)	Nycomed	Osteoporosis	2006 (EU)
Valtropin (rhGH produced in *S. cerevisiae*)	Biopartners	Growth failure/GH deficiency	2006 (EU) 2007 (USA)
Vectibix (panitumumab, a rh Mab that binds to hEGFR, produced in a CHO cell line)	Amgen	EGFR-expressing colorectal carcinoma	2006 (USA) 2007 (EU)
Abseamed (recombinant human erythropoietin alfa, a rhEPO produced in a CHO cell line)	Medice Arzneimittel Putter	Anemia associated with chronic renal failure	2007 (EU)

Binocrit (recombinant human erythropoietin alfa, a rhEPO produced in a CHO cell line)	Sandoz	Anemia associated with chronic renal failure	2007 (EU)
Cervarix (r, C-terminally truncated major capsid L 1 proteins from human papillomavirus types 16 and 18 produced in a baculovirus-based expression system)	GlaxoSmithKline	Prevention of cervical cancer	2007 (EU)
Epoetin alfa Hexal (recombinant human erythropoietin alfa, a rhEPO produced in a CHO cell line)	Hexal Biotech	Anemia associated with chronic renal failure	2007 (EU)
Mircera [methoxy poly(ethylene glycol)-epoetin beta, PEGylated rhEPO produced in a CHO cell line]	Roche	Anemia associated with chronic kidney disease	2007 (EU and USA)
Pergoveris (follitropin alfa/lutropin alfa; combination product containing rhFSH and rhLH, both produced in a CHO cell line)	Serono	Stimulation of follicular development in women with severe LH and FSH deficiency	2007 (EU)
Retacrit (epoetin zeta; a rhEPO produced in a CHO cell line)	Hospira enterprises	Anemia associated with chronic renal failure	2007 (EU)
Silapo (epoetin zeta; a rhEPO produced in a CHO cell line)	Stada Arzneimittel	Anemia associated with chronic renal failure	2007 (EU)
Soliris (eculizumab, a humanized IgG that binds human C5 complement protein, produced in a murine myeloma cell line)	Alexion	Paroxysmal nocturnal hemoglobinuria	2007 (EU and USA)

[a]Abbreviations: r = recombinant, rh = recombinant human, CHO = Chinese hamster ovary, VEGF = vascular endothelial growth factor, IGF-1 = insulin like growth factor 1, IGFBP-3 = insulin like growth factor binding protein, hGH = human growth hormone, TNF = tumor necrosis factor, EGF = epidermal growth factor, LH = luteinizing hormone, EPO = erythropoietin, FSH = follicle stimulating hormone, PDGF-BB = platelet derived growth factor BB.

occur only rarely (e.g., selenoylation, the co-translational incorporation of selenium into some proteins). Some PTMs are reversible (e.g., phosphorylation in many cases), while others are irreversible (e.g., proteolysis). Most but not all have an obvious function. Most are introduced into target proteins via specific enzymatic steps/pathways, while some (e.g., the oxidation of cysteine and methionine residues or the deamidation of asparagine or glutamine residues) arise non-enzymatically. Table 1.2

Table 1.2 Some of the more common post-translational modifications characteristic of proteins.

Modification	Description/comment
Acetylation	Acetylation (i.e., the incorporation of a $COCH_3$ group) of the N-terminal alpha-amine of polypeptides is a widespread modification in eukaryotes. It is enzymatically introduced by *N*-alpha-acetyltransferases. Little is actually known about the biological role of N-terminal acetylation, although acetylation of actin is important in the formation of actin filaments
Acylation	Acylation [attachment of a (RCO–) group] may help some polypeptides interact with/anchor in biological membranes
ADP-ribosylation	ADP-ribosylation describes the attachment of one or more ADP and ribose moieties to a polypeptide's backbone. This modification is achieved enzymatically by ADP-ribosyltransferase enzymes and can play various roles in cell signaling and control
Amidation	The enzymatic replacement of a polypeptide's C-terminal carboxyl group with an amide functional group is characteristic of many bioactive peptides/short polypeptides where it may influence their biological activity/stability
γ-Carboxyglutamate formation	The enzymatic conversion of target glutamate residues into γ-carboxyglutamate. Important in allowing some blood proteins bind calcium
Disulfide bond formation	Disulfide bonds are formed between the thiol (–SH) groups of cysteine residues and generally assist in the folding and conformational stabilization of some proteins
Glycosylation	The covalent attachment of a carbohydrate component to the polypeptide backbone. For some proteins glycosylation can increase solubility, influence biological half-life and/or biological activity
Hydroxylation	The enzymatic conversion of target aspartate residues into β-hydroxyaspartate or target asparagine residues into β-hydroxyasparagine. Important to the structural assembly of certain proteins
Phosphorylation	The covalent attachment of one or more phosphate groups to the polypeptide backbone. Often influences/regulates biological activity of various polypeptide hormones
Sulfation	The transfer of a SO_3^- group to target tyrosine residues. Influences biological activity of some neuropeptides and the proteolytic processing of some polypeptides

presents a selection of the more common protein PTMs, and further information may be obtained from the references [7, 8].

1.4
PTMs in the Context of Biopharmaceuticals

Only a subset of PTMs are generally associated with therapeutic proteins. Protein biopharmaceuticals, with very few exceptions, are recombinant/modified forms of native extracellular proteins. Some PTMs, such as acetylation, ADP ribosylation and phosphorylation, tend to regulate various intracellular processes, including gene expression, endosomal vesicle trafficking and signal transduction, and hence are rarely associated with biopharmaceuticals. PTMs most characteristic of biopharmaceuticals include glycosylation, proteolytic processing and disulfide bond formation, as well as (to a more limited extent) carboxylation, hydroxylation, sulfation and amidation.

Moreover, many biopharmaceuticals bear a combination of two or more PTMs. For example, activated protein C (trade name Xigiris) is carboxylated, hydroxylated, glycosylated, proteolytically processed and houses a disulfide linkage (Figure 1.1).

PTMs can be important, sometimes essential, to the biological activity of many biopharmaceuticals. The exact PTM detail of many recombinant biopharmaceuticals will differ from the PTM profile of the native or "natural" endogenous molecule. Indeed, even when we look at that native product, it is important to remember that its "natural" PTM details likely harbors modifications that occurred during its residence time in blood prior to purification, or which may have occurred during its purification. As such it is not necessary that a specific biopharmaceutical product harbor an exact replica of the PTM complement/detail of its endogenous human counterpart. What is necessary is that clinical trials prove the product (and hence its PTM complement) displays acceptable safety and efficacy, and that quality data prove that the manufacturing process results in the generation of product with consistent structural features, including its PTM profile.

The single largest influence upon PTM detail achieved during production of any recombinant protein is that of the producer cell line chosen in its manufacture. For example, the production of a normally glycosylated protein in *Escherichia coli* will result in an aglycosylated moiety, while expression in a yeast-based system can result

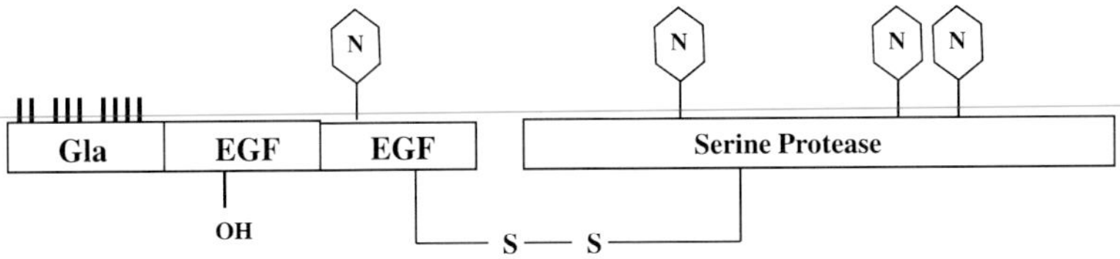

Figure 1.1 Schematic diagram of activated protein C. The protein consists of four domains as shown. The Gla domain houses nine γ-carboxylation sites. The first EGF domain houses a single β-hydroxylation site (Asp^{71}). Additional PTMs present include proteolytic processing, disulfide bond formation and the attachment of four N-linked carbohydrate side chains.

in the attachment of sugar side chains high in mannose content, which will negatively influence product serum half-life (Chapter 6). In contrast, production in plant-based systems can result in hyperglycosylated product containing xylose and fucose moieties that are immunogenic in man [9]. Consequently, the use of CHO or other mammalian cell lines is largely dictated by the latter's ability to undertake appropriate PTMs, generating product with acceptable therapeutic properties (Chapter 3). High expression levels may also overwhelm the PTM machinery, resulting in poor product quality, as shown in the case of some antibodies.

In addition to the production cell line chosen, exact upstream processing conditions employed during manufacture can also influence PTM characteristics. Temperature, growth rate and media composition have, for example, been shown to influence the glycoform profile of several biopharmaceuticals, including tPA and γ-interferon [10]. Downstream processing can also potentially compromise product integrity by, for example, selectively purifying or enriching a particular PTM product variant. Final product formulation can also have a potential effect by minimizing or preventing chemical-based modification of a PTM.

1.5 Some Specific PTMs

1.5.1 Glycosylation

Glycosylation represents the most complex of all PTMs in the context of therapeutic proteins. As such, the seven chapters in Part 1 of this book are devoted to this topic. Chapter 2 provides a comprehensive overview of the basic science of glycosylation, with subsequent chapters focusing upon the glycosylation characteristics of mammalian, yeast/fungal and insect cell lines and various case studies in the context of specific biopharmaceuticals.

It is estimated that up to 50% of all native human proteins are glycosylated and that somewhere between 1 and 2% of the human genome encodes proteins that contribute to glycosylation [11]. The exact glycosylation detail of a therapeutic protein can potentially effect a broad range of functionally significant characteristics:

- Glycosylation can aid in correct protein folding and assembly – for example the glycocomponent of gonadotrophic hormones such as FSH and LH has been implicated in proper protein folding, assembly and secretion).
- It can aid in the targeting and trafficking of a newly synthesized protein to its final destination, be it destined for extracellular secretion or destined for an intracellular organelle. For example, the removal of two or more of EPOs three N-linked glycosylation sites results in a product that is very poorly secreted from a producer cell.
- The glycocomponent can play a direct role in ligand binding, as for example in the case of some gonadotropins.

- The glycocomponent can play a direct role in triggering a biological activity upon ligand binding – for example removal of the N-52 glycocomponent of gonadotropins actually increases their receptor binding affinity but abolishes their ability to trigger signal transduction upon binding. In other instances it can more indirectly influence biological responses subsequent to ligand binding, as, for example, in the case of antibody triggered ADCC (Chapter 4).
- The glycocomponent may play a direct role in stabilizing the protein – for example α-galactosidase (trade name Fabrazyme), an enzyme used to treat Fabry's disease, is glycosylated at asparagine residue184. Removal of the sugar results in protein aggregation and precipitation.
- The glycocomponent often plays a role in regulating a protein's serum half-life. For example, the sialic acid content of EPO has a significant effect on its half-life. Highly branched, highly sialated glycoforms have the longest half-life, while de-sialation results in speedy product removal from the blood (removal of the sialic acid caps exposes terminal galactose residues, thus prompting rapid product removal from the blood via binding to galactose-specific receptors on liver cells).

Table 1.3 lists the therapeutic proteins approved (by the end of 2007) that are glycosylated.

In summary, there were 67 glycosylated products in total, representing one-third of all approved products. The single most significant category of glycosylated product are antibodies, of which there are 23 products listed – the only non-glycosylated antibody-based products are a few antigen-binding fragments approved for *in vivo* diagnostic purposes.

Table 1.3 Glycosylated biopharmaceuticals that had gained regulatory approval in the EU and/or USA by December 2007.

Product category	Products (by trade name)
Antibodies	Avastin, Bexxar, Erbitux, Herceptin, Humaspect, Humira, Mabcampath/Campath-H1, Mabthera/Rituxan, Mylotarg, Neutrospec, Oncoscint, Orthoclone OKT-3, Prostascint, Raptiva, Remicade, Simulect, Soliris, Synagis, Tysabri, Vectibix, Xolair, Zenapax, Zevalin
Blood factors, anticoagulants and thrombolytics	Activase, Advate, Atryn, Benefix, Bioclate, Helixate/Kogenate, Metalyse/TNKase, Novoseven, Recombinate, Refacto, Xigiris
Hormones	Gonal F, Luveris, Ovitrelle/Ovidrel, Pergoveris, Puregon/Follistim, Thyrogen
EPO and CSFs	Abseamed, Binocrit, Epoetin alfa hexal, Epogen/Procrit, Leukine, Mircera, Neorecormon, Nespo/Aranesp, Retacrit, Silapo
Interferons	Avonex, Rebif
Additional	Aldurazyme, Amevive, Cerezyme, Elaprase, Enbrel, Fabrazyme, Inductos, Infuse, Myozyme, Naglazyme, Orencia, Osigraft/OP-1 implant, Pulmozyme, Regranex, Replagal

1.5.2
Disulfide Bond Formation and Proteolytic Cleavage

Disulfide bond formation and proteolytic cleavage represent two prominent examples of PTMs that introduce structural change into proteins, and both are associated with multiple biopharmaceutical products [12, 13]. Some products such as therapeutic insulins display both modifications (Figure 1.2).

Disulfide bonds play an important role in the folding and stability of some proteins, usually proteins secreted into the extracellular medium. As cellular compartments are generally reducing environments, disulfide bonds are normally unstable in the cytosol. Table 1.4 presents examples of prominent therapeutic proteins displaying one or more disulfide bonds, and this PTM is considered in more detail in Chapter 11.

Proteolytic cleavage events are also a characteristic modification of many therapeutic and other proteins. Many proteins destined for export or for targeting to specific intracellular organelles are synthesized in an elongated "pro" form, with the "leader sequence" playing a critical role in targeting the protein to its destination. Typically specific proteolytic enzymes then cleave off the "pro" peptide, yielding the mature protein. Proteolysis can also serve as a mechanism of releasing biologically active protein from an inactive (or poorly active) precursor form. Prominent examples here would include several proteases functioning in the digestive tract, as well as the proteolytic processing of proinsulin, yielding the mature insulin product.

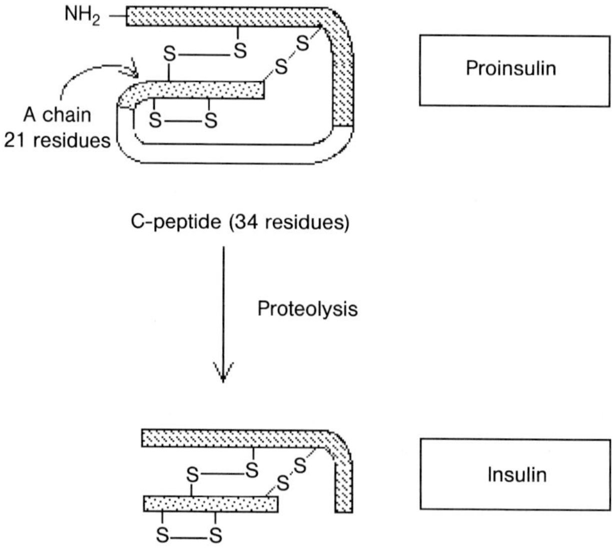

Figure 1.2 Schematic representation of the proteolytic processing of proinsulin, yielding mature insulin, as well as the positions of the disulfide linkages present. (Reproduced with permission from: G. Walsh, (2007) *Pharmaceutical Biotechnology*, John Wiley & Sons, Ltd., Chichester, UK.)

Table 1.4 Examples of some therapeutic proteins that display one or more disulfide linkages[a].

Protein	Disulfide linkage(s)	Medical application
IFN-β	Contains a single intrachain disulfide linkage	Multiple sclerosis
IL-2	Contains a single intrachain disulfide linkage	Treatment of certain cancers
TNF-α	Contains a single intrachain disulfide linkage	Treatment of soft tissue carcinoma
hGH	Contains two intrachain disulfide linkages	Treatment of some forms of dwarfism
Hirudin	Contains two intrachain disulfide linkages	Anticoagulant
Insulin	Contains two interchain and one intrachain disulfide linkages	Diabetes
Antibodies (IgG)	Contains two interchain and two intrachain disulfide linkages	Various
tPA	Contains 17 intrachain disulfide linkages	Thrombolytic agent

[a]Abbreviations: IFN = interferon, IL = interleukin, hGH = human growth hormone, tPA = tissue plasminogen activator.

Proteolytic processing tends not to be problematic during biopharmaceutical manufacture once an appropriate producer cell line has been chosen.

1.5.3 γ-Carboxylation and β-Hydroxylation

γ-Carboxylation and β-hydroxylation are PTMs characteristic of a small number of proteins, mainly a subset of proteins involved in blood coagulation [14, 15]. These modifications are undertaken by specific carboxylase and hydroxylase enzymes, with conversion of target glutamate residues in the protein backbone into γ-carboxyglutamate (Glu → Gla) and target aspartate residues into β-hydroxyaspartate (Asp → Hya). Asparagine residues can also rarely be hydroxylated, forming β-hydroxyasparagine (Asn → Hyn).

Both carboxylation and hydroxylation help mediate the binding of calcium ions, and these PTMs are important, in some cases essential, to the functioning of blood factors VII, IX and X, as well as activated protein C of the anticoagulant system.

In the context of approved biopharmaceuticals, therefore, carboxylation and/or hydroxylation is significant in the cases Novoseven (Novo's recombinant form of activated blood factor VIIa, indicated for the treatment of both hemophilia A and B), Wyeth's Benefix, (a recombinant version of blood factor IX, used to treat hemophilia B) and Xigiris (recombinant activated protein C, indicated for the treatment of severe sepsis). These PTM are considered in detail in Part 2 (Chapter 9).

1.5.4
Amidation and Sulfation

Amidation refers to the replacement of a protein's C-terminal carboxyl group with an amide group. This PTM is characteristic of some bioactive peptides/short polypeptides, including vasopressin, oxytocin, gastrin and calcitonin [16]. While widely associated with higher eukaryotes, amidation is not characteristic of yeast or prokaryotes. The exact biological role(s) of amidation remain to be fully elucidated. It may contribute to peptide stability and/or activity.

Most peptides used in the clinic are manufactured by direct chemical synthesis, including earlier preparations of salmon calcitonin. A recombinant version of this amidated 32 amino acid peptide hormone, however, is now approved for the treatment of Paget's disease and hypercalcemia of malignancy (Forcaltonin, Unigene, UK). Produced in *E. coli*, forcaltonin initially lacks this PTM and therefore must be subsequently amidated as part of downstream processing. This is achieved enzymatically, using an α-amidating enzyme (α-AE), which is itself produced by recombinant DNA technology using a CHO cell line. Amidation forms the subject matter of Chapter 10.

Sulfation is a PTM that entails the covalent attachment of a sulfate (SO_3^-) group, usually via selected tyrosine residues within the protein backbone [17]. Sulfation is performed by a sulfotransferase mediated co/post-translational process, occurring in the *trans* Golgi network, and is predominantly associated with secretory and membrane proteins. Tyrosine sulfated proteins/sulfotransferases are found in higher eukaryotic species but have not been reported in prokaryotes or in yeast.

Sulfation is apparently not dependant upon the occurrence of a target tyrosine within a strictly defined consensus sequence. However, 3–4 acidic amino acid residues are normally found within five residues of sulfated tyrosines and elements of local secondary structure also likely play a determinant role.

In many instances the function of sulfation remains unknown, although it often plays a role in protein–protein interactions. Generally, the absence of this PTM reduces rather than abolishes activity. In the context of biopharmaceuticals native hirudin and also blood factors VIII and IX are usually sulfated. Neither of the approved recombinant forms of hirudin (Revasc and Refludan, both produced in *Saccharomyces cerevisiae*) are sulfated and yet are therapeutically effective. Studies, however, have demonstrated that sulfated hirudin displays significantly tighter affinity for thrombin than do unsulfated forms.

While over 90% of native factor IX molecules are sulfated (at Tyr^{155}), less than 15% of the recombinant form approved (Benefix, produced in a CHO cell line) are, again with apparently little if any difference in therapeutic function.

1.6
Extending and Engineering PTM Profiles

Several approved therapeutic proteins are engineered post-synthesis. This approach normally entails the covalent attachment of a chemical group to the polypeptide's backbone, or the alteration of a pre-existing post-translational modification, such as a glycosylation pattern.

The covalent attachment of one or more molecules of poly(ethylene glycol) (PEG) to the polypeptide backbone represents a common form of such engineering [18]. PEGylation generally increases the plasma half-life of the protein drug by reducing the rate of systemic clearance. As a result less frequent dosage regimes are necessitated, with consequent economic savings and (usually) improved patient compliance and convenience. Several PEGylated biopharmaceuticals have gained regulatory approval over the last few years, most notably several interferon-based products. PEGylation is reviewed Part 3 (Chapter 14).

Levemir (trade name) is another therapeutic protein in which a "synthetic" PTM is introduced post-biological synthesis. This long-acting insulin analogue was developed for the treatment of diabetes by Novo. It differs from the native human molecule in that a 14-carbon fatty acid group is covalently attached via the lysine B29 side chain during downstream processing. The fatty acid promotes the binding of insulin to human serum albumin (which has three high-affinity fatty acid binding sites), both at the injection site and in plasma. This in turn ensures constant release of product, giving it duration of action of up to 24 hours [19].

Two approved biopharmaceuticals have been engineered by modification of their carbohydrate component. Cerezyme (trade name) is a carbohydrate-modified form of glucocerebrosidase, a lysosomal enzyme involved in the catalytic degradation of glycolipids. Gaucher's disease is a genetic condition caused by lack of lysosomal glucocerebrosidase activity, with tissue-based macrophages being amongst the most severely effected cell type. The product, which is naturally glycosylated, is produced by recombinant means in a CHO cell line and downstream processing includes an enzyme-based processing step using an exoglycosidase. The exoglycosidase removes sialic acid sugar residues that cap the oligosaccharide side chains. This exposes mannose residues underneath, facilitating specific uptake by macrophages via macrophage cell surface mannose receptors. In this way the product is specifically targeted to the cell type most affected by the disease. Unmodified glucocerebrosidase, if administered, is quickly removed from the bloodstream by the liver. Cerezyme is reviewed in Chapter 13.

Aranesp (Nespo) is a recombinant form of human erythropoietin (EPO) used to treat anemia associated with chronic renal failure. Produced in a recombinant CHO cell line, it displays an increased overall carbohydrate content when compared to native EPO. The natural molecule harbors three N-linked carbohydrate side chains whereas the recombinant product displays five such side chains. The increased carbohydrate content extends the product's half-life, facilitating once weekly (and in some circumstances once every second week) administration. Aranesp is discussed in detail in Chapter 12.

1.7 Conclusion

PTMs can have a potentially profound influence upon the stability, activity or pharmacokinetics of therapeutic proteins. Characterization of any PTM present, along with the establishment of its biological influence(s), represents a critical element of early stage biopharmaceutical drug development. The requirement to routinely manufacture product with an appropriate and reproducible PTM profile is a fundamental concern when developing manufacturing procedures. An increasing understanding of the link between PTM structure and function, particularly in the case of glycosylation, is now facilitating the development of second-generation products displaying PTMs engineered in order to tailor or optimize selected product functional attributes. The issue of post-translational modifications is clearly one of fundamental importance within the biopharmaceutical sector.

References

1 Lawrence, S. (2007) State of the biotech sector – 2006. *Nature Biotechnology*, **25**, 706.

2 Lawrence, S. (2005) Biotech drug market steadily expands. *Nature Biotechnology*, **23**, 1466.

3 Pavlou, A. and Belsey, M. (2005) The therapeutic antibody market to 2008. *European Journal of Pharmaceutics and Biopharmaceutics*, **59**, 389–396.

4 Pavlou, A. and Reichert, J. (2004) Recombinant protein therapeutics – success rates, market trends and values to 2010. *Nature Biotechnology*, **22**, 1513–1519.

5 Walsh, G. (2006) Biopharmaceutical benchmarks 2006. *Nature Biotechnology*, **24**, 769–776.

6 Rader, R. (2007) *Biopharmaceutical Products in the US and the European Markets*, 6th edn, Bioplan Associates, Rockville, MD.

7 Higgins, S. and Hames, B. (1999) *Post Translational Processing, a Practical Approach*, Oxford University Press.

8 Kannicht, C. (2002) *Post Translational Modification of Proteins*, Humana Press, USA.

9 Gomord, V. *et al.* (2004) Production and glycosylation of plant made pharmaceuticals: the antibodies as a challenge. *Plant Biotechnology Journal*, **2**, 83–100.

10 Butler, M. (2005) Animal cell culture: recent achievements and perspectives in the production of biopharmaceuticals. *Applied Microbiology and Biotechnology*, **68**, 283–291.

11 Walsh, G. and Jefferis, R. (2006) Post translational modifications in the context of therapeutic proteins. *Nature Biotechnology*, **24**, 1241–1252.

12 Steiner, D. *et al.* (1974) Proteolytic processing in the biosynthesis of insulin and other proteins. *Federation Proceedings*, **33**, 2105–2115.

13 Sevier, C.S. and Kaiser, C.A. (2002) Formation and transfer of disulphide bonds in living cells. *Nature Reviews Mollecular and Cellular Biology*, **3**, 836–847.

14 Hansson, K. and Stenflo, J. (2005) Post translational modifications in proteins involved in blood coagulation. *Journal of Thrombosis and Haemostasis*, **3**, 2633–2648.

15 Kaufman, R. (1998) Post translational modifications required for coagulation factor secretion and function. *Thrombosis and Haemostasis*, **79**, 1068–1079.

16 Bradbury, A. and Smyth, D. (1991) Peptide amidation. *Trends in Biochemical Science*, **16**, 112–115.

17 Moore, K. (2003) The biology and enzymology of protein tyrosine-O-sulfation. *Journal of Biological Chemistry*, **278**, 24243–24246.

18 Veronese, F. and Pasut, G. (2005) PEGylation, successful approaches to drug delivery. *Drug Discovery Today*, **10**, 1451–1458.

19 Goldman-Levine, J. and Lee, K. (2005) Insulin detemir – a new basal insulin analogue. *Annals of Pharmacotheraphy*, **39**, 502–507.

Part One
Glycosylation

Post-translational Modification of Protein Biopharmaceuticals. Edited by Gary Walsh

ISBN: 978-3-527-32074-5

2
Protein Glycosylation: The Basic Science

Susan A. Brooks

2.1
Introduction – Glycosylated Proteins

A glycoprotein is a protein containing one or more covalently linked carbohydrate groups. The carbohydrate, or glycan, can be either a single monosaccharide or be composed of many monosaccharides linked together as branched or linear chains, termed oligosaccharides. Glycoproteins contain polypeptide sequences that have one or more potential N-linked or O-linked glycosylation sites. Each site may or may not be glycosylated, and may be glycosylated with a range of different glycans. Thus, glycoproteins occur as a mixture of closely related glycosylation variants, an "ensemble of glycoforms" [1]. The range of different glycoforms exhibited by a single glycoprotein is specific to the species or tissue where it is located and may differ under varying physiological conditions, differentiation and disease states [2–5] and may be involved in regulating molecular interactions governing cellular responses. The possibility of glycosylating the same protein differently under different conditions means that functionally different molecules can be readily synthesized without the need for genetic alteration. Biologically, the heterogeneity in the way that a single glycoprotein is glycosylated may arise because of the way in which the enzymes of glycosylation act in a sequential, partially competitive manner in the Golgi apparatus, and where factors such as substrate availability and protein transit time through the endomembrane system have an influence. Alternatively, each individual cell may synthesize a specific glycoform, but this may differ subtly from that of its neighbors [6]. The factors that regulate and control glycoprotein glycosylation are not understood and this is a significant challenge for the biopharmaceutical industry.

The cells of different species often glycosylate their proteins differently and with glycan structures distinct from those synthesized by human cells. The repertoire of glycans synthesized by, for example, fungi, plants, yeasts, bacteria and invertebrates are extremely different from those synthesized by mammalian cells. This has strong implications for biotechnology, as recombinant glycoproteins synthesized by expression systems relying on cells from these types of sources may be glycosylated with glycans alien to human cells. This may have biological consequences,

Post-translational Modification of Protein Biopharmaceuticals. Edited by Gary Walsh

ISBN: 978-3-527-32074-5

including undesirable immunogenicity and reduced half-life in circulation (see References [7, 8]). Moreover, glycosylation of proteins produced by cultured cells will be unpredictably influenced by factors that include pH and the availability of nutrients and precursor molecules, thus giving rise to variable glycoforms. The present chapter describes glycoprotein glycosylation in human cells. For descriptions of glycosylation in other cell systems, see References [7–9].

The biological importance of glycosylation of proteins is apparent in that over 50% of naturally occurring proteins are glycoproteins [3, 10–12]. They exhibit great heterogeneity, complex and varied structures and widespread occurrence. Their biological roles are extremely diverse. They encompass structural and modulatory functions and recognition/receptor interactions and are involved in mechanisms of interest to immunology, neurobiology, hematology, ontogeny and metabolism. Changes in glycosylation are associated with disease [5]. Most glycoproteins are membrane-bound and the oligosaccharides associated with them are often physically bulky in comparison to the proteins to which they are attached. They therefore protrude significant distances from the cell membrane, hence rendering them well placed to mediate molecular recognition events as diverse as protein trafficking, fertilization, embryogenesis, intra- and intercellular signaling, cell–cell adhesion and infection processes of bacteria, viruses and parasites. Other glycoproteins are soluble and secreted and again have myriad and diverse biological functions, including, for example, endocrine function, involvement in the immune response and roles in blood clotting mechanisms [1, 13–16]. Glycans significantly alter protein conformation, functional activity, stability, binding interactions with other molecules and cellular signaling [17, 18]. There have been numerous reports implicating abnormal glycosylation in disease states, and glycosylation in health and disease is well reviewed [5, 16, 19]. For example, aberrant glycosylation has been recognized in pathological conditions, including rheumatoid arthritis, Crohn's disease, tuberculosis [13], and is linked with the metastatic progression of cancer [20–27]. This suggests that glycosylated proteins have potential in clinical diagnostics and as targets for developing therapeutics.

This chapter describes the way in which proteins are glycosylated in human cells and discusses the methods for their analysis.

2.2
Basic Building Blocks of Glycosylation in Human Cells

The basic chemistry of carbohydrates is well described in any standard biochemistry, cell biology or glycobiology textbook such as References [28, 29]. In human cells, glycans are composed of a combination of seven monosaccharides:

- Mannose (Man)
- Glucose (Glc)
- Fucose (Fuc)
- Galactose (Gal)
- *N*-Acetylgalactosamine (GalNAc)

- *N*-Acetylglucosamine (GlcNAc)
- Sialic acids or neuraminic acids (SA or NeuNAc).

They can exist as linear molecules but tend to occur in an energetically more favorable ring-shaped formation (Figure 2.1). They are very similar to each other in structure and in molecular weight, and this is one reason why the analysis of complex glycans, described later, is so technically challenging.

They occur as optical isomers (stereoisomerism), with the alpha (α) isomer being when the hydroxyl (–OH) group of the first carbon atom (C1) atom lies below the plane of the ring and the beta (β) isomer being when the hydroxyl group lies above the plane of the ring. The monosaccharides illustrated in Figure 2.1 are all α-isomers as the hydroxyl group of the first carbon in each case lies below the plane of the ring. Figure 2.2 illustrates both the α- and β-isomers of glucose.

The linear or branching glycan chains that are attached to human glycoproteins are composed of these seven monosaccharides linked together. Monosaccharides are covalently linked to each other by a reaction catalyzed by specific glycosyltransferases, resulting in the formation of disaccharides (consisting of two monosaccharide units), trisaccharides (consisting of three monosaccharide units) or oligosaccharides (which contain between three and ten monosaccharide units). These structures can be linear or branching chains.

2.3 Formation of Complex Glycan Structures

The formation of the bond between two monosaccharides occurs between hydroxyl groups in a condensation reaction that eliminates a molecule of water (Figure 2.2). The resulting acetal oxygen bond between the two units is a glycosidic bond. These bonds can occur between hydroxyl groups associated with different carbon atoms of the monosaccharide, so different linkages (e.g., a $1 \rightarrow 3$, $1 \rightarrow 6$ or $2 \rightarrow 3$ linkage) are possible.

2.3.1 α and β Glycosidic Bonds

The glycosidic linkage between two monosaccharide units can be either an α linkage or a β linkage. An α glycosidic linkage lies below the plane of the left-hand monosaccharide; a β glycosidic linkage lies above the plane of the left-hand monosaccharide. Thus, linkages between the first carbon of one monosaccharide and the fourth carbon of a second monosaccharide can be an $\alpha 1 \rightarrow 4$ linkage or a $\beta 1 \rightarrow 4$ linkage (Figure 2.2).

2.3.2 Structural Complexity of Glycoprotein Glycans

The features of carbohydrate chemistry – the seven monosaccharide building blocks, the possibility of linking them together through different carbon atoms, the choice of

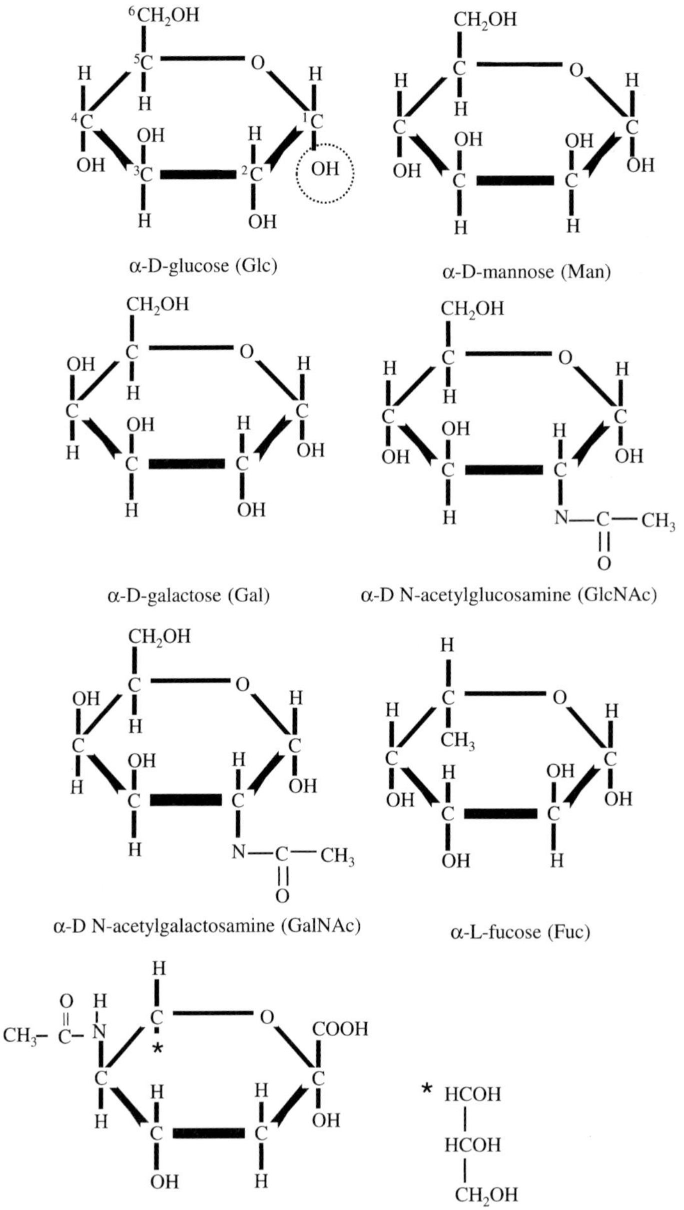

α-N-acetylneuraminic acid (NeuNAc), a sialic acid

Figure 2.1 The seven monosaccharides that make up the glycans of human glycoproteins. They are illustrated here in their ring-shaped conformation, drawn as Haworth projection formulae. All are presented here as α-isomers, the hydroxyl (–OH) group attached to the first carbon in the ring (C1) lies below the plane of the ring, as marked on the glucose molecule.

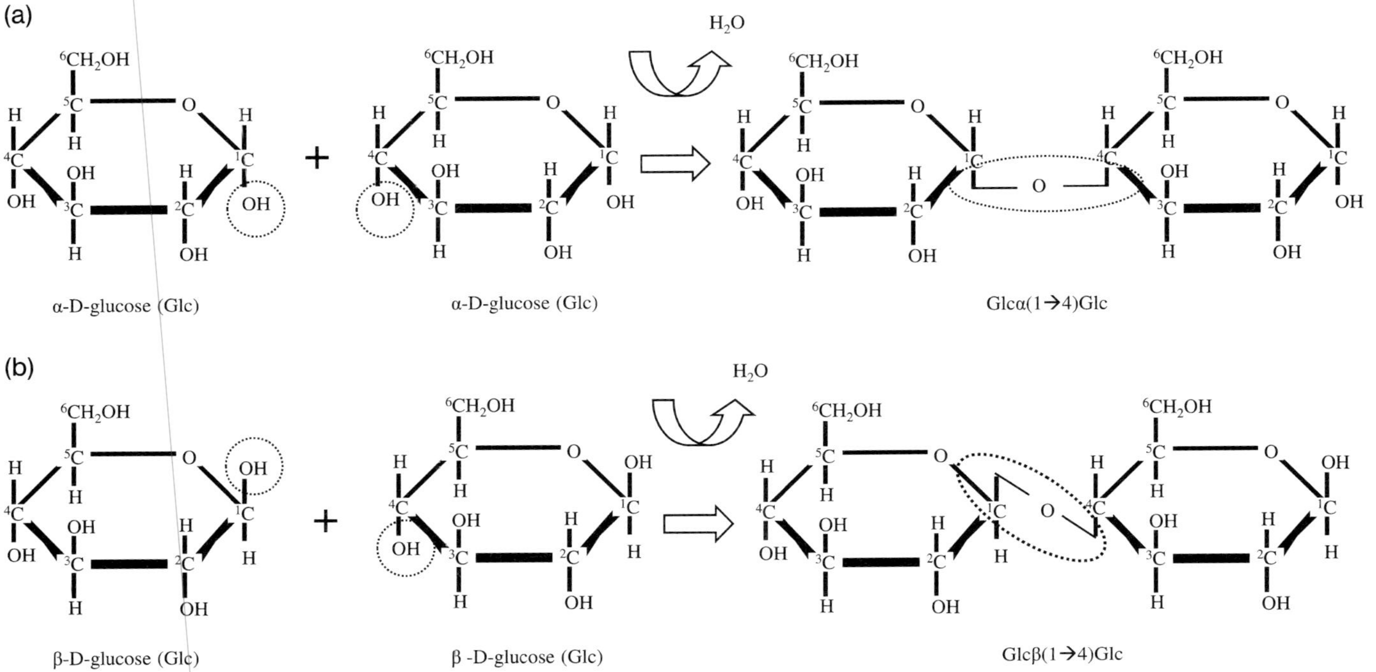

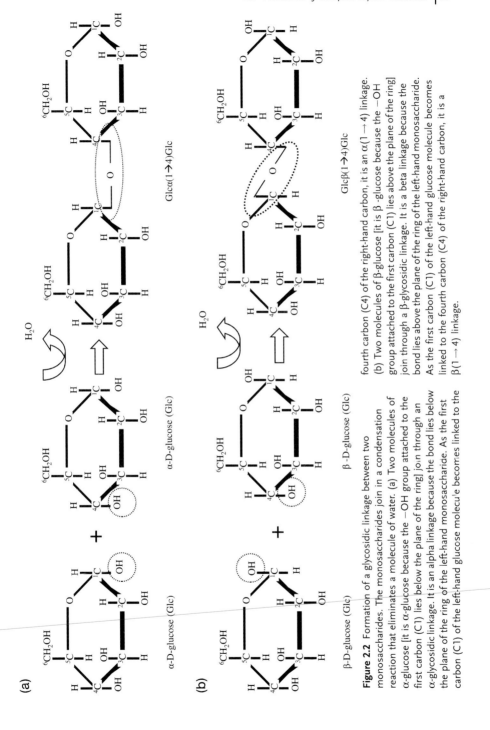

Figure 2.2 Formation of a glycosidic linkage between two monosaccharides. The monosaccharides join in a condensation reaction that eliminates a molecule of water. (a) Two molecules of α-glucose [it is α-glucose because the –OH group attached to the first carbon (C1) lies below the plane of the ring] join through an α-glycosidic linkage. It is an alpha linkage because the bond lies below the plane of the ring of the left-hand monosaccharide. As the first carbon (C1) of the left-hand glucose molecule becomes linked to the fourth carbon (C4) of the right-hand carbon, it is an α(1 → 4) linkage. (b) Two molecules of β-glucose [it is β -glucose because the –OH group attached to the first carbon (C1) lies above the plane of the ring] join through a β-glycosidic linkage. It is a beta linkage because the bond lies above the plane of the ring of the left-hand monosaccharide. As the first carbon (C1) of the left-hand glucose molecule becomes linked to the fourth carbon (C4) of the right-hand carbon, it is a β(1 → 4) linkage.

α and β glycosidic bonds and the possibility of building linear or branching chains of variable lengths and complexity – explain the extraordinary structural diversity of the glycan structures found in human cells. This diversity and the complexities of carbohydrate chemistry, including the physical similarities between the monosaccharide building blocks, also make analysis of glycans technically challenging in comparison to, for example, protein analysis. The development of technological approaches to glycan analysis, described below, is currently an area of intense research activity.

2.4 Glycan Synthesis is Catalyzed by Enzymes of Glycosylation – the "Glycozymes"

Nucleic acids and proteins are generated from instructions coded by DNA. In contrast, glycans are not primary gene products; their synthesis is not "template driven," rather they are constructed by the action of "building" enzymes (glycosyltransferases) and "trimming" enzymes (glycosidases), collectively sometimes referred to as "glycozymes," of the rough endoplasmic reticulum (RER) and Golgi apparatus (GA). Human cells have a repertoire of about 250 glycozymes, reflecting the diversity of oligosaccharide structures that can be built, and the co-ordinated action of as many as 30 different enzymes can be required for the synthesis of a single complex oligosaccharide [30]. Some 1–2% of the mammalian genome codes for these enzymes.

Glycosyltransferases are classified according to the monosaccharide they transfer, the linkage they form and the structure they recognize. For example, GalNAcα2-6-sialyltransferase catalyses transfer of a sialic acid residue to a GalNAc in an $\alpha 2 \rightarrow 6$ glycosidic linkage. Glycosidases and glycosyltransferases display highly discriminate substrate specificity [1, 20] and it was long believed that glycosyltransferases were each responsible for catalysis of a single specific linkage, sometimes referred to as the "one enzyme – one linkage" rule. However, this is not always the case. For example, a large family of polypeptide GalNAc-transferases (ppGalNAc-Ts) catalyze the transfer of GalNAc to Ser/Thr of the polypeptide [31] and some glycosyltransferases are able to catalyze more than one type of glycosidic linkage.

Glycosyltranferases are grouped into "families" according to what monosaccharide is transferred in the reactions they catalyze – for example, the sialyltransferase family catalyze the transfer of sialic acids. There is little sequence homology between members of different glycosyltransferase families, although common motifs are shared amongst related members of the same family. For example, different sialyltransferases share common "sialyl motifs." There are, however, physical similarities between apparently unrelated glycosyltransferases. Glycosyltransferases located in the GA have a common tertiary structure [32]. They are all type II transmembrane proteins with a short cytoplasmic amino terminal tail, a "stem" region that extends into the lumen of the GA and a "globular" carboxy-terminal catalytic domain [33]. A proteolytic site is sometimes located in the stem region, facilitating release of a soluble form of the enzyme into the GA lumen [34].

Glycosyltranferases catalyze transglycosylation reactions in which a monosaccharide is transferred from a high energy nucleotide sugar donor (e.g., UDP-GlcNAc or GDP-Man) to the glycan [33, 34]. Most glycosyltransferases require divalent cations such as Mg^{2+} or Mn^{2+} as cofactors to work optimally. The optimal pH for glycosyltransferase action ranges between 5 and 7, the pH of various parts of the RER and GA [33].

The rules that govern what glycan structures are synthesized and whether potential glycosylation sites on the polypeptide are occupied or unoccupied are not understood. However, they are influenced by several factors. These include the comparable levels of expression and activity of the glycosyltransferases, the availability of appropriate sugar-donor and acceptor and competition with other glycosyltransferases for available substrate [35]. Moreover, the transit time of the glycoprotein through the compartments of the GA may play a role, as rapid passage limits the time available for glycosylation reactions to take place. Glycosylation is under strict cellular control. Microarray approaches to detect transcript levels of different glycozymes may be helpful in future in providing predictive information on cellular glycosylation patterns [36], but at present it is not possible to do this.

The location of glycosyltransferases within the RER and GA secretory pathway roughly corresponds to the order in which they act. Differences in their localization occur in different cell types and also in some disease states and are associated with consequent differences in the glycosylation repertoire of the cell. How their positioning within the secretory pathway is determined and maintained is not understood. They appear to locate to regions where the pH of the environment and the characteristics of stiffness and thickness of the membrane in which they are embedded is appropriate for the structure of their transmembrane domain [37, 38]. Their retention within the GA seems to rely on several factors: (i) their transmembrane regions are too large to allow them to become incorporated into budding transport vesicles, (ii) they occur in aggregate complexes too large to enter transport vesicles and (iii) they interact with matrix proteins, which anchor them in place [39, 40].

2.5 Protein Glycosylation – Relationship Between N-linked and O-linked Glycoproteins

Glycans may be N-linked or O-linked to proteins. Both N- and O-linked glycoproteins share common features, and many, including most mucin-type glycoproteins, carry both N- and O-linked glycan chains attached to a single polypeptide. Furthermore, many glycosyltransferases, for example those that synthesize blood group ABO sugars, act on both N- and O-linked glycan chains, so that the terminal glycan structures, described later, are identical. However, N-linked and O-linked glycans also have unique features with regard to structure, function and biosynthesis. The synthesis of N- and O-linked glycans are reviewed in References [41–46].

2.6 N-linked Glycoproteins

2.6.1 Where Does N-linked Glycosylation of Proteins Take Place?

N-linked glycosylation of proteins begins in the RER during protein synthesis. It is thus a co-translational event. This is in contrast to O-linked glycosylation, described later, which begins after protein synthesis is completed and is thus entirely post-translational. *N*-Glycosylation continues during the transport of the protein from the RER to the GA, with glycosylation being completed prior to the protein leaving the *trans*-Golgi network. The enzymes of glycosylation necessary for N-linked glycosylation – glycosidases and glycosyltransferases – are transmembrane proteins located along the secretory pathway of RER and GA, and their location corresponds to the sequence in which they act. N-linked glycosylation has been recently well reviewed [47].

2.6.2 Why is it Called N-linked?

The N-linking of glycoproteins involves the formation of a bond between the first carbon (C1) of a GlcNAc monosaccharide in a α-*N*-glycosidic bond to a nitrogen (hence N-linked) of the amide group of an asparagine (Asn) residue on the polypeptide backbone.

2.6.3 *N*-Glycosylation Step-by-Step

2.6.3.1 Polypeptide Enters the RER

During protein synthesis, the nascent (i.e., still undergoing synthesis) polypeptide chain is guided to the RER membrane by a cytosolic signal recognition particle (SRP) binding to its RER signal sequence. The SRP is then recognized by a receptor embedded in the plasma membrane of the RER and this positions the polypeptide chain correctly on the RER membrane. The SRP is then released. The nascent polypeptide chain, as it is synthesized, feeds through a translocation channel in the RER membrane into the lumen and the addition of N-linked oligosaccharides begins immediately. Figure 2.3 illustrates the N-linked glycan structures.

2.6.3.2 Building and Positioning the Dolichol Oligosaccharide Precursor

An oligosaccharide precursor ($Glc_3Man_9GlcNAc_2$) is synthesized on a dolichol-pyrophosphate (lipid) carrier. The donor nucleotide sugars UDP-GlcNAc and GDP-Man are involved in reactions mediated by the sequential action of glycosyltransferases that transfer the first seven monosaccharides ($Man_5GlcNAc_2$) onto the lipid carrier on the cytoplasmic side of the RER (Figure 2.3a). This lipid-glycan structure is then "flipped" across to the luminal side of the RER [48, 49] and a further four Man and three Glc residues, from donor nucleotide sugars Dol-P-Man and Dol-P-Glc, are added to build the dolichol oligosaccharide precursor, $Glc_3Man_9GlcNAc_2$ (Figure 2.3b).

2.6.3.3 Attachment of the Dolichol Oligosaccharide Precursor to the Polypeptide Chain

The dolichol oligosaccharide precursor $Glc_3Man_9GlcNAc_2$ is then transferred *en bloc* to the amide nitrogen of an asparagine (Asn) residue on the nascent polypeptide by the action of an RER membrane resident multimeric enzyme, oligosaccharyltransferase (OST) [50, 51]. A consensus amino acid sequence Asn-X-Thr/Ser (X = any amino acid except proline) on the nascent polypeptide chain is a pre-requisite for N-linked oligosaccharide attachment [50].

2.6.3.4 Trimming the Dolichol Oligosaccharide Precursor in the RER

The $Glc_3Man_9GlcNAc_2$-Asn is then trimmed by the sequential action of α-glucosidase I, which cleaves the terminal Glc residue, α-glucosidase II, which cleaves the remaining two terminal Glc residues, and α-mannosidase I, which removes a terminal Man to yield $Man_8GlcNAc_2$-Asn (Figure 2.3c). The process of building a large glycan structure and then trimming it is a key part of protein folding and lysosomal trafficking. Chaperone molecules calnexin and calreticulin both recognize the Glc residues on the oligosaccharide precursor. The removal of the terminal Glc residues is part of the protein folding mechanism and improperly folded proteins are re-glucosylated by an α-glucosyltransferase located in the lumen of the RER and re-folded. Terminally improperly folded proteins are de-glucosylated, targeted to the lysosome and degraded. Correctly folded proteins are transferred to the GA for further processing [51–53].

2.6.3.5 Processing of N-linked Oligosaccharides in the Golgi Apparatus

When the glycoprotein leaves the RER and reaches the GA, the $Man_8GlcNAc_2$-Asn glycan is usually trimmed by class I α-mannosidases, which remove further Man residues, and then extended by the action of various glycosyltransferases. Some *N*-glycans with between five and nine Man remain, but most are processed to a $Man_5GlcNAc_2$-Asn glycan (Figure 2.3d) that becomes the basis for building a range of structurally diverse *N*-glycans. All N-linked glycans contain a common branched trimannosyl core, $Man_3GlcNAc_2$-Asn (Figure 2.3e), and the repertoire of N-linked structures that are synthesized depend on the type, order and linkages of the monosaccharide residues attached to it. N-linked glycoproteins are grouped into three main classes, as described below. They are highly heterogeneous in structure and can be bi-, tri- or tetra-antennary.

2.6.4 Three Classes of N-linked Glycoproteins

1. *Oligomannose or high mannose:* where the core contains between five and nine Man residues.

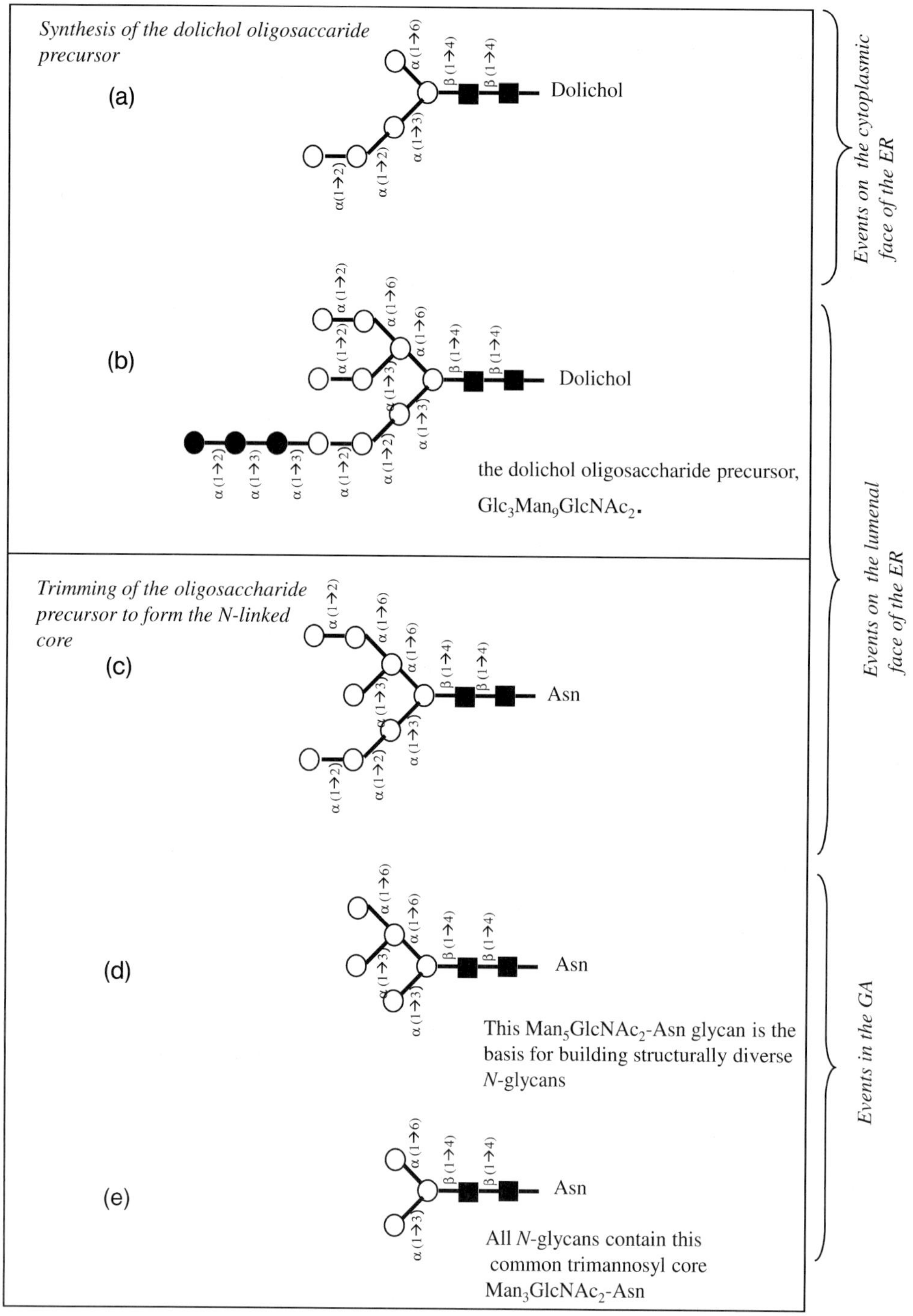

Synthesis of the dolichol oligosaccaride precursor
(a)
α(1→6)
α(1→3)
α(1→2)
α(1→2)
β(1→4)
β(1→4)
Dolichol
(b)
Dolichol
the dolichol oligosaccharide precursor,
$Glc_3Man_9GlcNAc_2$.
Trimming of the oligosaccharide precursor to form the N-linked core
(c)
Asn
(d)
Asn
This $Man_5GlcNAc_2$-Asn glycan is the basis for building structurally diverse *N*-glycans
(e)
Asn
All *N*-glycans contain this common trimannosyl core $Man_3GlcNAc_2$-Asn
Events on the cytoplasmic face of the ER
Events on the lumenal face of the ER
Events in the GA

2. *Complex*: which do not contain Man residues additional to those of the trimannosyl core and are often branched, resulting in a high-level of complexity. These first two are the most common and structurally diverse group of the N-linked glycans.
3. *Hybrid*: which possess structural features of both oligomannose and complex oligosaccharides.

2.6.5
What Determines Whether a Potential Site of *N*-Glycosylation on the Polypeptide Chain is Occupied or Not?

The consensus sequence Asn-X- Ser/Thr may occur many times along the polypeptide chain and each potential glycosylation site may or may not be glycosylated. The glycans are often a heterogeneous mixture of oligomannose, complex and hybrid types. The same glycoprotein may be differently glycosylated under different circumstances and the regulation of glycosylation pattern is not well understood. However, often, occupied *N*-glycosylation sites occur in a "loop" or "turn" in the tertiary structure of the protein, which is consistent with the need for their exposure during the initial steps of N-linked glycosylation. The addition of the oligosaccharide precursor to a glycosylation site on the growing polypeptide chain may also influence occupancy of other potential glycosylation sites down-stream. The protein environment surrounding occupied *N*-glycosylation sites has been modeled [54].

2.7
O-linked Glycoproteins

The *O*-glycans are usually smaller and less branched than *N*-glycans; they are commonly bi-antennary. However, as, unlike *N*-glycans, they do not share a common core, they are more heterogeneous in structure and their analysis is more challenging. Furthermore, *O*-glycosylation is not as well understood as *N*-glycosylation.

Figure 2.3 Early events in the N-linked glycosylation of proteins. (a) Two GlcNAc and five Man residues are attached to a dolichol carrier on the cytoplasmic face of the RER. (b) The dolichol-glycan structure is flipped into the RER lumen and a further four Man and three Glc residues added to build the dolichol oligosaccharide precursor. This is attached to an Asn residue on the polypeptide at the site of a consensus sequence: Asn-X-Thr/Ser. (c) The oligosaccharide precursor is trimmed of three terminal Glc residues and a Man, a process that is essential to protein folding and quality control. (d) On entering the GA, further Man residues are removed; some structures remain with between five and nine Man residues, but most are processed to $Man_5GlcNAc_2$-Asn and this is the basic structure on which many structurally diverse *N*-glycans are then built. (e) All N-linked glycans contain a common branched trimannosyl core. Key to symbols Glc: ●; Man: ○; GlcNAc: ■.

2.7.1
Why is it Called O-linked?

In O-linked-glycosylation the first monosaccharide of the oligosaccharide chain, usually a GalNAc, is attached through an α-*O*-glycosidic linkage to an oxygen molecule (hence "O-linked") of an amino acid residue, usually a serine (Ser) or threonine (Thr), on the polypeptide chain.

2.7.2
Different Types of O-linked Glycosylation

O-linked glycosylation begins with the attachment of a single monosaccharide to an amino acid residue on the polypeptide. Several different types of *O*-glycosylation occur and most are confined to certain species, tissues or to specific proteins [55]. GlcNAc O-linked to Ser/Thr residues that are also sites of phosphorylation are found in cytoplasmic and nuclear proteins in eukaryotes [56, 57]. Xylose (Xyl) O-linked to Ser/Thr occurs in proteoglycans of the extracellular matrix of animal tissues. Gal and Glc-Gal linked to the hydroxyl group of hydroxylysine are found in collagen and proteins containing a collagen-like domain. Xyl-Glc or Glc linked to Ser/Thr are seen in blood clotting factors. O-linked Fuc occurs in plasma glycoproteins, including tissue plasminogen activator [58, 59]. Man O-linked to Ser/Thr has been described, for example, in vertebrate brain proteoglycans. Arabinose linked to hydroxyproline is commonly found in plants [60] and may be analogous to the Gal/Glc-Gal linked to hydroxylysine in animal collagens, described above. Gal linked to Ser occurs in plants and Man linked to Ser/Thr is seen in fungi and yeasts [61].

By far the most common type of *O*-glycosylation in mammalian cells is initiated by the linking of a single GalNAc to a Ser or Thr residue on the polypeptide and this type of *O*-glycosylation is described in detail below. This type of O-linked glycosylation is found on many glycoproteins, and one type (i.e., mucins), which provide protective secretions in the body, is particularly heavily *O*-glycosylated [55, 62, 63]. For this reason, this type of O-linked glycosylation is sometimes referred to as "mucin-type" glycosylation.

2.7.3
Overview of O-linked Mucin-Type Protein Glycosylation

Unlike N-linked glycosylation, which, as described previously, is initiated by the attachment of a large, pre-formed oligosaccharide, O-linked glycosylation is initiated by the attachment of a single monosaccharide to the polypeptide. The glycan is then built, simply, step-by-step, by the sequential addition of monosaccharides, each addition catalyzed by a glycosyltransferase, to build several core structures (Figure 2.4). There is no trimming by glycosidases. O-linked glycosylation occurs entirely post-translationally (i.e., once the protein has been synthesized and folded). It is initiated in the cis-compartment of the GA and the addition of monosaccharides continues as the protein passes through more distal compartments. Thus, the amino acid residues to which the glycans are attached must be exposed and accessible in the complete and folded polypeptide. O-linked glycosylation has been well reviewed [64].

2.7.4 No Consensus Sequence for Mucin-Type O-linked Glycoprotein Glycosylation

There is no consensus amino acid sequence, comparable to the N-linked glycosylation consensus sequence Asn-X- Ser/Thr, required for O-linked glycan attachment to the polypeptide. However, glycosylated Ser/Thr residues are often located in proline-rich sequences, and there has been interest in understanding the factors that govern whether particular Ser or Thr residues are glycosylated. Sequence homology has been identified between regions of polypeptide featuring dense *O*-glycan attachment [64]. Occupancy of potential glycosylation sites appears to be influenced by complex relationships between the amino acid sequence and whether or not prior glycosylation of the polypeptide has occurred [65]. Predictive algorithms for *O*-glycosylation sites have been developed [66–70] but this remains an area that is incompletely understood.

2.7.5 Synthesis of O-linked Mucin-Type Glycan Core Structures Step-by-Step

2.7.5.1 Initiation – Synthesis of GalNAcα1 → Ser/Thr (Tn Epitope)

All mucin-type *O*-glycan synthesis begins with the attachment of a single GalNAc monosaccharide to a Ser or Thr residue on the polypeptide chain. This attachment is catalyzed by a large family of polypeptide GalNAc transferases (UDP-GalNAc: polypeptide α1,3-*N*-acetylgalactosaminyltransferase, ppGalNAc-Ts) [21, 31, 50, 71]. Attachment of GalNAc to the Ser/Thr of the polypeptide backbone yields GalNAcα1 → Ser/Thr, called the Tn epitope (Figure 2.4a). Further chain extension always occurs in healthy human cells, but Tn epitope often remains unelaborated and exposed in cancer cells, where it is considered a cancer-associated antigen [72, 73].

2.7.5.2 Synthesis of NeuNAc(α2 → 6)GalNAcα1 → Ser/Thr (Sialyl Tn)

Tn epitope can be acted upon by α-2,6-sialyltransferase (CMP-sialic acid: R_1-GalNAc-R α2,6-sialyltransferase I, ST2,6GalNAcI, ST6GalNAcI) to yield NeuNAc (α2 → 6)GalNAcα1 → Ser/Thr, also known as sialyl Tn or sTn (Figure 2.4b). This disaccharide can no longer act as an acceptor for other glycosyltransferases and is therefore always the final product. It is found commonly in submaxillary mucins and some other glycoproteins, and its presence is often increased in cancers where it has been a target for cancer vaccination therapy [74, 75].

2.7.5.3 Synthesis of Core 1, Gal(β1 → 3)GalNAcα-Ser/Thr, or the Thomsen–Friedenreich (T or TF) Antigen

The addition of (β1 → 3) linked Gal, catalyzed by core β1,3-galactosyltransferase (UDP-Gal: GalNAc-R β1,3-galactosyltransferase, β1,3 Gal-T or core 1 Gal-T), to the original Tn epitope yields Gal(β1 → 3)GalNAcα-Ser/Thr, the Thomsen–Friedenreich (T or TF) antigen, also called core 1 (Figure 2.4c). The same enzyme that sialylates Tn

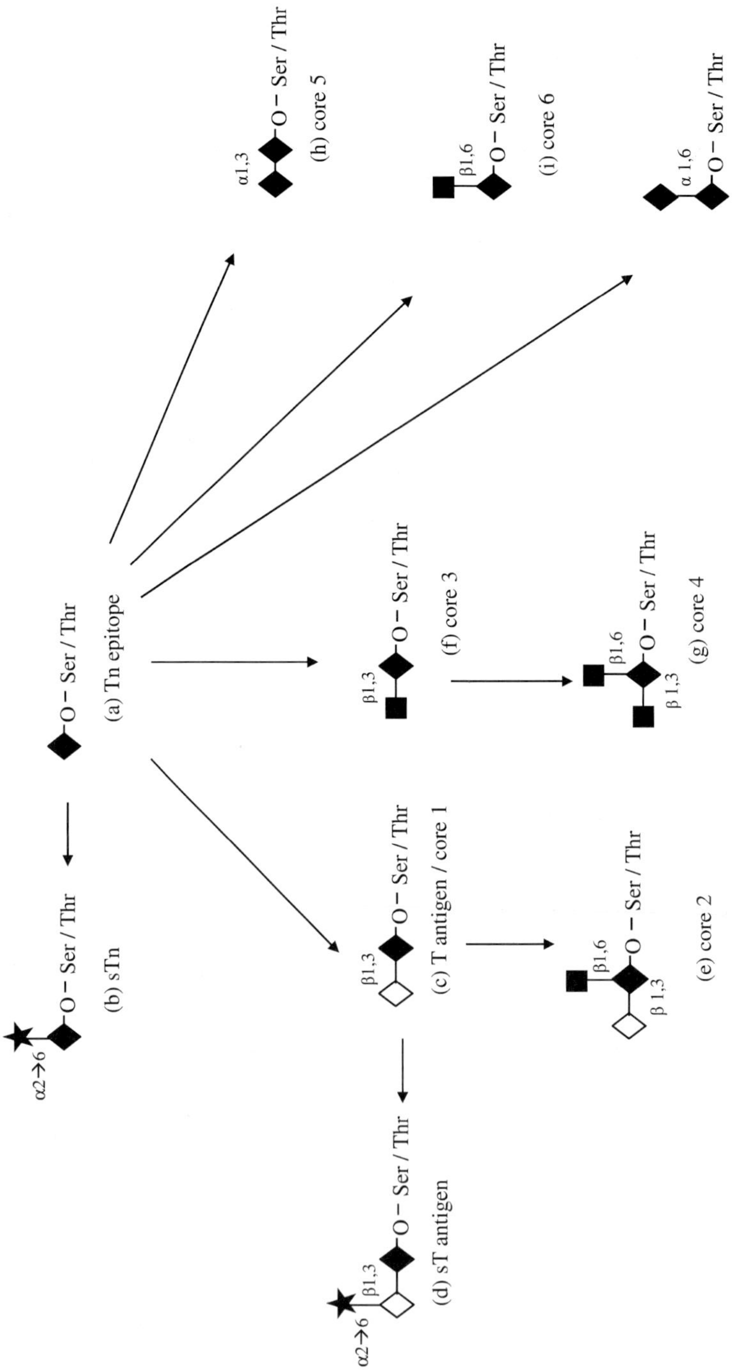
O – Ser / Thr
(a) Tn epitope
α2→6
O – Ser / Thr
(b) sTn
β1,3
O – Ser / Thr
(c) T antigen / core 1
α2→6
β1,3
O – Ser / Thr
(d) sT antigen
β1,6
β 1,3
O – Ser / Thr
(e) core 2
β1,3
O – Ser / Thr
(f) core 3
β1,6
β 1,3
O – Ser / Thr
(g) core 4
α1,3
O – Ser / Thr
(h) core 5
β1,6
O – Ser / Thr
(i) core 6
α 1,6
O – Ser / Thr
(j) core 7

epitope can also act on T antigen to yield sialyl T (sT or sTF) (Figure 2.4d); sialylation terminates further chain extension. T antigen is usually cryptic, being either sialylated or subject to further chain extension, but is frequently exposed in cancer [76, 77].

2.7.5.4 Synthesis of Core 2, GlcNAc(β1 → 6)[Galβ(1 → 3)]GalNAcα1-Ser/Thr

If core 1 remains unsialylated, it may be extended by the addition of further monosaccharides. It can be converted into the simple biantennary core 2 by the addition of a side branching (β1 → 6)GlcNAc (Figure 2.4e). This addition is catalyzed by the action of β1,6-*N*-acetylglucosaminyltransferase (UDP-GlcNAc: Gal(β1 → 3) GalNAc-R or (β1 → 6)GlcNAc transferase, core 2 β1,6 GlcNAc-T or β1,6 GlcNAc-T). Core 2 is GlcNAc(β1 → 6)[Galβ(1 → 3)]GalNAcα1-Ser/Thr. Core 2-based glycans are the most common type of O-linked oligosaccharides.

Further elongation of core 2 occurs in a similar manner to elongation of core 1, and is described below, but is slightly more complicated in that elongation can occur at both the Gal and the side branching GlcNAc residues of the core 2 structure.

2.7.5.5 Synthesis of Core 3, GlcNAc(β1 → 3)GalNAcα1 → Ser/Thr

Core 3 is formed from the Tn epitope by the action of β1,3-*N*-acetylglucosaminyltransferase (UDP-GlcNAc: GalNAc-R β1,3-GlcNAc transferase, core 3 β1,3 GlcNAc-T), which adds a (β1 → 3)GlcNAc to give GlcNAc(β1 → 3)GalNAcα1 → Ser/Thr (Figure 2.4f).

2.7.5.6 Synthesis of Core 4, GlcNAc(β1 → 6)[GlcNAc(β1 → 3)]GalNAcα1-Ser/Thr

Core 3 can be converted into the branching core 4 by the action of β1,6-*N*-acetylglucosaminyltransferase (UDP-GlcNAc: GlcNAc(β1 → 3)GalNAc-R or GlcNAc to GalNAc β1,6 GlcNAc transferase, core 4 β1,6 GlcNAc-T, or β1,6-GlcNAc-T), which adds a (β1 → 6) GlcNAc side branch to yield the simple bi-antennary core structure GlcNAc(β1 → 6)[GlcNAc(β1 → 3)]GalNAcα1-Ser/Thr (Figure 2.4g).

2.7.5.7 Less Common Core Structures

The most common O-linked glycan core structures are types 1 to 4 described above; however, other less commonly occurring core structures have been described [43, 55, 62]. These include:

Core 5: GalNAc(α1 → 3)GalNAcα1-Ser/Thr (Figure 2.4h), which has been reported on some human adenocarcinomas and on embryonic gut cells. This is

Figure 2.4 Mucin-type *O*-glycan core structures. Cores 1–4 are found commonly while cores 4–7 have only occasionally been described. (a) Mucin-type *O*-glycan synthesis begins with attachment of a single GalNAc to a Ser or Thr residue on the polypeptide chain to yield GalNAcα1 → Ser/Thr, Tn epitope; (b) Tn can be sialylated to yield NeuNAc(α2 → 6)GalNAcα1 → Ser/Thr, sialyl Tn. Sialylation terminates chain extension; (c) (β1 → 3) linked Gal can be added to Tn to yield Gal(β1 → 3)GalNAcα-Ser/Thr, the Thomsen–Friedenreich (T or TF) antigen, also called core 1; (d) T antigen can be sialylated to give sialyl T. Sialylation terminates further chain extension; (e) a side branching (β1 → 6)GlcNAc can be added to core 1 to give GlcNAc(β1 → 6) [Galβ(1 → 3)]GalNAcα-Ser/Thr, core 2; (f) addition of a (β1 → 3)GlcNAc to Tn yields GlcNAc(β1 → 3)GalNAcα1 → Ser/Thr, core 3; (g) addition of a (β1 → 6) GlcNAc side branch to core 3 yields GlcNAc(β1 → 6)[GlcNAc(β1 → 3)] GalNAcα-Ser/Thr, core 4; (h) core 5; (i) core 6; (j) core 7. Key to symbols: GalNAc: ♦; GlcNAc: ■; Gal: ◇; sialic acid: (★).

synthesized from GalNAcα-Ser/Thr by the action of a core 5 *N*-acetylgalactosaminyltransferase (core 5 α1,3 GalNAc-T).

Core 6: GlcNAc(β1 → 6)GalNAcα1-Ser/Thr (Figure 2.4i), which has been reported on human embryonic gut and ovarian cyst mucins. This is synthesized from GalNAcα-Ser/Thr by the action of a core 6 *N*-acetylglucosaminyltransferase (core 6β1,6 GlcNAc-T).

Core 7: GalNAc(α1 → 6)GalNAcα1-Ser/Thr (Figure 2.4i), which has been reported on bovine submaxillary mucins. This is synthesized from GalNAcα-Ser/Thr by the action of a core 7 *N*-acetylgalactosaminyltransferase (core 7 α1,6 GalNAc-T).

2.7.6
Regulation of *O*-Glycan Synthesis

Control of *O*-glycosylation is largely regulated by glycosyltransferase availability and activity, with these enzymes competing for available substrates. However, in addition, the availability of nucleotide sugar donors, acceptor substrates and cations, and subcellular membrane organization also influence the final glycosylation repertoire of the cell. Crucially, also, the rate at which potentially O-linked glycoproteins are synthesized, and their transit rate through the secretory pathway, will have an effect; if throughput is high, the machinery of glycosylation may not be able to cope with the demand, and incomplete synthesis of oligosaccharides will therefore result. Patterns of *O*-glycosylation of proteins commonly change during development, differentiation, growth and in disease, and cell or tissue types often feature characteristic *O*-glycans associated with their specialized biological functions.

2.8
O- and N-linked Glycan Chain Extension and Commonly Occurring Glycan Motifs

The *N*- and *O*-glycan core structures described previously can be extended by the stepwise addition of monosaccharides, the addition of each catalyzed by glycosyltransferases, to build commonly occurring glycan motifs, some of which are briefly described below and in References [23, 62].

2.8.1
Type I Chains or Neolactosamine Units, Gal(β1 → 3)GlcNAc/GalNAc

The addition of a (β1 → 3) linked Gal, catalyzed by a β1,3 galactosyltransferase, to a terminal GlcNAc yields Gal(β1 → 3)GlcNAc/GalNAc, the type 1 chain or neolactosamine unit. Type 1 chains are normally then extended with Fuc, GalNAc, Gal and SA residues, attached by various linkages, to yield a range of commonly occurring glycan motifs, including the ABH and Lewis blood group system antigens [78]. They are a feature of glycans associated with epithelial surfaces, such as the lining of the reproductive and gastrointestinal systems.

2.8.2 Type 2 Chains or Lactosamine Units, Gal(β1 → 4)GlcNAc

The most common core extension begins with the addition of a (β1 → 4) linked Gal, catalyzed by a β1,4 galactosyltransferase, to a terminal GlcNAc residue to yield Gal (β1 → 4) GlcNAc, the type II chain or lactosamine unit. Repeating sequences of lactosamines units, $[Gal(\beta1 \rightarrow 4)GlcNAc(\beta1 \rightarrow 3)Gal]_n$, referred to as polylactosamine chains, are a common feature of elongated O-linked glycans. The lactosamine unit can act as an acceptor substrate for many modifications (e.g., addition of Lewis and ABH glycans motifs [78]) to both the terminal Gal and the sub-terminal GlcNAc, yielding many diverse linear and branching structures.

2.8.3 Termination of Glycan Chains

Sialylation or fucosylation terminates chain extension and these residues are commonly found as the terminal monosaccharides in both *N*- and *O*-glycans. Further modifications of the glycans, including sulfation, methylation and phosphorylation, result in further heterogeneity in the range of structures possible.

Common sialylated terminal structures include NeuNAc(α2 → 6)GalNAc- (sialyl Tn), NeuNAc(α2 → 3)Gal(β1 → 3)GalNAc- and NeuNAc(α2 → 6)[NeuNAc(α2 → 3) Gal(β1 → 3)]GalNAc- (sialylated core 1).

Sialic acids, sulfated groups and Fuc(α1 → 3) or Fuc(α1 → 4) residues may also be attached as side branches to internal sugar residues.

2.8.4 Blood Group Antigens

Many O-linked and N-linked glycans terminate in glycan motifs such as ABO, Lewis (a family of α1–3-fucosylated glycan structures), Cad and S^d blood group antigens. These are described in Referencc [78].

2.9 Analytical Methodologies Developed to Detect and Characterize Glycosylation

2.9.1 Glycan Analysis is Complex and Requires a Number of Techniques

As described previously, the chemistry of glycans is complex. They are built from a range of chemically very similar monosaccharide units that can be linked together through their different carbon atoms in α or β linkages to form linear or branching chains of variable length and complexity and they exhibit great structural diversity. For this reason, their analysis is far more technically challenging than that of other classes of biological molecules, such as proteins or nucleic acids. One method rarely provides full details on oligosaccharide composition and linkage analysis of the constituent monosaccharides, and so several methods usually need to be employed in combination to achieve full oligosaccharide sequence information. Some of these techniques require expensive and specialist equipment and considerable technical expertise. There is currently interest in developing approaches to provide more sensitive, high throughput, less expensive and more broadly accessible approaches to analyzing these structurally heterogeneous molecules. This topic is reviewed in References [79–86].

The particular approaches adopted to analyze a particular sample will depend on several factors, including: the complexity of the sample, the amount of sample available for analysis, the equipment and expertise available and how much detail regarding composition is required. Recently formed core facilities provide databases, reagents and analytical services [87] and informatics resources are becoming more established (recently reviewed in Reference [88]).

2.9.2 Detection of Glycans

2.9.2.1 Periodic Acid-Schiff (PAS) Reaction

Traditionally, glycosylated molecules, including glycoproteins, have been identified in tissue sections or in electrophoretic gels using the periodic acid-Schiff (PAS) reaction, which, owing to the susceptibility of carbohydrates to periodate oxidation, stain these substances pink. The standard PAS reaction is fairly insensitive and commercial kits are now available that amplify the signal, allowing detection of as little as 5–10 ng of glycoprotein [89].

2.9.2.2 Recognition Through Lectin Binding

Glycoproteins can also be detected through their recognition by lectins. Lectins are naturally occurring proteins and glycoproteins that recognize glycan structures with a high degree of selectively [90, 91]. Around 200 different lectins are available commercially; some are well characterized with defined molecular structure and carbohydrate binding characteristics, while others are little explored. Many are defined simply by their binding preference for a particular monosaccharide – for example, the lectin Concanavalin A is usually said to be mannose-binding – but their naturally occurring binding partners, and therefore their binding preference, is usually for (often undetermined) more complex glycan structures [92–95]. They can discriminate between subtly different glycans. Lectins are useful tools for investigating glycosylation; for example, for purification of specific glycoconjugates, investigation of cellular glycosylation patterns and isolation of cell populations on that basis. They may be used in histochemical techniques to map glycan structures on cells and in tissue sections in a manner analogous to the way that antibodies are used to detect cell or tissue-bound antigens in immunohistochemistry, at the light, electron or confocal microscope level [92–95]. Lectins are also widely used in affinity purification of glycoproteins, glycopeptides and free glycans and can be employed bound to supporting matrices, including agarose, Sepharose or magnetic beads. Serial lectin affinity chromatography facilitates sequential fractionation of glycoconjugates on a series of affinity columns bearing lectins with differing carbohydrate binding characteristics. Lectins with a relatively broad binding repertoire can be used to recover pools of glycoproteins sharing common determinants – sometimes called a "glyco-catch" approach – and lectins with finer specificity can be used to differentiate glycoforms. Lectins have also been employed in arrays for quantitative and qualitative glycan profiling. Antibodies against specific glycan structures can, of course, also be used in these approaches, but the range of good commercially available antibodies is very limited. Lectins and antibodies are useful tools to probe glycosylation but are limited in their ability to reveal the intricacies of glycan structure.

2.9.2.3 Pulsed Amperometric Detection (PAD) of Unlabelled Free Glycans

Unlabelled monosaccharides and oligosaccharides chemically or enzymatically released from glycoproteins are difficult to detect. Unlike proteins and peptides, monosaccharides and oligosaccharides only weakly absorb light in the ultraviolet range but may be detected in native form using refractive index or pulsed amperometric detection (PAD) [96, 97].

2.9.2.4 Labeling Glycans with Radiolabels or Fluorescent Labels

For convenience and also for sensitivity of detection, glycans are commonly labeled. They may be radiolabeled metabolically by isotope incorporation during synthesis either *in vitro* or *in vivo* and then detected by conventional methods. However, derivatization with a fluorescent label by reductive amination is now the most common approach to labeling glycans. Examples of labels include 2-aminobenzamide, [2-AB], 2-aminopyridine, [2-AP], 2-amino-antranillic acid [2-AA] and 8-amino-naphthalene-1,3,6-trisulfonic acid, [ANTS]. Fluorescent tagging with these types of label allows quantitative detection in the sub-picomolar range. This level of sensitivity is important as glycans for analysis are often present in tiny amounts (e.g., see References [98–100]). The use of 2,6-diaminopyridine (DAP) to generate fluorescently labeled glycans that contain a primary amine, facilitating further conjugation for immobilization on solid supports and quantification of very tiny samples, has been described recently [101].

2.9.3 Profiling Glycans Using Microarrays

There has recently been interest in using "chips" arrayed with lectins or antibodies to explore the glycosylation of biomolecules, cells, biological fluids or cell lysates [102–105]. The glycosylated samples are usually fluorescently labeled and the pattern of their binding to the arrays indicates the range/profile of glycan structures present. Such an approach provides a moderate amount of structural information in a convenient and high-throughput format. This type of glycosylation profiling can be performed on as little as 1 μg of sample and is possible using conventional scanners and technology available in many research laboratories [103, 105]. N- and O-linked glycans do not need to be analyzed separately, as they do with many other methods.

At present, this approach is limited by the range of (defined) glycan epitopes for which lectins or antibodies are available and the cross-specificity of some lectins that recognize several glycan structures (although this can sometime be advantageous). Driven by the need for this type of technology, there is currently much work being done to more clearly define lectin binding characteristics and in future recombinant lectins may become available, overcoming the problems of using necessarily variable reagents derived from natural sources [102, 106]. Furthermore, while at present the data yielded by this type of approach need to be analyzed by hand, in future, computer-assisted algorithms may streamline the interpretation. As the technology matures, it is likely to become an important, powerful and readily accessible tool for high-throughput analysis of glycomics.

2.9.4
One- and Two-Dimensional Gel Electrophoresis – Exploring the Glycome

Glycoproteins can be separated by SDS-PAGE and identified in the gel by staining using the PAS reaction. They may alternatively be transferred by electroblotting onto nitrocellulose or other membrane supports and labeled for the binding of specific antibodies or lectins [89, 107]. Glycoproteins separated in two-dimensional (2D) gels [separation according to molecular weight by SDS-PAGE in one dimension, then by isoelectric point (pI) in a second dimension] can also be treated in a similar manner to reveal information regarding glycosylation profiles [108–112]. Owing to their heterogeneous glycosylation patterns, which influences their molecular mass and isoelectric point, glycoproteins often separate as broad, poorly resolved bands on SDS-PAGE gels and as a "train" of spots on 2D gels, and some large molecular weight mucins may be too large to enter conventional separating gels. Another limitation of these type of approaches is the under-representation of membrane glycoproteins, which are difficult to solubilize completely.

Glycoproteins, when successfully separated and identified as being of interest by these types of methods, can be extracted from the gel or blotting membrane and the glycans released and analyzed by other methods, such as HPLC or mass spectroscopy (described below). Detailed information on the glycosylation of proteins present in complex mixtures, such as cellular or tissue extracts of protein, can be obtained. This sort of approach, that of releasing glycans from glycoproteins separated by SDS-PAGE prior to their subsequent analysis by HPLC, is, for example, described in References [113, 114].

2.9.5
Chemical Release and Analysis of Monosaccharides

Identification of the constituent monosaccharides and analysis of their relative proportion present in a glycan or glycan mixture enables at least preliminary identification of the class of oligosaccharide to be determined. Acid hydrolysis is used to cleave the glycosidic linkages between monosaccharides, breaking the glycan chain into its component parts. The free monosaccharides can then analyzed by high-performance liquid chromatography (HPLC) or by gas–liquid chromatographic separation with, usually, mass spectrometry detection, as described below [115].

2.9.6
Chemical Release of Intact Oligosaccharides

Intact oligosaccharides can be released from glycoproteins, even from cell or tissue samples, by the action of anhydrous hydrazine [116]. Hydrazinolysis removes the acetyl groups from acetylated monosaccharides – so chemical procedures need to be adapted to re-*N*-acetylate the oligosaccharide following hydrazinolysis. This approach can be adapted to optimally release O- or N-linked glycans, or mixtures of both. O-linked oligosaccharides can also be released from glycoproteins by mild alkali treatment, which is also called beta-elimination.

2.9.7
Enzymatic Release of Intact Oligosaccharides

N-linked oligosaccharides can be enzymatically released from glycoproteins by the enzyme peptide *N*-glycosidase F (PNGase F) or PNGase A (for example, see [117]). Endo H may be used for the selective release of high-mannose and hybrid type N-linked structures. This approach can be used to release N-linked oligosaccharides from glycoproteins separated in SDS-PAGE or 2D gel electrophoresis without the need to extract the glycoprotein from the gel before release [118].

2.9.8
Sequential Exoglycosidase Digestions to Provide Monosaccharide Sequence and Linkage Data

To obtain information regarding the sequence of monosaccharides in a glycan chain and the linkages between them, sequential exoglycosidase digestion can be used to cleave monosaccharides from oligosaccharides in a stepwise manner. For example, if a sample is treated with a sialidase that specifically cleaves $\alpha 2 \rightarrow 3$ linked NeuNAc and the position of the peak corresponding to that sample then shifts position on an HPLC or mass spectrometry analysis then that constitutes evidence that the glycan contained a terminal $\alpha 2 \rightarrow 3$ linked NeuNAc. The sample would then be treated with a second specific glycosidase and a corresponding peak shift sought, and so on. An alternative to sequential exoglycosidase digestions is exoglycosidases "arrays" (called the "reagent array analysis method," RAAM), a relatively fast approach to characterizing monosaccharide sequences in a glycan. Here different cocktails of exoglycosidases are applied to parallel samples simultaneously and analysis of the peak shift allows determination of the original sequence/linkage of the monosaccharides in the glycan. This type of approach can provide good data on relatively small glycans but is time consuming and its scope is limited by the specific exoglycosidases available.

2.9.9 Oligosaccharide Separation and Mapping by High-Performance Liquid Chromatography (HPLC)

One of the most commonly employed approaches for the analysis of N- and O-linked glycans released from glycoprotein preparations, either chemically or enzymatically as described above, is high-performance liquid chromatography (HPLC). Different types of HPLC separate glycans on the basis of different properties and it is common for more than one approach to be applied sequentially to a single sample. Moreover, HPLC is commonly employed in tandem with other approaches, such as mass spectrometry or nuclear magnetic resonance, to yield full structural analysis. Glycans may be analyzed either in their native state or after conjugation to a label.

Recently, an automated approach to analyzing femtomole quantities of N-linked glycans released from glycoproteins from biological samples, such as serum or tissue extracts, has been described [119]. Samples are first immobilized on 96-well plates, glycans are released and labeled fluorescently, subjected to exoglycosidase digestions and then analyzed by quantitative HPLC analysis. Structures are then assigned using web-based software accessing a database of *N*-glycan structures.

2.9.9.1 Normal Phase HPLC (NP-HPLC)

Normal phase HPLC (NP-HPLC) [80] separates glycans on the basis of differences in their hydrophilicity. Larger glycans are retained on the column for longer because they are more hydrophilic and this has the effect of separating glycans according to their size. There is also some separation of different isomers. Commonly, an oligomeric mixture of glucose polymers termed a "dextran ladder" is run alongside experimental samples for the purpose of calibration. The size of experimental glycans separated by this method is then described in "glucose units" (GU) based on their elution positions relative to the dextran ladder. Putative structures can be assigned to experimental glycans based on comparison of GU values with those of known standards. Some data on linkages and monosaccharide sequences of experimental glycans can be obtained by sequential digestion with specific exoglycosidases, as described previously, as the enzymes cleave monosaccharides from the non-reducing end of the glycan, causing a predictable shift in the GU value.

2.9.9.2 Weak Anion Exchange High-Performance Liquid Chromatography (WAX-HPLC)

In weak anion exchange high-performance liquid chromatography (WAX-HPLC) [120] glycans are separated by their electrical charge, which relates to the number of (negatively charged) sialic acid residues present. Retention times are also influenced by the overall size of the glycan, such that smaller monosialylated structures are retained on the column longer than larger monosialylated oligosaccharides. Linkage data can be obtained by digestion of samples with neuraminidases that cleave specific linkages, such as $\alpha 2 \rightarrow 3$ linkages, followed by re-analysis to seek a peak shift.

2.9.9.3 High-Performance (or High pH) Anion Exchange Chromatography (HPAEC)

High-performance anion exchange chromatography, also known as high pH anion exchange chromatography (HPAEC) can be used to separate glycans. Separation is performed in strong alkaline solutions, usually high molarity sodium hydroxide solution [121, 122]. Columns are available for the separation of mono- and oligosaccharides. HPAEC has often been used in conjunction with pulsed amperometric detection (HPAEC-PAD) for direct analysis of unlabeled free glycans. Contaminating peptides can be derivatized with 2,4,6-trinitrobenzene-1-sulfonate (TNBS), which makes them strongly hydrophobic. They can then be separated from the glycans as they are retained strongly on a graphitized carbon column [123] and this can be a useful initial clean-up step prior to further analysis (e.g., by mass spectrometry).

2.9.10 Separation and Mapping of Oligosaccharides by Fluorophore-Assisted Carbohydrate Electrophoresis (FACE)

Fluorophore-assisted carbohydrate electrophoresis (FACE) is also sometimes termed polyacrylamide gel electrophoresis of fluorophore labeled saccharides (PAGEFS) [124]. Fluorescently-labeled glycans are separated electrophoretically in high-density polyacrylamide gels (30–60% acrylamide) with cooling, visualized under ultraviolet light and images captured on film or by camera with a charge-coupled device. Multiple samples of both N- and O-linked glycans released from glycoproteins can be analyzed and the technique has the advantage that it uses low cost equipment that is commonly available in many laboratories. The sensitivity of detection is in the sub-picomolar range.

2.9.11 Separation and Mapping of Oligosaccharides by Capillary Electrophoresis (CE)

Capillary electrophoresis (CE) separates glycans on the basis of their electrical charge. It is performed under an electric field of several hundred volts per cm in fused-silica capillaries 20–50 cm long and of approximately 50 μm internal diameter. Glycans are first derivatized with a compound containing sulfonic groups, which impart a net negative charge, enabling their migration in the electric field and their subsequent detection. Fluorescently labeled oligosaccharides are detected in nanomolar quantities and separation is possible in a short timescale, usually less than half an hour. It is often used in conjunction with mass spectrometry approaches.

In capillary affinity electrophoresis (CAE), fluorescently-labeled glycans are first separated by CE. They are then subjected to CE in the presence of a lectin with defined specificity in the electrolyte. If the lectin recognizes glycan in the sample, the appearance of a peak is delayed in comparison to the original run. If the lectin does not recognize glycan in the sample, the peak appears in the same position as in the absence of the lectin. The procedure is then repeated in the presence of a series of lectins with known specificities, allowing categorization of glycans in the sample [125].

2.9.12 Oligosaccharide Analysis by Nuclear Magnetic Resonance (NMR)

Nuclear magnetic resonance (NMR) [81] analysis is based on the extent to which a material (in this case a glycan) distorts under a magnetic field. It is a very powerful tool for glycan analysis. It is non-destructive and is often therefore used as a first analysis technique because samples can be subsequently subjected to other, destructive, approaches. It is the only technique currently available that can provide full structural information, including anomericity of monosaccharides, in a complex glycan. However, usually, 2D NMR, such as correlated spectroscopy (COSY) and total correlation spectroscopy (TOCSY) for ^{1}H, then heteronuclear single-quantum coherence (HSQC) for ^{13}C, need to be performed to determine full structural analysis. 2D ^{1}H NMR spectroscopy, using through-space effects (nuclear Overhauser effects, NOEs) can also yield full structural analysis. Milligram amounts of sample are, however, required for these methods. The use of this technique is further limited by both the high capital cost of equipment and the requirement for specialist training in its use.

2.9.13 Determining the Mass of an Oligosaccharide Using Mass Spectrometry (MS)

Mass spectrometry (MS) enables the determination of the mass weight of tiny samples (pico- to femto-mole range) of glycans with great accuracy and allows distinctions to be made between oligosaccharides with very similar properties that are difficult to separate chromatographically. Typically, glycans are chemically or enzymatically released from the protein and derivatized and several chromatography steps are then carried out to obtain a sample appropriate for analysis. This is a complex technique, a detailed description of which goes beyond the scope of this chapter.

These purification steps are often the most cumbersome and time consuming part of the analysis and recently approaches have been developed using carbohydrate capture agents to clean up samples from biological sources, such as serum samples, prior to MS analysis [126, 127]. A potential advantage of this type of approach is that the chemistry allows the incorporation of a mass label [127, 128]. It is anticipated that such developments will further MS technology for high-throughput analysis.

Following sample preparation and labeling, N-links or O-links are usually analyzed separately, and the techniques for analyzing O-links are generally less well developed. Another shortcoming of MS is that sample preparation techniques often remove modifications such as sulfation and methylation.

Initial analysis gives data regarding the proportion of different types of monosaccharides (e.g., hexoses such as Glc, Man, and so on, the *N*-acetylated hexoses

GlcNAc and GalNAc, and sialic acids). Mass numbers of many component monosaccharides are the same – for example, Glc, Gal and Man are identical – and they therefore cannot be distinguished. Further analysis on sub-fractions of interest can then be performed. Methods for analyzing protein glycosylation by MS have recently been reviewed [85, 129]. Protocols for glycomic profiling of cells and tissues using MS have also been described recently [130–132]. Annotation and interpretation of the mass spectra obtained is complex, requires specialist knowledge and is time consuming. Computer-assisted databases and algorithms are being developed to streamline this part of the analysis process (e.g., References [133–136]).

Matrix-assisted laser desorption/ionization mass spectrometry (MALDI-MS) [137] is the most widely available and straightforward type of MS. It yields relatively uncomplicated spectra, presenting information regarding mass weight and mass-to-charge ratio (*m*/*z*) of glycans which, if the machine is calibrated with known standards, enables putative compositional and thus structural assignations to be made by comparison with known database values. MALDI-MS is often coupled with time of flight (TOF) analysis and this approach can resolve glycoforms of glycoproteins with limited glycosylation sites [138].

Fast atom bombardment (FAB) and electrospray ionization (ESI) [139] MS can be used to yield information regarding monosaccharide linkage and position from ionization fragments of the oligosaccharides. MALDI-MS with NP-HPLC and online electrospray ionization (LC-ESI-MS) have been used in combination with other chromatographic techniques to provide very complete composition and structural information on glycans [140, 141]. In recent years, MS instruments have become much more affordable and bench-top models are commonly seen in laboratories; however, this type of approach still requires considerable expertise in instrument operation, sample preparation and interpretation of results. MALDI and negative ion nanospray mass spectrometry have recently been reported to be among the most useful mass spectrometric techniques yet developed for N-linked glycan analysis [142, 143].

2.9.14
Gas–Liquid Chromatography (GLC)/MS for Determining the Linkage Position of Monosaccharides in Oligosaccharides

Determining the linkage and position of monosaccharides in a complex glycan is difficult and multiple iterative methods are often required. Methylation analysis in combination with gas–liquid chromatography (GLC) and MS is the most widely used approach to obtain information on both the monosaccharides present and their linkages to one another in the glycan. However, the anomericity (α or β configuration) of the monosaccharide linkages can not be determined and, also, this approach does not provide sequence information. The principle of methylation analysis is to introduce a methyl group in place of each free hydroxyl group of the native oligosaccharide. The glycosidic linkages are then cleaved to yield a mixture of individual monosaccharide residues with newly-exposed hydroxyl groups at the positions that were previously involved in a glycosidic bond. The monosaccharides are then labeled and analyzed by GLC-MS. Identification is based on a combination of their retention time on the GLC column and their electron impact-MS fragmentation pattern in comparison to known standards. Original oligosaccharide samples of >10 picomole are required for this type of approach.

2.10 Conclusion

Most human proteins are glycosylated in various subtly different glycoforms. The biological control of glycosylation is not understood but appears to rely on factors including competition between the enzymes of glycosylation that act sequentially on proteins as they pass through the RER and GA and is different in different cells and tissues and under different physiological conditions. Glycoproteins have myriad biological functions and changes in their glycosylation pattern influence their biological activity. Human cells glycosylate proteins very differently to the cells of other species. The chemistry of glycosylation is complex and results in astonishing diversity of glycan structures. Their analysis is technically challenging and usually requires several approaches to yield full structural information. A major challenge for the biotechnology industry is understanding – and ultimately controlling – the regulation of glycosylation by cells, especially those cell types used in expression systems that glycosylate their proteins very differently to human cells. There is also a pressing need to develop new, simpler, high-throughput technologies for analyzing glycans derived from complex biological samples that do not require expensive capital equipment and sophisticated technical expertise.

References

1 Dwek, R.A. (1996) Glycobiology: toward the understanding of the function of sugars. *Chemical Reviews*, **96**, 683–720.
2 Rudd, P.M. and Dwek, R.A. (1991) Glycobiology: a coming of age. *Chemistry & Industry*, **2**, 660–663.
3 Rudd, P.M. (1993) Oligosaccharides in human biology. *Glyco News*, **3**, 1–8.
4 Fukuda, M. (1994) Cell surface carbohydrates: cell-type specific expression, in *Molecular Glycobiology* (eds M. Fukuda and O. Hindsgaul), IRL Press, Ch 1, pp. 1–52.
5 Taylor, M.E. and Drickamer, K. (2006) Glycosylation and disease, in *Introduction to Glycobiology*, 2nd edn, Oxford University Press, Ch 12, pp. 218–235.
6 Varki, A. *et al.* (1999) Historical background and overview, in *Essentials of Glycobiology*, Cold Harbor Laboratory Press, New York, Ch 1, pp. 1–16.
7 Brooks, S.A. (2004) Appropriate glycosylation of recombinant proteins for human use: implications of choice of expression system. *Molecular Biotechnology*, **28**, 241–256.
8 Brooks, S.A. (2006) Protein glycosylation in diverse cell systems: implications for modification and analysis of recombinant proteins. *Expert Review of Proteomics*, **3**, 345–359.

9 Brooks, S.A., Dwek, M.V. and Schumacher, U. (2002) Carbohydrate biotechnology, in *Functional and Molecular Glycobiology*, Bios Scientific Publishers, Ltd, Oxford, UK, Ch 18, pp. 329–345.

10 Apweiler, R., Hermjakob, H. and Sharon, N. (1999) On the frequency of protein glycosylation as deduced from analysis of the SWISS-PROT database. *Biochimica et Biophysica Acta*, **1473**, 4–8.

11 Hagglund, P. *et al.* (2004) A new strategy for identification of I-glycosylated proteins and unambiguous assignment of their glycosylation sites using HILIC enrichment and partial deglycosylation. *Journal of Proteome Research*, **3**, 556–566.

12 Kameyama, A. *et al.* (2006) Strategy for simulation of CID spectra of N-linked oligosaccharids towards glycomics. *Journal of Proteome Research*, **5**, 808–814.

13 Rademacher, T.W., Parekh, R.B. and Dwek, R.A. (1998) Glycobiology. *Annual Review of Biochemistry*, **57**, 785–838.

14 Delves, P.J. (1998) The role of glycosylation in autoimmune disease. *Autoimmunity*, **27**, 239–253.

15 Haltiwanger, R.S. and Lowe, J.B. (2004) Role of glycosylation in development. *Annual Review of Biochemistry*, **73**, 491–537.

16 Ohtsubo, K. and Marth, J.D. (2006) Glycosylation in cellular mechanisms of health and disease. *Cell*, **126**, 855–867.

17 Rudd, P.M., Wormald, M.R. and Dwek, R.A. (2004) Sugar-mediated interactions in the immune system. *Trends in Biotechnology*, **22**, 524–530.

18 Roseman, S. (2001) Reflections on glycobiology. *The Journal of Biological Chemistry*, **276**, 41527–41542.

19 Brooks, S.A., Dwek, M.V. and Schumacher, U. (2002) Disease processes in which carbohydrates are involved, in *Functional and Molecular Glycobiology*, Bios Scientific Publishers, Ltd, Oxford, UK, Ch 16, pp. 287–312.

20 Dennis, J.W. (1992) Oligosaccharides in carcinogenesis and metastasis. *Glyco News*, **2**, 1–4.

21 Dall'Olio, F. (1996) Protein glycosylation in cancer biology: an overview. *Journal of Clinical Pathology*, **49**, M126–M135.

22 Fukuda, M. (1996) Possible roles of tumour associated carbohydrate antigens. *Cancer Research*, **56**, 2237–2244.

23 Ho, S.B. and Kim, Y.S. (1997) Glycoproteins and glycosylation changes in cancer, in *Encyclopaedia of Cancer*, Vol II, Academic Press, pp. 744–759.

24 Kannagi, R. (1997) Carbohydrate-mediated cell adhesion involved in hematogenous metastasis of cancer. *Glycoconjugate Journal*, **14**, 577–584.

25 Kim, Y.J. and Varki, A. (1997) Perspectives on the significance of altered glycosylation of glycoproteins in cancer. *Glycoconjugate Journal*, **14**, 569–576.

26 Price, J.T., Bonovich, M.T. and Kohn, E.C. (1997) The biochemistry of cancer dissemination. *Critical Reviews in Biochemistry and Molecular Biology*, **32**, 175–253.

27 Dube, D.H. and Bertozzi, C.R. (2005) Glycans in cancer and inflammation – potential for therapeutics and diagnostics. *Nature Reviews. Drug Discovery*, **4**, 477–488.

28 Varki, A. *et al.* (1999) Saccharide structure and nomenclature, in *Essentials of Glycobiology*, Cold Harbor Laboratory Press, New York, Ch 2, pp. 17–30.

29 Brooks, S.A., Dwek, M.V. and Schumacher, U. (2002) An introduction to carbohydrate chemistry, in *Functional and Molecular Glycobiology*, Bios Scientific Publishers, Ltd, Oxford, UK, Ch 1, pp. 1–30.

30 Van den Eijnden, D.H. and Joziasse, D.H. (1993) Enzymes associated with glycosylation. *Current Opinion in Structural Biology*, **3**, 711–721.

31 Ten Hagen, K.G., Fritz, T.A. and Tabak, L.A. (2003) All the family: the UDP:GalNAc:polypeptide

N-acetylgalactosaminyltransferases. *Glycobiology*, **13**, 1R–16.

32 Kleene, R. and Berger, E.G. (1993) The molecular and cell biology of glycosyltransferases. *Biochimica et Biophysica Acta*, **1154**, 283–325.

33 Brockhausen, I. and Schachter, H. (1997) Glycosyltransferases involved in *N*- and *O*-glycan synthesis, in *Glycosciences – Status and Perspectives* (eds H.-J. Gabius and S. Gabius), Chapman and Hall, Weinheim, Ch 5, pp. 79–111.

34 Joziasse, D.H. (1992) Mammalian glycosyltranferases: genomic organisation and protein structure. *Glycobiology*, **9**, 271–277.

35 Drickamer, K. and Taylor, M.E. (1998) Evolving views of protein glycosylation. *Trends in Biochemical Sciences*, **23**, 321–324.

36 Cornelli, E.M. *et al.* (2006) A focussed microarray approach to functional glycomics: transcriptional regulation for the glycome. *Glycobiology*, **16**, 117–131.

37 Bretscher, M.S. and Munro, S. (1993) Cholesterol and the Golgi apparatus. *Science*, **261**, 1280–1281.

38 Masibay, A.S. *et al.* (1993) Mutational analysis of the Golgi retention signal of bovine β−1,4-galactosyltransferase. *The Journal of Biological Chemistry*, **268**, 9908–9916.

39 Machamer, C.E. (1991) Golgi retention signals: do membranes hold the key? *Trends in Cell Biology*, **1**, 141–144.

40 Nilsson, I.M. and von Heijne, G. (1993) Determination of the distance between the oligosaccharyltransferase active site and the endoplasmic reticulum membrane. *The Journal of Biological Chemistry*, **268**, 5798–5801.

41 Schachter, H. (2000) The joys of HexNAc. The synthesis and function of *N*- and *O*-glycan branches. *Glycoconjugate Journal*, **17**, 465–483.

42 Varki, A. *et al.* (1999) *N*-glycans, in *Essentials of Glycobiology*, Cold Harbor Laboratory Press, New York, Ch 7, pp. 85–100.

43 Varki, A. *et al.* (1999) O-glycans, in *Essentials of Glycobiology*, Cold Harbor Laboratory Press, New York, Ch 8, pp. 101–114.

44 Brooks, S.A., Dwek, M.V. and Schumacher, U. (2002) N-linked glycoproteins, in *Functional and Molecular Glycobiology*, Bios Scientific Publishers, Ltd, Oxford, UK, Ch 4, pp. 73–88.

45 Brooks, S.A., Dwek, M.V. and Schumacher, U. (2002) O-linked (mucin-type) glycoproteins, in *Functional and Molecular Glycobiology*, Bios Scientific Publishers, Ltd, Oxford, UK, Ch 5, pp. 89–115.

46 Brooks, S.A. *et al.* (2008) Altered glycosylation of proteins in cancer: what is the potential for new anti-tumour strategies? *Anti-Cancer Agents in Medicinal Chemistry*, **8**, 2–21.

47 Yan, A. and Lennarz, W.J. (2005) Unraveling the mechanism of protein *N*-glycosylation. *The Journal of Biological Chemistry*, **280**, 3121–3124.

48 Hirschberg, C.B. and Snider, M.D. (1987) Topology of glycosylation in the rough endoplasmic reticulum and Golgi apparatus. *Annual Review of Biochemistry*, **56**, 63–87.

49 Helenius, J. and Aebi, M. (2002) Transmembrane movement of dolichol linked carbohydrates during *N*-glycoprotein biosynthesis in the endoplasmic reticulum. *Cell & Developmental Biology*, **13**, 171–178.

50 Sears, P. and Wong, C.H. (1998) Enzyme action in glycoprotein synthesis. *Cellular and Molecular Life Sciences*, **54**, 223–252.

51 Dempski, R.E. Jr, and Imperiali, B. (2002) Oligosaccharyl transferase: gatekeeper to the secretory pathway. *Current Opinion in Chemical Biology*, **6**, 844–850.

52 Parodi, A.J. (2000) Protein glucosylation and its role in protein folding. *Annual Review of Biochemistry*, **69**, 69–93.

53 Ito, Y. *et al.* (2005) Structural approaches to the study of oligosaccharides in glycoprotein quality control. *Current Opinion in Structural Biology*, **15**, 481–489.

54 Petrescu, A.J. *et al.* (2004) Statistical analysis of the protein environment of *N*-glycosylation sites: implications for occupancy, structure and folding. *Glycobiology*, **14**, 103–114.
55 Hanisch, F.-G. (2001) *O*-glycosylation of the mucin type. *Biological Chemistry*, **382**, 143–149.
56 Haltiwanger, R.S. *et al.* (1992) Glycosylation of nuclear and cytoplasmic proteins is ubiquitous and dynamic. *Biochemical Society Transactions*, **20**, 264–269.
57 Hart, G.W. (1997) Dynamic O-linked glycosylation of nuclear and cytoskeletal proteins. *Annual Review of Biochemistry*, **66**, 315–335.
58 Nishimura, H. *et al.* (1992) Human factor IX has a tetrasaccharide *O*-glycosidically linked to serine 61 through the fucose residue. *The Journal of Biological Chemistry*, **267**, 17520–17525.
59 Harris, R.J. *et al.* (1991) Tissue plasminogen activator has an O-linked fucose attached to threonine-61 in the epidermal growth factor domain. *Biochemistry*, **30**, 2311–2314.
60 Kieliszewski, M.J. *et al.* (1995) Tandem mass spectrometry and structural elucidation of glycopeptides from a hydroxyproline-rich plant cell wall glycoprotein indicate that contiguous hydroxyproline residues are the major sites of hydoxyproline *O*-arabinosylation. *The Journal of Biological Chemistry*, **270**, 2541–2549.
61 Duman, J.G. *et al.* (1998) *O*-Mannosylation of *Pichia pastoris* cellular and recombinant proteins. *Biotechnology and Applied Biochemistry*, **28**, 39–45.
62 Brockhausen, I. (1995) Biosynthesis of O-glycans of the N-acetylgalactosamine-α-Ser/Thr linkage type, in *Glycoproteins* (eds J. Montreuil, H. Schacter and J.F.G. Vliegenthart), Elsevier Science BV, Amsterdam, Ch 5, pp. 201–259.
63 Dekker, J. *et al.* (2002) Review: the MUC family – an obituary. *Trends in Biochemical Sciences*, **27**, 126–131.
64 Van den Steen, P. *et al.* (1998) Concepts and principles of O-linked glycosylation. *Critical Reviews in Biochemistry and Molecular Biology*, **33**, 151–208.
65 Brockhausen, I. *et al.* (1996) Specificity of *O*-glycosylation by bovine colostrums UDP-GalNAc: polypeptide alpha-*N*-acetylgalactosaminyltransferase using synthetic glycopeptide substrates. *Glycoconjugate Journal*, **13**, 849–856.
66 Hansen, J.E. *et al.* (1995) Prediction of *O*-glycosylation of mammalian proteins: specificity patterns of UDP-GalNAc:poly- peptide *N*-acetylgalactosaminyltransferase. *The Biochemical Journal*, **308**, 801–813.
67 Hansen, J.E. *et al.* (1998) NetOglyc: prediction of mucin type *O*-glycosylation sites based on sequence context and surface accessibility. *Glycoconjugate Journal*, **15**, 115–130.
68 Gupta, R. *et al.* (1999) O-GLYCBASE version 4.0: a revised database of *O*-glycosylated proteins. *Nucleic Acids Research*, **27**, 370–372.
69 Thanka Christlet, T.H. and Veluraja, K. (2001) Database analysis of *O*-glycosylation sites in proteins. *Biophysical Journal*, **80**, 952–960.
70 Julienus, K. *et al.* (2005) Prediction, conservation analysis, and structural characterisation of mammalian mucin-type *O*-glycosylation sites. *Glycobiology*, **15**, 153–164.
71 Pavelka, M. (1997) Topology of glycosylation – a histochemist's view, in *Glycosciences – Status and Perspectives*, (eds H.-J. Gabius and S. Gabius), Chapman and Hall, Weinheim, Ch 6, pp. 115–120.
72 Springer, G.F. (1984) T and Tn, general carcinoma auto-antigens. *Science*, **224**, 1198–1206.
73 Springer, G.F. (1989) Tn epitope (*N*-acetylgalactosamine α-*O*-serine/threonine) density in primary breast carcinoma: a functional predictor of aggressiveness. *Molecular Immunology*, **26**, 1–5.

74 Ibrahim, N.K. and Murray, J.L. (2003) Clinical development of the STn-KLH vaccine (Theratope). *Clinical Breast Cancer*, **3** (Suppl. 4), S139–S143.

75 Holmberg, L.A. and Sandmaier, B.M. (2004) Vaccination with Theratope (STn-KLH) as treatment for breast cancer. *Expert Review of Vaccines*, **3**, 655–663.

76 Hakomori, S. (2002) Glycosylation defining malignancy: new wine in an old bottle. *Proceedings of the National Academy of Sciences of the United States of America*, **99**, 10231–10233.

77 Kobata, A. and Amano, J. (2005) Latered glycosylation of proteins produced by malignant cells, and applications in the diagnosis and immunotherapy of tumors. *Immunology and Cell Biology*, **83**, 429–439.

78 Brooks, S.A., Dwek, M.V. and Schumacher, U. (2002) Glycan chain extension and some common and important glycan structures, in *Functional and Molecular Glycobiology*, Bios Scientific Publishers, Ltd, Oxford, UK, Ch 10, pp. 187–202.

79 Dwek, R.A. *et al.* (1993) Analysis of glycoprotein-associated oligosaccharides. *Annual Review of Biochemistry*, **62**, 65–100.

80 Rudd, P.M. *et al.* (1997) Oligosaccharide sequencing technology. *Nature*, **388**, 205–207.

81 Rudd, P.M. and Dwek, R.A. (1997) Rapid, sensitive sequencing of oligosaccharides from glycoproteins. *Current Opinion in Biotechnology*, **8**, 488–497.

82 Anumula, K.R. (2000) High-sensitivity and high-resolution methods for glycoprotein analysis. *Analytical Biochemistry*, **283**, 17–26.

83 Wuhrer, M., Deelder, A.M. and Hokke, C.H. (2005) Protein glycosylation analysis by liquid chromatography – mass spectrometry. *Journal of Chromatography B*, **825**, 124–133.

84 Geyer, H. and Geyer, R. (2006) Strategies for analysis of glycosylation. *Biochim Biophys Acta – Proteins and Proteomics*, **1764**, 1853–1869.

85 Mechref, Y. and Novotny, M.V. (2006) Miniturized separation techniques in glycomic investigations. *Journal of Chromatography B*, **841**, 65–78.

86 Pilobello, K.T. and Mahal, L.K. (2007) Deciphering the glycocode: the complexity and analytical challenge of glycomics. *Current Opinion in Chemical Biology*, **11**, 300–305.

87 Raman, R. *et al.* (2005) Glycomics: an integrated systems approach to structure-function relationships of glycans. *Nature Methods*, **2**, 817–824.

88 Mamitsuka, H. (2007) Informatic innovations in glycobiology: relevance to drug discovery. *Drug Discovery Today*, **13**, 118–123.

89 Patton, W.F. (2002) Fluorescence detection of glycoproteins in gels and on electroblots. *Current Protocols in Cell Biology*, John Wiley & Sons, Chapter 6, Unit 6.8.

90 Rudiger, H. and Gabius, H.J. (2001) Plant lectins: occurrence, biochemistry, functions and applications. *Glycoconjugate Journal*, **18**, 589–613.

91 Sharon, N. and Lis, H. (2004) History of lectins: from hemagglutinins to biological recognition molecules. *Glycobiology*, **14**, 53–62.

92 Schumacher, U., Brooks, S.A. and Leathem, A.J. (1991) Lectins as tools in histochemical techniques: a review of methodological aspects, in *Lectin Reviews*, Vol 1 (eds D.C. Kilpatrick, E. Van Driessche and T.C. Bog-Hansen), Sigma Chemical Company, St Louis, MO, USA, pp. 195–201.

93 Brooks, S.A., Leathem, A.J.C. and Schumacher, U. (1997) *Lectin Histochemistry – A Concise Practical Handbook*, Bios Scientific Publishers, Oxford, UK.

94 Brooks, S.A. and Hall, D.M.S. (2001) Lectin histochemistry to detect altered glycosylation in cells and tissues, in *Metastasis Research Protocols, Vol I: Analysis of Cells and Tissues* (eds S.A.

Brooks and U. Schumacher), Humana Press, Ch 4, pp. 49–66.

95 Carter, T.M. and Brooks, S.A. (2006) Detection of aberrant glycosylation in breast cancer using lectin histochemistry, in *Breast Cancer Research Protocols* (eds S.A. Brooks and A. Harris), Humana Press, Ch 16, pp. 201–216.

96 Cataldi, T.R., Campa, C. and De Benedetto, G.E. (2000) Carbohydrate analysis by high-performance anion-exchange chromatography with pulsed amperometric detection: the potential is still growing. *Fresenius' Journal of Analytical Chemistry*, **368**, 739–758.

97 Rohrer, J.S. (2000) Analyzing sialic acids using high-performance anion-exchange chromatography with pulsed amperometric detection. *Analytical Biochemistry*, **283**, 3–9.

98 Bigge, J.C. *et al.* (1995) Nonselective and efficient fluorescent labeling of glycans using 2-amino benzamide and anthranilic acid. *Analytical Biochemistry*, **230**, 229–238.

99 Zhuang, Z. *et al.* (2007) Electrophoretic analysis of *N*-glycans on microfluidic devices. *Analytical Chemistry*, **79**, 7170–7175.

100 Anumula, K.R. (2008) Unique anthranilic acid chemistry facilitates profiling and characterization of Ser/Thr-linked sugar chains following hydrazinolysis. *Analytical Biochemistry*, **373**, 104–111.

101 Xia, B. *et al.* (2005) Versatile fluorescent derivatization of glycans for glycomic analysis. *Nature Methods*, **2**, 845–850.

102 Hirabayshi, J. (2004) Lectin-based structural glycomics: glycoproteomics and glycan profiling. *Glycoconjugate Journal*, **21**, 35–40.

103 Pilobello, K.T. *et al.* (2005) Development of a lectin microarray for the rapid analysis of protein glycopatterns. *ChemBioChem*, **6**, 985–989.

104 Hsu, K.L., Pilobello, K.T. and Mahal, L.K. (2006) Analyzing the dynamic bacterial glycome with a lectin microarray approach. *Nature Chemical Biology*, **2**, 153–157.

105 Hsu, K.L. and Mahal, L.K. (2006) A lectin microarray approach for the rapid analysis of bacterial glycans. *Nature Protocols*, **1**, 543–549.

106 Manimala, J.C. *et al.* (2005) Carbohydrate array analysis of anti-Tn antibodies and lectins reveals unexpected specificities: implications for diagnostic and vaccine development. *Chem Bio Chem*, **6**, 2229–2241.

107 Osborne, C. and Brooks, S.A. (2006) SDS-PAGE and Western blotting to detect proteins and glycoproteins of interest in breast cancer research, in *Breast Cancer Research Protocols* (eds S.A. Brooks and A. Harris), Humana Press, Ch 17, pp. 217–230.

108 Wilson, N.L. *et al.* (2002) Sequential analysis of *N*- and O-linked glycosylation of 2D-PAGE separated glycoproteins. *Journal of Proteome Research*, **1**, 521–529.

109 Dwek, M.V. and Rawlings, S.L. (2002) Current perspectives in cancer proteomics. *Molecular Biotechnology*, **22**, 139–152.

110 Dwek, M.V. and Rawlings, S.L. (2006) Breast cancer proteomics using two-dimensional electrophoresis: studying the breast cancer proteome, in *Breast Cancer Research Protocols* (eds S.A. Brooks and A. Harris), Humana Press, Ch 18, pp. 231–244.

111 Dwek, M.V. and Alaiya, A.A. (2003) Proteome analysis enables separate clustering of normal breast, benign breast and breast cancer tissues. *British Journal of Cancer*, **89**, 305–307.

112 Block, T.M. *et al.* (2005) Use of targeted glycoproteomics to identify serum glycoproteins that correlate with liver cancer in woodchucks and humans. *Proceedings of the National Academy of Sciences of the United States of America*, **102**, 779–784.

113 Rudd, P.M. *et al.* (2001) A high-performance liquid chromatography based strategy for rapid, sensitive sequencing of N-linked oligosaccharide modifications to proteins in sodium dodecyl sulphate polyacrylamide electrophoresis gel bands. *Proteomics*, **1**, 285–294.

114 Royle, L. *et al.* (2006) Detailed structural analysis of *N*-glycans released from glycoproteins in SDS-PAGE gel bands using HPLC combined with exoglycosidase array digestions. *Methods in Molecular Biology (Clifton, NJ)*, **347**, 125–143.

115 Manzi, A.E. (2001) Total compositional analysis by high-performance liquid chromatography or gas-liquid chromatography. *Current Protocols in Molecular Biology*, John Wiley & Sons, Chapter 17, Unit17, 19A.

116 Patel, T.P. and Parekh, R.B. (1994) Release of oligosaccharides from glycoproteins by hydrazinolysis. *Methods in Enzymology*, **230**, 57–66.

117 Freeze, H.H., Varki, A. and Endo-glycosidase, F. (1986) peptide *N*-glycosidase F release the great majority of total cellular N-linked oligosaccharides: use in demonstrating that sulfated N-linked oligosaccharides are frequently found in cultured cells. *Biochemical and Biophysical Research Communications*, **140**, 967–973.

118 Kuster, B. *et al.* (1998) Structural determination of N-linked carbohydrates by matrix-assisted laser desorption/ionization-mass spectrometry following enzymatic release within sodium dodecyl sulphate-polyacrylamide electrophoresis gels: application to species-specific glycosylation of alpha1-acid glycoprotein. *Electrophoresis*, **19**, 1950–1959.

119 Royle, L. *et al.* (2008) HPLC-based analysis of serum *N*-glycans on a 96-well plate platform with dedicated database software. *Analytical Biochemistry*, **376**, 1–12.

120 Guile, G.R., Wong, S.Y.C. and Dwek, R.A. (1994) Analytical and preparative separation of anionic oligosaccharides by weak anion exchange high performance liquid chromatography on an inert polymer column. *Analytical Biochemistry*, **222**, 231–235.

121 Lee, Y.C. (1990) Review: high performance anion-exchange chromatography for carbohydrate analysis. *Analytical Biochemistry*, **189**, 151–162.

122 Lee, Y.C. (1996) Review: carbohydrate analyses with high performance anion-exchange chromatography. *Journal of Chromatography. A*, **720**, 137–149.

123 Nakano, M., Kakehi, K. and Lee, Y.C. (2003) Sample clean-up method for analysis of complex N-glycans released from glycopeptides. *Journal of Chromatography A*, **1005**, 13–21.

124 Jackson, P. (1991) Polyacrylamide gel electrophoresis of reducing saccharides labelled with the fluorophore 2-aminoacridone: subpicomolar detection using an imaging system based on a cooled charge-coupled device. *Analytical Biochemistry*, **196**, 238–244.

125 Nakajima, K. *et al.* (2003) Capillary affinity electrophoresis for the screening of post-translational modification of protein with carbohydrates. *Journal of Proteome Research*, **2**, 81–88.

126 Shimaoka, H. *et al.* (2007) One-pot solid-phase glycoblotting and probing by transoximization for high-throuput glycomics and glycoproteomics. *Chemistry (Weinheim an der Bergstrasse, Germany)*, **13**, 1664–1673.

127 Lohse, A. *et al.* (2006) Solid phase oligosaccharide tagginf (SPOT): validation on glycolipid-derived structures. *Angewandte Chemie, International Edition*, **45**, 4167–4172.

128 Uematsu, R. *et al.* (2005) High throughput quantitative glycomics and glycoform-focused proteomics of murine dermis and epidenrmis. *Molecular & Cellular Proteomics*, **4**, 1977–1989.

129 Budnik, B.A., Lee, R.S. and Steen, J.A.J. (2006) Global methods for protein glycosylation analysis by mass spectrometry. *Biochim Biophys Acta – Proteins and Proteomics*, **1764**, 1870–1880.

130 Jang-Jee, J. *et al.* (2006) Glycomic profiling of cells and tissues by mass spectrometry: fingerprinting and sequencing methodologies. *Methods in Enzymology*, **415**, 59–86.

131 Morelle, W. and Michalski, J.C. (2005) Glycomics and mass spectrometry. *Current Pharmaceutical Design*, **11**, 2615–2645.

132 Morelle, W. and Michalski, J.C. (2007) Analysis of protein glycosylation by mass spectrometry. *Nature Protocols*, **2**, 1585–1602.

133 Joshi, H.J. *et al.* (2004) Development of a mass fingerprinting tool for automated interpretation of oligosaccharide fragmentation data. *Proteomics*, **4**, 1640–1664.

134 Lohmann, K.K. and von der Leith, C.W. (2004) GlycoFragment and GlycoSearchMS: two tools to support the interpretation of mass spectra of complex carbohydrates. *Nucleic Acids Research*, **32**, W261–W266.

135 Goldberg, D. *et al.* (2005) Automatic determination of *O*-glycan structure form fragmentation spectra. *Journal of Proteome Research*, **5**, 1429–1434.

136 von der Leith, C.W., Lutteke, T. and Frank, M. (2006) The role of informatics in glycobiology research with special emphasis on automatic interpretation of MS spectra. *Biochimica et Biophysica Acta*, **1760**, 568–577.

137 Karas, M. and Hillenkamp, F. (1988) Laser desorption ionization of proteins with molecular masses exceeding 10,000 daltons. *Analytical Chemistry*, **60**, 2299–2301.

138 Wada, Y. *et al.* (1994) Diagnosis of carbohydrate-deficient glycoprotein syndrome by matrix-assisted laser desorption time-of-flight mass spectrometry. *Biological Mass Spectrometry*, **23**, 108–109.

139 Fenn, J.B. *et al.* (1989) Electrospray ionization for mass spectrometry of large biomolecules. *Science*, **246**, 64–71.

140 Harvey, D.J. (1999) Matrix-assisted laser desorption/ionization mass spectrometry of carbohydrates. *Mass Spectrometry Reviews*, **18**, 349–450.

141 Dell, A. and Morris, H.R. (2001) Glycoprotein structure determination by mass spectrometry. *Science*, **291**, 2351–2356.

142 Park, Y. and Lebrilla, C.B. (2005) Application of Fourier transform ion cyclotron resonance mass spectrometry to oligosaccharides. *Mass Spectrometry Reviews*, **24**, 232–264.

143 Harvey, D.J. *et al.* (2008) Structural and quantitative analysis of N-linked glycans by matrix-assisted laser desorption ionization and negative ion nanospray mass spectrometry. *Analytical Biochemistry*, **376**, 44–60.

3
Mammalian Cell Lines and Glycosylation: A Case Study

Michael Butler

3.1
Introduction

Animal cell cultures are used in bioprocesses for the production of biopharmaceuticals because of their capabilities of post-translational modification of proteins, including glycosylation [1]. The glycosylation profile of these proteins is essential to ensure structural stability and biological and clinical activity. However, the ability to control the glycosylation in a bioprocess is limited by our understanding of the parameters that affect the heterogeneity of glycan structures. The basic protein structures can be controlled and directed by the expression of appropriate genetic sequences. However, controlling the pool of glycan structures (glycomics) that occupy a recombinant protein is still difficult. The glycoform profile of a recombinant glycoprotein expressed by a cell line in culture is affected by various parameters, including the host cell line [2–5], the method of culture [6–8], extracellular environment or the protein structure [9, 10].

From a process development perspective there are many culture parameters that can be considered to enable a consistent glycosylation profile to emerge from each batch culture [11]. Such consistency is essential for the approval of a process for a therapeutic product. A further, but more difficult, goal is to control the culture conditions to enable the enrichment of specific glycoforms identified with desirable biological activities. This chapter discusses the cellular metabolism associated with protein glycosylation and reviews the attempts to manipulate, control or engineer this metabolism to allow the expression of a glycoprotein in producer lines such as genetically engineered Chinese hamster ovary (CHO) cells.

3.2
The Choice of Cell Line for Glycoprotein Production

The selection of a high-producing animal cell line is key to the initial stages of the development of a cell culture bioprocess [12, 13]. Chinese hamster ovary (CHO) cells

Post-translational Modification of Protein Biopharmaceuticals. Edited by Gary Walsh

ISBN: 978-3-527-32074-5

have become the standard mammalian host cells used in the production of recombinant proteins, although the mouse myeloma (NS0), baby hamster kidney (BHK), murine C127 cells, human embryonic kidney (HEK-293) or human retina-derived (PER-C6) cells are alternatives. All these cell lines have been adapted to grow in suspension culture and are well-suited for scale-up in stirred tank bioreactors. The advantage of CHO and NSO cells is that there are well-characterized platform technologies that allow for transfection, amplification and selection of high producer clones. Transfection of cells with the target gene along with an amplifiable gene such as dihydrofolate reductase (DHFR) [14, 15] or glutamine synthetase (GS) [16] have offered effective platforms for expression of the required proteins.

The pattern of protein glycosylation is dependent upon the expression of various glycosylation enzymes (glycosidases and glycotransferases) that are present in the host cell (Reference [17] and Chapter 2). Differences in the relative activity of these enzymes among species can account for significant variations in structure. In a systematic study of glycan structures of IgG produced from cells of 13 different species significant heterogeneity was found [18]. Extensive variations were found in the proportion of terminal galactose, core fucose and bisecting GlcNAc. For example, 90% of sheep IgG is galactosylated whereas the equivalent figure in rat cells is only 10%. Such variation may be accounted for by differential enzyme activities in the glycosyltransferase reactions in the Golgi that may not go to completion. Some species variation may be of particular concern when choosing a host cell line for the production of biotherapeutics. For example, the *N*-glycolylneuraminic acid groups that may be transferred to glycoproteins in mouse cells are potentially immunogenic to humans.

CHO cells produce predominantly complex and oligomannosyl *N*-glycan structures. Polylactosamine and polysialic structures have also been observed as minor species. The glycosylation pathways of CHO cells have been well characterized through a panel of mutants that have been selected by resistance to the cytotoxicity of specific plant lectins [19]. The mutants were derived by random mutagenesis of parental CHO cells and resulted in loss-of-function (lec) or gain-of-function (LEC) at specific points of the glycosylation pathway. Although these pathways are similar to those identified in human cells there are some notable differences. CHO cells do not express the ST6Galα2,6-sialyltransferase that is present in human cells [20]. This means that CHO-produced proteins only contain the 2,3 between sialic acid and galactose rather than the combination of 2,3 and 2,6 typical of human proteins. Core-fucosylation of proteins occurs in CHO cells through the α1,6-fucosyltransferase but the α1,2-, α1,3- or α1,4-fucosyltransferases are absent, thus preventing the possibility of peripheral fucosylation of a glycan [21]. Interestingly, some gain-of-function CHO mutants have been obtained that are capable of fucosylation of the peripheral lactosamine of glycans. Such mutants gain α1,3-fucosyltransferase activity through a mechanism of gene rearrangement [19]. GlcNAc transferase III that transfers bisecting *N*-acetylglucosamine to the mannose core structure is also absent in CHO cells.

The *O*-glycans found from CHO cells include core 1-type mucin structures containing up to four monosaccharides but generally no core 2 structures (Figure 3.1). The most common is the core 1 structure (Galβ1 $\rightarrow$ 3GalNAc) that

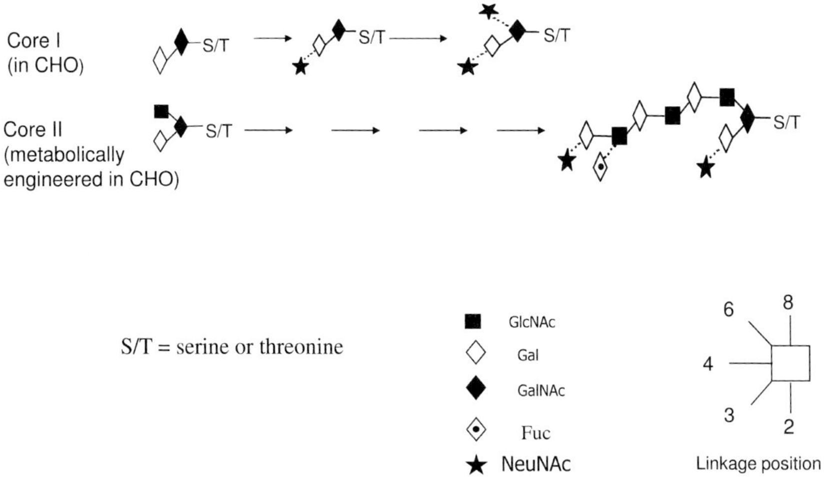

Figure 3.1 O-linked glycans in CHO cells.

may be monosialylated or disialylated [22]. Prati *et al.* [23] transformed *O*-glycan synthesis in CHO cells by metabolic engineering by simultaneously up-regulating and down-regulating selected enzymes to alter the metabolic pathway (Figure 3.2). They co-expressed the core 2 GlcNAc transferase (C2GnT) and an antisense fragment

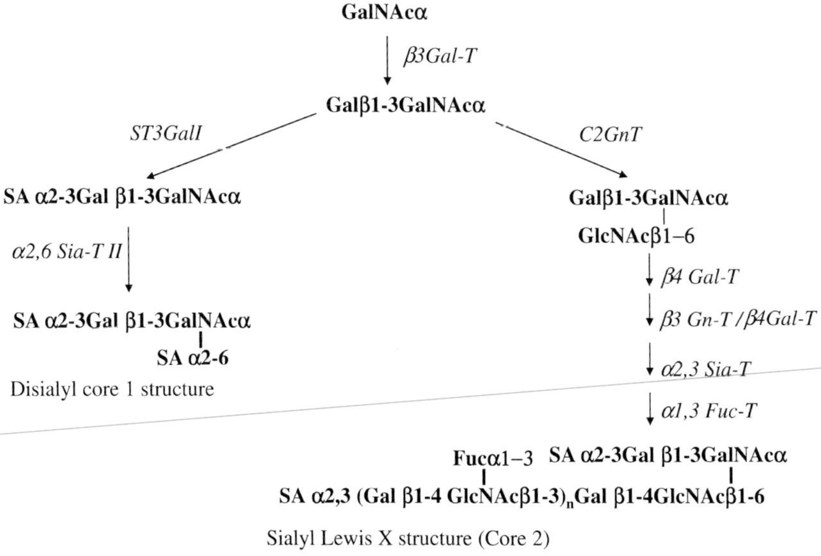

Figure 3.2 The O-linked glycosylation pathway, showing the branching point for the synthesis of core 1 or core 2 structures. (Data from Reference [23].)

of the sialyltransferase (ST3Gal I) enzyme in a CHO cell line already transfected with a fucosyltransferase. The effect of this change diverted the *O*-glycosylation pathway from the formation of core 1 glycans to the formation of core 2 glycans, which are the well-studied sialyl-Lewis X glycan structures that mediate cell–cell adhesion.

3.3
Effect of Growth and Protein Production Rate on Glycosylation

N-Glycosylation occurs co-translationally in the ER over a limited time period in which the *N*-glycan sequon is exposed to the active site of the oligosaccharyl transferase (OST) enzyme. This means that the glycan site occupancy may be dependent upon the rate of the elongation step of protein synthesis. There are several studies that support this. Shelikoff *et al.* [24] showed that C127 cells that expressed recombinant prolactin enhanced glycosylation of the secreted protein from 20 to 80% in the presence of cycloheximide. Cycloheximide is highly specific in its inhibition of protein elongation and so could extend the time available for glycosylation by reducing the elongation rate.

Human tissue-type plasminogen activator (t-PA) has three *N*-glycan sites, one of which is variably occupied (Asn-184). This gives rise to the production of a mixture of type 1 (three glycans) and type 2 (two glycans) from the culture of transfected CHO cells. The relative levels of each of these two forms of t-PA is dependent upon various factors, including the cellular growth rate, and it is suggested that site-occupancy could vary with the fraction of cells in the G0/G1 phase of the cell cycle [25].

This suggests a mechanism by which glycosylation efficiency improves at a reduced rate of protein translation. A decrease in growth rate produced by supplementation of butyrate or lowering the culture temperature resulted in increased site occupancy and higher levels of type 2 t-PA [25]. This also supports the hypothesis that the lowering of the rate of protein elongation would increase the exposure time of the glycan site to the oligosaccharide transferase enzyme in the ER.

The residence time of proteins in the Golgi can also be important in determining the extent of exposure to glycosyltransferase enzymes. Wang *et al.* [26] incubated cells at 21 °C to decrease the flow of glycoproteins through the Golgi. They found that the glycans produced under these conditions had 100% more *N*-acetyllactosamine repeats than the controls, suggesting that glycan processing increases with increasing residence time in the Golgi.

Nabi and Dennis [27] found, similarly, that the extent of polylactosamine formation in lysosomal membrane glycoprotein (LAMP-2) was dependent upon the Golgi residence time. Polylactosamine consists of repeated Gal β1 → 4GlcNAc disaccharide units that are produced by the repeated action of two transferases, β1 → 3GlcNAc transferase and β1 → 4Gal transferase. Low temperature culture was used to increase the residence time of the synthesized protein in the Golgi and enhance its glycosylation by increased exposure to the limiting transferase enzymes.

3.4
Enzymes Associated with Glycan Heterogeneity

3.4.1
N-Acetyl Glucosaminyltransferases

Significant heterogeneity of glycan structures is a result of variable activity of a series of *N*-acetyl glucosaminyltransferase (GnT) enzymes that are found within the Golgi. These enzymes are important in converting a high mannose glycan structure into a complex structure. They introduce variable antennarity onto the core mannose structure and also may allow a bisecting GlcNAc to occur between two antennae. *N*-Acetyl glucosaminyltransferase1 (GnT1) is the first of these enzymes and is responsible for the transfer of GlcNAc from the nucleotide-sugar donor (UDP-GlcNAc) to the α1,3 mannose arm of the high mannose structure, M5. This creates a hybrid structure, M_5Gn, which is converted into M_3Gn by the sequential action of a mannosidase. The free α1,6 mannose arm of the core structure is then extended by GnTII to produce M_3Gn_2, which is a complex biantennary glycan. Both arms may be extended by the sequential action of galactosyl transferase and sialyl transferase enzymes to produce fully sialylated biantennary structures. However, the other GnTs can allow further branching. GnTIV may add GlcNAc in a β1,4 linkage to the α1,3 mannose arm of the core structure and leads to M_3Gn_3, which is a triantennary complex glycan. GnTV has equivalent activity for addition of a GlcNAc in a β1,6 linkage on the α1,6 mannose arm of the core structure. Thus the combined activity of GnTIV and GnTV leads to a tetra-antennary structure, M_3Gn_4 (Figure 3.3). The degree of branching of cell surface glycans partially regulates growth control and

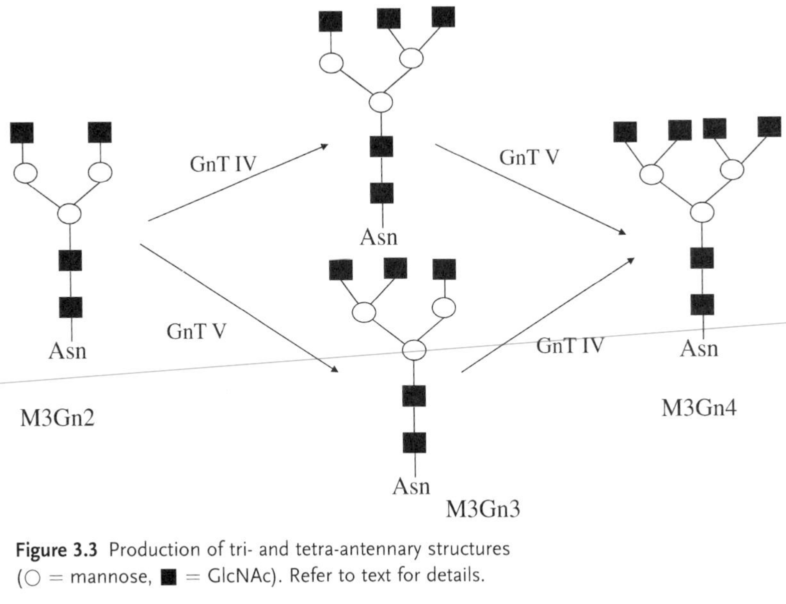

Figure 3.3 Production of tri- and tetra-antennary structures (○ = mannose, ■ = GlcNAc). Refer to text for details.

differentiation of cells via a cell signaling system [28]. The ability to produce high antennary glycan structures is also highly dependent upon the intracellular nucleotide-sugar, UDP-GlcNAc.

A bisecting *N*-acetylglucosamine (GlcNAc^b) may also be added via a β1,4 linkage to the core mannose through the activity of GnTIII. This GlcNAc^b cannot be extended further because of steric interference that does not allow access of the galactosyl transferase. However, GnTIII is not expressed in CHO or BHK cells and therefore the bisecting GlcNAc is normally absent in glycoproteins produced from these cell lines.

The importance of each of the GnT enzymes is emphasized by the determination of a set of abnormal glycan structures found in a human patient with a genetic mutation in the GnTII enzyme that leads to a severe pathological condition characterized as a Congenital Disorder of Glycosylation type II [29]. This led to the biosynthesis of a series of hybrid glycan structures that appeared to be present in all serum proteins.

This complex of reactions that involve the GnTs has been modeled mathematically and described as the central reaction network of glycosylation [30]. The authors based the model on kinetic constants and mass balances associated with the formation of 33 different glycan structures, and then used the model to predict the effect of GnTIII on the glycosylation metabolism of CHO cells, in which the enzyme is normally absent (Figure 3.4). The ability of CHO cells transfected with the GnTIII gene to produce bisected glycans was shown for β-interferon and an immunoglobulin [31, 32]. In the

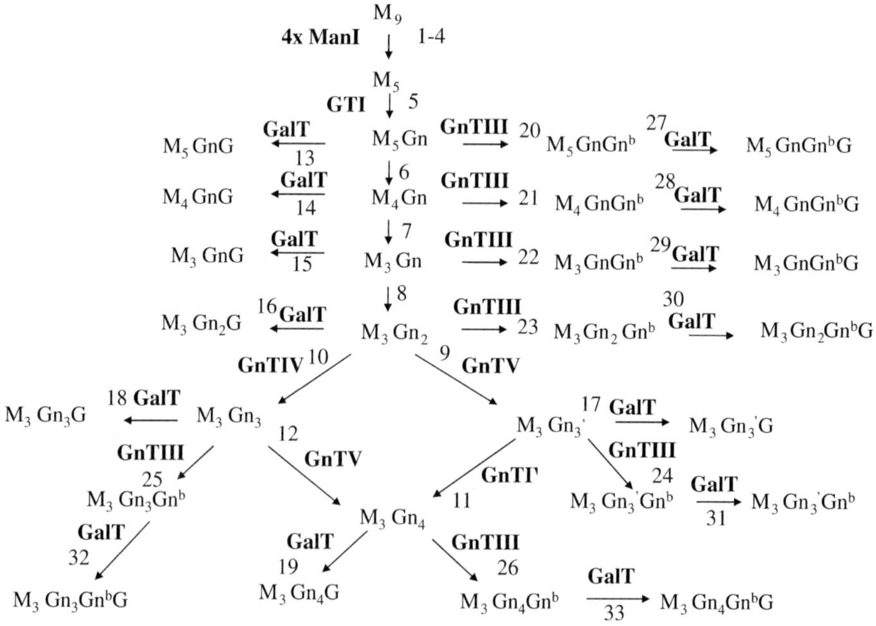

Figure 3.4 Reaction network for N-linked glycosylation. Refer to text for details. (From Reference [30].)

case of immunoglobulin it was shown that the presence of a bisected GlcNAc[b] resulted in a significant enhancement of effector function through antibody-dependent cell-mediated cytotoxicity (ADCC) [32].

3.4.2
Fucosylation

The core-fucosylation pathway includes the synthesis of the fucose donor GDP-fucose, the transport of the nucleotide-sugar from the cytosol into the Golgi, and the transfer of fucose to the nascent glycoprotein by the enzyme α1,6 fucosyltransferase (α1,6FT) [33, 34]. α 1,6 Fucosyltransferase is a glycosyltransferase enzyme that catalyzes the transfer of fucose from GDP-fucose to the Asn-linked GlcNAc residue [35, 36]. The enzyme is a type II transmembrane protein, consisting of a short amino-terminal cytoplasmic tail, a transmembrane domain and a large intraluminal carboxyl-terminal catalytic region [37].

Although the α1,6 fucose residue is frequently found in the *N*-glycans of various glycoproteins and it has been shown that this modification is widespread in various tissues [38], little is known about the function of the core fucose residue. Fucose-containing glycoproteins are removed from the blood into the liver and fucose/mannose receptors exist on the surface of macrophages and mediate phagocytosis [39]. Some studies have found that core fucose residues play an important role in defining oligosaccharide conformation needed for specific carbohydrate–protein interaction [40] and they can regulate the activity of some glycoproteins [41].

A congenital defect in the activity of the enzymes involved in the de novo synthesis of GDP-fucose from GDP-mannose has been reported in humans [42]. This involves the enzymes GDP-mannose 4–6 dehydratase (GDM) and FX protein, a NADP(H)-binding protein that apparently catalyzes a combined epimerase and NADPH-dependent reductase reaction [43, 44]. Altered core-fucosylation occurs in many glycoproteins of tumor cells [39, 45]. A high α1,6 fucosyltransferase activity has been associated with hepatocellular carcinoma [46, 47], ovarian adenocarcinoma [48] and several other human cell lines, including myeloma, pancreatic cancer, lung cancer and gastric cancer [49]. This increase in the α1,6 fucosyltransferase activity in tumor cells was shown to be at the level of transcription [35].

There are reports of altered fucosylation of glycoproteins in patients with cystic fibrosis; this has been attributed to a change in the intravacuolar pH of the Golgi apparatus [50]. Therefore, a change in the glycosylation of a glycoprotein could occur because of an altered activity of the glycosyltransferases as a result of a perturbation in the pH or due to a change in the proper localization of glycosyltransferases [51]. Furthermore, because of the high specificity of the enzyme α1,6 fucosyltransferase [52], a modification in the glycan substrate might cause a significant change in the resulting fucosylation.

Fucosylation may be particularly important to the biological activity of some biopharmaceutical proteins, such as IgG, in which the fucose plays an important role in the determination of the binding capacity of the protein to Fc receptors (Section 3.7).

3.4.3 Sialylation

A high level of terminal sialylation of glycans is important in therapeutic glycoproteins to avoid the effects of asialoglycoprotein receptors present in the liver and macrophages that cause the removal of the glycoprotein from the circulatory system [53]. This interaction with the receptors reduces therapeutic efficacy by decreasing the effective half-life of the glycoprotein in the circulatory system.

Sialylation is the last intracellular stage of the glycosylation process that takes place in the *trans*-Golgi. This involves the enzymatic transfer of sialic acid from the nucleotide sugar precursor, CMP-sialic acid, to an available galactose on the emerging glycan structure that is attached to the newly synthesized protein. Possible limitations to the process that might cause incomplete sialylation include the availability of CMP-sialic acid and the activity of the sialyl transferase enzyme.

The pool of CMP-sialic acid (CMP-NANA) in the *trans*-Golgi is generated from a well-defined biosynthetic pathway in which glucose is the original precursor but the unique portion of the pathway derives from UDP-GlcNac (Figure 3.5). The metabolic conversion from UDP-GlcNac into CMP-NANA proceeds via five enzymic steps and one transport step into the Golgi that includes the formation of a nine-carbon sialic acid from the fusion of a hexose and a three-carbon pyruvate structure. There is a metabolic control network in this pathway governed through two key steps – the UDP GlcNAc 2-epimerase enzyme and the transporter II. The activity of the epimerase enzyme is regulated allosterically through the cytoplasmic concentration of CMP-NANA [54]. This feedback mechanism is important to prevent excessive accumulation of CMP-NANA. It has been shown that the human hereditary disease sialuria, characterized by the accumulation of excessive sialic acid, arises through a mutation

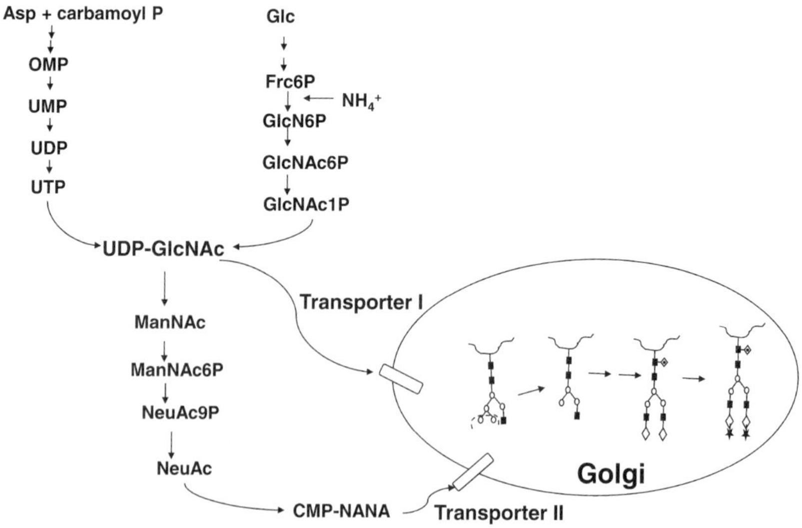

Figure 3.5 Substrates for sialylation. Refer to text for details.

in the allosteric site of this enzyme [55]. A further regulatory mechanism associated with the transporter (II in Figure 3.5) involves competitive inhibition of CMP-NANA transport into the Golgi if the cytoplasmic level of UDP-GlcNAc is too high. This may occur in the presence of excess ammonia, which is required for the formation of glucosamine 6-phosphate, a precursor of UDP-GlcNAc [56].

One of the strategies that can be used to enhance the intracellular concentration of CMP-NANA is to supply one of its metabolic precursors. Because of its location in the metabolic pathway and high cell membrane permeability, *N*-acetylmannosamine (ManNAc) is the best candidate for this, as sialic acid and CMP-sialic acid have poor membrane permeability. Gu and Wang [57] showed that the supplementation of 20 mM ManNAc to CHO cell culture increased the intracellular pool of CMP-sialic acid ×30 and enhanced the sialylation of γ-interferon (Figure 3.6). However, the enhancement of sialylation was selective and limited. One *N*-glycan site (Asn^{25}) was unaltered at a high level of sialylation (90%) whereas in the second site (Asn^{97}) the proportion of incompletely sialylated biantennary structures was reduced from 35 to 20%. This suggests that in this system there were other limiting factors, which were probably related to the rate of the transport system or the relative accessibility of the glycan sites to the transferase enzymes. Results from the use of this strategy in other systems have been variable. Baker *et al.* [58] have reported an increase of intracellular CMP-NeuAc as a result of ManNAc feeding to either CHO or NS0 cells but no increase in overall sialylation for a recombinant protein (TIMP 1) containing two *N*-glycan sites. However, this strategy did change the ratio of *N*-glycolylneuraminic acid to *N*-acetylneuraminic acid from 1 : 1 to 1 : 2 in the NS0 cells. In a further study using this type of feeding strategy Follstad [59] showed a 36% increase in protein sialylation from the addition of 4 mM each of ManNAc, fructose, galactose and mannose.

Analogues of ManNAc may also be incorporated into the glycosylation pathway, leading to non-natural sialylated glycans that may be incorporated into secreted

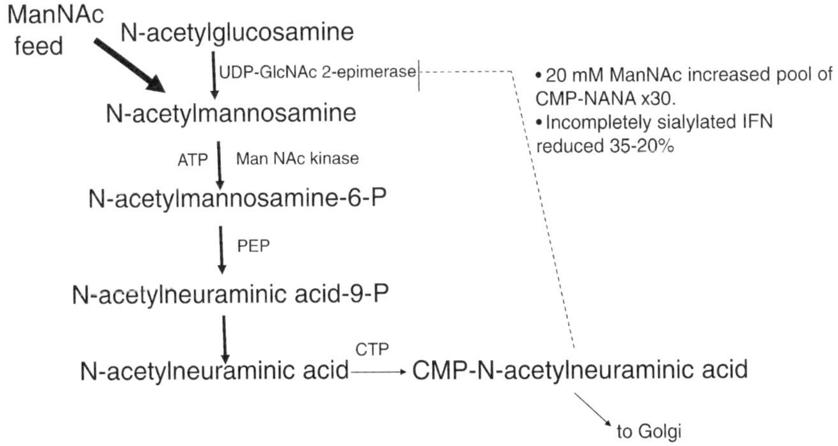

Figure 3.6 Improved sialylation by feeding *N*-acetylmannosamine (ManNAc). Refer to text for details. (Reference [57].)

proteins or cell surface proteins [60]. This strategy has potential in sialic acid engineering that may lead to novel glycoproteins with unique properties.

Sialylation can be maximized by enhancing the activity of specific glycosyltransferase enzymes. Weikert *et al.* [61] have used this approach with CHO cells that had already been selected for the secretion of different recombinant glycoproteins and transfected them with human genes to allow the overexpression of both β1,4-galactosyltransferase and α2,3-sialyl transferase. The predominant structures (>90%) secreted by these re-engineered cells were fully sialylated bi-, tri- and tetra-antennary glycans. This decreased the heterogeneity of glycoforms produced by the cells and the more highly sialylated structures were shown to have improved pharmacokinetics as tested in animals.

The structure of sialic acid varies between species, with *N*-glycolyl-neuraminic acid (NGNA) found in goat, sheep and cows rather than the *N*-acetyl-neuraminic acid (NANA) found in humans. NGNA is the predominant sialic acid in mice but CHO-produced glycoproteins have predominantly NANA, although a small proportion (up to 15%) of NGNA can occur [58]. These differences in glycan structure are important for the effectiveness of biotherapeutics because of the potential immunogenicity of these structures in humans.

Eighteen different types of sialyltransferases have been cloned from various animal species and they give rise to various terminal linkages [62]. The two that are most important for human glycoproteins are the α2,6 and α2,3 sialyl linkages to galactose, each of which is catalyzed by a specific enzyme – the α2,3 or α2,6 sialyl transferase (ST). In humans both enzymes compete for the same substrate and this results in glycoproteins with either linkage type, depending upon the cells. The linkage and expression levels of terminal sialic acids have important effects in human cells that include interactions with the extracellular matrix and susceptibility to apoptosis [63]. The sialyl transferase enzyme, (α2,6 ST) is absent in the hamster and so CHO and BHK cells produce glycoproteins with exclusively α2,3 terminal sialic acid residues. However, these differences in sialylation profiles between CHO and human cells do not appear to result in glycoproteins that are immunogenic. Natural human erythropoietin (EPO) consists of a mixture of sialylated forms – 60% are 2,3 linked and 40% are 2,6 linked. Because of the restricted sialylation capacity of CHO cells, commercially available EPO is sialylated entirely via the α2,3 linkages [64]. Nevertheless, recombinant EPO produced from CHO cells has proven to be a highly effective therapeutic agent with no evidence of an adverse physiological effect due to the structural differences in terminal sialylation.

The sialic acid pattern of glycoproteins from producer cell lines can be humanized by transfecting with appropriate glycosyltransferase enzymes. Attempts have been made to enable α2,6 sialylation in hamster cells by transfection of the α2,6 sialyl transferase gene. BHK cells produced proteins containing a mixture of α2,3 and α2,6 sialyl structures following transfection with an α2,6 STgene [65]. CHO cells that were co-transfected with genes for γ-interferon (IFNγ) and α2,6 ST produced sialylated IFNγ, 40% of which was in the α2,6 form. This form of IFNγ was shown to have improved pharmacokinetics in clearance studies compared to IFNγ produced from a normal CHO host [66].

3.5
Immunogenicity of Non-Human Glycans

Mouse cells express the enzyme α1,3 galactosyltransferase that generates Gal α1,3-Galβ1,4-GlcNAc residues that are highly immunogenic in humans [9].

In one study of NS0 murine cells, 30% of glycan antennae of a recombinant protein terminated with the Gal α1,3-Gal motif [58]. The NSO cells and their derived hybridomas are used extensively for the production of IgG monoclonal antibodies. However, in the immunoglobulin structure terminal galactosylation of the conserved glycan site at Asn_{297} of the Fc region is restricted by steric hindrance [67] and so the extent of terminal α1,3-Gal is likely to be much lower [4].

Fortunately, the α1,3 galactosyltransferase enzyme appears to be inactive in CHO and BHK cells, which are the most commonly used cell lines for the production of recombinant proteins. However, both CHO and BHK show differences in their potential for glycosylation compared to human cells. The absence of a functional α1,3 fucosyltransferase in CHO cells prevents the addition of peripheral fucose residues, and the absence of *N*-acetylglucosaminyltransferase III (Gn TIII) prevents the addition of bisecting GlcNAc to glycan structures [6].

The sialic acid variant – *N*-glycolylneuraminic acid – is prevalent in mouse-derived cells [18] and is potentially immunogenic in humans [68]. For NS0-derived glycoproteins, the NGNA can be >50% of the total sialylation, whereas in CHO cells the proportion is lower but not insignificant; 1–5% for one study [69] and 15% in another [58].

3.6
Culture Parameters that may Affect Glycosylation

3.6.1
Nutrient Depletion

In a batch culture there is a continuous depletion of nutrients from the medium and also a gradual accumulation of metabolic by-products. This gives rise to changing conditions for the cells and may cause the extent of glycosylation to decrease over time. Figure 3.7 shows the reduction of glycan site occupancy leading to macroheterogeneity of γ-interferon during the batch culture of CHO cells [70]. This is probably to be due to the continuous depletion of nutrients, particularly glucose or glutamine, which have been shown to limit the glycosylation process [6, 71, 72].

A second example shows a decrease in the proportion of terminal galactose in an antibody over the course of a batch culture. This microheterogeneity was alleviated by a suitable feeding regime of galactose (Figure 3.8) [73]. These examples of glycan heterogeneity in batch cultures are of major concern in trying to produce consistent biopharmaceuticals. It can lead to significant batch to batch variation in the production process and diminished therapeutic efficacy.

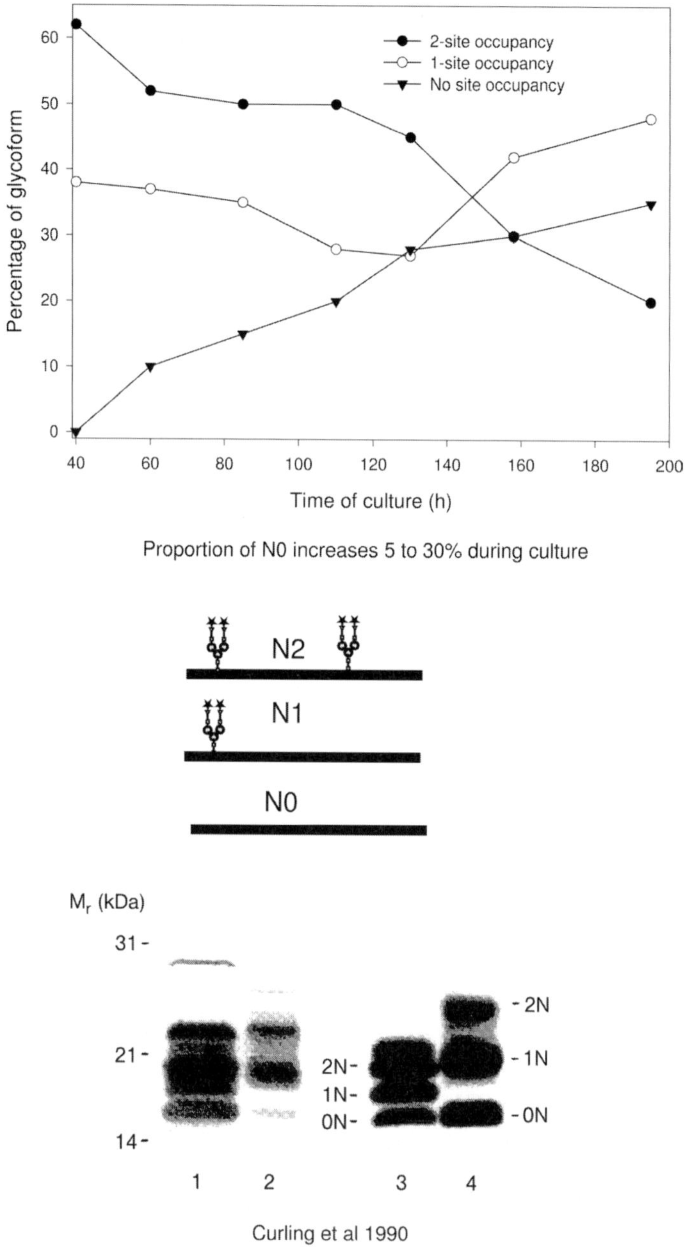

Figure 3.7 Production of γ-interferon in extended batch culture [70] and separation of glycoforms by SDS-PAGE [71]. Lanes 1 and 2 show intracellular γ-interferon from Sf-9 and CHO cells respectively. Lanes 3 and 4 show secreted γ-interferon from the same cells.

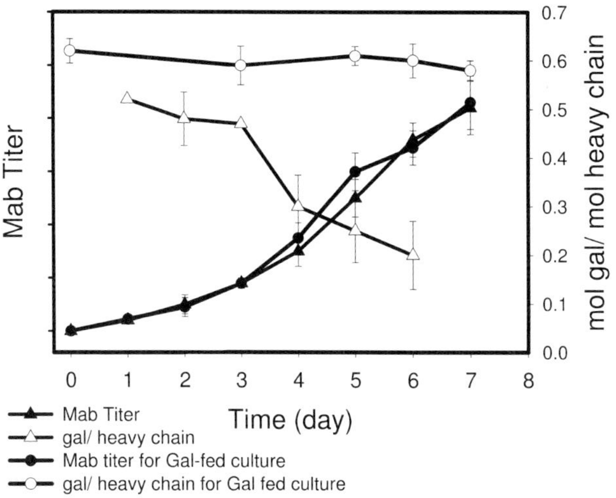

Figure 3.8 Feeding galactose during batch culture prevents the decline in galactosylation during the culture. (With kind permission of Robert Kiss, [73].)

3.6.2
Fed-Batch Cultures and Supplements

Feeding strategies in batch cultures have been highly successful for bioprocess development, allowing product yields in excess of 5 g L^{-1} [12]. The principle is that a slow feed of essential nutrients such as glucose and glutamine can allow an efficient cellular metabolism with minimal accumulation of by-products. This allows cells to reach a high cell density (>10^7 mL^{-1}) that can be maintained over a prolonged period. From this, target proteins can be secreted into the culture medium over an extended time period. Critical to this strategy is the maintenance of nutrients at a set-point of a low concentration. This can be achieved by several methods. For example, stoichiometric feeding can be based upon the projected cell growth over daily time points (12–24 h). Alternatively, a dynamic nutrient feeding regime may be based upon regular sample analysis over shorter time periods (1–2 h). The choice of method will directly affect the variability of nutrient concentration about the predefined set-point. However, it is important to recognize that the critical nutrient concentrations needed to maximize product titers may be different from those that affect glycosylation. Also, significant fluctuations around the set-point could cause variability of product glycosylation (Figure 3.9).

Fed-batch strategies may be designed to ensure that the concentrations of these key nutrients do not decrease below a critical level that could compromise protein glycosylation [74]. From a series of fed-batch culture studies, the lower levels of nutrients for the production of γ-interferon from CHO cells were found to be 0.1 mM glutamine and 0.7 mM glucose [75]. Nutrient levels below these critical concentrations led to decreased sialylation and an increase in hybrid and high mannose-type glycans. Reduced site-occupancy of *N*-glycans was also shown in immunoglobulin synthesis from mouse myeloma cells at low glucose concentration (<0.5 mM) [76].

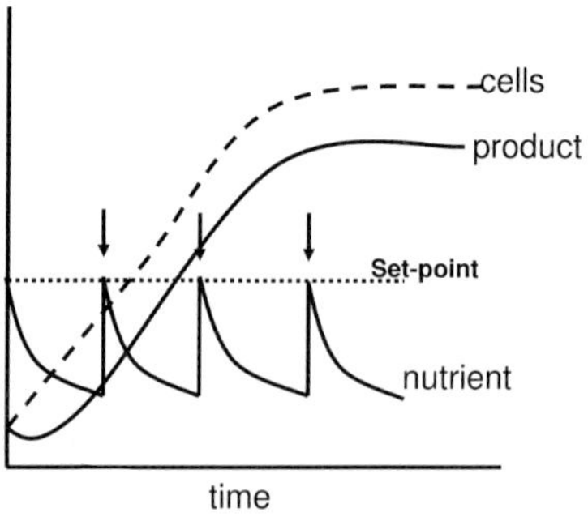

Figure 3.9 Kinetics of nutrient consumption and product accumulation in a controlled fed-batch culture.

Davidson and Hunt [77] showed under-glycosylation and the presence of abnormal truncated glycans in viral proteins derived from CHO cells deprived of glucose.

Glucose starvation may result in an intracellular depleted state or a shortage of glucose-derived precursors of glycans [78]. This was shown to give rise to a higher proportion of high mannose structures. Hayter *et al.* [71] showed that in a glucose-limited chemostat culture CHO cells produced an increase in the proportion of non-glycosylated γ-interferon, although pulsed additions restored normal levels of glycosylation rapidly.

A plausible mechanism for reduced site-occupancy of a recombinant protein produced by glucose-depleted or glutamine-depleted CHO cell cultures was offered by Nyberg *et al.* [72]. They showed that in both cases the low level of glycosylation was related to a decreased intracellular concentration of UDP-GlcNAc but from different metabolic causes. Metabolic flux analysis showed that glutamine depletion was likely to decrease significantly the formation of glucosamine phosphate via the glutamine: fructose 6-phosphate amidotransferase (GFAT) enzymic reaction. On the other hand, glucose-depletion affected the synthesis of UTP, which was found at a low intracellular concentration. UTP and glucosamine-phosphate are the key precursors of UDP-GlcNAc, which in turn is required for glycosylation of proteins.

Tachibana *et al.* [79] have shown that by replacing glucose with GlcNAc in the media of a human hybridoma they were able to change the glycosylation profile of the hypervariable region of the light chain of the antibody, with a tenfold increase in affinity binding to its antigen. This change was associated with a lower level of sialylation of the light chain glycans. The role of UDP-GlcNAc in the metabolic network for glycosylation is shown in Figure 3.5.

3.6.2.1 **Glucosamine as a Supplement**

Glucosamine is a precursor for UDP-GlcNAc, which is an important intracellular nucleotide-sugar that is the substrate for a range of GlcNAc transferases present in the Golgi. Supplementation of cultures with glucosamine leads to an elevated level of intracellular UDP-GlcNAc, particularly if added in conjunction with uridine [58]. The UDP-GlcNAc requires transport into the Golgi lumen before it can be acted upon by the GlcNAc transferases (GlcNAc T). The elevated level of UDP-GlcNAc has been shown to enhance the antennarity of glycan structures produced in baby hamster kidney (BHK) cells, probably through a stimulation of the specific GlcNAc TIV and TV [56, 80, 81]. However, the phenomenon is not universal for all cell lines. Baker *et al.* [58] found enhanced antennarity following glucosamine supplementation in CHO cells but not NS0 cells that produced the same recombinant glycoprotein. However, in all cases an elevated UDP-GlcNAc appears to cause a decrease in sialylation, which may be explained metabolically by the inhibition of CMP-sialic acid transport [82].

3.6.2.2 Galactose as a Supplement

Terminal galactosylation of glycans of recombinant antibodies exhibits significant variability, dependent upon the state of the medium. Andersen [73] has shown that feeding cultures with galactose up to a concentration of 36 mM can ensure high levels of terminal galactosylation, as shown in the production of several antibodies. Galactose feeding was shown to increase the UDP-galactose pool in the cell up to 20× that of control levels and corresponded to a concentration of 7 fmol per 10^5 viable cells. However, in a separate study, Clark *et al.* [83] showed that the sialic acid content of a glycoprotein was not increased by galactose feeding. They attributed this to an enhanced intracellular sialidase activity in the galactose-fed cultures that increased the potential for desialylation. Figure 3.10 shows the metabolic network in which UDP-Gal is a precursor transported into the Golgi by a specific transporter prior to addition to the *N*-glycan on the protein.

3.6.3 Ammonia

Ammonia (NH_3) or the ammonium ion (NH_4^+) accumulates in culture as a by-product from cellular glutamine metabolism (glutaminolysis) or from the non-

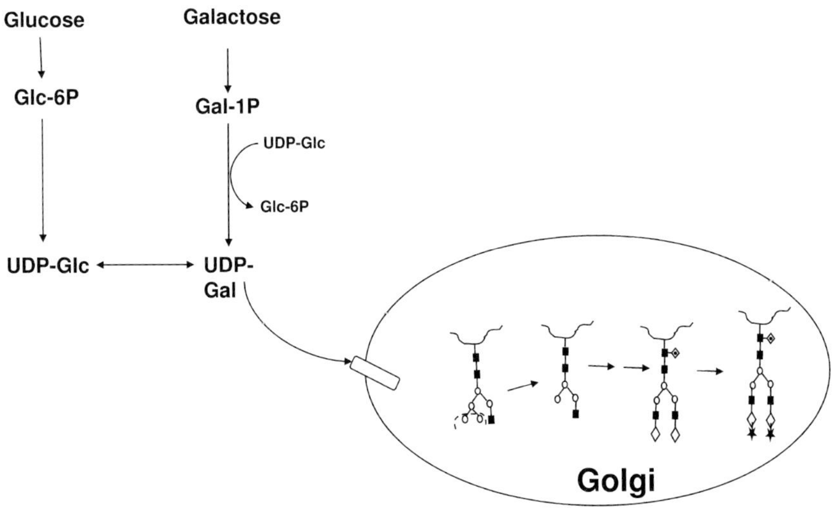

Figure 3.10 Pathway for galactosylation from galactose feeding.

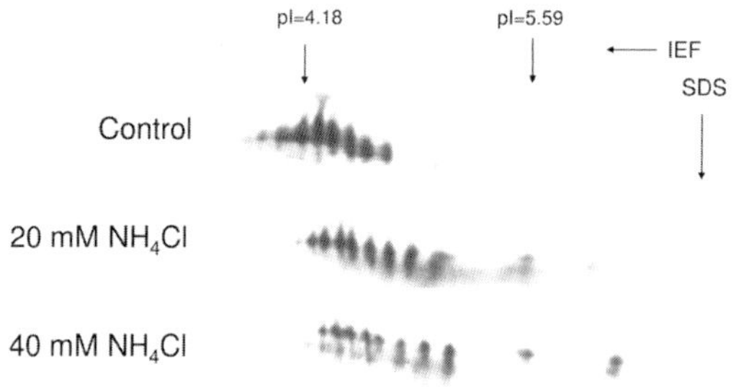

Figure 3.11 EPO heterogeneity analysis by 2D electrophoresis. Refer to text for details. (Reference [88].)

enzymatic decomposition of glutamine in the medium. It has been known for some time that the accumulated ammonia is inhibitory to cell growth [84], an effect that is greater at high pH [85].

The major affect that ammonia exerts on glycosylation is to decrease terminal sialylation, a phenomenon observed in the production of various recombinant proteins [86–88].

2D gel electrophoresis shows the shift of EPO glycoforms to higher pI values in the presence of high levels of ammonia (Figure 3.11). Andersen and Goochee [89] have reported that even a low level of ammonia (2 mM) could affect the sialylation of *O*-glycans. There are two possible mechanisms to explain this effect. The first is the observed increase in the UDP-GlcNAc/UTP ratio that is brought about by the enhanced incorporation of ammonia into glucosamine, a precursor for UDP-GlcNAc. This nucleotide-sugar competes with the transport of CMP-NANA into the Golgi and would therefore decrease the available substrate concentration for sialylation. The second plausible mechanism is that the ammonia raises the pH of the Golgi, thereby shifting from the optimal pH of the sialyl transferase enzymes [56].

3.6.4
pH

Under adverse external pH conditions the internal pH of the Golgi is likely to change, resulting in a reduction of the activities of key glycosylating enzymes. The pH of the medium was shown to have some effect on the distribution of glycoforms of IgG secreted by a murine hybridoma [90]. Borys *et al.* [91] have related the extracellular pH to the specific expression rate and glycosylation pattern of recombinant mouse placental lactogen-I (mPL-I) by CHO cells. They observed that the maximum specific mPL-I expression rates occurred between pH 7.6 and 8.0. The level of site occupancy was maximum within this pH range, decreasing at lower (<6.9) and higher (>8.2) pH.

3.6.5
Oxygen

Control of the dissolved oxygen (DO) level is important to maintain optimal metabolism and growth of producer cells in bioprocesses [92, 93]. The effect of DO

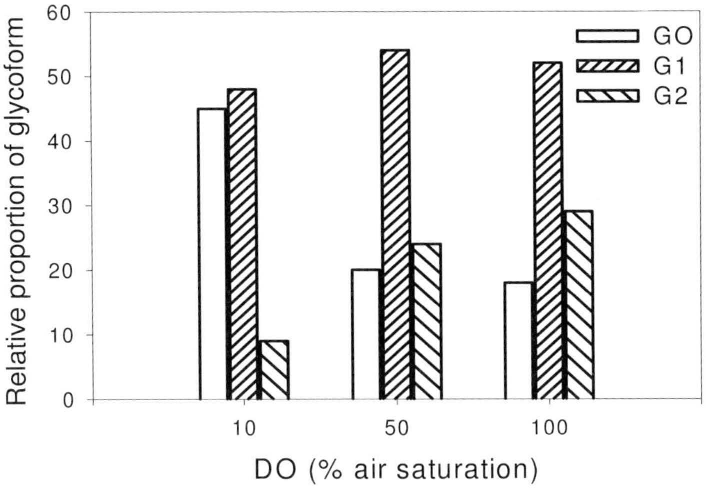

Figure 3.12 Effect of dissolved oxygen on the glycoform profile of IgG. (Data from Reference [95].)

on the glycosylation of a recombinant protein from CHO cells has been observed by a changing glycoform profile [94]. In particular, an increase in sialyltransferase was observed at high oxygen levels that translated into increased sialylation of recombinant follicle stimulating hormone (FSH).

By controlling DO set-points between a range of 1 to 100% air saturation the terminal galactosylation of an immunoglobulin (IgG) was changed significantly, with a gradual decrease in the digalactosylated glycans (G2) from 30% at the higher oxygen level to around 12% under low oxygen conditions (Figure 3.12) [95]. The mechanism for the effect of DO is unclear but it is unlikely to be due to a change in the activity of the transferase enzyme [96]. One explanation is that reduced DO causes a decline in the availability of UDP-Gal, which might occur through reduced synthesis or reduced transport into the Golgi lumen. A second explanation is that galactosylation might be sterically impeded by the early formation of an inter-heavy chain disulfide. It has been proposed previously that the timing and rate of formation of the disulfide bond in the hinge region of IgG is critical to the extent of galactosylation [97]. The redox environment of the ER or the Golgi may perturb the pathway of interchain disulfide bond formation.

3.7 Functional Glycomics

The production of specific protein glycoforms may allow the possibility of even more efficacious biopharmaceuticals [98]. Functional glycomics is an expanding area of science that attempts to understand the physiological function of specific carbohydrate groups. This approach established the importance of the sialylation of EPO with the discovery that the removal of sialic acid groups from the glycans resulted in a significantly reduced half-life in the blood stream [99]. Therapeutic recombinant EPO is normally enriched with highly sialylated glycoforms by selection using ion-exchange chromatography of the product secreted from CHO cells. Protein engineering has allowed the creation of a modified EPO with two extra glycan attachment sites and with the potential to incorporate eight extra sialic acid groups per molecule. This has led to a new generation EPO called "darbepoetin," which has a three times higher drug half-life [100]. This strategy of enhancing the half-life of a biotherapeutic has also been successful for other recombinant proteins such as follicle stimulating hormone [101] and thyroid stimulating hormone [102].

Structural changes of glycans can also be brought about by metabolic engineering of the host cell line. This includes gene-knockout of already expressed glycosyltransferases or the insertion of novel activities [61]. The presence of a bisecting *N*-acetylglucosamine [32, 103] or the absence of fucose [104–106] in the conserved glycan of an IgG antibody has been shown to enhance attachment to Fc receptors and result in an increase in antibody-dependent cell-mediated cytotoxicity (ADCC). Afucosylated antibodies can be produced by metabolic engineering. RNAi has been used to decrease the FUT8 mRNA transcripts that reduced the activity of α1,6-fucosyl transferase [107]. This has been of value in the design of antibody therapeutics. For example, work with Herceptin, which is a novel humanized antibody approved for the treatment of breast cancer, has shown that a glycoform with no fucose has a 53× higher binding capacity to an Fc receptor that triggers its therapeutic activity [105]. This enhancement of ADCC allows the antibody to be effective at lower doses.

An alternative method of generating glycans with minimal fucose is by the use of the specific inhibitor, kifunesine. This is an alkaloid isolated from the actinomycete *Kitasatosporia kifunense* and shown to inhibit the enzyme α-mannosidase I, resulting in the accumulation of high mannose glycans, Man8 and Man9 [108]. This addition of kifunesine to CHO cells or hybridomas resulted in the production of anti-CD20 monoclonal antibodies containing oligomannose-type glycans. These were shown to have high affinity for FcγRIIIA, which is key to high ADCC activity [109].

Complete glycosylation of recombinant proteins is usually associated with maximization of galactosylation and sialylation. Often these two processes are incomplete and this gives rise to considerable glycan structural variation. CHO cells can be engineered with a combination of human β1,4-galactosyltransferase and α2,3-sialyltransferase to ensure high activities of these enzymes. The recombinant proteins produced by these cells exhibited greater homogeneity than controls and increased terminal sialic acid residues [61]. An alternative approach involves glycoengineering of the proteins *in vitro* [110]. Preparations of these terminal transferase enzymes can be immobilized so that glycoproteins can be galactosylated and sialylated in the presence of appropriate galactose and sialic acid donors.

3.8 Conclusion

Glycan structures have a major effect on the biological activities of glycoproteins and so the control of glycosylation is extremely important for the production of biopharmaceuticals. The strategic control of a bioprocess is essential to ensure high yields of product but also to ensure consistency of quality manifested by a consistent glycosylation profile. The study of functional glycomics is rapidly revealing examples of desirable biological activities that are expressed only by specific glycoforms. This is likely to lead to the future design of bioprocesses that allow synthesis of glycoprotein with only pre-selected glycans. However, before this becomes a reality there is a need for further fundamental understanding of the parameters that control protein glycosylation in host cells as well as in bioreactors.

Acknowledgment

The Natural Science and Engineering Research Council (NSERC) of Canada is gratefully acknowledged for financial support for the study of protein glycosylation through a series of Discovery and Collaborative Research and Development (CRD) grants.

References

1 Butler, M. (2004) *Animal Cell Culture and Technology*, 2nd edn, Bios Scientific, Oxford.

2 Goto, M., Akai, K., Murakami, A., Hashimoto, C., Tsuda, E., Ueda, M., Kawanishi, G., Takahashi, N., Ishimoto, A., Chiba, H. and Sasaki, R. (1988) Production of recombinant human erythropoietin in mammalian cells: host cell dependency of the biological activity of the cloned glycoprotein. *Bio/Technology*, **6**, 67–71.

3 Goochee, C.F. (1992) Bioprocess factors affecting glycoprotein oligosaccharide structure. *Developments in Biological Standardization*, **76**, 95–104. Review.

4 Sheeley, D.M., Merrill, B.M. and Taylor, L.C. (1997) Characterization of monoclonal antibody glycosylation: comparison of expression systems and identification of terminal alpha-linked galactose. *Analytical Biochemistry*, **247**, 102–110.

5 Kagawa, Y., Takasaki, S., Utsumi, J., Hosoi, K., Shimizu, H., Kochibe, N. and Kobata, A. (1988) Comparative study of the asparagine-linked sugar chains of natural human interferon-beta 1 and recombinant human interferon-beta 1 produced by three different mammalian cells. *The Journal of Biological Chemistry*, **263**, 17508–17515.

6 Jenkins, N. and Curling, E.M. (1994) Glycosylation of recombinant proteins: problems and prospects. *Enzyme and Microbial Technology*, **16**, 354–364.

7 Gawlitzek, M., Valley, U., Nimtz, M., Wagner, R. and Conradt, H.S. (1995) Characterization of changes in the glycosylation pattern of recombinant proteins from BHK-21 cells due to different culture conditions. *Journal of Biotechnology*, **42**, 117–131.

8 Schweikart, F., Jones, R., Jaton, J.C. and Hughes, G.J. (1999) Rapid structural characterisation of a murine monoclonal IgA alpha chain: heterogeneity in the oligosaccharide structures at a specific site in samples produced in different bioreactor systems. *Journal of Biotechnology*, **69**, 191–201.

9 Jenkins, N., Parekh, R.B. and James, D.C. (1996) Getting the glycosylation right: implications for the biotechnology industry. *Nature Biotechnology*, **14**, 975–981.

10 Reuter, G. and Gabius, H.J. (1999) Eukaryotic glycosylation: whim of nature or multipurpose tool? *Cellular and Molecular Life Sciences*, **55**, 368–422. Review.

11 Restelli, V. and Butler, M. (2002) The effect of cell culture parameters on protein glycosylation, in *Glycosylation*, vol 3 (ed. M. Al-Rubeai), Kluwer, Dordrecht, pp. 61–92.

12 Wurm, F.M. (2004) Production of recombinant protein therapeutics in cultivated mammalian cells. *Nature Biotechnology*, **22**, 1393–1398.

13 Andersen, D.C. and Krummen, L. (2002) Recombinant protein expression for therapeutic applications. *Current Opinion in Biotechnology*, **13**, 117–123.

14 Gasser, C.S., Simonsen, C.C., Schilling, J.W. and Schimke, R.T. (1982) Expression of abbreviated mouse dihydrofolate reductase genes in, cultured hamster cells. *Proceedings of the National Academy of Sciences of the United States of America*, **79**, 6522–6526.

15 Lucas, B.K., Giere, L.M., DeMarco, R.A., Shen, A., Chisholm, V. and Crowley, C.W. (1996) High-level production of recombinant proteins in CHO cells using a dicistronic DHFR intron expression vector. *Nucleic Acids Research*, **24**, 1774–1779.

16 Bebbington, C.R., Renner, G., Thomson, S., King, D., Abrams, D. and Yarranton, G.T. (1992) High-level expression of a recombinant antibody from myeloma cells, using a glutamine synthetase gene as an amplifiable selectable marker. *Bio/Technology (Nature Publishing Company)*, **10**, 169–175.

17 Rudd, P.M. and Dwek, R.A. (1997) Glycosylation: heterogeneity and the 3D structure of proteins. *Critical Reviews in Biochemistry and Molecular Biology*, **32**, 1–100.

18 Raju, T.S., Briggs, J.B., Borge, S.M. and Jones, A.J. (2000) Species-specific variation in glycosylation of IgG: evidence for the species-specific sialylation and branch-specific galactosylation and importance for engineering recombinant glycoprotein therapeutics. *Glycobiology*, **10**, 477–486.

19 Patnaik, S.K. and Stanley, P. (2006) Lectin-resistant CHO glycosylation mutants. *Methods in Enzymology*. 2006, **416**, 159–182. Review.

20 Sasaki, H., Bothner, B., Dell, A. and Fukuda, M. (1987) Carbohydrate structure of erythropoietin expressed in Chinese hamster ovary cells by a human erythropoietin cDNA. *The Journal of Biological Chemistry*, **262** (25), 12059–12076.

21 Howard, D.R., Fukuda, M., Fukuda, M.N. and Stanley, P. (1987) The GDP-fucose:*N*-acetylglucosaminide 3-alpha-L-fucosyltransferases of LEC11 and LEC12 Chinese hamster ovary mutants exhibit novel specificities for glycolipid substrates. *The Journal of Biological Chemistry*, **262** (35), 16830–16837.

22 Backstrom, M., Link, T., Olson, F.J., Karlsson, H., Graham, R., Picco, G., Burchell, J., Taylor-Papadimitriou, J., Noll, T. and Hansson, G.C. (2003) Recombinant MUC1 mucin with a breast cancer-like *O*-glycosylation produced in large amounts in Chinese-hamster ovary cells. *The Biochemical Journal*, **376**, 677–686.

23 Prati, E.G., Matasci, M., Suter, T.B., Dinter, A., Sburlati, A.R. and Bailey, J.E. (2000) Engineering of coordinated up- and down-regulation of two glycosyltransferases of the *O*-glycosylation pathway in Chinese hamster ovary (CHO) cells. *Biotechnology and Bioengineering*, **68** (3), 239–244.

24 Shelikoff, M., Sinskey, A.J. and Stephanopoulos, G. (1994) The effect of protein synthesis inhibitors on the glycosylation site occupancy of recombinant human prolactin. *Cytotechnology*, **15**, 195–208.

25 Allen, S. Naim, H.Y. Bulleid, N.J. 1995 Intracellular folding of tissue-type plasminogen activator. Effects of disulfide bond formation on N-linked glycosylation and, secretion *The Journal of Biological Chemistry,* **270**, 4797–4804. Andersen, D.C. Bridges, T. Gawlitzek, M. Hoy, C. 2000 Multiple cell culture factors can affect the glycosylation of Asn-184 in CHO-produced tissue-type plasminogen activator. *Biotechnology and Bioengineering,* **70**, 25–31.

26 Wang, W.C., Lee, N., Aoki, D., Fukuda, M.N. and Fukuda, M. (1991) The poly-*N*-acetyllactosamines attached to lysosomal membrane glycoproteins are increased by the prolonged association with the Golgi complex. *The Journal of Biological Chemistry,* **266**, 23185–23190.

27 Nabi, I.R. and Dennis, J.W. (1998) The extent of polylactosamine glycosylation of MDCK LAMP-2 is determined by its Golgi residence time. *Glycobiology,* **8**, 947–953.

28 Lau, K.S., Partridge, E.A., Grigorian, A., Silvescu, C.I., Reinhold, V.N., Demetriou, M., Dennis, J.W. (2007) Complex N-glycan number and degree of branching cooperate to regulate cell proliferation and differentiation. *Cell,* **129**(1): 123–34.

29 Butler, M., Quelhas, D., Critchley, A.J., Carchon, H., Hebestreit, H.F., Hibbert, R.G., Vilarinho, L., Teles, E., Matthijs, G., Schollen Argibay, P., Harvey, D.J., Dwek, R.A., Jaeken, J. and Rudd, P.M. (2003) Detailed glycan analysis of serum glycoproteins of patients with congenital disorders of glycosylation indicates the specific enzyme defect and, coupled with proteomics, provides an insight into pathogenesis. *Glycobiology,* **13**, 601–622.

30 Umana, P. and Bailey, J.E. (1997) A mathematical model of N-linked glycoform biosynthesis. *Biotechnology and Bioengineering,* **55**, 890–908.

31 Sburlati, A.R., Umana, P., Prati, E.G. and Bailey, J.E. (1998) Synthesis of bisected glycoforms of recombinant IFN-beta by overexpression of beta-1.4-*N*-acetylglucosaminyltransferase III in Chinese hamster ovary cells. *Biotechnology Progress,* **14** (2), 189–192.

32 Umana, P., Jean-Mairet, J., Moudry, R., Amstutz, H. and Bailey, J.E. (1999) Engineered glycoforms of an antineuroblastoma IgG1 with optimized antibody-dependent cellular cytotoxic activity. *Nature Biotechnology,* **17**, 176–180.

33 Hirschberg, C.B. (2001) Golgi nucleotide sugar transport and leukocyte adhesion deficiency II. *The Journal of Clinical Investigation,* **108**, 3–6.

34 Freeze, H.H.J. (2002) Sweet solution: sugars to the rescue. *Cell Biology,* **158**, 615–616.

35 Narhi, L.O. Arakawa, T. Aoki, K.H. Elmore, R. Rohde, M.F. Boone, T. Strickland, T.W. 1991 The effect of carbohydrate on the structure and stability of erythropoietin *The Journal of Biological Chemistry,* **266**, 23022–23026. Noda, K. Miyoshi, E. Uozumi, N. Gao, C.X. Suzuki, K. Hayashi, N. Hori, M. Taniguchi, N. 1998 High expression of alpha-1-6 fucosyltransferase during rat hepatocarcinogenesis. *International Journal of Cancer,* **75**, 444–450.

36 Voynow, J.A., Kaiser, R.S., Scanlin, T.F. and Glick, M.C. (1991) Purification and characterization of GDP-L-fucose-*N*-acetyl beta-D-glucosaminide alpha 1-6 fucosyltransferase from cultured human skin fibroblasts. Requirement of a specific biantennary oligosaccharide as substrate. *The Journal of Biological Chemistry,* **266**, 21572–21577.

37 Breton, C., Oriol, R. and Imberty, A. (1998) Conserved structural features in eukaryotic and prokaryotic fucosyltransferases. *Glycobiology,* **8**, 87–94.

38 Srikrishna, G., Varki, N.M., Newell, P.C., Varki, A. and Freeze, H.H. (1997) An IgG monoclonal antibody against Dictyostelium discoideum glycoproteins specifically recognizes Fucalpha1.6GlcNAcbeta in the core of N-linked glycans. Localized expression of core-fucosylated glycoconjugates in human tissues. *The Journal of Biological Chemistry,* **272**, 25743–25752.

4
Antibody Glycosylation

Roy Jefferis

4.1
Introduction

Recombinant antibody (rMAb) therapeutics are exemplars of translational medicine. The rMAbs currently licensed represent a significant success in terms of clinical benefit delivered and revenue (profit) generated within the biopharmaceutical industry. Additionally, it is estimated that ~30% of new drugs likely to be licensed during the next decade will be based on antibody products [1–4]. High volume production with the maintenance of structural and functional fidelity of these large biological molecules results in high "cost of goods" (CoG) and, consequent, "high cost of treatment" (CoT) that can limit their availability to patients, due to the strain it puts on national and private health budgets. This can result in international, national and regional differences in drug availability with "bio-selection" of patients determining which products are made available, for which disease indications and to which patient groups. The perceived benefits that could flow from lower CoG/CoT have acted as an incentive for innovation in the development of rMAbs with improved efficacy and novel (low cost?) production vehicles. An added incentive is provided by the increasing capability of low labor cost economies to compete with and undercut production costs of high labor cost economies.

Post-translational modifications can critically influence product equivalence and immunogenicity – issues that are central to the debate of whether biopharmaceuticals can ever be classified as "generic" products [5–7]. Possibly the most frequent and diverse PTM is glycosylation since it is estimated that ~50% of genes encode for proteins expressing the Asn-X-Ser/Thr motif, where X may be any amino acid except proline [8]; this is the motif required, though not sufficient, for N-linked glycosylation. Glycoprotein complexity can be further extended by the addition of O-linked glycans; however, their potential presence cannot be predicted from gene or protein sequences. Defects in genes contributing to N- and O-linked glycosylation pathways can result in congenital disorders of glycosylation (CDG), with serious medical

Post-translational Modification of Protein Biopharmaceuticals. Edited by Gary Walsh

ISBN: 978-3-527-32074-5

consequences [9]. Changes in the glycosylation profiles of specific proteins may serve as disease markers [10–14] whilst the significance of other disease related changes are yet to be elucidated [15].

It will be evident that a recombinant protein should, ideally, exhibit the same PTMs as the endogenous protein product; however, it is important to recognize that the structure determined for an endogenous protein is that of molecules that have had a residence time in a body compartment/fluid prior to being subject to multiple isolation and purification protocols. The structure of this purified product could differ from that of the nascent molecule secreted from its tissue of origin. Similarly, recombinant proteins are synthesized in an "alien" tissue (CHO, NS0, Sp2/0 cells, etc.), are exposed to the culture medium, products of the host cell line and subject to rigorous downstream and formulation processes. Lack of structural fidelity can impact on function, stability and immunogenicity; an immune response may impact therapeutic efficacy and/or result in harmful reactions (side effects) [16]. Glycosylation and other PTMs have been shown to be species, tissue and gender specific [17–21]. Current antibody therapeutics have a shelf life of 18–24 months, which is testament to a lack of structural stability and/or sub-optimal formulation.

Considerable success has been reported for increased productivity of antibody in CHO cell lines, with levels of 5 $g\,L^{-1}$ being achieved and 10 $g\,L^{-1}$ being set as a goal [22]. However, high production levels may overwhelm the PTM machinery, resulting in poor product quality; it is essential, therefore, to characterize product at an early stage in clone selection to optimize both productivity and quality. Essential nutrients may also compromise product quality – for example, glycation through the non-enzymatic addition of glucose [18, 23, 24] or oxidation of methionine side chains [18, 24]. The mammalian CHO, NS0 and Sp2/0 cell lines produce an endogenous carboxypeptidase-b that differentially cleaves the C-terminal lysine residues from antibody heavy chains, adding structural and charge heterogeneity [18, 24].

4.2
Antibodies

4.2.1
Basic Structure/Function

Early studies of serum components defined the soluble albumin fraction and the less soluble globulin fraction. Subsequently, it was shown that humoral immune protection was mediated by components of the globulin fraction – hence, immunoglobulins. Antibodies are, formally, immunoglobulins of defined antigen specificity; however, the terms antibody and immunoglobulin are widely used interchangeably, as synonyms. Antibodies are often referred to as adaptor molecules that function as a "bridge" between humoral and cellular immunity. Initially, antibodies are expressed as receptor molecules on the cell membrane of B-lymphocytes, which, following productive engagement with antigen, undergo proliferation and differentiation to generate non-proliferating plasma cells that are specialized antibody producing and secreting "factories." In humans, five classes of secreted immunoglobulin (Ig) are defined that each exhibit unique structural and functional features: IgG, IgM, IgA, IgD and IgE [25–28]. Initial stimulation of the adaptive immune system results in a primary response characterized by the secretion of specific IgM antibody, by plasma cells and the generation of immunological "memory." Subsequent contact with the antigen results in a rapid, heightened secondary response exemplified by "class switching" and the production of plasma cells secreting IgM, IgG, IgA, IgE or IgD antibodies.

Antibody of the IgG class predominates in human blood and extra-vascular fluids; it is the only antibody class that is transported across the placenta to provide passive antibody protection for the human newborn. The IgG molecule exemplifies the basic four-chain antibody structural unit; composed of two identical light chains and two identical heavy chains (Figure 4.1a). Examination of the amino acid sequence reveals repeating structural motifs or homology regions, each of ~110 amino acid residues. Light chains consist of two homology regions and heavy chains four (IgG, IgA, IgD) or five (IgM, IgE). Each homology region consists of two anti-parallel β-sheets that form a β-barrel structure stabilized by the orientation of hydrophobic amino acid side chains towards the interior and the presence of an inter-sheet disulfide bridge (Figure 4.1b). This structure, referred to as the immunoglobulin fold or domain, is structurally robust whilst the β-turns allow for essentially infinite structural

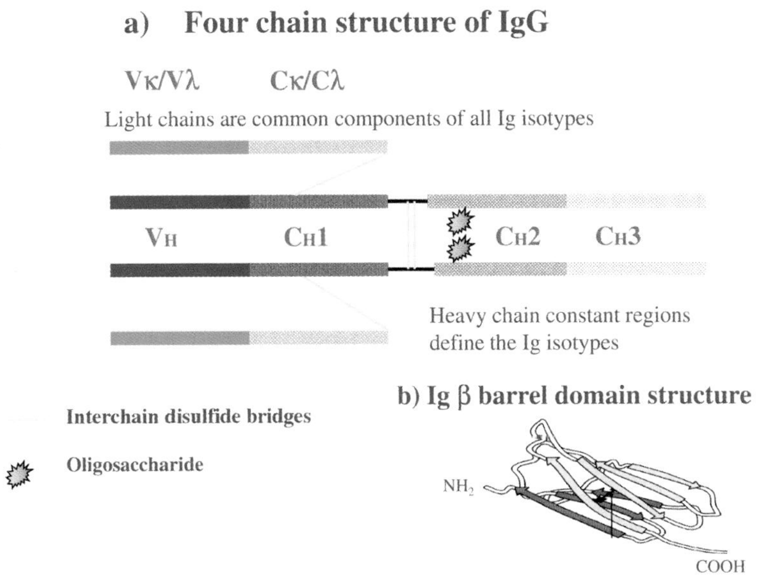

Figure 4.1 (a) The basic four-chain structure of the IgG molecule, showing the variable regions of the heavy (VH) and light (Vκ/Vλ) chains and the constant homology domains (CL, CH1, CH2, CH3); (b) the β barrel domain structure.

variation. The immunoglobulin fold, first defined for IgG molecules, is the structural paradigm for molecules of the immunoglobulin supergene family. The N-terminal domains of heavy and light chains are variable in sequence and determine the antigen specificity of the antibody molecule, referred to as VH and VL, respectively. The other light and heavy chain domains are referred to as the "constant domains" (CL, CH1, CH2, CH3, etc.) since they have a conserved sequence that is characteristic for a light chain type or heavy chain isotype [25–28].

In humans there are two types of light chain, kappa (κ) and lambda (λ), defined by the sequences of their variable and constant domains. Kappa light chains are encoded on chromosome 2 by a unique "library" of Vκ gene segments and a single Cκ gene; lambda chains are encoded on chromosome 22 by a unique "library" of Vλ gene segments and multiple Cλ genes. The heavy chain isotypes are encoded on chromosome 14 by a unique "library" of VH variable region gene segments and single copies of genes for the heavy chain constant regions; designated μ, δ, γ, α and ε for IgM, IgD, IgG, IgA and IgE, respectively. Rearrangements within each V gene segment library generate unique sequences, and random pairing of VH and VL domains, referred to as "combinatorial pairing," provides the diversity of the primary antibody repertoire. The secondary immune response is characterized by class switching and somatic hypermutation of the rearranged V gene segments; whilst most mutations will be deleterious a minority have increased specificity and affinity for antigen and are positively selected to provide enhanced humoral protection. The mechanism for generation of the antibody repertoire, by random recombination of gene segments and somatic mutation, is essential if the individual is to mount a protective immune response to new pathogens arising in the environment, also resulting from random mutations and selection. The individual must, therefore, be able to mount a protective antibody response to any external antigen but not against self-antigens; when this occurs it gives rise to autoimmunity; self-reactive B or T cells are generated but are normally eliminated to establish self-tolerance. Since individual B cells and plasma cell clones synthesize only one light and one heavy chain sequence (i.e., antibody of a unique antigen specificity), an individual antibody molecule bears either two kappa chains or two lambda chains, never one of each.

The gross three-dimensional structure of the IgG molecule is of three globular regions linked through a flexible "hinge" region (Figure 4.2). The IgG-Fab (fragment antigen binding) and IgG-Fc (fragment crystallizable) regions are defined as fragments released following exposure to the enzyme papain. Within an individual antibody molecule the IgG-Fab regions are of identical sequence and antigen binding specificity whilst the IgG-Fc region mediates effector functions resulting in the clearance and destruction of immune complexes. The compact structure of IgG-Fab is due to the "pairing" of hydrophobic surfaces of VH/VL, CH1/CL domains, respectively; CH3 domains similarly "pair;" however, CH2 domains do not pair but the hydrophobic surface is overlain by complex diantennary oligosaccharides structures (Figures 4.2 and 4.3). The flexible nature of the hinge region allows each IgG-Fab arm to bind antigen and for the IgG-Fc to be mobile and accessible to engage effector ligands, for example, IgG-Fc receptors (FcγR) and the C1 component of

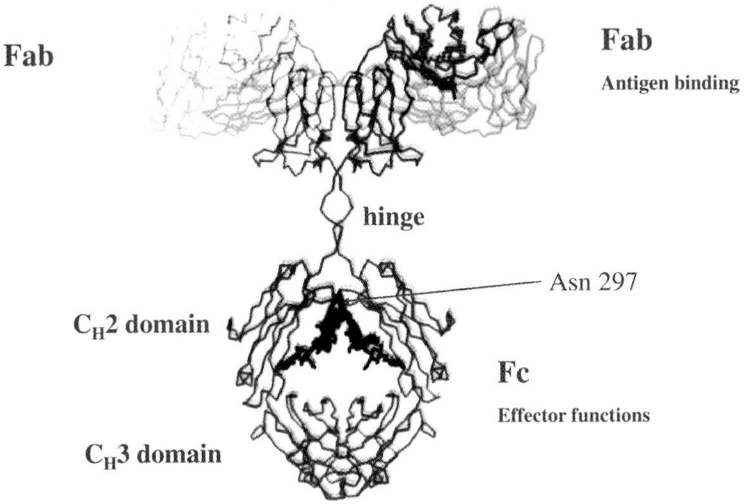

Figure 4.2 The alpha carbon backbone structure of the IgG molecule illustrates the sequence of β-pleated sheet homology domains. The oligosaccharide is integral to the protein structure and has defined conformation.

complement (see below). The complex diantennary oligosaccharide moiety (Figure 4.3) attached to the CH2 domain of IgG has been shown to be essential for recognition and engagement of these ligands and subsequent activation of effector functions [10, 11, 25–29].

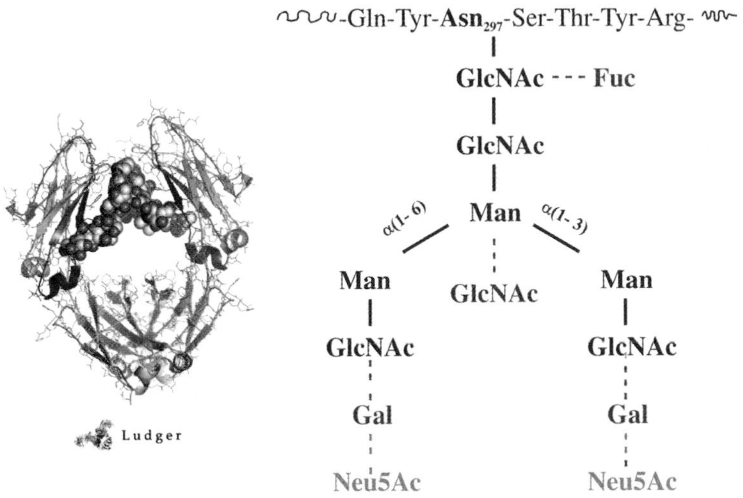

Figure 4.3 The complex diantennary oligosaccharide structures released from normal human IgG-Fc. A "core" heptasaccharide (blue) is identified and heterogeneity arises from variable addition of outer arm sugars (red).

4.2.2
Antibody (Immunoglobulin) Isotypes

Structural and serological studies define five human antibody (immunoglobulin) classes: IgM, IgG, IgA, IgD and IgE, and within IgG and IgA four and two subclasses, respectively. The subclasses are designated IgG1, IgG2, IgG3 and IgG4 and IgA1 and IgA2 according to their relative concentrations in serum; approximate proportions are IgG1 (60%), IgG2 (25%), IgG3 (10%), IgG4 (5%) and IgA1 (60%) and IgA2 (40%). Each class or subclass is defined by the amino acid sequence of the heavy chain constant regions and collectively constitute nine immunoglobulin isotypes that are present in all normal individuals. The sequence homology between heavy chains of different Ig class is $\sim$30%, whilst between the subclasses of IgG and IgA it is >90%, reflecting recent gene duplication and selection events in evolution. Each immunoglobulin heavy chain isotype expresses a unique profile of effector ligand recognition and consequently biological functions [25–28]. The ability of IgG to activate its effector ligands is critically dependent on the IgG-Fc glycosylation [10, 11, 25–28].

4.3
Glycosylation

4.3.1
Glycosylation of Normal Human IgG

The oligosaccharides released from normal polyclonal human IgG are of the complex diantennary type and exhibit considerable heterogeneity. Potentially, there are 32 [10, 11] different complex diantennary structures that could generate $(32 \times 32 = 1024)/2$, that is, 512 IgG-Fc glycoforms, given random "pairing" of heavy chain glycoforms; the apparent total is divided by 2 to account for the symmetry of the molecule. Analysis of oligosaccharides released from normal IgG-Fc and myeloma IgG proteins have defined a "core" heptasaccharide (blue in Figure 4.3), with heterogeneity arising from variable addition of outer arm sugar residues (red in Figure 4.3). The oligosaccharides released from IgG-Fc reveal a paucity of sialylation (<10%) with twelve of the sixteen possible neutral oligosaccharides predominating (Figures 4.4 and 4.5) [30]; however, minor glycoforms present in polyclonal IgG may predominate in IgG myeloma proteins. The 16 neutral oligosaccharides can generate a total of 128 glycoforms, assuming random pairing of heavy chain glycoforms [31, 32].

Unfortunately, there are several systems of nomenclature currently in use to represent oligosaccharide structures and, consequently, antibody glycoforms [33–36]. Carbohydrate chemists and specialist mass spectrometry scientists, amongst others, have developed different nomenclatures. A shorthand system has evolved from investigation IgG-Fc galactosylation – particularly with respect to variations encountered in rheumatoid arthritis and some other inflammatory diseases. Initially, IgG-Fc oligosaccharides devoid of galactose were designated G0 (zero galactose) and oligosaccharides bearing one or two galactose residues were designated G1 and G2,

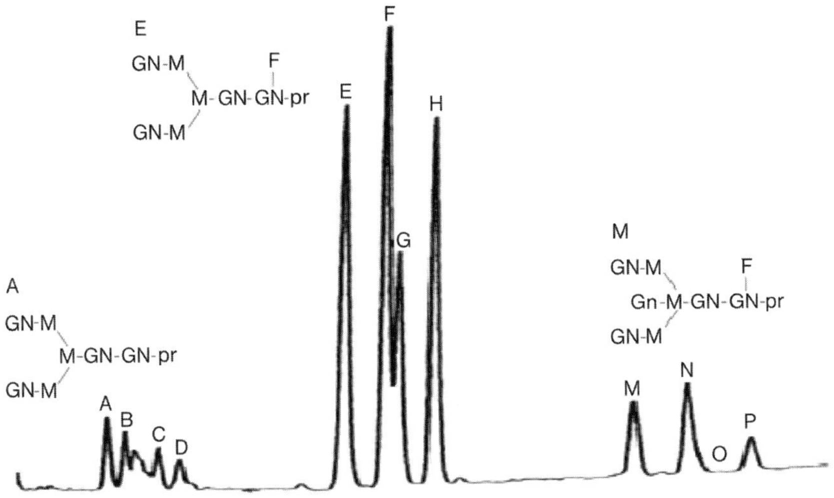

Figure 4.4 HPLC profile of neutral oligosaccharides released from normal human IgG-Fc.

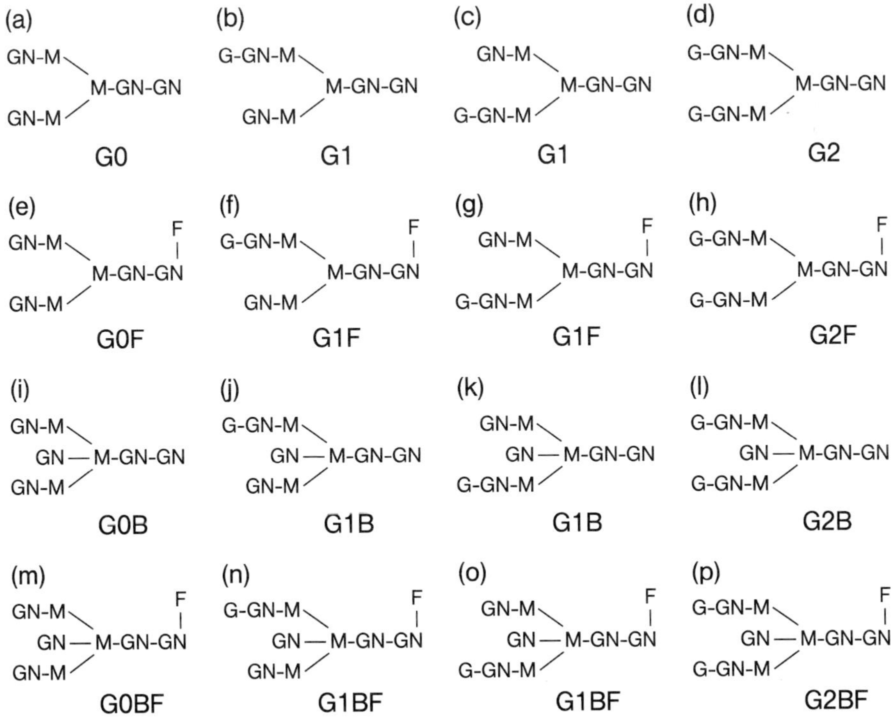

Figure 4.5 Potential "library" of neutral complex diantennary oligosaccharides – see Figure 4.4.

respectively. At that time the presence of fucose was presumed to be a constant; however, more recently the presence or absence of fucose has been shown to exert a major influence on biologic activity. The preferred shorthand nomenclature for the heptasaccharide + fucose ± galactose should be: G0F, G1F and G2F and the non-fucosylated forms as G0, G1, G2; when bisecting *N*-acetylglucosamine is present a B is added, for example, G0B, G0BF, and so on. Figure 4.5 illustrates this shorthand nomenclature; the two G1 forms of oligosaccharide reflect differential addition of galactose to the $\alpha(1 \rightarrow 3)$ or $\alpha(1 \rightarrow 6)$ arm. The extent of IgG-Fc galactosylation has been shown to vary with age, gender and pregnancy [37–39]. It is similarly important to define the glycoform of the whole IgG molecule, (G0)2, (G0F)2, (G0BF)2, and so on since it has been hypothesized that enhanced ADCC may be observed for IgG in which only one heavy chain bears oligosaccharide devoid of fucose [40]; thus, the IgG-Fc glycoform (G0/G0F) could be as potent as the (G0)2 glycoform. This shorthand nomenclature will be adopted within this chapter. Mass spectroscopic analysis of intact recombinant antibodies reveals symmetric and asymmetric glycoforms [31, 32]. Since the IgG molecule is fully assembled before entering the Golgi apparatus, asymmetrically glycosylated molecules result from differential processing of heavy chains within an individual molecule; it is not determined whether pairing is totally random or whether one heavy chain glycoform influences the processing of the "companion" heavy chain glycoforms.

Addition of galactose to the $\alpha(1 \rightarrow 6)$ arm is seen to predominate over addition to the $\alpha(1 \rightarrow 3)$ arm for normal polyclonal IgG-Fc; however, this reflects the quantitative predominance of the IgG1 subclass (60%). Analysis of myeloma proteins, of each of the four IgG subclasses, reveals the influence of subtle structural differences. Whilst for IgG1 and IgG4 proteins preference for addition to the $\alpha(1 \rightarrow 6)$ arm is observed, for IgG2 myeloma proteins the preference is for addition to the $\alpha(1 \rightarrow 3)$ arm. For IgG3 myeloma proteins the arm preference varies with the allotype of the heavy chain, possibly correlating with a phenylalanine/tyrosine interchange at residue 296 [41]. Further evidence of the influence of structure on oligosaccharide processing is provided by the profiles of oligosaccharides released from a series of mutant IgG3 proteins in which amino acids contributing to interactions with the oligosaccharide were replaced, sequentially, by alanine. Increased accessibility to the β 1→4 galactosyltransferase and α 2→3 sialyltransferase was evident since each mutant protein was extensively galactosylated and sialylated [42].

4.3.2
Impact of Glycosylation on Structure

The crystal structure of an IgG-Fc fragment, composed of residues 216–446, revealed electron density for residues 238–443 only [29]; thus residues 216–237, which constitute the hinge and lower hinge region, were assumed to be mobile and without defined structure. This finding was not consistent with the later proposal that residues in the lower hinge region were critical to the differential recognition of the three human FcγR [10, 11, 43, 44]. It was proposed, therefore, that the hinge/lower hinge region was composed of multiple structural conformers, in equilibrium, amongst which are particular conformers that recognize individual effector ligands [10, 11]. All crystal structures reveal a distinct conformation for the oligosaccharide that is integral to the three-dimensional protein structure and overlays the hydrophobic surface of the CH2 domain, making a ~2400 Å^2 contact [29, 43–45]. Each α(1 → 6) arm is directed towards the CH2/CH3 interface region and, in addition to covalent attachment at Asn-297 (Eu sequence [46] in a cis conformation, has the potential to make 85 non-covalent contacts with the protein backbone and amino acid side chains, including at least six hydrogen bonds [29, 43, 45]. Sugar residues of the α(1 → 3) arm extend into the internal space and, in some IgG-Fc crystal structures, contact the α(1 → 3) arm of the opposing CH2 domains. The integration of the oligosaccharide into the tertiary and quaternary structure of IgG-Fc suggests that the observed heterogeneity in galactosylation and the paucity of sialylation may be due to limited access for the β(1 → 4) galactosyltransferase and α(2 → 6) sialyltransferase enzymes.

4.3.3 Impact of Glycosylation on Stability

Differential scanning micro-calorimetry (DSC) has been employed to compare the stability of a series of homogeneous, normal and truncated glycoforms of IgG1-Fc with the aglycosylated form [47]. Fully galactosylated, (G2)2 and agalactosylated (G0)2 glycoforms of IgG-Fc exhibited two transition temperatures, *Tm*1 and *Tm*2, of 71.4 and 82.2 °C, respectively. The *Tm*1 and *Tm*2 transitions represent the unfolding of the CH2 and CH3 domains, respectively. These data suggest that whilst it is proposed that the galactose residue on the α(1 → 6) arm has substantial contacts with the protein structure it does not impact CH2 domain stability. Sequential removal of the terminal GlcNAc and the two arm mannose residues, generating a glycoform bearing only the $(GlcNAc)_2Man$ trisaccharide, resulted in destabilization of the CH2 domain and a lowering of *Tm*1 to 67.7 °C whilst *Tm*2 was unchanged. The thermodynamic parameters describing CH2 thermal denaturation of all IgG-Fc glycoforms was consistent with cooperative unfolding. By comparison the unfolding of the CH2 domain of deglycosylated IgG1-Fc was non-cooperative, involving at least one intermediate. It was proposed that this intermediate is a partially unfolded CH2 domain pair possessing hinge proximal disordered/unfolded loops that may account for the compromised functional activities of deglycosylated IgG and IgG-Fc [10, 11, 47–49]. Thus, the (GlcNAc)2Man trisaccharide is sufficient to confer a degree of order/structural stability on the CH2 domain and the generation of lower hinge conformers compatible with FcγRI and C1 binding and activation. It is proposed that, in addition to the covalent bond, the $(GlcNAc)_2Man$ trisaccharide has the potential to form 31 non-covalent contacts with the protein, with at least three being hydrogen bonds [45]. These data are consistent with X-ray crystal data obtained for the same truncated IgG-Fc glycoforms that showed progressive increases in the temperature factors for the CH2 domain, as evidence of progressive structural disorder (destabilization) [47]. Interestingly, in this study all attempts to obtain a crystal form of deglycosylated IgG-Fc failed. This may be indicative of the structural destabilization that results from complete removal of oligosaccharides. Truncation of the sugar residues results in the mutual approach of Cγ2 domains with the generation of a "closed" conformation; in contrast to the "open" conformation observed for the fully galactosylated IgG-Fc [49]. A minimal requirement of an $(GlcNAc)_2$ oligosaccharide for an element of structural and functional activity is suggested by the demonstration that a truncated glycoform bearing only a fucosylated primary GlcNAc was reported not to bind soluble recombinant FcγRIIIa and NMR studies indicated conformational alterations in the lower hinge and CH2 hinge proximal region that contributes to FcγR binding sites [50].

Protein engineering has been applied to generate a mutant, S239D/A330L/I332E, IgG1-Fc that has a 100-fold increase in affinity for FcγRIII; however, DSC revealed a considerable destabilization of the CH2 domain [51]. Thus, the melting temperature for the CH2 domain, *Tm*1, was reported as 46 °C, in contrast to 68 °C for the normal, unmutated IgG-Fc. The melting temperature for the CH3 domain, *Tm*2, was unchanged at 73 °C. Data for the normal IgG-Fc were in good agreement with the previously reported values of *Tm*1, 71.4 °C and *Tm*2, 82.2 °C [47]. The residues replaced do not make contacts with the oligosaccharide; however, modeling suggests that the introduced amino acid residues may form additional hydrogen and hydrophobic bonds at the interface with FcγRIII. The X-ray crystal structure does not reveal conformational change within the CH2 domain; however, there is a "slight opening" of the IgG-Fc structure. This may be, in part, due to the formation of hydrogen bonds between aspartic acid 265 and the introduced glutamic acid residue 332.

4.4 IgG-Fc Effector Functions

4.4.1 Inflammatory Cascades

The flexible hinge region of the IgG molecule allows each Fab region to engage its cognate antigen with the potential to form large immune complexes. Such complexes can activate downstream functional activities that result in their removal and destruction. These inflammatory processes are mediated by leucocytes expressing receptors with specificity for the Fc region of the IgG heavy chain (γ); they are termed Fcγ receptors (FcγR). Similarly, such complexes can activate the complement cascade through initial binding to the C1q component; this induces a conformational change with the generation of an enzyme that precipitates the cascade. Importantly, monomeric IgG is monovalent for FcγR and C1q and, therefore, whilst IgG, C1q and cells bearing FcγR co-exist in blood and tissue and may interact, transiently, such engagement does not result in activation. In humans the family of structurally homologous FcγR (FcγRI, FcγRII and FcγRIII) are constitutively but differentially expressed on human leucocytes and expression may be upregulated or induced by inflammatory cytokines [25–28, 52, 53]. There are multiple isoforms and polymorphic variants of FcγR that differ in IgG subclass recognition and activation or inhibitory functions [52, 53]. The FcγRs are transmembrane glycoproteins and there is evidence that the glycoform profile of the receptor may influence the outcome of encounters with IgG immune complexes, in a tissue-specific manner [40, 54, 55]. There is emerging data suggesting that certain glycoforms of IgG, and other immunoglobulin classes, may activate complement through the lectin pathway mediated by mannan binding lectin (MBL), a C1q homologue [56–60]. Similarly, some glycoforms may bind the mannose receptor (MR) that is expressed on phagocytic antigen presenting cells [61, 62]. Many parameters impact on the nature of immune complexes formed, for example, IgG subclass, antibody/antigen ratio, epitope density, glycoform, and so on, and consequently the inflammatory cascades activated; these include antibody-dependent cellular cytotoxicity (ADCC), secretion of inflammatory mediators, enhanced antigen presentation, regulation of antibody production, the oxidative burst and phagocytosis [63–65]. Since a polyclonal antibody response to an environmental antigen may consist of multiple isotypes it is virtually impossible to predict or monitor the effector functions activated. In addition, pathogens have evolved mechanisms that may subvert immune protection; thus bacteria produce IgG-Fc binding proteins – for example, SpA and SpG and virus may encode for membrane bound IgG-Fc receptor homologues that aid cellular infection [10, 16, 29, 66]. *Streptococcus pyogenes* produces an endoglycosidase, EndoS, that selectively removes the IgG-Fc oligosaccharides to generate IgG bearing only a fucosylated primary GlcNAc that is unable to activate FcγR or complement [67].

It is established that glycosylation of the IgG-Fc is essential for optimal expression of biological activities mediated through FcγRI, FcγRII, FcγRIII and the C1q component of complement [10, 11]. The association constant of aglycosylated IgG1 or IgG3 binding for FcγRI is reduced by two orders of magnitude, relative to that observed for the normally glycosylated form; however, aglycosylated IgG3 antibody can mediate ADCC if a high level of target cell sensitization is achieved; cellular activation through FcγRII and FcγRIII appears to be completely ablated [48, 68]. The association constant for C1q binding to aglycosylated IgG is also reduced by one order of magnitude and results in loss of complement dependent cytotoxicity (CDC) [48, 69, 70]. The binding of aglycosylated IgG-Fc to FcRn is only marginally affected so that the long catabolic half-life is maintained; similarly, binding to bacterial proteins, for example, SpA, SpG, to human IgG-Fc is essentially unaffected by the presence or absence of oligosaccharides [10, 11, 25–28].

Protein engineering, employing alanine scanning, has been used to "map" amino acid residues deemed to be critical for FcγR and C1q binding. These studies "map" the binding site for all four of these ligands to the hinge proximal or lower hinge region of the CH2 domain (Figure 4.6) [43, 44, 71–73]. The proposal that the hinge region of the IgG molecule is not without structure but is composed of multiple conformers in equilibrium is supported by the X-ray crystal structure of the complex of IgG-Fc with a recombinant, soluble form of FcγRIII [43, 44]. The interaction site on the IgG-Fc is seen to include asymmetric binding to discrete conformations of the lower hinge residues of each heavy chain. Other residues of the hinge proximal region of the CH2 domain also form contacts with the FcγRIII. Interestingly, one structure reveals a possible contribution of the primary *N*-acetylglucosamine residue

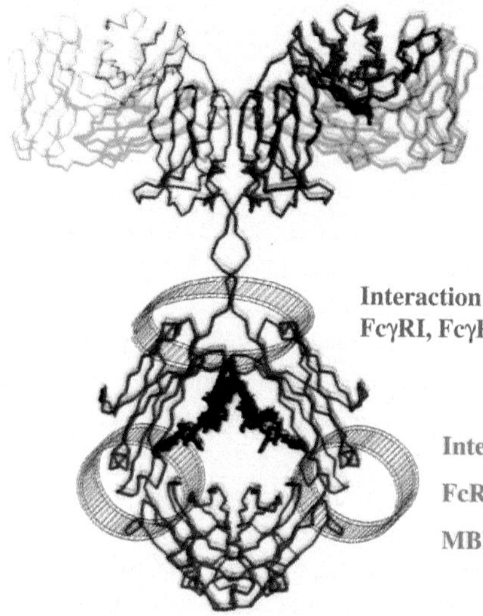

Figure 4.6 Interaction sites have been localized to the hinge and hinge proximal region of the CH2 domain for FcγRI, FcγRII, FcγRIII and C1q, and to the CH2/CH3 interface for FcRn, SpA, SpG, MBL and MR.

to binding [43] whilst the other primary publication held that there is no direct contact [44]; in a subsequent review the authors of the latter publication stated that the oligosaccharides contributes $\sim$100 Å^2 to the contact interface [74]. This conclusion resulted from refinement of the crystal data (P. Sun – personal communication). Taken together these data suggest that any direct contribution of the oligosaccharide to binding is minimal and that the oligosaccharide contributes indirectly through its influence on protein conformation. The asymmetric binding to both heavy chains provides a structural explanation for an essential requirement – that the IgG should be univalent for the FcγR; if it were divalent monomeric IgG could crosslink cellular receptors and hence constantly activate inflammatory reactions [10, 11, 25–27]. The X-ray crystal structures of IgG-Fc in complex with SpA, SpG and the autoantibody rheumatoid factor (RF) show that each of these ligands interacts with sites embracing residues at the junction of the CH2 and CH3 domain and the IgG-Fc is divalent for these ligands (Figure 4.6) [29, 75, 76].

4.4.2
Catabolism, Pharmacokinetics and Placental Transport

Catabolism and placental transport of human IgG is mediated by interactions with the neonatal IgG-Fc receptor FcRn – so-named because it was first identified in the rat where it mediates transport of IgG across the neonatal gut wall – effected through pathways of pinocytosis followed by exocytosis or transcytosis, respectively [27, 77]. It is a homologue of MHC class I molecules and in humans is expressed as a transmembrane protein on endothelial, epithelial, trophoblast, dendritic and monocyte cells. The high concentration of IgG in blood and intravascular fluids leads to cellular internalization during the natural process of non-specific pinocytosis. The pinocytotic vesicles fuse with endosomes that express membrane-bound FcRn. At the pH within the endosome, ~pH 6.5, IgG binds FcRn and is protected from enzymatic degradation; however, IgG remaining in the fluid phase is degraded. At exocytosis the IgG/FcRn complexes are exposed to extracellular fluid at physiological, pH 7.2, and the IgG is released. The catabolic half-life of human IgG1, IgG2 and IgG4 is ~23 days, the longest of any serum protein, whilst for IgG3 it is 7 days. Present evidence suggests that the catabolic half-life of IgG is independent of the natural glycoform and is essentially unaltered for aglycosylated IgG [78, 79]; there is no data available for placental transport of aglycosylated IgG. Several groups have reported enhanced FcRn binding [27, 80] and extended catabolic half-life for protein engineered IgG and Fc-fusion proteins [27, 81]. The potential benefit of increased half-life is less frequent attendance to the clinic for re-dosing and hence lowered CoT.

The long catabolic half-life of IgG is being exploited to increase the *in vivo* activity of therapeutic proteins that naturally have a short half-life (e.g., cytokines) by the generation of cytokine/IgG-Fc fusion proteins. An exciting recent development has been the demonstration that lung epithelial cells express FcRn and can mediate transcytosis with the delivery of locally dosed drug-IgG-Fc fusion proteins to the systemic circulation; bioavailability of 70% was reported for an erythropoietin/IgG-Fc fusion protein [81]. Abnormal glycoforms of IgG may be catabolized by alternate pathways – for example, cells expressing the mannose receptor may clear high mannose glycoforms. Notably, whilst circulating IgG bears terminal galactose residues it is not cleared in the liver by cells expressing the asialylglycoprotein receptor (ASGPR).

4.5 Individual IgG-Fc Glycoforms

4.5.1 Individual IgG-Fc Glycoforms and Effector Activities

Since glycosylation of IgG-Fc is shown to be essential to recognition and activation of numerous effector ligands quantitative and/or qualitative functional differences between glycoforms might be anticipated. Whilst it is virtually impossible to analyze the structural and functional diversity of a polyclonal antibody response to an individual target antigen, analysis of the oligosaccharides released from serum IgG has revealed significant differences associated with several inflammatory diseases, including rheumatoid arthritis [13, 82–84], inflammatory bowel disease [85], vasculitis [86], coeliac disease [87], periodontal disease [88], and so on; in addition, differentiation amongst rheumatic disease(s) is claimed for “sugar mapping” [13]. Studies employing recombinant monoclonal antibodies have established that individual glycoforms can have a very significant impact on functional activity. This leads one to question whether the immune system responds not only by production of an optimal isotype but also an optimal glycoform.

4.5.2
Sialylation of IgG-Fc Oligosaccharides

A minority of oligosaccharides released from polyclonal IgG-Fc are sialylated whilst a majority bear one or two terminal galactose sugar residues [19, 30, 41, 89]. The paucity of sialylation is presumed to reflect the intimate integration of the oligosaccharides within the IgG-Fc structure such that the steric/spatial requirements of the $\alpha(2 \rightarrow 6)$ sialyltransferase cannot be met. This conclusion is supported by the fact that when the IgG-Fab is glycosylated it bears highly galactosylated and sialylated structures, demonstrating that the sialylation pathway is fully functional. The presence of terminal galactose residues does not result in clearance through the ASGPR, possibly due to a lack of accessibility of the galactose residues. As discussed above, catabolism of IgG is mediated through FcRn and is independent of glycoform, including absence of IgG-Fc oligosaccharides. The situation for IgG-Fab glycoforms is explored below.

The balance between structure and accessibility is graphically illustrated for a panel of IgG-Fcs in which amino acid residues contributing to contacts with the oligosaccharide were replaced, sequentially, by alanine. In each case replacement of a single contact residue resulted in the production of hypergalactosylated and highly sialylated glycoforms [42]. This again demonstrated that hypogalactosylation and paucisialylation are not due to deficits in the processing machinery of the Golgi apparatus. Recent studies suggest that sialylated human IgG-Fc may be anti-inflammatory, relative to asialylated IgG-Fc, by virtue of increased binding to the inhibitory mouse FcγRIIb relative to mouse FcγRIIa/FcγRIV receptors [90, 91]; reduced affinity for human FcγRIIIa was manifest as reduced NK cell-mediated ADCC activity [92].

4.5.3
Influence of Galactosylation on IgG-Fc Activities

It can be appreciated from Figure 4.4 that differences in galactosylation are a major source of IgG-Fc glycoform heterogeneity. The distribution for normal adult serum derived IgG-Fc is agalactosyl (G0 + G0F + G0B + G0BF), 20–25%; mono-galactosyl (G1 + G1F + G1B + G1BF), 35–45%; di-galactosyl (G2 + G2F + G2B + G2BF), 10–20% [10, 11, 30, 89]. Lower levels of IgG-Fc galactosylation are observed for young children and for older adults; there is also a small but significant gender difference [37]. Hypogalactosylation of IgG-Fc is reported for several inflammatory states associated with autoimmune diseases [82–88]. An increase in IgG-Fc galactosylation occurs over the course of a normal pregnancy with levels returning to the adult norm following parturition [38, 39]. The extent of IgG-Fc galactosylation observed for monoclonal human myeloma IgG proteins is highly variable, indicating that the level of IgG-Fc galactosylation is a clonal property [41]. The antibody products of CHO, Sp2/0 and NS0 cell lines used in commercial production of recombinant antibody are generally fucosylated but hypogalactosylated, relative to normal human IgG-Fc [10, 11, 93–96] and consist of (G0 + G0F), 60–80%; mono-galactosyl (G1 + G1F), 20–40%; di-galactosyl (G2 + G2F), 10–20%; these cells do not add the bisecting *N*-acetylglucosamine residue. It is essential, therefore, to consider the impact of differential IgG-Fc galactosylation on functional activity.

Several studies have probed the influence of the presence or absence of galactose residues on IgG-Fc structure and function and published conflicting reports. A NMR study of galactosylated (G1F, G2F) and agalactosylated (G0F) glycoforms of IgG-Fc reported the mobility of the glycan to be comparable to that of the backbone polypeptide chain, with the exception of the galactose residue on the $\alpha(1 \rightarrow 3)$ arm, which was highly mobile; it was concluded that agalactosylation does not induce any significant change in glycan mobility and that it remains "buried" within the protein structure [50]. This report is consistent with binding and stability studies showing minimal differences between G0F and G2F glycoforms [47, 48, 97] and crystal structures [49] of a series of truncated glycoforms of IgG-Fc. A sophisticated NMR study probed changes in local environments on the binding of soluble recombinant FcγRIIIa to G2F and G0F glycoforms of IgG1-Fc and reported chemical shift differences of >0.2 ppm for Lys-248 and Val-308 residues [50]; this is a very localized change distant from the interaction site for the FcγRIIIa moiety. The finding of a changed environment for these residues is interesting since they were not predicted, from the crystal structure, to make contacts with the $\alpha(1 \rightarrow 6)$ arm galactose residue; small perturbations for the predicted contact residues Lys-246, Asp-249 and Thr-256 were also reported. Conflicting data were reported from a NMR study that concluded that glycans with $\alpha(1 \rightarrow 6)$ galactose residues had the same relaxation time as the protein backbone, whilst in the absence of $\alpha(1 \rightarrow 6)$ arm galactose the glycan had relaxation rates 30× slower, indicating high mobility and freedom from interactions with the protein structure [98].

The conflicting reports of the impact of galactosylation on IgG-Fc effector functions may be due, in part, to the fact that most studies were performed before our understanding of the impact of core fucosylation on FcγRIII binding and function [40, 47–50, 55]. Early studies reported only "slight" differences in binding/adherence to FcγRI, FcγRII or FcγRIII bearing cells for highly galactosylated anti-D antibody, compared to pauci-galactosylated anti-D; however, consideration of the lytic activity of an extensive panel of recombinant anti-D antibodies did not allow for a rational structural explanation [99]. No difference in binding or receptor-mediated signaling of FcγRII was reported for G0F and G2 glycoforms of the humanized IgG1 Campath-1H antibody [100].

As previously stated recombinant IgG antibody therapeutics secreted by cell lines adapted for commercial production are hypogalactosylated, relative to normal IgG-Fc. The possible consequences for *in vivo* activity are extrapolated from *in vitro* assays and animal experiments. Removal of terminal galactose residues from Campath-1H was shown to reduce CDC but to be without effect on FcR mediated functions [101]. Similarly, the ability of Rituximab to kill tumor cells by CDC was shown to be reduced by a factor of two for the (G0)2 glycoform in comparison to the (G2)2 glycoform [102]. The product that gained licensing approval consisted of ~25%

of the G1 glycoform; therefore, regulatory authorities required that galactosylation of the manufactured product be controlled to within a few % of this value.

In the absence of sialic acid and galactose the terminal sugar residues of IgG-Fc are GlcNAc. The serum protein mannan binding lectin (MBL) and the cellular mannose receptor (MR) can each recognize and bind arrays of GlcNAc [56–61]. It is possible, therefore, that immune complexes composed of G0/G0F IgG-Fc glycoforms may engage and activate these lectin molecules. MBL is a structural homologue of the C1q molecule that forms a complex with MASP-1, MASP-2 and MASP-3 molecules that are the homologues of C1s and C1r. The MBL/MASP complex circulates in the blood and when activated triggers the complement cascade through the initial binding and cleavage of C4, as for the C1 complex. A degalactosylated form of IgG1 was shown to bind and activate the MBL/MASP complex and the consequent cascade [56]. A G0F glycoform of IgG4-Fc was also shown to bind MBL [58]. These findings have been cited to suggest that in inflammatory diseases characterized by increased levels of G0F IgG-Fc glycoforms that activation of the lectin pathway may contribute to inflammation [103, 104].

The MR is a C-type lectin, expressed at the surface of macrophages, endothelial cells and dendritic cells, that recognizes arrays of mannose and *N*-acetylglucosamine residues [105]. Thus, immune complexes of G0F antibody glycoforms may aid their uptake by these cell types. Of particular interest is uptake by dendritic cells since they are "professional" antigen presenting cells and may be implicated in the development of autoimmunity. Experimental evidence is provided by the mouse collagen model of arthritis in which monoclonal anti-collagen type II antibodies were passively administered as complexes of collagen with G0F or G2F glycoforms of anti-collagen type II antibodies. The onset and severity of induced arthritis was greater for complexes formed with G0F glycoform of anti-collagen type II antibodies [61]. This finding may be relevant to the generation of immune responses to monoclonal antibody therapeutics; thus, the predominance of G0F glycoforms may similarly result in immune complex uptake by dendritic cells and the presentation of non-self peptides generated from the antibody, for example, mouse V regions, idiotypic determinants, and so on. This role for MR has been disputed and there is evidence that enhanced antigen presentation is through the processing of glycoproteins taken up by DC-SIGN, another C-type lectin molecule [106, 107].

4.5.4
Influence of Fucose and Bisecting *N*-Acetylglucosamine on IgG-Fc Activities

Antibody produced in rat-derived Y0 cells has been shown to be more active in ADCC than the same antibody produced in CHO (hamster) or NS0 cells (mouse) [108]. Analysis of the oligosaccharide profiles of these antibodies showed a possible correlation of ADCC activity with the ability of Y0 cells to produce IgG glycoforms bearing a bisecting *N*-acetylglucosamine residue. This rationale appeared to be vindicated by the demonstration of increased FcγRIII-mediated ADCC for antibody produced in CHO cells that had been transfected with the human $\beta(1\rightarrow4)$ *N*-acetylglucosaminyltransferase III (GnTIII) gene to produce antibody bearing bisecting *N*-acetylglucosamine residues [109, 110]. A profound increase in FcγRIII-mediated ADCC was also reported for antibody produced in a mutant CHO cell line that failed to add either bisecting GlcNAc or fucose [111]. Comparison of the ability of non-fucosylated, fucosylated and bisecting GlcNAc glycoforms forms of IgG antibody to mediate ADCC led to the conclusion that it is the absence of fucose rather than the presence of bisecting GlcNAc that accounts for increased FcγRIII-mediated ADCC [112]. Further studies showed that in CHO cells the expression of GnTIII competes with the endogenous $\alpha(1 \rightarrow 6)$-fucosyltransferase and inhibits addition of fucose [113]. Thus, the increased ADCC was in fact due to the absence of fucose. When transfected with genes for chimeric transferases that localize the GnTIII transferase to an earlier Golgi compartments increased addition of bisecting GlcNAc and increased inhibition of the addition of fucose resulted [114]. It is difficult to reconcile these data with the conclusion of Davies *et al.* [110], who credited the improved ADCC to an IgG glycoform bearing bisecting GlcNAc and fucose. A possible explanation for this discrepancy could be that only the major glycoforms were positively characterized in this study and that the ADCC activity observed was due to the presence of a minor but increased population of non-fucosylated glycoforms. Due to the asymmetry of the IgG-Fc/FcγRIII interface the lack of fucose on one heavy chain should be sufficient to provide improved binding and activation. The IgG-Fc glycoform bearing bisecting GlcNAc in the absence of fucose is a minor component of normal polyclonal human IgG and was not identified in the analysis represented in Figure 4.4.

Surface plasmon resonance studies show that non-fucosylated IgG-Fc binds soluble recombinant FcγRIII with higher affinity than does the fucosylated form, whilst aglycosylated IgG-Fc shows no evidence of binding [113, 114]. A similar study of the binding to different glycoforms of FcγRIIIa showed that glycosylation at asparagine-162 of the receptor influenced IgG-Fc binding; aglycosylation at this site resulting in increased affinity for the normal fucosylated glycoform of IgG-Fc [55]. This residue is at the interface between the FcγRIIIa receptor and IgG-Fc and it was suggested that the presence of fucose on IgG-Fc might result in steric inhibition of glycosylated srFcγRIIIa binding [114]. It was concluded that high affinity IgG-Fc/FcγRIIIa binding requires an interaction of sugar residues attached at Asn-162 with surface structures of the non-fucosylated IgG-Fc glycoform and that due to the asymmetry of the IgG-Fc/FcγRIIIa interaction non-fucosylation of one heavy chain would be sufficient for tight binding. Interestingly, it was shown that the increased affinity for the non-fucosylated glycoform of IgG-Fc was negated when FcγRIIIa was not glycosylated at Asn-162 [114]. The presence or absence of fucose was shown not to influence the binding affinity of IgG-Fc for the inhibitory FcγRIIb receptor and it was suggested that IgG antibody glycoform might be a sensitive modulator of FcγRIIIa-mediated ADCC – through tissue-specific production of different Asn-162 glycoforms of FcγRIII [40, 54]. Although most studies were conducted with IgG1, subclass protein increased ADCC was also demonstrated for non-fucosylated IgG3 and IgG4 antibodies; some activity was also observed for IgG2 [115]. Increased ADCC activity was also reported for a non-fucosylated glycoform of CH1/CL deleted fusion protein and could, presumably, be extended to IgG-Fc fusion proteins [116].

As stated previously, crystal structures for IgG-Fc in complex with aglycosylated soluble recombinant FcγRIIIa were initially interpreted to show that there is no direct contact between the oligosaccharide and the FcγRIIIa protein moiety [43, 44]. However, a subsequent review article [74] claimed that, following model refinement, the carbohydrate on IgG-Fc was seen to contribute 100 $Å^2$ to receptor contact (P. Sun – personal communication). These investigators also demonstrated that whilst the binding of deglycosylated IgG-Fc to aglycosylated soluble FcγRIIIa was undetectable the binding of deglycosylated whole IgG was only decreased, 10–15-fold [117]. This serves to remind us to exercise great caution when attempting to extrapolate from *in vitro* experimental data to predict *in vivo* biological outcomes.

Non-fucosylated oligosaccharides account for ~10% of those released from normal polyclonal IgG-Fc. Given random pairing between different heavy chain glycoforms a maximum of ~10% of assembled IgG molecules may be anticipated to contain one non-fucosylated heavy chain, with variable galactosylation. Studies of human IgG myeloma proteins, however, show that antibody producing plasma cell clones can secret predominantly fully fucosylated or non-fucosylated IgG glycoforms [30]. Thus, polyclonal IgG may similarly contain populations of IgG composed of two non-fucosylated heavy chain glycoforms. Enhanced ADCC has been demonstrated for non-fucosylated "Herceptin," *in vitro*, using peripheral mononuclear cells obtained from patients undergoing treatment for breast cancer [118].

4.6
IgG-Fab Glycosylation

It has been established that 15–20% of polyclonal human IgG molecules bear N-linked oligosaccharides within the IgG-Fab region, in addition to the conserved glycosylation site at Asn-297 in the IgG-Fc [10, 11, 119, 120]. There are no consensus sequences for N-linked oligosaccharide within the constant domains of either the kappa or lambda light chains or the CH1 and CH3 domains of heavy chains; therefore, when present they are attached in the variable regions of the kappa (Vκ), lambda (Vλ) or heavy (VH) chains – and sometimes both. In the immunoglobulin sequence database ~20% of IgG V regions have N-linked glycosylation consensus motifs (Asn-X-Thr/Ser; where X can be any amino acid except proline). Interestingly, these consensus sequences are mostly not germline encoded but result from somatic mutation – suggesting positive selection for improved antigen binding [120]. The functional significance for IgG-Fab glycosylation of polyclonal IgG has not been fully evaluated but data emerging for monoclonal antibodies suggest that Vκ, Vλ or VH glycosylation can have a neutral, positive or negative influence on antigen binding [10, 11, 93, 94].

Analysis of polyclonal human IgG-Fab reveals the presence of diantennary oligosaccharides that are extensively galactosylated and substantially sialylated, in contrast to the oligosaccharides released from IgG-Fc [10, 14, 119, 120]. This is somewhat surprising since "random" mutations could result in the generation of glycosylation motifs at many sites within a polyclonal antibody population and as a consequence site-specific glycosylation would be observed with varied incorporation of complex diantennary, high mannose, triantennary oligosaccharides, and so on. Indeed, the application of high sensitivity analysis of oligosaccharides released from polyclonal IgG has reported the presence of very low levels of tri- and tetra-antennary oligosaccharides [121]; a caveat might be entered here, to the effect that the IgG was a commercial source of undefined purity.

A similar glycoform profile has been observed for a human monoclonal IgG1 (myeloma) protein that is glycosylated within the VL region. Whilst both the IgG-Fc and IgG-Fab bore diantennary structures the IgG-Fc oligosaccharides were predominantly non-fucosylated G0 and G1 whilst the IgG-Fab oligosaccharides were predominantly fucosylated, galactosylated and sialylated [41]. Interestingly, whilst the IgG-Fc oligosaccharides could be quantitatively released on exposure to PNGase F the IgG-Fab oligosaccharides were refractory, but could be released on exposure to Endo F, in contrast to the IgG-Fc oligosaccharides that could not be released with this enzyme [32].

The glycoform profile of the therapeutic antibody Cetuximab (Erbitux) that is licensed for the treatment of colon, head and neck cancer reveals both IgG-Fc and IgG-Fab glycosylation. The antibody is produced in Sp2/0 cells and bears an N-linked oligosaccharide at Asn-88 of the VH region; interestingly, there is also a glycosylation motif at Asn-41 of the VL but it is not occupied [94]. The oligosaccharides could be released on exposure to PNGase F; however, the glycoform profiles at IgG-Fc and IgG-Fab are profoundly different. Thus, whilst the IgG-Fc glycoform profile is of the familiar diantennary G0F/G1F type the IgG-Fab consisted of 21 oligosaccharide structures that differ in degree of sialylation with *N*-glycolylneuraminic acid and extent of galactosylation, including mono- and di- gal $\alpha(1\text{–}3)$ gal residues; hybrid and tri-antennary oligosaccharides are also present.

Unfortunately, some patients have experienced severe hypersensitivity reactions on first exposure to Cetuximab [122] and it has been demonstrated that most of these patients have pre-existing IgE antibodies specific for the gal $\alpha(1 \rightarrow 3)$ gal epitope; the *in vitro* assay employed did not detect any reaction for "Cetuximab" produced in CHO cells. It might be concluded, therefore, that the Sp2/0 cell line is not suitable for the production of IgG-Fab glycosylated antibodies; however, it remains to be seen whether CHO cells can provide human type IgG-Fab glycoforms.

A similar experience has been reported for a detailed analysis of the glycoforms of a humanized IgG antibody, expressed in Sp2/0 cells, bearing oligosaccharides at Asn-56 of the VH and Asn-297. The expected IgG-Fc oligosaccharides profile of predominantly fucosylated G0F/G1F oligosaccharides was observed; however, eleven oligosaccharides were released from the IgG-Fab, including triantennary, and other oligosaccharides not observed for normal human IgG [93]. Clearance rates, in mice, were independent of IgG-Fc and nine of the IgG-Fab glycoforms but marginally accelerated clearance was observed for two IgG-Fab glycoforms. All IgG-Fab oligosaccharides were extensively sialylated with *N*-glycolylneuraminic acid, rather than *N*-acetylneuraminic acid. The IgG-Fab oligosaccharide linkages were refractory to release by PNGase F. Several other examples of recombinant antibodies bearing IgG-Fab glycosylation have been reported and the presence of a diantennary oligosaccharide resistant to removal by PNGase F has been a repeated observation, except for Cetuximab. Partial occupancy of IgG-Fab but not IgG-Fc has also been observed. Presentations made by biopharmaceutical companies at meetings have reported similar findings of conserved IgG-Fc oligosaccharide profiles and more heterogeneous IgG-Fab glycosylation; the latter showing relatively high levels of galactosylation and sialylation.

The influence of IgG-Fab glycosylation on antigen binding has been the subject of several reports. Three antibodies with specificity for $\alpha(1 \rightarrow 6)$ dextran, differing only in potential *N*-glycosylation sites at Asn-54, -58 or -60 in the VH CDR2 region, were evaluated for antigen binding affinity. The Asn-54 and -58 molecules each bore a complex diantennary oligosaccharide rich in sialic acid, were equivalent in antigen binding and the glycosylated forms had a 10–50-fold higher affinity for antigen compared with aglycosylated forms. In contrast the Asn-60 molecule bore a high mannose oligosaccharide and had a lower affinity for antigen [123]. A significant proportion of IgG-Fab oligosaccharides bore Gal $\alpha(1 \rightarrow 3)$ Gal structures [124]. By contrast humanization of a mouse anti-CD33 antibody with concomitant removal of a potential glycosylation site at Asn-73 of the VH resulted in higher affinity for antigen; subsequent deglycosylation of the original mouse antibody similarly resulted in increased affinity [125]. Increased affinity for ovomucoid was reported for the deglycosylated form of a mouse antibody bearing N-linked oligosaccharide in the light chain CDR2 [126]. A multi-specific human monoclonal antibody, produced in mouse-human heterohybridoma cells, has been reported to bear both di- and tetra-antennary oligosaccharides attached at Asn-75 of the VH region and to include antigenic *N*-glycolylneuraminic acid sugar residues [127]. IgG-Fab glycosylation can impact differentially on the structural and functional characteristics of IgG. It may be exploited to increase the solubility and stability of antibodies; limiting aggregation and hence immunogenicity. However, given the essential demand for product consistency it offers an additional challenge to the biopharmaceutical industry.

A further intriguing feature of IgG-Fab glycosylation is being revealed in studies of human B cell lymphoproliferative disease. Whilst ~10% of normal B cells bear surface Ig glycosylated within VH or VL the frequency amongst patients with sporadic Burkitt's lymphoma, endemic Burkitt's lymphoma and follicular lymphoma is 42%, 82% and 94%, respectively [128, 129]. Sequence analysis revealed multiple glycosylation motifs; one sequence encoding four VH and two VL motifs. Significantly, whilst the normal IgG-Fc glycoform profile was maintained IgG-Fab sites were shown to bear high mannose glycans that have potential to engage mannose-binding proteins/receptors that may be relevant to the aetiology of these diseases.

4.7
Recombinant Monoclonal Antibodies for Therapy

It is projected that 30–40% of new drugs entering the market over the next 10 years will be based on antibodies and that the market value will be $30 billion in 2010. The efficacy of recombinant antibodies results from their specificity for the target antigen and the biological activities (effector functions) activated by the immune complexes formed. To date all licensed therapeutic antibodies have been of the IgG class. Numerous parameters impact on the effector function profile (e.g., IgG subclass, glycoform, etc.) and their definition offers opportunities for optimizing an antibody for a given disease indication. It is increasingly appreciated that the choice of IgG subclass is a critical decision. In oncology, a major disease indication, it would seem beneficial to maximize the potential to induce ADCC and CDC to eliminate targeted cancer cells [11, 130–132]. However, in chronic diseases neutralization of a soluble target (e.g., a cytokine) may be the central objective and excessive effector activity could be detrimental [11, 132, 133] for instance where the target is also expressed as a membrane protein on certain cells – for example, the targeting of TNFα by Infliximab in patients with rheumatoid arthritis. Consequently, IgG1 may not be the automatic choice and the other IgG subclasses are now being evaluated, both in native and engineered forms. Some companies have selected IgG4 as the preferred alternative subclass and two IgG4 antibodies have been licensed. However, recent studies suggest that IgG4 may not be the best alternative since it is has been shown that it can activate inflammatory reactions through cellular IgG-Fc receptors [11, 114, 134, 135]. Notably, the "super-agonistic" anti-CD28 antibody TGN 1412, which induced a "cytokine storm" in healthy volunteers [136, 137], is a humanized IgG4 recombinant antibody.

To date all licensed therapeutic antibodies have been produced in CHO, NSO or Sp2/0 cells [138, 139]. The glycoform profile of the product can vary widely from clone to clone and is dependent on the mode of production and culture conditions. Under non-optimal conditions CHO, NS0 and Sp2/0 cells can produce several abnormally glycosylated products, including partial occupancy, high mannose forms, the addition of galactose $\alpha(1 \rightarrow 3)$ galactose and *N*-glycolylneuraminic acid structures [18, 19]. These glycoforms may compromise potency and potentiate immunogenicity [140]. The licensed antibody products are characteristically hypogalactosylated, relative to normal polyclonal IgG, and lack the presence of glycoforms bearing bisecting *N*-acetylglucosamine residues; abnormal glycoforms are deemed to be at an acceptably low level. Alternatives to mammalian cell culture for the production of glycosylated whole length antibodies are now reaching maturity [141, 142]; antibody fragments that are not glycosylated are produced in *Escherichia coli* [143].

The glycoform profiles of the currently licensed antibodies were determined rather late in the clone selection process; however, the lessons learned are being applied to clone selection at a much earlier stage and monitored throughout development and production, as an integral element of process analytical technology (PAT) [144, 145]. There have been quite dramatic increases in the productivity of antibody-producing mammalian cells with concentrations of 1–5 g L^{-1} being reported; however, when selecting for high specific protein production particular attention has to given to product quality. High levels of antibody production can compromise glycosylation, resulting in partial occupancy, unacceptable levels of high mannose, and so on. However, clones producing antibody at high levels and acceptable quality have been achieved [139]. A fully human cell line, PerC.6, has been approved by the FDA that yields a human IgG glycoform profile and has recently been reported to produce $\sim 15\,g\,L^{-1}$ of antibody [146].

4.8 Conclusions and Future Perspectives

The body of this chapter demonstrates that the functional activity of recombinant monoclonal antibody molecules is influenced by the structure of the oligosaccharides attached at Asn-297. Notably, however, almost all studies have employed antibodies of the IgG1 subclass. It remains to be determined whether the glycoform similarly impacts on the activities of the other subclasses; however, the absence of fucose has been shown to influence FcγRIIIa for all four subclasses [115]. It is particularly interesting to note that glycosylation at Asn-162 of FcγRIIIa also regulates functionality and that differential tissue glycosylation of FcγRIIIa has been reported [54], allowing for subtle variations in functionality. It is anticipated that FcγRI and FcγRII receptors may interface similarly with IgG-Fc but glycosylation motifs homologous to Asn-162 of FcγRIIIa are not present in these receptors.

It remains to be established whether the glycoform profile of a natural antibody response is regulated by the immune system to optimize and vary functionality – between initial inflammatory and later anti-inflammatory activities [90]. Yeast and plant cell lines have been developed that allow for the glycoform profile of an antibody therapeutic to be selected in advance; this allows for the oligosaccharide to be employed as "rheostat" to optimize functional activity for a given disease indication [147, 148]. A significant advantage of glycoform selection is that natural glycoforms are employed; protein engineering always carries the possibility of introducing immunogenicity. A present and future challenge is to develop means of monitoring mechanisms activated by antibody glycoforms *in vivo* and to anticipate variations in individual patient responses, determined by the interaction of multiple polymorphisms resident in an out-bred human population – it is called "systems biology."

References

1 Baker, M. (2005) Upping the ante on antibodies. *Nature Biotechnology*, **23**, 1065–1072.

2 Reichert, J.M. *et al.* (2005) Monoclonal antibody successes in the clinic. *Nature Biotechnology*, **23**, 1073–1078.

3 Carter, P.J. (2006) Potent antibody therapeutics by design. *Nature Reviews*, **6**, 343–357.

4 Satoh, M., Iida, S., Shitara, K. (2006) Non-fucosylated therapeutic antibodies as next-generation therapeutic antibodies. *Expert Opinion on Biological Therapy*, **6**, 1161–1173.

5 Ledford, H. (2007) Biotechs go generic: the same but different. *Nature*, **449**, 274–276.

6 Woodcock, J. *et al.* (2007) The FDA's assessment of follow-on protein products: a historical perspective. *Nature Reviews. Drug Discovery*, **6**, 437–442.

7 Schellenkens, H. (2008) The first biosimilar epoetin: but how similar is it? *Clinical Journal of the American Society of Nephrology*, **3**, 174–178.

8 Liang, P.H. *et al.* (2008) Glycan arrays: biological and medical applications. *Current Opinion in Chemical Biology*, **12**, 86–92.

9 Freeze, H.H. (2007) Genetic defects in the human glycome. *Nature Reviews. Genetics*, **7**, 537–551.

10 Jefferis, R. (2005) Glycosylation of recombinant antibody therapeutics. *Biotechnology Progress*, **21**, 11–16.

11 Jefferis, R. (2007) Antibody therapeutics: isotype and glycoform selection. *Expert Opinion on Biological Therapy*, **7**, 1401–1413.

12 Vanhooren, V. *et al.* (2007) *N*-Glycomic changes in serum proteins during human aging. *Rejuvenation Research*, **10**, 521–531.

13 Alavi, A. and Axford, J.S. (2006) The pivotal nature of sugars in normal physiology and disease. *Wiener Medizinische Wochenschrift*, **156**, 19–33.

14 Holland, M. *et al.* (2006) Differential glycosylation of polyclonal IgG, IgG-Fc and IgG-Fab isolated from the sera of patients with ANCA associated systemic vasculitis. *Biochimica et Biophysica Acta*, **1760**, 669–677.

15 Poland, D.C. *et al.* (2005) Activated human PMN synthesize and release a strongly fucosylated glycoform of {alpha}1-acid glycoprotein, which is transiently deposited in human myocardial infarction. *Journal of Leukocyte Biology*, **78**, 453–461.

16 Smalling, R. *et al.* (2004) Drug-induced and antibody-mediated pure red cell aplasia: a review of literature and current knowledge. *Biotechnology Annual Review*, **10**, 37–50.

17 Sinclair, A.M. and Elloitt, S. (2005) Glycoengineering: the effect of glycosylation on the properties of therapeutic proteins. *Journal of Pharmaceutical Sciences*, **94**, 1626–1635.

18 Walsh, G. and Jefferis, R. (2006) Post-translational modifications in the context of therapeutic proteins. *Nature Biotechnology*, **24**, 1241–1252.

19 Raju, T.S. *et al.* (2000) Species-specific variation in glycosylation of IgG: evidence for the species-specific sialylation and branch-specific galactosylation and importance for engineering recombinant glycoprotein therapeutics. *Glycobiology*, **10**, 477–486.

20 Van den Nieuwenhof, I.M. *et al.* (2000) Recombinant glycodelin carrying the same type of glycan structures as contraceptive glycodelin-A can be produced in human kidney 293 cells but not in Chinese hamster ovary cells. *European Journal of Biochemistry*, **267**, 4753–4762.

21 Lienard, D. *et al.* (2007) Pharming and transgenic plants. *Biotechnology Annual Review*, **13**, 115–1147.

22 Birch, J.R. and Racher, A.J. (2006) Antibody production. *Advanced Drug Delivery Reviews*, **58**, 671–685.

23 Quan, C. *et al.* (2008) A study in glycation of a therapeutic recombinant humanized monoclonal antibody: where it is, how it got there, and how it affects charge-based behavior. *Analytical Biochemistry*, **373**, 179–191.

24 Harris, R.J. (2005) Heterogeneity of recombinant antibodies: linking structure to function. *Developmental Biology*, **122**, 117–127.

25 Burton, D.R. and Woof, J.M. (1992) Human antibody effector function. *Advances in Immunology*, **51**, 1–84.

26 Woof, J.M. and Burton, D.R. (2004) Human antibody-Fc receptor interactions illuminated by crystal structures. *Nature Reviews. Immunology*, **4**, 89–99.

27 Nezlin, R. and Ghetie, V. (2004) Interactions of immunoglobulins outside the antigen-combining site. *Advances in Immunology*, **82**, 155–215.

28 Arnold, J.N. *et al.* (2007) The impact of glycosylation on the biological function and structure of human

immunoglobulins. *Annual Review of Immunology*, **25**, 21–50.
29 Deisenhofer, J. (1981) Crystallographic refinement and atomic models of a human Fc fragment and its complex with fragment B of protein A from Staphylococcus aureus at 2.9- and 2.8-Å resolution. *Biochemistry*, **20**, 2361–2370.
30 Routier, F.H. *et al.* (1998) Quantitation of human IgG glycoforms isolated from rheumatoid sera: A critical evaluation of chromatographic methods. *Journal of Immunological Methods*, **213**, 113–130.
31 Masuda, K. *et al.* (2000) Pairing of oligosaccharides in the Fc region of immunoglobulin G. *FEBS Letters*, **473**, 349–357.
32 Mimura, Y. *et al.* (2007) Contrasting glycosylation profiles between Fab and Fc of a human IgG protein studied by electrospray ionization mass spectrometry. *Journal of Immunological Methods*, **326**, 116–126.
33 http://glycomics.scripps.edu/CFGnomenclature.pdf. Accessed January 5th 2009.
34 Crispin, M., Stuart, D.I., Jones, E.Y. (2007) Building meaningful models of glycoproteins. *Nature Structural & Molecular Biology*, **14**, 354–355.
35 http://www.functionalglycomics.org. Accessed January 5th 2009.
36 www.proglycan.com. Accessed January 5th 2009.
37 Yamada, E. *et al.* (1997) Structural changes of immunoglobulin G oligosaccharides with age in healthy human serum. *Glycoconjugate Journal*, **14**, 401–405.
38 Williams, P.J., Arkwright, P.D. and Rudd, P.M. (1995) Short communication: Selective transport of maternal IgG to the foetus. *Placenta*, **16**, 749–756.
39 Kibe, T. *et al.* (1996) Glycosylation and placental transport of immunoglobulin G. *Journal of Clinical Biochemistry and Nutrition*, **21**, 57–63.
40 Ferrara, C. *et al.* (2006) The carbohydrate at FcgammaRIIIa Asn-162. An element required for high affinity binding to non-fucosylated IgG glycoforms. *The Journal of Biological Chemistry*, **281**, 5032–5036.
41 Farooq, M. *et al.* (1997) Glycosylation of antibody molecules in multiple myeloma. *Glycoconjugate Journal*, **14**, 489–492.
42 Lund, J. *et al.* (1996) Multiple interactions of IgG with its core oligosaccharide can modulate recognition by complement and human FcγRI and influence the synthesis of its oligosaccharide chains. *Journal of Immunology*, **157**, 4963–4969.
43 Sondermann, P. *et al.* (2000) The 3.2-A crystal structure of the human IgG1 Fc-FcγRIII complex. *Nature*, **406**, 267–273.
44 Radaev, S. *et al.* (2001) The structure of human type FcγIII receptor in complex with Fc. *The Journal of Biological Chemistry*, **276**, 16469–16477.
45 Padlan, E.A. (1990) *Fc Receptors and the Action of Antibodies* (ed. H. Metzger), American Society for Microbiology, Washington D.C., pp. 12–30.
46 Edelman, G.M. *et al.* (1969) The covalent structure of an entire IgG molecule. *Proceedings of the National Academy of Sciences of the, United States of America* **63**, 78–85.
47 Mimura, Y. *et al.* (2000) The influence of glycosylation on the thermal stability and effector function expression of human IgG1-Fc: properties of a series of truncated glycoforms. *Molecular Immunology*, **37**, 697–706.
48 Mimura, Y. *et al.* (2001) The role of oligosaccharide residues of IgG1-Fc in FcγIIb binding. *The Journal of Biological Chemistry*, **276**, 45539–45547.
49 Krapp, S. *et al.* (2003) Structural analysis of human IgG glycoforms reveals a correlation between oligosaccharide content, structural integrity and Fcγ-receptor affinity. *Journal of Molecular Biology*, **325**, 979–989.
50 Yamaguchi, Y. *et al.* (2006) Glycoform-dependent conformational alteration of the Fc region of human immunoglobulin G1 as revealed by NMR spectroscopy. *Biochimica et Biophysica Acta*, **1760**, 693–700.

51 Oganesyan, V. *et al.* (2008) Structural characterization of a mutated, ADCC-enhanced human Fc fragment. *Molecular Immunology*, **45**, 1872–1882.

52 Nimmerjahn, F. and Ravetch, J. (2007) Fc-receptors as regulators of immunity. *Advances in Immunology*, **96**, 179–204.

53 Nimmerjahn, F. and Ravetch, J. (2008) Fcgamma receptors as regulators of immune responses. *Nature Reviews. Immunology*, **8**, 34–47.

54 Edberg, J.C. and Kimberly, R.P. (1997) Cell type-specific glycoforms of Fc gamma RIIIa (CD16): differential ligand binding. *Journal of Immunology*, **159**, 3849–3857.

55 Drescher, B., Witte, T., Schmidt, R.E. (2003) Glycosylation of FcgammaRIII in N163 as mechanism of regulating receptor affinity. *Immunology*, **110**, 335–340.

56 Malhotra, R. *et al.* (1995) Glycosylation changes of IgG associated with rheumatoid arthritis can activate complement via the mannose-binding protein. *Nature Medicine*, **1**, 237–243.

57 Abadeh, S. *et al.* (1997) Remodelling the oligosaccharide of human IgG antibodies: effects on biological activities. *Biochemical Society Transactions*, **25**, S661.

58 Wright, A. and Morrison, S.L. (1998) Effect of C2-associated carbohydrate structure on Ig effector function: studies with chimeric mouse-human IgG1 antibodies in glycosylation mutants of Chinese hamster ovary cells. *Journal of Immunology*, **160**, 3393–3402.

59 Arnold, J.N. *et al.* (2005) Human serum IgM glycosylation: identification of glycoforms that can bind to mannan-binding lectin. *The Journal of Biological Chemistry*, **280**, 29080–29087.

60 Terai, I. *et al.* (2006) Degalactosylated and/or denatured IgA, but not native IgA in any form, bind to mannose-binding lectin. *Journal of Immunology*, **177**, 1737–1745.

61 Dong, X., Storkus, W.J., Salter, R.D. (1999) Binding and uptake of agalactosyl IgG by mannose receptor on macrophages and dendritic cells. *Journal of Immunology*, **163**, 5427–5434.

62 Wright, A. *et al.* (2000) In vivo trafficking and catabolism of IgG1 antibodies with Fc associated carbohydrates of differing structure. *Glycobiology*, **10**, 1347–1355.

63 Garred, P., Michaesen, T.E., Aase, A. (1989) The IgG subclass pattern of complement activation depends on epitope density and antibody and complement concentration. *Scandinavian Journal of Immunology*, **30**, 379–382.

64 Lucisano Valim, Y.M. and Lachmann, P.J. (1991) The effect of antibody isotype and antigenic epitope density on the complement-fixing activity of immune complexes: a systematic study using chimaeric anti-NIP antibodies with human Fc regions. *Clinical and Experimental Immunology*, **84**, 1–8.

65 Voice, J.K. and Lachmann, P.J. (1997) Neutrophil Fc gamma and complement receptors involved in binding soluble IgG immune complexes and in specific granule release induced by soluble IgG immune complexes. *European Journal of Immunology*, **27**, 2514–2523.

66 Armour, K.L. *et al.* (2002) The contrasting IgG-binding interactions of human and herpes simplex virus Fc receptors. *Biochemical Society Transactions*, **30**, 495–500.

67 Nandakumar, K.S. *et al.* (2007) Endoglycosidase treatment abrogates IgG arthritogenicity: importance of IgG glycosylation in arthritis. *European Journal of Immunology*, **37**, 2973–2982.

68 Tao, M.H. and Morrison, S.L. (1989) Studies of aglycosylated chimeric mouse-human IgG. Role of carbohydrate in the structure and effector functions mediated by the human IgG constant region. *Journal of Immunology*, **143**, 2595.

69 Pound, J.D., Lund, J., Jefferis, R. (1993) Aglycosylated chimeric human IgG3 can trigger the human phagocyte respiratory burst. *Molecular Immunology*, **30**, 469–478.

70 Lund, J. *et al.* (1991) Human FcγRI and FcγRII interact with distinct but overlapping sites on human IgG. *Journal of Immunology*, **147**, 2657–2662.

71 Sarmay, G. *et al.* (1992) Mapping and comparison of the interaction sites on the Fc region of IgG responsible for triggering antibody dependent cellular cytotoxicity (ADCC) through different types of Fcγ receptor. *Molecular Immunology*, **29**, 633–639.

72 Shields, R.L. *et al.* (2001) High resolution mapping of the binding site on human IgG1 for FcγRI, FcγRII, FcγRIII, and FcRn and design of IgG1 variants with improved binding to the FcγR. *The Journal of Biological Chemistry*, **276**, 6591–6604.

73 Desjarlais, J.R. *et al.* (2007) Optimizing engagement of the immune system by anti-tumor antibodies: an engineer's perspective. *Drug Discovery Today*, **12**, 898–910.

74 Radaev, S. and Sun, P. (2002) Recognition of immunoglobulins by Fcgamma receptors. *Molecular Immunology*, **38**, 1073–1083.

75 Sauer-Eriksson, A.E. *et al.* (1995) Crystal structure of the C2 fragment of streptococcal protein G in complex with the Fc domain of human IgG. *Structure*, **3**, 265–278.

76 Corper, A.L. *et al.* (1997) Structure of human IgM rheumatoid factor Fab bound to its autoantigen IgG Fc reveals a novel topology of antibody-antigen interaction. *Nature Structural Biology*, **4**, 374–381.

77 Brambell, F.W.R., Hemmings, W.A. and Morris, I.G. (1964) A theoretical model of gammaglobulin catabolism. *Nature*, **203**, 1352–1355.

78 Jones, A.J. *et al.* (2007) Selective clearance of glycoforms of a complex glycoprotein pharmaceutical caused by terminal *N*-acetylglucosamine is similar in humans and cynomolgus monkeys. *Glycobiology*, **17**, 529–540.

79 Stork, R. *et al.* (2008) *N*-glycosylation as novel strategy to improve pharmacokinetic properties of bispecific single-chain diabodies. *The Journal of Biological Chemistry*, **283**, 7804–7812.

80 Hinton, P.R. *et al.* (2006) An engineered human IgG1 antibody with longer serum half-life. *Journal of Immunology*, **176**, 346–356.

81 Bitonti, A.J. and Dumont, J.A. (2006) Pulmonary administration of therapeutic proteins using an immunoglobulin transport pathway. *Advanced Drug Delivery Reviews*, **58**, 1106–1118.

82 Parekh, R.B. *et al.* (1985) Association of rheumatoid arthritis and primary osteoarthritis with changes in the glycosylation pattern of total serum IgG. *Nature*, **316**, 452–457.

83 Axford, J.S. *et al.* (2003) Rheumatic disease differentiation using immunoglobulin G sugar printing by high density electrophoresis. *The Journal of Rheumatology*, **12**, 2540–2546.

84 Flogel, M. *et al.* (1998) Fucosylation and galactosylation of IgG heavy chains differ between acute and remission phases of juvenile chronic arthritis. *Clinical Chemistry and Laboratory Medicine*, **36**, 99–102.

85 Go, M.F., Schrohenloher, R.E., and Tomana, M. (1994) Deficient galactosylation of serum IgG in inflammatory bowel disease: correlation with disease activity. *Journal of Clinical Gastroenterology*, **18**, 86–87.

86 Holland, M. *et al.* (2002) Hypogalactosylation of serum IgG in patients with ANCA-associated systemic vasculitis. *Clinical and Experimental Immunology*, **29**, 183–190.

87 Cremata, J.A., Sorell, L. and Montesino, R. (2003) Hypogalactosylation of serum IgG in patients with coeliac disease. *Clinical and Experimental Immunology*, **133**, 422–429.

88 Novak, J. *et al.* (2005) Heterogeneity of IgG glycosylation in adult periodontal disease. *Journal of Dental Research*, **84**, 897–901.

89 Mizouchi, T. *et al.* (1982) Structural and numerical variations of the carbohydrate moiety of immunoglobulin G. *Journal of Immunology*, **129**, 2016–2020.

90 Kaneko, Y., Nimmerjahn, F. and Ravetch, J. (2006) Anti-inflammatory activity of immunoglobulin G resulting from Fc sialylation. *Science*, **313**, 670–673.
91 Jefferis, R. (2006) A sugar switch for anti-inflammatory antibodies. *Nature Biotechnology*, **24**, 1230–1231.
92 Scallon, B. *et al.* (2007) Higher levels of sialylated Fc glycans in immunoglobulin G molecules can adversely impact functionality. *Molecular Immunology*, **44**, 1524–1534.
93 Huang, L. *et al.* (2006) Impact of variable domain glycosylation on antibody clearance: An LC/MS characterization. *Analytical Biochemistry*, **349**, 197–207.
94 Qian, J. *et al.* (2007) Structural characterization of N-linked oligosaccharides on monoclonal antibody cetuximab by the combination of orthogonal matrix-assisted laser desorption/ionization hybrid quadrupole-quadrupole time-of-flight tandem mass spectrometry and sequential enzymatic digestion. *Analytical Biochemistry*, **364**, 8–18.
95 Sethuraman, N. and Stadheim, T.A. (2006) Challenges in therapeutic glycoprotein production. *Current Opinion in Biotechnology*, **17**, 341–346.
96 Werner, R.G., Kopp, K., Schleuter, M. (2007) Glycosylation of therapeutic proteins in different production systems. *Acta Paediatrica (Oslo, Norway: 1992) Supplement*, **96**, 17–22.
97 Maenaka, K. *et al.* (2001) The human low affinity Fcgamma receptors IIa, IIb, and III bind IgG with fast kinetics and distinct thermodynamic properties. *The Journal of Biological Chemistry*, **276**, 44898–44904.
98 Wormald, M.R. *et al.* (1997) Variations in oligosaccharide-protein interactions in immunoglobulin G determine the site-specific glycosylation profiles and modulate the dynamic motion of the Fc oligosaccharides. *Biochemistry*, **36**, 1370–1380.
99 Kumpel, B.M. (2007) Efficacy of RhD monoclonal antibodies in clinical trials as replacement therapy for prophylactic anti-D immunoglobulin: more questions than answers. *Vox Sanguinis*, **93**, 99–111.
100 Groenink, J. *et al.* (1996) On the interaction between agalactosyl IgG and Fc gamma receptors. *European Journal of Immunology*, **26**, 1404–1407.
101 Boyd, P.N. *et al.* (1995) The effect of the removal of sialic acid, galactose and total carbohydrate on the functional activity of Campath-1H. *Molecular Immunology*, **32**, 1311–1318.
102 http://www.fda.gov/cder/biologics/review/ritugen112697-r2.pdf. Accessed January 5th 2009.
103 Garred, P. *et al.* (2000) Two edged role of mannose binding lectin in rheumatoid arthritis: a cross sectional study. *The Journal of Rheumatology*, **27**, 26–34.
104 Saevarsdottir, S., Vikingsdottir, T. and Valdimarsson, H. (2004) The potential role of mannan-binding lectin in the clearance of self-components including immune complexes. *Scandinavian Journal of Immunology*, **60**, 23–29.
105 Taylor, M.E. and Drickamer, K. (2003) Structure-function analysis of C-type animal lectins. *Methods in Enzymology*, **363**, 3–16.
106 Napper, C.E. and Taylor, M.E. (2004) The mannose receptor fails to enhance processing and presentation of a glycoprotein antigen in transfected fibroblasts. *Glycobiology*, **14**, 7C–12C.
107 Cambi, A. and Figdor, C.G. (2003) Levels of complexity in pathogen recognition by C-type lectins. *Current Opinion in Cell Biology*, **2003**, **15**, 539–546.
108 Lifely, M.R., Hale, G. and Boyse, S. (1995) Glycosylation and biological activity of CAMPATH-1H expressed in different cell lines and grown under different culture conditions. *Glycobiology*, **5**, 813–822.
109 Umana, P. *et al.* (1999) Engineered glycoforms of an anti-neuroblastoma. IgG1 with optimized antibody-dependent cellular cytotoxic activity. *Nature Biotechnology*, **17**, 176–180.

110 Davies, J. *et al.* (2001) Expression of GTIII in a recombinant anti-CD20 CHO production cell line: expression of antibodies of altered glycoforms leads to an increase in ADCC thro' higher affinity for FcRIII. *Biotechnology and Bioengineering*, **74**, 288–294.

111 Shields, R.L. *et al.* (2002) Lack of fucose on human IgG1 N-linked oligosaccharide improves binding to human FcγRIII and antibody-dependent cellular toxicity. *The Journal of Biological Chemistry*, **277**, 26733–26740.

112 Shinkawa, T. *et al.* (2003) The absence of fucose but not the presence of galactose or bisecting *N*-acetylglucosamine of human IgG1 complex-type oligosaccharides shows the critical role of enhancing antibody-dependent cellular cytotoxicity. *The Journal of Biological Chemistry*, **278**, 3466–3473.

113 Okazaki, A. *et al.* (2004) Fucose depletion from human IgG1 oligosaccharide enhances binding enthalpy and association rate between IgG1 and FcgammaRIIIa. *Journal of Molecular Biology*, **336**, 1239–1249.

114 Ferrara, C. *et al.* (2006) Modulation of therapeutic antibody effector functions by glycosylation engineering: influence of Golgi enzyme localization domain and co-expression of heterologous beta1, 4-*N*-acetylglucosaminyltransferase III and Golgi alpha-mannosidase II. *Biotechnology and Bioengineering*, **93**, 851–861.

115 Niwa, R. *et al.* (2005) IgG subclass-independent improvement of antibody-dependent cellular cytotoxicity by fucose removal from Asn297-linked oligosaccharides. *Journal of Immunological Methods*, **306**, 151–160.

116 Natsume, A. *et al.* (2005) Fucose removal from complex-type oligosaccharide enhances the antibody-dependent cellular cytotoxicity of single-gene-encoded antibody comprising a single-chain antibody linked the antibody constant region. *Journal of Immunological Methods*, **306**, 93–103.

117 Radaev, S. and Sun, P.D. (2001) Recognition of IgG by Fcgamma receptor. The role of Fc glycosylation and the binding of peptide inhibitors. *The Journal of Biological Chemistry*, **276**, 16478–16483.

118 Suzuki, E. *et al.* (2007) A nonfucosylated anti-HER2 antibody augments antibody-dependent cellular cytotoxicity in breast cancer patients. *Clinical Cancer Research*, **13**, 1875–1882.

119 Youings, A. *et al.* (1996) Site-specific glycosylation of human immunoglobulin G is altered in four rheumatoid arthritis patients. *The Biochemical Journal*, **314**, 621–630.

120 Dunn-Walters, D.K., Boursier, L. and Spencer, J. (2000) Effect of somatic hypermutation on potential *N*-glycosylation sites in human immunoglobulin heavy chain variable regions. *Molecular Immunology*, **37**, 107–113.

121 Harvey, D.J. *et al.* (2008) Differentiation between isomeric triantennary N-linked glycans by negative ion tandem mass spectrometry and confirmation of glycans containing galactose attached to the bisecting (beta1–4-GlcNAc) residue in *N*-glycans from IgG. *Rapid Communications in Mass Spectrometry*, **22**, 1047–1052.

122 Chung, C.H. *et al.* (2008) Cetuximab-induced anaphylaxis and IgE specific for galactose-α-1,3-galactose. *The New England Journal of Medicine*, **358**, 1109–1117.

123 Wright, A. and Morrison, S.L. (1997) Effect of glycosylation on antibody function: implications for genetic engineering. *Trends in Biotechnology*, **15**, 26–32.

124 Gala, F.A. and Morrison, S.L. (2004) V region carbohydrate and antibody expression. *Journal of Immunology*, **172**, 5489–5494.

125 Co, M.S. *et al.* (1993) Genetically engineered deglycosylation of the variable domain increases the affinity of an anti-CD33 monoclonal antibody. *Molecular Immunology*, **30**, 1361–1367.

126 Fujomura, Y. *et al.* (2004) Antigen binding of an ovomucoid-specific antibody is affected by a carbohydrate chain located on the light chain variable region. *Bioscience, Biotechnology, and Biochemistry*, **64**, 2298–2305.

127 Leibiger, H. *et al.* (1999) Variable domain-linked oligosaccharides of a human monoclonal IgG: structure and influence on antigen binding. *The Biochemical Journal*, **338**, (Pt 2) 529–538.

128 Radcliffe, C.M. *et al.* (2007) Human follicular lymphoma cells contain oligomannose glycans in the antigen-binding site of the B-cell receptor. *The Journal of Biological Chemistry*, **282**, 7405–7415.

129 McCann, K.J. *et al.* (2008) Remarkable selective glycosylation of the immunoglobulin variable region in follicular lymphoma. *Molecular Immunology*, **45**, 1567–1572.

130 Weiner, G.J. (2007) Monoclonal antibody mechanisms of action in cancer. *Immunologic Research*, **39**, 271–278.

131 Siberil, S. *et al.* (2007) FcγR, the key to optimize therapeutic antibodies? *Critical Reviews in Oncology/Hematology*, **62**, 26–33.

132 Congy-Jolivet, N. *et al.* (2007) Recombinant therapeutic monoclonal antibodies: mechanisms of action in relation to structural and functional duality. *Critical Reviews in Oncology/Hematology*, **64**, 226–233.

133 Furst, D.E. *et al.* (2007) Updated consensus statement on biological agents for the treatment of rheumatic diseases, 2007. *Annals of the Rheumatic Diseases*, **66**, (Suppl 3) iii2–iii22.

134 Holland, M. *et al.* (2004) Anti-neutrophil cytoplasm antibody IgG subclasses in Wegener's granulomatosis: a possible pathogenic role for the IgG4 subclass. *Clinical and Experimental Immunology*, **138**, 183–192.

135 van der Neut Kolfschoten, M. *et al.* (2007) Anti-inflammatory activity of human IgG4 antibodies by dynamic Fab arm exchange. *Science*, **317**, 1554–1557.

136 Nada, A. and Somberg, J. (2007) First-in-Man (FIM) clinical trials post-TeGenero: a review of the impact of the TeGenero trial on the design, conduct, and ethics of FIM trials. *American Journal of Therapeutics*, **14**, 594–604.

137 Dayan, C.M. and Wraith, D.C. (2008) Preparing for first-in-man studies: the challenges for translational immunology post-TGN1412. *Clinical and Experimental Immunology*, **151**, 231–234.

138 Birch, J.R. and Racher, A.J. (2006) Antibody production. *Advanced Drug Delivery Reviews*, **58**, 671–685.

139 Farid, S.S. (2006) Established bioprocesses for producing antibodies as a basis for future planning. *Advances in Biochemical Engineering/Biotechnology*, **101**, 1–42.

140 Mirik, G.R. *et al.* (2004) A review of human anti-globulin antibody (HAGA, HAMA, HACA, HAHA) responses to monoclonal antibodies. Not four letter words. *Quarterly Journal of Nuclear Medicine and Molecular, Imaging* **48**, 251–257.

141 Sparrow, P.A. (2007) Pharma-Planta: road testing the developing regulatory guidelines for plant-made pharmaceuticals. *Transgenic Research*, **16**, 147–161.

142 Gasser, B. and Mattanovitch, D. (2007) Antibody production with yeasts and filamentous fungi: on the road to large scale? *Biotechnology Letters*, **29**, 201–212.

143 Humphreys, D.P. (2003) Production of antibodies and antibody fragments in Escherichia coli and a comparison of their functions, uses and modification. *Current Opinion in Drug Discovery & Development*, **6**, 188–196.

144 Kourti, T. (2006) The Process Analytical Technology initiative and multivariate process analysis, monitoring and control. *Analytical and Bioanalytical Chemistry*, **384**, 1043–1048.

145 http://www.fda.gov/CDER/GUIDANCE/6419fnl.htm. Accessed January 5th 2009.

146 http://www.percivia.com/index.html. Accessed January 5th 2009.

147 Hamilton, S.R. and Gerngross, T.U. (2007) Glycosylation engineering in yeast: the advent of fully humanized yeast. *Current Opinion in Biotechnology*, **18**, 387–392.

148 Cox, K.M. *et al.* (2006) Glycan optimization of a human monoclonal antibody in the aquatic plant Lemna minor. *Nature Biotechnology*, **24**, 1591–1597.

5
Gonadotropins and the Importance of Glycosylation

Alfredo Ulloa-Aguirre, James A. Dias, and George R. Bousfield

5.1
Introduction

The gonadotropins, luteinizing hormone (LH), follicle-stimulating hormone (FSH) and choriogonadotropin (CG) are the hormones that regulate gonadal function. Luteinizing hormone and FSH are produced by the gonadotropes of the anterior pituitary gland [1–4]. Although CG is mainly synthesized by the placental trophoblasts, there is evidence indicating that this gonadotropin is also normally produced by the anterior pituitary [5]. Choriogonadotropin is found only in primates and equids, whilst the pituitary gonadotropins are found in all mammals and in some non-mammalian species [6]. Both LH and FSH play an essential role in regulating gonadal function; their coordinated secretion and action allow for an extremely precise control of ovarian and testicular function. Granulosa and Sertoli cells are the targets for FSH; in these cells, FSH promotes and/or supports the synthesis of estrogens and several non-steroidal factors as well as gametogenesis. More recently, it has been proposed that FSH may also play an important role in osteoclast function [7] and oocyte development [8]. The main target tissues for LH are the testicular Leydig cells and ovarian thecal cells, in which this gonadotropin regulates the local and systemic concentrations of sex steroids, mainly androgens and progesterone, as well as some non-steroidal factors; LH is also essential for follicular rupture and the subsequent release of the oocyte. The placental gonadotropin, CG, is synthesized and secreted by differentiated syncytiotrophoblast cells early in pregnancy and maintains steroid hormone production (mainly progesterone) by the corpus luteum until the placenta is functionally competent and assumes this steroidogenic function later in pregnancy.

The initial event in glycoprotein hormone action begins with binding of hormone to highly specific receptors located in the cell surface membrane of the target cell. Glycoprotein hormone receptors belong to the superfamily of G protein-coupled receptors, specifically the family of rhodopsin-like receptors (family A) [3, 9–11]. These receptors consist of a single polypeptide chain of variable length that begins

Post-translational Modification of Protein Biopharmaceuticals. Edited by Gary Walsh

ISBN: 978-3-527-32074-5

with an extracellular domain and then enters the membrane, continuing to thread back and forth across the lipid bilayer seven times, forming characteristic α-helical transmembrane domains connected by alternating extracellular and intracellular loops, ending with an intracellular carboxyl-terminal segment (Figure 5.1). Follicle-stimulating hormone binds to its cognate receptor, the FSHR. It has been demonstrated that the FSHR exists in three forms, called R1, R2 and R3 [12]. The relative

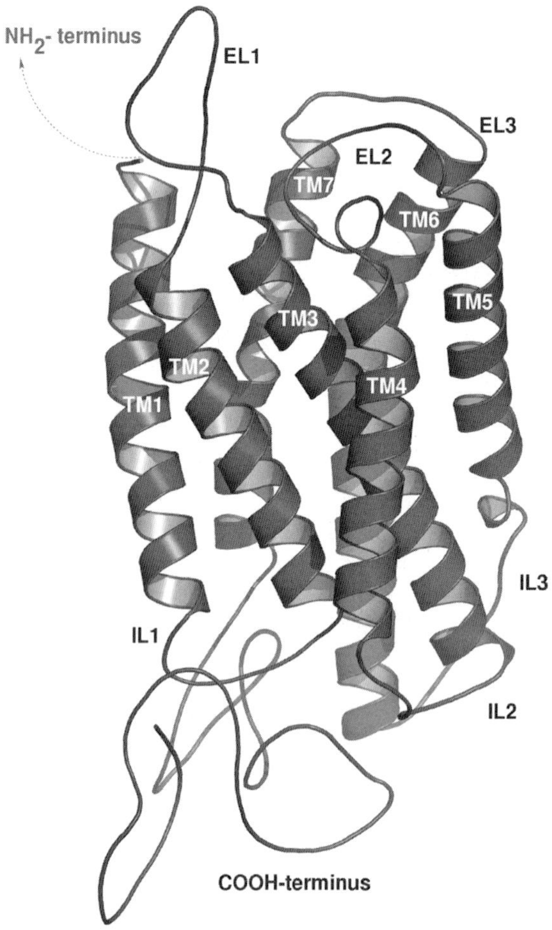

Figure 5.1 Predicted structure of human FSHR based on homology modeling with the structure of bovine rhodopsin (Protein Data Bank ID 1U19A). The coiled structures represent the antiparallel α-helices of transmembrane domains 1–7 connected by the extra- (EL) and intracellular (IL) loops of the receptor. The structure of the ILs and the COOH-terminus were obtained after 100 ns of molecular dynamics simulations performed with a partial solvation approach that optimize the conformational sampling of local protein regions with low computational cost [17]. The NH_2-terminus of the receptor is not shown for clarity. (The model was generated with the molecular visualization program PyMOL 0.99 (DeLano Scientific, San Francisco, CA, USA) and is reproduced by courtesy of Dr. Eduardo Jardón-Valadez, from the Instituto Mexicano del Seguro Social and the National University of Mexico.)

levels of each of these forms of the receptor are not clear. The splice variant R3 appears to be expressed in ovarian cancer cells [13]. In addition, FSH appears to act on osteoclast cells [7, 14]. Luteinizing hormone and CG bind to the same receptor (the LHCGR). Both the FSHR and the LHCGR transduce intracellular signals via coupling to heterotrimeric G proteins, mainly the G_s protein. Activation of gonadotropin receptors leads to production of the second messenger cAMP, which in turn activates directly or indirectly several protein kinases and pathways that eventually lead to an array of biological responses [15, 16].

5.2 Structure of Gonadotropins

Luteinizing hormone, FSH and CG as well as thyroid-stimulating hormone (TSH) – produced by a distinct anterior pituitary cell type, the thyrotrope – are heterodimers consisting of a common α-subunit non-covalently bound to a β-subunit [2, 18]. Within a given animal species, the α-subunit of the four glycoproteins is encoded by the same gene and, therefore, its amino acid sequence is identical; in contrast, the β-subunits arise from separate genes and confer to each hormone a high degree of biological specificity [5, 18–22]. The presence of both subunits is required for the expression of full biological activity. Each β-subunit may combine with the common α-subunit [23], thus indicating that some regions within the three-dimensional structure of the various β-subunits are very similar among the glycoprotein hormones [24–26]. Interestingly and unexpectedly, both the conserved residues as well as non-identical residues play a role in specificity of binding [27].

The primary structure of the common human (h) α-subunit encoded by cDNA is a protein 116 amino acids in length, including a signal peptide of 24 amino acids that is cleaved during the processing of the subunit to yield a mature protein of 92 amino acids. In the human, the absence of four amino acids corresponding to residues 6–9 in other species yields a gap in the comparative Cys-aligned sequence of the α-subunit [18]; this gap, however, does not alter the loops formed by any of the α disulfide bridges. The primary structure of the β-subunit of FSH encoded by cDNA is a 120-amino acid protein that after cleaving of the 19-amino acid signal peptide results in a mature protein 111 amino acids long (FSHβ). However, apparently due to use of an alternative signal peptidase cleavage site, 80% of pituitary hFSHβ and 50% of pituitary equine FSHβ consist of only residues 3–111 [28]. The gene organization of the β-subunit of LH (LHβ) is closely related to that of its homologous gene in CG, since apparently both genes are derived from a common ancestor [18, 20]. There are seven sets of *LH-CG* β-subunit genes in humans; six of these genes contain the coding sequence for CG β-subunit (CGβ) and one contains the sequence for LHβ. Only one of the *CG* β-subunit genes is capable of being expressed as a functional mRNA. There are several differences between the *LH* β-subunit and *CG* β-subunit genes; the latter is only present in primates and equids, whereas the former is present in all vertebrate and in some invertebrate species [18, 20]. Whilst the *LH* β-subunit gene codes for a protein with 121 amino acid residues, with a 24-residue signal

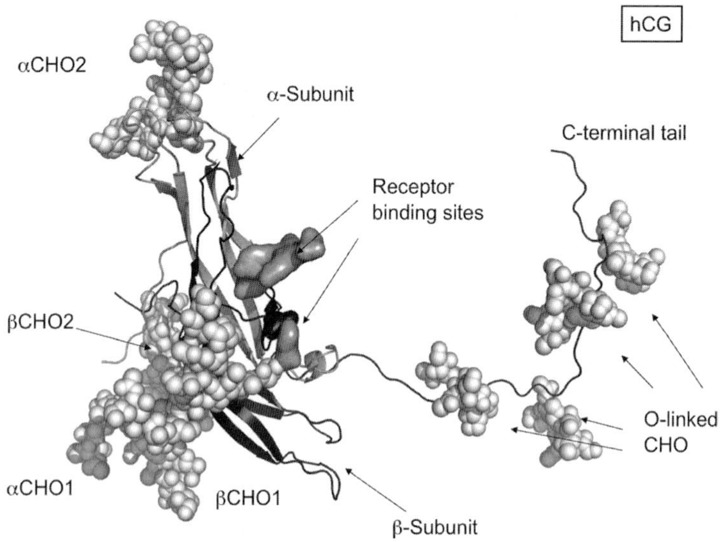

Figure 5.2 A rendering of the hCG crystal structure (1HCN) with glycan moieties added to the reported structure. Glycan structures were attached to hCG structure glycans through alignment of the GlcNAc residue on each structure using Weblab Viewer Pro (Accelrys, San Diego, CA, USA). Attachment of the glycans is for illustrative purposes only, to give the reader a sense of the contribution of glycan moieties to both the overall glycoprotein size and potential steric hindrance to formation of the hormone–receptor interaction.

peptide, the mature CGβ molecule exhibits 145 amino acids due to a mutation in the terminal codon of the ancestral gene. This accident of nature gives rise to an extension that is heavily *O*-glycosylated (Figure 5.2), prolonging the half-life of CG in blood, compared to other glycoprotein hormones. Furthermore, this extended COOH-terminus has recently been used to create fusion proteins of FSHβ that have prolonged half-life in circulation, and that are currently employed in clinical trials for *in vitro* fertilization protocols (see below) [29–32].

The secondary, tertiary and quaternary structures of the gonadotropins have been revealed by the three-dimensional structures of hFSH and hCG [24–26, 33, 34]. The three-dimensional structures of hFSH and hCG reveal that they belong to the family of cystine knot proteins, which include activin, bone morphogenetic protein, transforming growth factor beta, platelet-derived growth factor, nerve growth factor and vascular endothelial growth factor (Figure 5.2) [35]. The cystine knot proteins are generally dimeric in nature, some being homodimers, some heterodimers. Their signature folding motif is one where six cysteines form three interlinked (knot) disulfide bridges [36]. The glycoprotein hormone secondary structure contains three antiparallel beta sheet elongated loops joined at the center by the cystine knot. The first and third loop associate into a tertiary structure whereas the second loop appears as a long loop in β-subunit or, as in the α-subunit, a partial α-helix. Although the specificity of binding of the glycoprotein hormones was for years considered to be due to the β-subunit, the crystal structure of single chain hFSH in complex with the hFSHR1-250 revealed that the α-subunit also plays a role in binding specificity [37].

The two subunits assemble into a quaternary structure in a head to tail orientation, such that the first and third loops of one subunit are juxtaposed to the second loop of the other. When the hCG and hFSH crystal structures are aligned via the α-subunits, there is very little deviation between the α-subunits but the β-subunit conformation differs slightly, suggesting some flexibility [25]. Comparison of the receptor bound form of hFSH with free hFSH reinforces this notion that there is conformational flexibility of the long loop (loop 2) in FSHβ [37]. Additionally, there are notable differences in the conformation of the β-subunits of hCG and FSH when the α-subunit of each of the heterodimers are aligned [25]. It is important to remember that crystal structures represent the conformation of proteins in a crystalline state. In contrast, solution structures of the α-subunit demonstrate a range of extensive conformational space [38]. In this regard, it is worth noting that the conformation of the β-subunit loop 2 of hFSH differs when the FSH structure is compared to the hFSH:hFSHR1-250 structure. However, the key residue contacts that form the high-affinity binding site appear similar in the crystal structure of the complex when compared to the structure–function studies of hFSH. This indicates that conformational flexibility may play a key role in formation of the high-affinity binding site interaction by sampling conformational space before formalizing key contacts.

Crystal structures of cystine knot proteins in complex with their receptors have revealed that the association of FSH with FSHR1-250 differs from, for example, the association of bone morphogenetic protein-7 with its receptor. In the former case, the association appears to involve a complex of 1 : 1, whereas in the latter the ratio is 2 : 1, with one ligand coordinating two receptors. These observations indicate that the mode of ligand-provoked intracellular signaling may differ. In the case of a 2 : 1 complex, a high-affinity interaction with ligand and one receptor may lead to a secondary low-affinity interaction with an additional receptor. Alternatively, the formation of a high-affinity site may require the presence of two closely associated receptors. In the case of activin, the former mode appears to be operative. Additional work needs to be done to determine if the FSH receptor is truly a 1 : 1 association with its ligand. It has recently been demonstrated that the FSHR is a dimer from birth [in the endoplasmic reticulum (ER)] to maturity (in the plasma membrane) [39]. This observation suggests that the extracellular domain could play a role in formation of the FSH binding site. In this vein, it has also been demonstrated that negative cooperativity is at play in the binding of TSH to its receptor and to a lesser extent hCG and FSH binding to their receptors [40].

The subunits of all glycoprotein hormones contain several N-linked heterogeneous oligosaccharide structures (Figure 5.3). Carbohydrates are an important structural component of the gonadotropin molecules, making nearly 20–30% of the hormone's mass (Figure 5.2). Each primary sequence of the common α-subunit and of FSHβ and CGβ contains two *N*-glycosidically-linked oligosaccharides, while LHβ and TSH β-subunit have only one [6, 41]. In addition to the two consensus sites for N-linked glycosylation, CGβ exhibits four *O*-glycosidically-linked oligosaccharide structures in its carboxyl terminal extension (Figures 5.2 and 5.3) [41]. In the pituitary free α-subunit, which is found in all glycoprotein hormone-producing tissues, an O-linked glycosylation site at position Thr43 has been described [42]; glycosylation

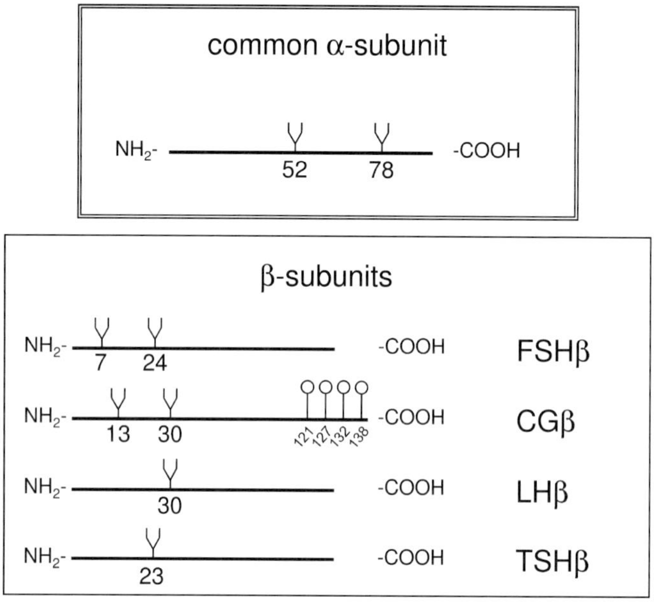

Figure 5.3 Glycosylation sites on the human glycoprotein hormone α- and β-subunits. The subunit proteins are indicated by the horizontal bars and the N-linked oligosaccharides by the branched-like structures. The "lollipop" indicates O-linked glycosylation sites in CGβ. The amino acid residue to which the oligosaccharide is attached is indicated by the number under each glycosylation site. (Reproduced from Reference [4] with permission from Elsevier Science.)

at this site prevents its association with the β-subunit [43]. In contrast, human placental free-α lacks *O*-glycans, and 3- to 4- branch glycans attached to Asn52 block association with the β-subunit [44]. A novel hFSH glycosylation variant, which possesses only α-subunit oligosaccharides, has been recently reported [28, 45]; this particular variant is significantly more active *in vitro* than the tetra-glycosylated form of the hormone. As in other multicellular eukaryote glycoproteins, oligosaccharide structures on glycoprotein hormones are highly variable [41, 46–52] (Figures 5.4 and 5.5) and play an important role in determining several key hormonal functions [53–57]. For example, in human and equine FSH more than 35 glycans may be identified using mass spectrometry of isolated glycopeptides (Figure 5.5) [52, 58]; ~90% of the total glycans in hFSH are sialylated or sulfated [48], and the heterogeneity of this glycoprotein is primarily determined by the variability in these negatively charged species (see below) [59, 60].

5.3 Glycosylation of Gonadotropins and Structural Microheterogeneity

Glycosylation of glycoprotein hormones begins in the rough ER with the co-translational transfer of a dolichol-linked oligosaccharide precursor to asparagine

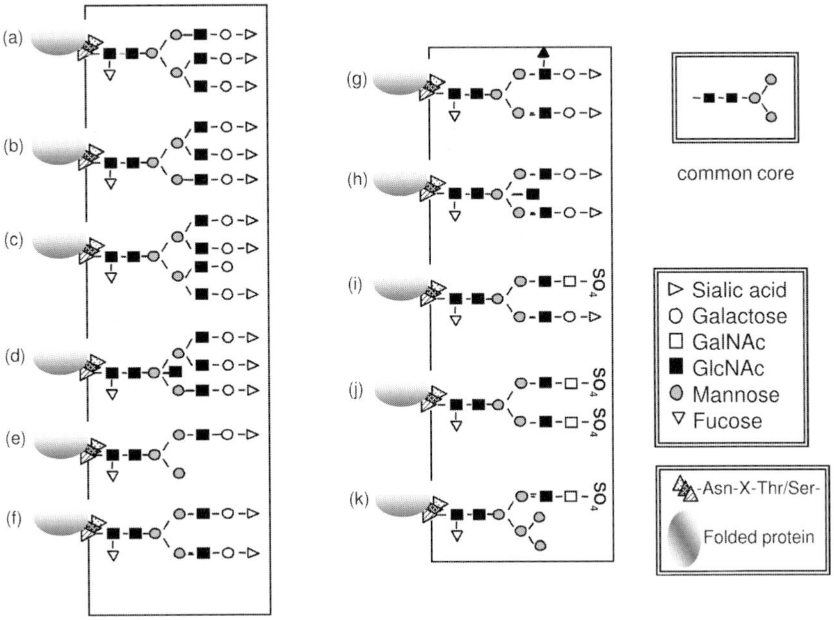

Figure 5.4 Some of the N-linked oligosaccharides structures found in ovine, bovine and human glycoprotein hormones. Only the complete structures are represented. Many other oligosaccharides may be found that are incomplete versions of these structures, mainly lacking terminal residues such as sulfate, sialic acid and fucose. GalNAc = *N*-acetyl galactosamine; GlcNAc = *N*-acetyl glucosamine; Asn = asparagine. Structures with bisecting GlcNAc residues (structures d and h), sialic acid bound in the α2,6 position and terminal sulfates are missing in recombinant gonadotropin preparations produced by Chinese hamster ovary cells.

residues at *N*-glycosylation consensus sites (Asn-X-Ser/Thr) [41, 62] (Figure 5.6, and see also Chapter 2). This precursor is further modified by exoglucosidases (exoglucosidases I and II) and by α-mannosidases, yielding a common core composed of two *N*-acetyl glucosamine (GlcNAc) residues linked to three mannose (Man) residues (Figure 5.6). Dimer formation and trimming of glucose and mannose residues to a $Man_5GlcNAc_2$ in the rough ER and *cis*-Golgi apparatus are followed by extensive processing of the oligosaccharides attached to the protein core of the hormone in the medial and *trans*-Golgi, resulting in the formation of mature glycans (Figure 5.6) [62]. Several enzymes are involved in the glycosylation of glycoprotein hormones; these include glucosidases and mannosidases as well as a group of glycosyltransferases [*N*-acetyl-galactosamine (GalNAc)-transferases, *N*-acetyl glucosamine-transferases, galactosyl- or galactose (Gal)-transferases, sialyltransferases and sulfotransferases] that markedly influence glycoprotein hormone glycosylation and sulfation [63–66]. Some of these enzymes require a specific recognition site in the primary sequence of the protein to add a new carbohydrate residue to the nascent oligosaccharide chain, while others may add it in a non-specific manner [64–66]. A particular GalNAc-transferase has been identified in pituitary membranes; it recognizes

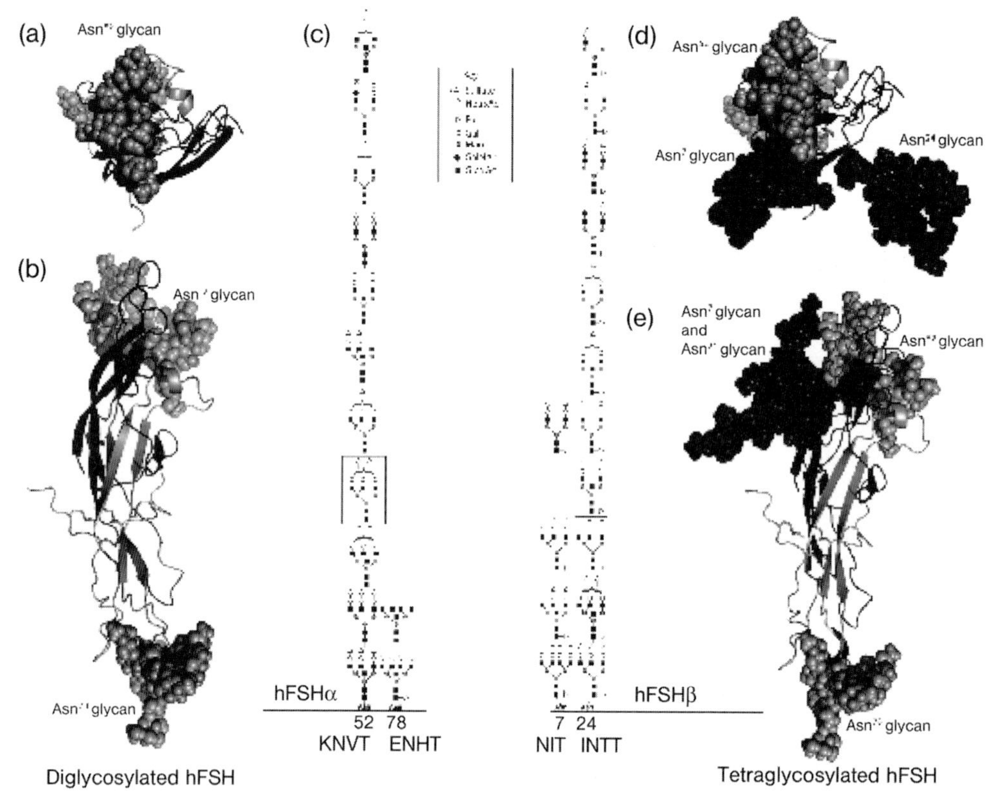

Figure 5.5 Human FSH (hFSH) glycoform glycosylation patterns. The α-subunit is represented in gray and the β-subunit in black. The glycan populations associated with each N-linked glycosylation site were determined by glycopeptide mass spectrometry. Structures associated with the strongest signals for Asn-7, -24, and -52 glycans (boxed, not enough information for Asn-78) were created using the Biopolymer module of InsightII (Accelrys, Inc., San Diego, CA), attached to the 3D structure of hFSH (1FL7) [25], and the model subjected to energy minimization using the Amber force field. Molecules were depicted with MacPyMol (DeLano Scientific LLC, San Francisco, CA). The orientations of the molecules show the bottom oriented toward the cell membrane as suggested in the co-crystal structure for hFSH bound to the extracellular domain of the FSH receptor [37]. This is consistent with monoclonal antibody binding studies involving CHO cells expressing the hFSHR [61]. (a) Top view of di-glycosylated hFSH (only α-subunit glycans); (b) side view of di-glycosylated hFSH; (c) glycan populations at each site for glycopeptides (sequence of the most abundant peptide sequence) derived by proteinase K digestion of reduced, carboxymethylated FSH [52, 58]; (d) top view of tetra-glycosylated hFSH (both α and β glycans); (e) tetra-glycosylated hFSH side view. The symbols are the same as in Figure 5.4.

the -Pro-Xaa-Arg/Lys- motif in the α-subunit when associated as an LHα/β dimer, adding a GalNAc residue to its glycans, allowing for a terminally sulfated sequence [64, 66]. The particular peptide sequence for this GalNAc-transferase is present in the α-subunit as well as in LHβ and CGβ; in hFSHβ, which is minimally sulfated [52], a Pro-Leu-Arg sequence is present but appears to be ignored by the

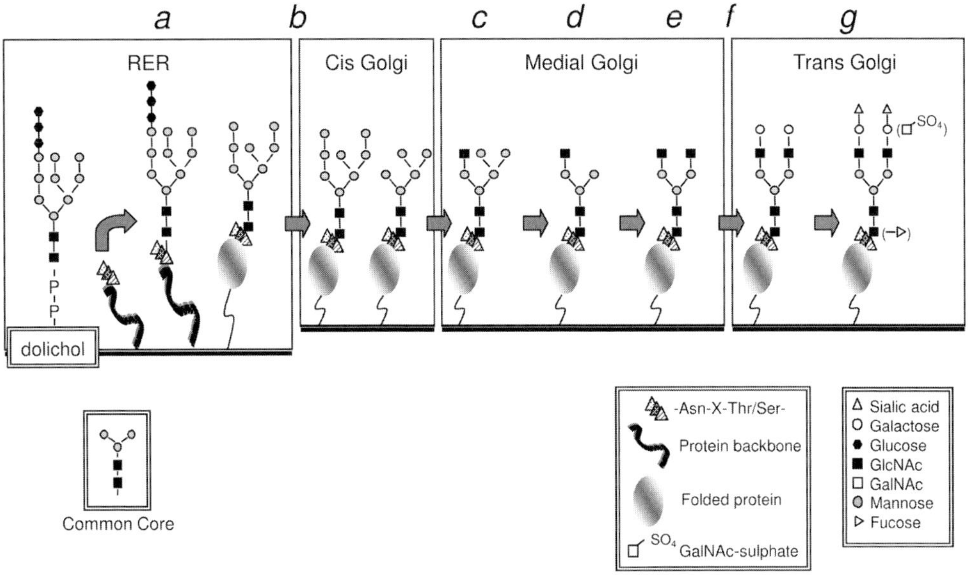

Figure 5.6 Summary of the biosynthetic pathway for N-linked oligosaccharides. Some of the key steps, starting with *en bloc* transfer of the common core from a lipid-linked precursor (dolichol), are shown. These steps include: Oligosaccharyltransferase-catalyzed transfer of dolichol-linked $Glc_3Man_9GlcNAc_2$ oligosaccharide to asparagine at glycosylation site –Asn-Xaa-Thr/Ser (step a); α-glucosidase I and II remove terminal glucose residues, endoplasmic reticulum (ER) α-1,2-mannosidase removes one mannose residue and Golgi α-mannosidase I removes another three mannose residues (step b); *N*-acetylglucosamine-transferase adds *N*-acetylglucosamine, forming precursor for a structure having only one completed branch (structure *k* in Figure 5.4) (step c) or leading to continued processing in which Golgi α-mannosidase II removes two mannose residues (step d) and *N*-acetylglucosamine-transferase II adds a second *N*-acetylglucosamine residue (step e), leading to further processing (steps f and g), which eventually results in the oligosaccharides shown in Figure 5.4.

GalNac-transferase, probably because its location differs from that of LHβ and CGβ. This may not be the case for more sulfated FSHβ from other species [41, 48, 67]. The Pro-Leu-Arg sequences in hCG and hFSH differ in the orientation of the Arg residue, which probably accounts for the fact that the GalNAc-transferase appears to bind hCGα, but not hFSHα. In the α-subunit, this sequence increases the catalytic efficiency of the enzyme ~500-fold. This specific GalNAc-transferase has been detected in the pituitary but not in placental cells; hence, CG produced by the pituitary differs from that synthesized by placental trophoblasts in that placental CG does not bear terminal sulfate residues [41]. Ovine and human LH oligosaccharides contain the GalNAc-sulfate terminal sequence that results from the consecutive actions of a GalNAc-transferase and a sulfotransferase, which act as a catalyst for terminal sulfate transfer [68]. The regulation of the remarkably different distribution of sialylated and sulfated oligosaccharides in LH and FSH is of particular interest considering that both glycoprotein hormones are synthesized within the same cell [69, 70]. In fact, it has been shown that the β-subunits influence the oligosaccharide processing of the bound α-subunit, particularly of the αAsn52-linked oligosaccharide, thus accounting in part for the differences in N-linked glycosylation between these gonadotropin molecules [71, 72]. In this regard, it has been proposed that the presence of the β-subunit may hinder the action of some processing enzymes or, alternatively, that conformational changes between free and combined α-subunits may be distinctly recognized by glycosidases and glycotransferases [71]. Although the presence of FSHβ may potentially mask the recognition site for GalNAc-transferase on the FSH α-subunit, hence facilitating FSH glycan termination predominantly in a galactose-sialic acid sequence, only hFSH (in which ~9% of its oligosaccharides are sulfated exclusively at Asn24 in FSHβ [58]) conforms completely to this model since nearly 30% of oligosaccharides on ovine FSH and 22–27% in equine FSH bear terminal sulfates [41, 67].

Numerous carbohydrate intermediates formed through post-translational processing eventually become final forms of the glycans attached to the protein core, which conform the repertoire of distinct oligosaccharides structures that characterize all glycoprotein hormones [6, 41, 59]. Human FSH and CG contain relatively high amounts of sialic acid-enriched oligosaccharides in their corresponding mono-, di-, tri- or tetra-antennary structures, thus conferring an overall negative charge to the molecule [5, 6, 41, 50, 51]. Human LH is sialylated to a lesser extent and thus is less negatively charged than hFSH and hCG [73–75]. A largely overlooked feature of the glycosylation differences between hLH and the other gonadotropins (CG and FSH) results from only three two-branch glycans (maximum charge contribution –6) in LH, whereas urinary CG possesses four N-linked glycans and four O-linked glycans for a maximum charge contribution of −11; in hFSH with two main glycoforms [28], but an average of three branches per glycan, the maximum charge contribution can be −6 to −12. Differences in the extent of sialylation and sulfation among these gonadotropins are of paramount importance for determining the *in vivo* bioavailability and net bioactivity of the hormone; terminal sialic acid prolongs whilst sulfate residues shortens the half-life of the glycoprotein in blood (see below). Fucose is another sugar that can functionally be as important as sulfate or sialic acid since it influences binding of the hormone to the appropriate receptor [76] and also, as in hTSH, may determine activation of particular intracellular signaling pathways (e.g., inositol phosphate accumulation) [77]. Interestingly, in this vein, fucosylation of αAsn56 in equine LH (homologous to human αAsn52) does not activate the inositol phosphate-mediated signaling pathway [78].

The degree of charge heterogeneity of the gonadotropins has been estimated largely on the basis of how gonadotropin fractionates during isoelectric focusing or chromatofocusing [59]. The separation of gonadotropin into these discrete fractions has been assumed to be largely dependant on the amount of terminal sialic acid and sulfate residues present in their glycans, and to vary according to the origin of the preparation analyzed. For example, the spectrum of charged variants of hFSH detected in anterior pituitary extracts and separated by charged-based procedures includes approximately 20 isoform fractions exhibiting both basic and extremely acidic pH values [59, 79]; although the mixture of charge isoforms in serum contains most of the isoforms detected in pituitary extracts, the strongly acidic isoforms predominate due to the rapid clearance of their less sialylated counterparts in circulation. Recombinant analogs of LH and FSH may show distinct glycosylation patterns depending on their cell of origin [73], and this has led to the hypothesis that their *in vitro* and *in vivo* effects may differ from naturally occurring analogs [59, 80–84]. The pattern of charge heterogeneity of both gonadotropins also differs according to sex, age and in the case of females the phase of the menstrual cycle, thus indicating that gonadotropin glycosylation is under tight endocrine control [59, 60, 85–89]. Recently, the power of mass spectrometry analysis of hFSH glycopeptides has been brought to bear on this microheterogeneity issue [52]. Remarkably, it has been found that nearly identical forms of gonadotropin can exist in each charge isoform fraction yielded by chromatofocusing of human pituitary FSH [58]. This has significance not only from a methodological point of view but also from a practical (pharmacological) point of view, since there has been a long discussion about whether the recombinant FSH produced by Chinese hamster ovary (CHO) cells is more akin than urinary FSH to the isoform profile of cycling women (see below) [3, 59, 60, 85, 90, 91]. In addition, although the presence of sialic acid capping galactose certainly will affect clearance in the liver [92], it is unclear how glycan structure beyond GlcNAc2 or perhaps Man1,2 will affect hormone binding. Most of the conformational constraints imposed by carbohydrate, if significant, will arise from GlcNAc1 interactions with the peptide backbone. Nevertheless, steric hindrance at the level of the receptor, in terms of forming a weak collisional complex, could be affected by the larger carbohydrate structures containing sialic acid.

The existence of partially glycosylated hFSHβ subunits suggests selective inhibition of oligosaccharyl transferase activity in the anterior pituitary [6, 28]. First reported in recombinant bovine FSH preparations intended for pharmaceutical use [93], partial glycosylation was subsequently reported at Asn7 in horse pituitary FSHβ [6], and at Asn24 in insect cell-expressed recombinant hFSHβ [25]. Sodium dodecyl sulfate-polyacrylamide gel electrophoresis (SDS-PAGE) and matrix-assisted laser desorption/ionization time-of-flight mass spectrometry (MALDI-TOF-MS) confirmed the existence of equine FSHβ possessing one or two N-linked glycans [6]. However, the 24 kDa and 21 kDa hFSHβ subunit bands revealed by Western blot subsequently were shown to represent an all-or-none pattern of hFSHβ N-linked glycosylation where the 21 kDa band lacked both glycans, while the 24 kDa band possessed both. Complementary automated Edman degradation confirmed the all-or-none hFSHβ N-linked glycosylation and the presence of >PhNCS (phenylthiohydantoin)-Asn in cycles corresponding to residues 7 and 24, indicating that carbohydrate had never been attached to these particular sites [28]. Human FSH possessing the 21 kDa β-subunit has been designated di-glycosylated FSH because it possesses only the two α-subunit N-linked glycans (Figure 5.5). Although initially dismissed as a minor variant, analysis of individual human pituitaries revealed that di-glycosylated FSH was more abundant (52–73%) than the classic tetra-glycosylated FSH in 21–24-year-old cycling females. Di-glycosylated FSH abundance dropped to 16% in individual postmenopausal pituitaries and pooled postmenopausal urinary gonadotropin preparations. Recombinant di-glycosylated hFSH was reported to be cleared from the circulation more rapidly than wild-type recombinant FSH (presumably mostly tetra-glycosylated FSH) [94]. Wide and colleagues reported that pituitary FSH from young, cycling women was cleared more rapidly than FSH from postmenopausal women [95]. In ensemble, these studies suggest that the difference in clearance rates probably reflects, at least in part, the high abundance of more rapidly cleared di-glycosylated FSH in young, cycling women and its low abundance in postmenopausal women. The initial di-glycosylated hFSH preparation was isolated by chromatofocusing and resided in the least acidic fraction [28]. Since this included the highest levels of LH contamination and because it was significantly more biologically active than the more acidic isoform preparations, it resembled the less acidic hFSH isoforms reported by others [96, 97].

As with other anterior pituitary glycoprotein hormones, CG is synthesized as different glycosylation variants and thus is structurally highly heterogeneous [98–100]. This gonadotropin also exhibits variants related to peptide cleavage (nicked forms of CG, in which β-subunits have bond cleavages within loop 2) and the so-called β-core fragment urinary metabolite [101–105], which may also be detected as a variant of LH in the pituitary and urine [106–109]. Free CG α-subunit is a major product of the placenta during pregnancy and is differentially glycosylated than CGα [110]. Oligosaccharide structures on CG vary throughout normal and abnormal pregnancy as well as in trophoblastic diseases [99, 100, 111–114]. In normal pregnancy, changes in the degree of branching, sialylation and core fucosylation on CG and free-CG α-subunit oligosaccharides and concomitantly in the plasma half-life, *in vivo* biological potency and *in vitro* B/I ratio of CG occur as pregnancy progresses [110–115]. Hyperglycosylated CG is an overexpressed, high molecular weight variant of hCG that is produced and independently secreted by stem cytotrophoblast cells [116, 117]. This important and unique CG variant exhibits an independent function to the regular CG produced by differentiated syncytiotrophoblast cells and directly relates to autocrine control of trophoblast invasion and development of malignancy, particularly choriocarcinoma [118, 119]. The principal difference between hyperglycosylated CG expressed in choriocarcinoma and pregnancy CG lies in the four O-linked oligosaccharides; the predominant O-linked glycan structure in the latter is a trisaccharide with the structure NeuAc2 → 3Galβ1 → 3GalNAc-O-Ser (were NeuAc is sialic acid), whilst in choriocarcinoma it is a double size hexasaccharide with the structure NeuAc2 → 3Galβ1 → 3 (NeuAc2 → 3Galβ1 → 4GlcNAcβ1 → 6)GalNAc-O-Ser [99, 120]. Some differences in N-linked glycans between normal pregnancy CG and CG from trophoblastic diseases are also apparent. The α-subunit from normal pregnancy CG mainly contains non-fucosylated mono- and biantennary N-linked structures as well as fucosylated biantennary and triantennary oligosaccharides in a lesser amount, whereas in choriocarcinoma CG α-subunit the amounts of fucosylated biantennary glycans are increased and the biantennary structures that predominate in normal pregnancy are decreased [99]. In its β-subunit, normal pregnancy CG predominantly exhibits N-linked fucosylated and non-fucosylated biantennary structures, accompanied by distinct levels of mono- and triantennary glycans; in contrast, CGβ from trophoblastic disease exhibits higher proportions of N-linked fucosylated triantennary oligosaccharides [99, 121]. Finally, the α-subunit from normal pregnancy CG bears a high proportion of N-linked sialylated antennae (nearly 100%), whilst in CG from trophoblastic diseases it varies from 79 to 83% [99, 121].

5.4 Role of Glycosylation in the Function of Gonadotropins

Glycans in glycoprotein hormones play an important role in many of the functional characteristics of the molecule [60, 62, 122]. They are important not only for folding, heterodimerization, quality control, sorting, transport and conformational maturation of the protein within the cell but also for enhancing the efficiency of heterodimer secretion and determining the metabolic fate and interaction with its cognate receptor [53–56, 62, 123–126]. Further, differential terminal glycosylation allows fine-tuning of the protein properties at the target cell *without* the need for a change in primary sequence. Glycans also play a role in the physicochemical properties of glycoprotein hormones [127]: increased subunit dissociation [128], alterations in thermal stability [124] and alterations in the antigenic structure [128, 129] of the glycoproteins have been observed when those structures are modified.

5.4.1 Role in Folding, Subunit Assembly and Secretion

The observation that the β-subunit of hCG can be expressed in bacteria and refolded as an unglycosylated subunit that associates with the α-subunit initially suggested that carbohydrate was unnecessary for folding of the β-subunit of this glycoprotein [130, 131]. However, in follow up work it was shown that carbohydrate at position 30 and to a lesser extent at position 13 enhanced the rate of folding of CGβ, but that the size of the glycan was not important [132]. Indeed, consistent with the observation that disulfide bond formation was critical to the folding process, it was found that oligosaccharide in the β-subunit facilitated disulfide bond formation [133]. In addition, the folding and secretion of CG appears to be dependent upon association of the free β-subunit with ER resident chaperones [134]. These kinetic and biochemical studies have proven very useful for helping us to understand the relationship of primary structure and post-translational modifications on CG subunit folding and assembly. The extent to which these observations are relevant to LHβ is clouded by the fact that LHβ is not efficiently folded and secreted in the absence LHα, whereas CGβ is efficiently secreted in the absence of CGα [57]. Follicle-stimulating hormone glycosylation variants are folded and secreted, illustrating that carbohydrate at each site is not essential for these processes [135, 136]. However, if the glycans of the α-subunit of CG become highly branched then they prevent association of CG α- and β-subunits, resulting in the secretion of free alpha subunit [44, 137].

In summary, glycosylation of the gonadotropin subunits may facilitate, but is not essential for, folding, assembly and secretion.

5.4.2
Metabolic Clearance Rate

The metabolic clearance rate and the *in vivo* biological potency of the gonadotropins are highly dependent on the specific type of terminal residues present in their oligosaccharide structures. As described above, oligosaccharides present in hFSH and CG predominantly terminate in sialic acid; this sugar content, and particularly the number of exposed terminal galactose residues, are essential in determining the hormone's survival in the circulation [138]. Exposure of terminal galactose residues dramatically increases glycoprotein clearance from plasma through a mechanism that involves hepatocyte receptors for the asialo-galactose-terminated complex molecules. However, this mechanism appears to be irrelevant for CG clearance, as blocking the hepatocyte receptors with asialoglycoproteins has no effect on intact CG clearance [139] and apparently kidney clearance appears to be a more important mechanism [140]. Heavily sialylated glycoproteins circulate for longer periods than those with less sialylated or sulfated glycans [3, 111, 141–143], whereas sulfated glycans, such as those present in LH and minimally in FSH, or oligosaccharides bearing terminal mannose or GlcNAc accelerate clearance of the molecule by specific receptors present in hepatic endothelial and Kupffer cells [144, 145]. However, this concept has been challenged by *in vivo* whole organism imaging, which indicated that porcine LH was cleared by the kidneys, although very little was recovered from urine [146]. The plasma half-life of FSH and CG is therefore longer than that of LH [3]. Recombinant asialo-gonadotropin variants as well as deglycosylated FSH are rapidly cleared from the circulation and are practically inactive *in vivo* when compared to the corresponding intact variants [147, 148]. Conversely, recombinant hFSH containing a hybrid β-subunit composed of the β-subunit of FSH and the carboxyl-terminal peptide of CGβ [(FSH-CTP), which is rich in O-linked sialylated glycans], or to which additional glycosylation sites have been added, exhibits a prolonged plasma half-life circa two to four times longer than intact recombinant FSH [149–152]. The half-life of the different charge variants of hFSH and hLH has been analyzed in some detail. In FSH, the more acidic/sialylated variants exhibit higher circulatory survival than the less acidic counterparts [3, 11, 59, 60, 143]. The plasma disappearance of endogenous LH is slower in postmenopausal than in young women [153], which is likely due to the increased sialylation of the gonadotropin that occurs after the menopause [79, 86, 154]. The mechanism for this shift is likely more complicated than the sole addition of GalNAc instead of Gal to partially synthesized glycans because sialylated GalNac has been reported for hLH [50]. Differences in half-lives among the different glycosylation variants that comprise circulating FSH also explains why *in vivo* administration of a more acidic/sialylated mix of FSH isoforms is more effective to stimulate follicular maturation and 17β-estradiol production than a poorly sialylated preparation [155].

Oligosaccharides on the β-subunits play a major role in determining the metabolic clearance rate of gonadotropins [94], whereas the oligosaccharide in position αAsn52 is essential for the activation of the receptor/signal transducer (G proteins) system and the subsequent biological response [124, 125, 156–158]. A potential role for the αAsn78 glycan in FSH-mediated signal transduction has been revealed by two mutagenesis studies [122, 135].

5.4.3 Binding and Signal Transduction

Removal of αAsn52 oligosaccharide attachment by either chemical methods or site-specific mutagenesis results in gonadotropin molecules exhibiting an increased capacity to bind to the receptor, but with a significantly reduced signal transduction capability [52, 78, 122, 124, 125, 135, 156–158]. There has been no elucidation of the underpinning of why glycosylation at this sequon is critical for signal transduction activity but not receptor-binding activity. One possibility is that deglycosylation may prevent hormone-provoked aggregation of the receptor; in fact, is has been reported that antibodies against CG can restore the signal transduction defect caused by CG deglycosylation [159]. Moreover, ovine LHα Asn56 glycans inhibited CG-induced fluorescence resonance energy transfer between LHCGR–green fluorescent protein and LHCGR–yellow fluorescent protein chimeras [78]. A second possibility is that the length of the Man(α1–6)Man oligosaccharide branch at αAsn52 of the α-subunit may provide steric hindrance, decreasing receptor–hormone affinity, and allowing conformational flexibility necessary for signal transduction [160]. A third and simpler explanation is that the signal transduction defect caused by removing glycosylation at CG αAsn52 is due to instability of the heterodimeric gonadotropin at 37 °C, as suggested by studies in which the effects caused by removing glycosylation at this position were abrogated by covalently linking the α- and β-subunits via intersubunit disulfide bonds [161]; this, however, was shown by Bousfield and colleagues [158, 162] not to be the case as an αAsn56-deglycosylated equine LH derivative showed no instability during a 72-h preincubation period at 37 °C, no significant loss of receptor-binding activity and no detectable dissociation into subunits. Further, recent studies, describing the three-dimensional structure of fully glycosylated FSH [24, 25] and the extracellular domain of the FSH receptor ($FSHR_{HB}$) in complex with FSH [37], predict contacts via hydrogen bonding [24] between the αAsn52-proximal GlcNAc and βTyr58, which is likely to contribute to stabilization of the dimer. Thus, a slight conformational shift could reposition the critical residues involved in binding, which are, incidentally, in exactly the same region as the carbohydrate attachment site (Figure 5.2). Also notable is that these variants always have a higher affinity constant, suggesting that the formation of a tighter complex may alter the final conformation of the otherwise signal productive hormone–receptor complex.

Studies on other possible functions of carbohydrates linked to the β-subunits are somewhat contradictory. In one study [135], site-specific deletion of the oligosaccharide in position at FSHβ Asn7 resulted in a 2.5-fold increase in receptor binding activity and a twofold decrease in its ability to promote intracellular signal transduction. In another study, the activity of the variant was unaffected by deleting this particular glycosylation site [122]. Further studies have confirmed the observation that oligosaccharides in the β-subunit play a minor role in both gonadotropin binding and signal transduction [157].

The structure of fully glycosylated single-chain FSH as well as that the extracellular domain of the hFSH receptor in complex with FSH [24, 25, 37, 163] reveal that carbohydrate has little influence on the overall structure of the protein and that there is no carbohydrate in the binding interface of the FSH-FSHR$_{HB}$, which explains why deglycosylation of FSH does not abrogate binding but profoundly affects signal transduction. However, as noted above, carbohydrate is definitely near the binding site and it is not known what accommodations the FSHR1-250 structure had to make to accept FSH binding. In addition, the receptor and single-chain FSH were both deglycosylated prior to crystallization [37, 163], and it is not known whether the conformation and the carbohydrate on single-chain FSH is the same as in heterodimeric FSH. Nevertheless, this concept suggests that ligands with antagonist properties may be developed that either bind to the receptor with very limited or null potency to evoke a biological effect [78, 162] or stabilize a conformation of the receptor that is signal transduction incompetent. In the latter case, since glycosylation variants of this gonadotropin may provoke diverse effects on signal transduction and end responses *in vitro* [59, 60, 85, 164], it is possible that some rearrangements in the transmembrane structure of the receptor should occur upon FSH binding.

5.4.4
LH and FSH Glycoforms and Gonadotropin Function

The functional characteristics of the glycosylation variants of human pituitary gonadotropins have been studied extensively [59, 85]. Using a variety of techniques, mainly zone electrophoresis, preparative isoelectric focusing and chromatofocusing [59], several gonadotropin isoform fractions can be separated (Figure 5.7). These techniques separate the different isoforms on the basis of charge, which in the gonadotropins is mainly determined by the amounts of sialic acid and/or sulfated $\beta 1 \rightarrow 4$-linked *N*-acetylgalactosamine residues present as terminal groups in the glycans attached to these hormones (Figure 5.4). The pattern of charge distribution of these isoforms varies, depending on the purity and source of the gonadotropin preparation as well as on the endocrine status present in the donor at the time of tissue or sample collection (Figure 5.7). In general, less acidic gonadotropin isoform fractions exhibit higher biological potencies than more acidic variants when tested either in single-cell-type or in complex cell-type *in vitro* assay systems [60, 85, 97, 165, 166]. Further, it has been shown that the charge variants of hFSH may exert differential and even unique effects on cell function when several intermediate and/or end-products are measured [85, 167, 168].

The results from these studies, however, should be taken with caution because of the possibility of aggregate formation during the electrofocusing procedures employed to separate crude or partially purified extracts [58], which may lead to retention of immunological activity but loss of receptor-binding activity. In addition, charge-based procedures distinguish populations of glycosylation variants differing in sialic acid and/or sulfate content only, but not in overall glycan structure (e.g., degree of branching) or subunit glycosylation (e.g., presence or absence deglycosylated FSHβ). In FSH, with two to four N-linked glycans per molecule, many combinations of

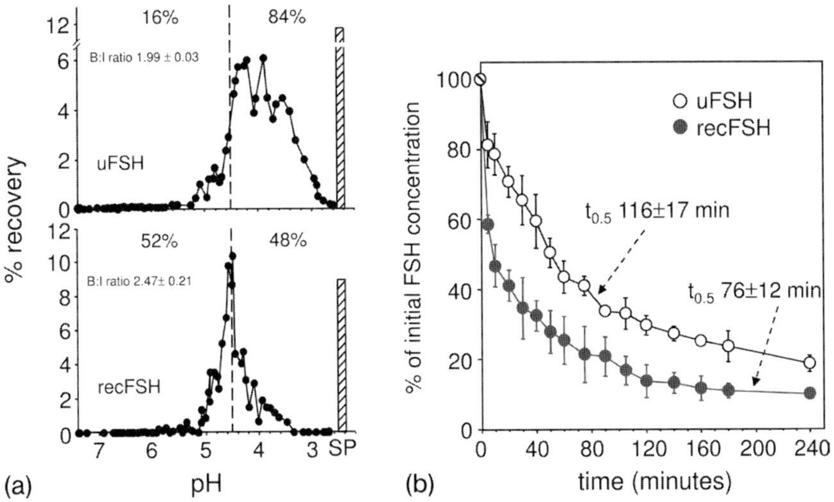

Figure 5.7 Chromatofocusing separation (a) and plasma disappearance curves (b) of recombinant human FSH (recFSH) produced by CHO cells transfected with the cDNA of both FSH subunits, and highly purified urinary FSH (uFSH). Urinary FSH present in this preparation is more acidic than recFSH. According to the pH distribution pattern exhibited by the FSH preparations, 84% of total uFSH immunoactivity was recovered at pH ≤4.5 [including the salt peak (SP, pH <3.5)], whereas in recFSH 48% of the total was detected within these values. Vertical broken lines separate isoforms with pH >4.5 and <4.5. The percentage recoveries within each pH window are indicated at the top of each graph. The FSH preparations (~60 μg of immunoactive uFSH or recFSH) shown in (a) were administrated i.v. (via a jugular vein) to adult male rats and blood samples were taken at 5 and 10 min after the injection and subsequently at regular time intervals during the ensuing 4 h. The less acidic preparation (recFSH) disappeared faster from the circulation than the more acidic/sialylated compound (uFSH); $t_{0.5}$, plasma half-life. Values are the mean ± SEM of three to five independent studies. (Reproduced from Reference [96] with permission from the European Society for Human Reproduction and Embryology and Oxford University Press.)

charges distributed over its glycosylation sites can provide the same isoelectric points, with common glycans in several charge isoform fractions (see above) [58, 169]. Nevertheless, when the *in vitro* biological activity of recombinant (CHO cells-derived) and urinary or pituitary preparations is compared, the recombinant preparations usually show higher activities than their naturally occurring counterparts (Figure 5.7) [3]. This may, in part, reflect the fact that pituitary (and probably urinary as well) hFSH preparations include aggregated and dissociated forms as well as the intact heterodimer, while recombinant preparations are virtually exclusively heterodimers. As shown in Figure 5.7, the pH distribution pattern of urinary FSH is more acidic than in recombinant FSH, which may additionally explain the lower *in vitro* bioactivity but longer plasma half-life of the former preparation. In this regard, it is important here to distinguish between the *in vitro* and *in vivo* bioactivity of glycoprotein hormones; the latter is concerned mainly with the half-life of the hormone whereby carbohydrates affect clearance, whereas the former is more related to the bioresponse as disclosed by measurements of activity at the receptor/target cell level. In addition, the *in vivo* response may vary, depending on the particular assay employed and end-response measured [3, 85, 96, 97, 165], whilst in the *in vitro* bioresponse several aspects may influence the estimated parameter, including potentially large non-uniformities among the immunoassays employed to quantitate the amount of the gonadotropin (when the potency is expressed as an estimate of apparent bioactive to immunoreactive ratio) and variations in the amount of glycoprotein recovered after extraction when amino acid analysis is used to determine the protein mass [74, 170, 171].

The impact of variations in LH and FSH glycan structure on receptor function and their role in the selective activation of intracellular signaling pathways is still uncertain. Studies with highly purified isoforms of known glycan structure will undoubtedly provide useful insights into the physiological role of gonadotropin microheterogeneity.

5.4.5
Chorionic Gonadotropin Glycoforms and Function

As mentioned previously, CG is essential to maintain progesterone production by the corpus luteum, until the placenta takes over this and other functions. Chorionic gonadotropin is heavily sialylated and several structural aspects of this residue, including carbon side chain, an intact ring structure and its position relative to other carbohydrate residues, are important for full biological activity [172]. Owing to differing degrees in sialylation, CG variants may also be fractionated by charge-based procedures [111–114]. Using either zone electrophoresis in agarose suspension or chromatofocusing, it has been shown that the pattern of pH distribution of serum CG changes through normal gestation; an initial shift towards less acidic isoforms is observed between 11 and 15 weeks of gestation [111, 112, 114], and a second shift during the third trimester [111]. The first shift in glycosylation pattern results from decreased production of hyperglycosylated CG by cytotrophoblasts gradually shifting to CG production by syncytiotrophoblasts [173]. Less acidic/sialylated CG isoforms exhibit higher *in vitro* biological potency per immunological unit than the more acidic analogs and, as expected, the plasma half-life of serum CG from the first trimester is longer than that presented by circulating CG from the second and third trimesters of pregnancy [111, 172]. Both changes in peptide backbone (see above) and carbohydrate structure of CG during pregnancy contribute to the maintenance of the low serum CG concentrations and decreased bioactivity observed late in pregnancy. In pregnancy, hyperglycosylated CG is produced at the time of implantation and in the two weeks that follow [117, 174]. The biological effect of this CG variant is mainly related to invasive events such as blastocyst implantation and development of trophoblast diseases, including malignancy. Isolated normal pregnancy cytotrophoblast cells and JAR choriocarcinoma cells (invasive cytotrophoblast cells) are highly invasive in *in vitro* models, and invasion is promoted by hyperglycosylated CG but not by CG produced by differentiated trophoblasts [175]. Further, treatment of JAR cells with CG α- or β-subunit antisense cDNA [176, 177], or mice with transplanted JEG-3 tumors with antibody against hyperglycosylated hCG [175], inhibited cell growth, and blocked invasion *in vitro* and tumorigenesis *in vivo*. Hyperglycosylated CG has also been detected in the serum or urine of patients suffering from other malignant conditions, including cervical and ovarian [178], colon [179], bladder [180–182] and lung cancer [183]. Whether hyperglycosylated CG variants have any role in the pathogenesis of these malignancies or only represent biochemical markers for the presence of such tumors, remains to be investigated.

5.5 Regulation of Gonadotropin Glycosylation

Our knowledge of the factors that regulate pituitary gonadotropin glycosylation is still limited. Nevertheless, evidence derived from experimental and clinical studies has provided useful information on this process and allowed the identification of some specific structural and endocrine factors that may potentially influence the glycan structure of and their attachment on these glycoproteins. Some of these studies demonstrated that, at least in rodents and in primates, gonadal steroids strongly influence gonadotropin glycosylation, thus raising the interesting possibility of an endocrine-governed mechanism for precise control of gonadotropin structure and potency.

5.5.1 Effects of Estrogens

Some of the cellular mechanisms whereby estrogens may influence the pituitary biosynthetic mechanisms involved in gonadotropin glycosylation have been explored [4, 63, 68, 85, 184]. Pituitary gonadotropins, particularly LH, bear oligosaccharides terminating in the SO_4-4-GalNAcβ1,4GlcNAcβ1,2Manα sequence (Figure 5.4). The synthesis of sulfated oligosaccharides on LH reflects the action of two highly specific glycosyltransferases; addition of GalNAc to oligosaccharide acceptors on LH is mediated by a GalNAc-transferase that recognizes a tripeptide motif located 6–9 amino acid residues NH_2-terminal to an Asn glycosylation site (see above). The subsequent addition of sulfate is completed by a sulfotransferase that does not recognize specific sequences in the peptide but that requires the presence of the trisaccharide sequence GalNAcβ1,4GlcNAcβ1,2Manα [64–66]. A remarkable feature of both the β1,4 GalNAc-transferase and the GalNAc-4-sulfotransferase is that their levels in the gonadotropes may be regulated by the concerted action between LH and estrogens. When serum estrogen concentrations decrease and the synthesis of LH rises, the β1,4-GalNAc-transferase levels, and to a lesser extent those of the GalNAc-4-sulfotransferase, concomitantly increase, thus favoring the production of more oligosaccharides terminating with the sequence GalNAc-4-SO_4 [68]. As mentioned earlier, changes in the degree of terminal sulfation and sialylation may potentially alter the circulatory survival of the gonadotropins and consequently their biological effects *in vivo*. In FSH, it has been shown that the mRNA expression of the pituitary Galβ1,3[4]GlcNAc α-2,3-sialyltransferase (one of the enzymes responsible for the incorporation of terminal sialic acid residues on the oligosaccharide chains of the glycoprotein hormones) changes according to the time and day of the rat estrous cycle; the dynamics of these changes in the enzyme mRNA levels correlate with variations in serum estradiol concentrations [4, 63]. Nevertheless, the finding that in ovariectomized rats the response to exogenous estradiol and antiestrogen administration on the enzyme mRNA levels was considerably attenuated when compared with that observed in the intact animals suggests that other extrapituitary factors may be additionally involved in the control of this particular glycosyltransferase [136]. The observations derived from these molecular studies correlate with reported changes in charge isoform distribution and glycan attachment of FSH and/or LH in pituitary and serum samples from women under different endocrine conditions [28, 45, 59, 60, 85, 154, 185, 186] and underline the critical role of estrogens in defining the structure and determining the biological potency of the gonadotropin signal delivered to the gonad during a given physiological condition.

5.5.2
Effects of Androgens

Less extensive research has been conducted to establish the role of androgens on gonadotropin glycosylation [59, 187–190]. These studies, however, do not establish whether androgens influence gonadotropin glycosylation directly at the gonadotrope level or indirectly through changes in hypothalamic gonadotropin-releasing hormone (GnRH) secretion and consequently in the rate of processing of the glycoprotein (see below). In fact, in both hypothyroid and euthyroid individuals, exogenous thyrotropin-releasing hormone (TRH) may provoke release of TSH molecules with an enhanced core fucosylation [77]; moreover, *in vitro* studies have shown that TRH may also regulate the branching pattern and relative sialylation of TSH carbohydrate chains [191, 192]. Orchidectomy leads to changes in the production of FSH glycoforms, although the trend varies in accordance with the animal species studied. In monkeys, changes in charge distribution of intrapituitary gonadotropin isoforms have been observed after orchidectomy [193], whereas in castrated men significant shifts towards less sialylated serum FSH isoforms were observed only after estrogen treatment [187]. Circulating LH in extremely obese individuals contains an increased proportion of more basic LH isoforms when compared to normal men [188]; although the elevated proportions of more basic LH isoforms in obese men may be due to the increased pituitary exposure to endogenous estrogens, reduced serum testosterone levels may also account for this altered gonadotropin glycosylation. In fact, androgen treatment of wethers and castrated male rats increased the percentage of less basic intrapituitary LH isoforms as well as the *in vitro* bioactivity of the gonadotropin [194, 195], and, in humans, serum testosterone concentrations correlate with more acidic circulating LH isotypes [189, 196]. Additional evidence supporting the possibility that androgens influence gonadotropin glycosylation directly at the pituitary level results from studies showing that, in both non-suppressed and GnRH antagonist-suppressed male rats, treatment with testosterone leads to parallel changes in the charge distribution profile of intrapituitary FSH [197] and that exposure of cultured pituitary cells to this sex steroid increases the relative proportion of more sialylated forms of secreted FSH [198]. Although these studies strongly suggest a direct role for androgens in gonadotropin glycosylation, additional studies are still required to define the molecular mechanisms whereby androgens may influence this process.

A series of clinical and experimental studies on another structurally-related glycoprotein hormone, TSH, might also illustrate how glycosylation of anterior pituitary glycoprotein hormones is regulated by target gland end-products under the control of a determined trophic hormone. In both experimental animals and in humans, thyroid hormone deficiency modifies sialylation and sulfation (and thus the functional properties) of secreted TSH [199–204]. In rodents, exogenously induced hypothyroidism leads to a dramatic increase in the ratio of sialylated to sulfated TSH oligosaccharides, affecting consequently its metabolic clearance rate and bioactivity *in vivo* and *in vitro* [205, 206]. Patients with subclinical and overt primary hypothyroidism have a markedly increased proportion of circulating TSH glycoforms exhibiting terminal sialic acid residues as well as a compromised *in vitro* biological potency [202, 203]. Studies performed at the molecular level have shown that the α-1,4-galactosyl-transferase and α-mannosidase II mRNA levels are increased in hypothyroid, propylthiouracil-treated mice when compared to euthyroid animals [199]; in this experimental model, a significant increase in α2,6-sialyltransferase mRNA synthesis and/or stability has also been documented [200], thus indicating that sialic acid incorporation into TSH oligosaccharide chains is also hormonally regulated.

5.5.3
Effects of Gonadotropin-Releasing Hormone

There is considerable variance in the results from studies exploring the role of GnRH on FSH glycosylation. In one study, treatment of pubertal children with GnRH agonists led to qualitative changes in the isoforms of both FSH and LH [88] and, in another, GnRH administration to testosterone-treated or untreated men favored release of less acidic/sialylated glycoforms [207]. Other studies, however, have found that the charge distribution profile and the *in vitro* B/I ratio of secreted FSH do not change after administration of low or high doses of exogenous GnRH to adult women [89, 208] and that pulsatile administration of GnRH to nutritionally growth-restricted ovariectomized lambs (a hypogonadotropic state related to central inhibition of GnRH secretion [209]) failed to alter the patterns of isoform distribution of intrapituitary and secreted FSH [210]. Whether GnRH modulates post-translational processing of the FSH oligosaccharides is an issue that needs further investigation.

In contrast to FSH, evidence derived from several experimental and clinical studies indicates that GnRH plays a role in the regulation of LH glycosylation. Early studies showed that the rate of carbohydrate incorporation on LH oligosaccharides is affected by GnRH treatment of anterior pituitary tissue [211, 212]. In men and normally cycling women, GnRH administration results in the preferential release of bioactive LH, with a consequent increase in the plasma LH biological to immunological ratio (this derived parameter has been used as an indirect index of the relative biopotency of gonadotropins, at least in relation to any one immunological estimate) [186]. This distinctive effect of GnRH upon gonadotropin bioactivity appears important particularly considering the crucial role of this releasing peptide in the regulated secretion of LH. The mechanism(s) whereby GnRH influences LH glycosylation may involve changes in the expression level and/or activity of particular glycosyltransferases or may be just a consequence of the GnRH-triggered alterations in the rate of synthesis, intracellular transport and/or residence time of the glycoprotein. Therefore, it appears that the nature of the final product is a result of the interplay between secretion and stimulation of gonadotropin synthesis occurring under the GnRH drive.

5.6
Therapeutic Applications of Gonadotropins

Various preparations containing FSH, LH or CG from urinary or recombinant origin have been used successfully to restore reproductive function in hypogonadotropic men and women as well as for stimulating multiple follicular development in assisted reproduction programs [213, 214]. Although some differences in glycosylation patterns and pharmacodynamic responses among the various preparations have been detected [215–217], in practice they do not significantly impact the final outcome [80, 81, 83, 218–221]. In general, gonadotropins from urinary source are highly sialylated [3, 59, 90, 215, 222]; recombinant preparations (LH, FSH and CG) produced by CHO cells bear complex type N-linked glycans with di-, tri-, tetra- and minor penta-antennary species with various levels of sialylation. The level of sialylation of recombinant hFSH at N-linked glycans is higher than for the recombinant LH and CG [216]. Other differences in glycosylation between recombinant gonadotropins and their pituitary counterparts include (i) oligosaccharides in CHO cell-derived gonadotropins are sialylated only in an $\alpha 2 \rightarrow 3$ conformation, (ii) CHO cell-derived gonadotropins do not contain bisecting GlcNAc moieties and (iii) sulfated terminal glycosylation is missing [215, 216, 223–225]. Notably, human embryonic kidney cells transfected with the human α- and β-subunit FSH genes produce FSH isoforms that represent the full spectrum found within the pituitary gland and in the circulation [226]; recombinant FSH from this particular source can not be produced on a large scale and thus is not available for clinical use.

In men, spermatogenesis is initiated and maintained by FSH and testosterone, respectively. In fact, men who have a mutation in FSHβ are infertile [227]. Some evidence indicates that, once initiated, spermatogenesis can be maintained by LH stimulated testosterone alone [228]. In male patients with hypogonadotropic hypogonadism (i.e., with complete to partial failure to produce and secrete pituitary gonadotropins in sufficient amounts to attain full sexual maturation, including reproductive competence), the induction and maintenance of spermatogenesis can be achieved by administering a combination of FSH and LH. The regimen most commonly used in clinical practice involves injections of FSH-containing preparations [e.g., human menopausal gonadotropin or hMG (a preparation from urinary source that contains equal FSH and LH activities as assessed by *in vivo* bioassays)] or either urinary or recombinant FSH alone, two or three times per week, as well as urinary CG given as a surrogate for LH [229–231]. After prolonged treatment, spermatogenesis can be stimulated and pregnancies achieved; nevertheless, the differences in circulatory half-life of FSH and CG usually lead to unphysiological fluctuations in serum FSH concentrations and relatively high constant concentrations of CG. Administration of recombinant FSH-CTP, which contains the α-subunit of hFSH and a hybrid β-subunit composed of the β-subunit of FSH and a carboxyl-terminal peptide of CG (a region in CGβ that is important for maintaining the prolonged plasma half-life of hCG dimer) [30], results in an increased circulatory half-life of FSH (2–3× longer than that exhibited by recombinant FSH), and in more constant concentrations of FSH, which may be more convenient because it could be administered less frequently than regular recombinant or urinary FSH [150]. Recombinant gonadotropins that exhibit dual FSH and LH activities [232, 233] represent highly promising therapeutic strategies for hypogonadal men.

In women, gonadotropins are used for ovulation induction and for stimulation of multiple follicular development. For the induction of monofollicular development in anovulatory women with low or normal gonadotropin levels, FSH with or without concomitant administration of LH is necessary to promote follicular growth and estrogen biosynthesis, whereas ovulation is induced by administration of CG. For this purpose, hMG plus urinary CG or the recombinant preparations may be employed [214, 220, 234–237]. Different regimens have been used for this purpose, including those that are designed to reproduce the changes in concentrations of FSH that occur in the normal cycle where there is a single ovulatory follicle [213, 214, 235, 236]. Stimulation of multiple follicles with exogenous gonadotropins is widely used in assisted reproduction clinics. In this situation, the normal physiological mechanism that selects a single ovulatory follicle is overridden by administration of high doses of FSH until the desired number of follicles have been stimulated [235, 238–240]. To avoid the occurrence of a premature rise in endogenous LH, suppression of endogenous gonadotropins is usually performed by the co-administration of GnRH analogs (agonists or antagonists). Finally, a single injection of urinary or recombinant CG is administered to stimulate resumption of oocyte meiotic maturation [241–244]. Although some studies have found potential advantages of a given preparation over others (i.e., among recombinant FSH, highly purified urinary FSH and hMG) [80, 81, 83, 221, 245, 246], which may be due, in part, to differences in glycosylation pattern among the various compounds [3, 59, 83, 141, 222] or presence of variable amounts of LH, in practice both preparations yield similar results in terms of pregnancy or live birth rates [83, 220]. Addition of recombinant LH administration as a supplementation for FSH stimulation has not shown any consistent or substantial improvement in pregnancy rates in normogonadotropic women [247, 248], except probably in those showing poor responses to conventional treatment [249]; in contrast, in hypogonadotropic women LH supplementation is necessary for optimal follicle maturation [236, 250]. Long-acting FSH analogs (e.g., FSH-CTP) have been used to stimulate multiple follicular maturation [32, 251]; in a study involving a small cohort of patients, the effects of a single dose of FSH-CTP to induce multiple follicular growth were comparable to those showed by conventional treatment with FSH alone [32].

5.7
Conclusions

From the previous discussion, it is clear glycans are crucial for proper synthesis, secretion and activity of glycoprotein hormones and that the biosynthesis of their oligosaccharide structures is under the subtle control of several factors, including hypothalamic-releasing factors and end products from their respective target tissues. Alterations in glycoprotein hormone glycosylation (and thus in the hormone's properties), as a result of end-organ failure or physiological variations in end-organ hormonal production, may represent adaptive responses that provide the target cell with optimal trophic signals. Genome sequencing efforts have unveiled the various enzymes involved in glycoprotein processing and a detailed understanding of the regulation of the biosynthetic pathways leading to mature glycoprotein hormones is deemed necessary; this is most relevant for the judicious design and choice of recombinant analogs optimal for therapeutic purposes, particularly considering that glycan structures from the currently available recombinant glycoprotein hormones are different from their natural counterparts. In this vein, new gonadotropin analogs with agonist or antagonist activity are being created using recombinant DNA technology. These include single- and double-chain chimeras that elicit both dual FSH and LH activities *in vivo* [232, 233], single-chain FSH analogs containing variable numbers of additional N- or O-linked carbohydrates that increase the *in vivo* potency of the hormone [151, 152, 252], and crosslinked bifunctional gonadotropin analogs that bind the FSH and LH receptors but exhibit reduced efficacy to activate the receptor and stimulate intracellular signaling [253]. These novel gonadotropins will undoubtedly enrich the arsenal of strategies useful for therapeutic purposes.

The functional significance of gonadotropin microheterogeneity is still uncertain. The possibility exists that the changing abundance of different gonadotropin glycoforms with particular circulatory survival and potency observed in different physiological conditions [85] may serve as a mechanism to refine the gonadal response to the gonadotropic stimulus.

Acknowledgments

The studies performed in the authors' laboratories have been supported by grant 45991M from the Consejo Nacional de Ciencia y Tecnología (CONACyT), Mexico (to A.U-A); grants HD18407 (to J.A.D.), P20 RR16475 and P20 RR017708 (INBRE Program and COBRE Program of the National Center for Research Resources, respectively) (to G.R.B.) from the National Institutes of Health (NIH), Bethesda, Maryland, USA; grant EPS-9874732 from the National Science Foundation, USA (to G.R.B.); and matching support from the State of Kansas (to G.R.B.). A.U-A is recipient of a Career Development Award from the Fundacion IMSS, México.

References

1 Pierce, J.G. and Parsons, T.F. (1979) Glycoprotein hormones: similar molecules with different functions. *UCLA Forum in Medical Sciences*, **21**, 99–117.

2 Pierce, J.G. and Parsons, T.F. (1981) Glycoprotein hormones: structure and function. *Annual Review of Biochemistry*, **50**, 465–495.

3 Ulloa-Aguirre, A. and Timossi, C. (2000) Biochemical and functional aspects of gonadotrophin-releasing hormone and gonadotrophins. *Reproductive Biomedicine Online*, **1**, 48–62.

4 Ulloa-Aguirre, A. *et al.* (2001) Endocrine regulation of gonadotropin glycosylation. *Archives of Medical Research*, **32**, 520–532.

5 Birken, S. *et al.* (1996) Isolation and characterization of human pituitary chorionic gonadotropin. *Endocrinology*, **137**, 1402–1411.

6 Bousfield, G.R. *et al.* (1996) Structural features of mammalian gonadotropins. *Molecular and Cellular Endocrinology*, **125**, 3–19.

7 Sun, L. *et al.* (2006) FSH directly regulates bone mass. *Cell*, **125**, 247–260.

8 Meduri, G. *et al.* (2002) Follicle-stimulating hormone receptors in oocytes? *The Journal of Clinical Endocrinology and Metabolism*, **87**, 2266–2276.

9 Ulloa-Aguirre, A. and Conn, P.M. (1998) G protein-coupled receptors and the G protein family, in *Handbook of Physiology-Endocrinology* (ed. P.M. Conn), Oxford University Press, New York, Section 7, Cellular Endocrinology, pp. 87–141.

10 Ulloa-Aguirre, A. *et al.* (1999) Structure-activity relationships of G protein-coupled receptors. *Archives of Medical Research*, **30**, 420–435.

11 Ulloa-Aguirre, A. and Timossi, C. (1998) Structure-function relationship of follicle-stimulating hormone and its receptor. *Human Reproduction Update*, **4**, 260–283.

12 Khan, H., Yarney, T.A. and Sairam, M.R. (1993) Cloning of alternately spliced mRNA transcripts coding for variants of ovine testicular follitropin receptor lacking the G protein coupling domains. *Biochemical and Biophysical Research Communications*, **190**, 888–894.

13 Li, Y. *et al.* (2007) FSH stimulates ovarian cancer cell growth by action on growth factor variant receptor. *Molecular and Cellular Endocrinology*, **267**, 26–37.

14 Iqbal, J. *et al.* (2006) Follicle-stimulating hormone stimulates TNF production from immune cells to enhance osteoblast and osteoclast formation. *Proceedings of the National Academy of Sciences of the United States of America*, **103**, 14925–14930.

15 Richards, J.S. *et al.* (2002) Novel signaling pathways that control ovarian follicular development, ovulation, and luteinization. *Recent Progress in Hormone Research*, **57**, 195–220.

16 Ulloa-Aguirre, A. *et al.* (2007) Multiple facets of follicle-stimulating hormone receptor function. *Endocrine*, **32**, 251–263.

17 Uribe, A. *et al.* (2008) Functional and structural roles of conserved cysteine residues in the carboxyl-terminal domain of the follicle-stimulating hormone receptor in human embryonic kidney 293 cells. *Biology of Reproduction*, **78**, 869–882.

18 Gharib, S.D. *et al.* (1990) Molecular biology of the pituitary gonadotropins. *Endocrine Reviews*, **11**, 177–199.

19 Fiddes, J.C. and Talmadge, K. (1984) Structure, expression, and evolution of the genes for the human glycoprotein hormones. *Recent Progress in Hormone Research*, **40**, 43–78.

20 Albanese, C. *et al.* (1996) The gonadotropin genes: evolution of distinct mechanisms for hormonal control. *Recent Progress in Hormone Research*, **51**, 23–58, discussion 59–61.

21 Gordon, D.F., Wood, W.M. and Ridgway, E.C. (1988) Organization and nucleotide sequence of the mouse alpha-subunit gene of the pituitary glycoprotein hormones. *DNA*, **7**, 679–690.

22 Mercer, J.E. and Chin, W.W. (1995) Regulation of pituitary gonadotrophin gene expression. *Human Reproduction Update*, **1**, 363–384.

23 Pierce, J.G. *et al.* (1971) Biologically active hormones prepared by recombination of the alpha chain of human chorionic gonadotropin and the hormone-specific chain of bovine thyrotropin or of bovine luteinizing hormone. *The Journal of Biological Chemistry*, **246**, 2321–2324.

24 Dias, J.A. and Van Roey, P. (2001) Structural biology of human follitropin and its receptor. *Archives of Medical Research*, **32**, 510–519.

25 Fox, K.M., Dias, J.A. and Van Roey, P. (2001) Three-dimensional structure of human follicle-stimulating hormone. *Molecular Endocrinology*, **15**, 378–389.

26 Lapthorn, A.J. *et al.* (1994) Crystal structure of human chorionic gonadotropin. *Nature*, **369**, 455–461.

27 Dias, J.A. (2005) Endocrinology: fertility hormone in repose. *Nature*, **433**, 203–204.

28 Walton, W.J. *et al.* (2001) Characterization of human FSH isoforms reveals a nonglycosylated beta-subunit in addition to the conventional glycosylated beta-subunit. *The Journal of Clinical Endocrinology and Metabolism*, **86**, 3675–3685.

29 Balen, A.H. *et al.* (2004) Pharmacodynamics of a single low dose of long-acting recombinant follicle-stimulating hormone (FSH-carboxy terminal peptide, corifollitropin alfa) in women with World Health Organization group II anovulatory infertility. *The Journal of Clinical Endocrinology and Metabolism*, **89**, 6297–6304.

30 Fares, F.A. *et al.* (1992) Design of a long-acting follitropin agonist by fusing the C-terminal sequence of the chorionic gonadotropin beta subunit to the follitropin beta subunit. *Proceedings of the National Academy of Sciences of the United States of America*, **89**, 4304–4308.

31 LaPolt, P.S. *et al.* (1992) Enhanced stimulation of follicle maturation and ovulatory potential by long acting follicle-stimulating hormone agonists with extended carboxyl-terminal peptides. *Endocrinology*, **131**, 2514–2520.

32 Devroey, P. *et al.* (2004) Induction of multiple follicular development by a single dose of long-acting recombinant follicle-stimulating hormone (FSH-CTP, corifollitropin alfa) for controlled ovarian stimulation before in vitro fertilization. *The Journal of Clinical Endocrinology and Metabolism*, **89**, 2062–2070.

33 Wu, H. *et al.* (1994) Structure of human chorionic gonadotropin at 2.6 A resolution from MAD analysis of the selenomethionyl protein. *Structure*, **2**, 545–558.

34 Lustbader, J.W. *et al.* (1998) Structural and molecular studies of human chorionic gonadotropin and its receptor. *Recent Progress in Hormone Research*, **53**, 395–424.

35 Krystek, S.R. Jr and Dias, J.A. (2005) Glycoprotein hormones tied but not tethered like other cysteine-knot cytokines. *Trends in Pharmacological Sciences*, **26**, 439–442.

36 Sun, P.D. and Davies, D.R. (1995) The cystine-knot growth-factor superfamily. *Annual Review of Biophysics and Biomolecular Structure*, **24**, 269–291.

37 Fan, Q.R. and Hendrickson, W.A. (2005) Structure of human follicle-stimulating hormone in complex with its receptor. *Nature*, **433**, 269–277.

38 Erbel, P.J. *et al.* (1999) Solution structure of the alpha-subunit of human chorionic gonadotropin. *European Journal of Biochemistry*, **260**, 490–498.

39 Thomas, R.M. *et al.* (2007) Follice-stimulating hormone receptor forms oligomers and shows evidence of carboxyl-terminal proteolytic processing. *Endocrinology*, **148**, 1987–1995.

40 Urizar, E. *et al.* (2005) Glycoprotein hormone receptors: link between receptor homodimerization and negative cooperativity. *EMBO Journal*, **24**, 1954–1964.

41 Baenziger, J.U. and Green, E.D. (1988) Pituitary glycoprotein hormone oligosaccharides: structure, synthesis and function of the asparagine-linked oligosaccharides on lutropin, follitropin and thyrotropin. *Biochimica et Biophysica Acta*, **947**, 287–306.

42 Parsons, T.F., Bloomfield, G.A. and Pierce, J.G. (1983) Purification of an alternate form of the alpha subunit of the glycoprotein hormones from bovine pituitaries and identification of its O-linked oligosaccharide. *The Journal of Biological Chemistry*, **258**, 240–244.

43 Parsons, T.F. and Pierce, J.G. (1984) Free alpha-like material from bovine pituitaries. Removal of its O-linked oligosaccharide permits combination with lutropin-beta. *The Journal of Biological Chemistry*, **259**, 2662–2666.

44 Blithe, D.L. (1990) N-linked oligosaccharides on free alpha interfere with its ability to combine with human chorionic gonadotropin-beta subunit. *The Journal of Biological Chemistry*, **265**, 21951–21956.

45 Bousfield, G.R. *et al.* (2007) All-or-none *N*-glycosylation in primate follicle-stimulating hormone beta-subunits. *Molecular and Cellular Endocrinology*, **260–262**, 40–48.

46 Green, E.D., Boime, I. and Baenziger, J.U. (1986) Differential processing of Asn-linked oligosaccharides on pituitary glycoprotein hormones: implications for biologic function. *Molecular and Cellular Biochemistry*, **72**, 81–100.

47 Renwick, A.G. *et al.* (1987) The asparagine-linked sugar chains of human follicle-stimulating hormone. *Journal of Biochemistry*, **101**, 1209–1221.

48 Green, E.D. and Baenziger, J.U. (1988) Asparagine-linked oligosaccharides on lutropin, follitropin, and thyrotropin. II. Distributions of sulfated and sialylated oligosaccharides on bovine, ovine, and human pituitary glycoprotein hormones. *The Journal of Biological Chemistry*, **263**, 36–44.

49 Green, E.D. and Baenziger, J.U. (1988) Asparagine-linked oligosaccharides on lutropin, follitropin, and thyrotropin. I. Structural elucidation of the sulfated and sialylated oligosaccharides on bovine, ovine, and human pituitary glycoprotein hormones. *The Journal of Biological Chemistry*, **263**, 25–35.

50 Weisshaar, G. *et al.* (1991) NMR investigations of the N-linked oligosaccharides at individual glycosylation sites of human lutropin. *European Journal of Biochemistry*, **195**, 257–268.

51 Weisshaar, G., Hiyama, J. and Renwick, A.G. (1991) Site-specific *N*-glycosylation of human chorionic gonadotrophin–structural analysis of glycopeptides by one- and two-dimensional 1H NMR spectroscopy. *Glycobiology*, **1**, 393–404.

52 Dalpathado, D.S. *et al.* (2006) Comparative glycomics of the glycoprotein follicle stimulating hormone: glycopeptide analysis of isolates from two mammalian species. *Biochemistry*, **45**, 8665–8673.

53 Matzuk, M.M. and Boime, I. (1988) Site-specific mutagenesis defines the intracellular role of the asparagine-linked oligosaccharides of chorionic gonadotropin beta subunit. *The Journal of Biological Chemistry*, **263**, 17106–17111.

54 Matzuk, M.M. and Boime, I. (1988) The role of the asparagine-linked oligosaccharides of the alpha subunit in the secretion and assembly of human chorionic gonadotrophin. *The Journal of Cell Biology*, **106**, 1049–1059.

55 Matzuk, M.M. and Boime, I. (1989) Mutagenesis and gene transfer define site-specific roles of the gonadotropin oligosaccharides. *Biology of Reproduction*, **40**, 48–53.

56 Matzuk, M.M. *et al.* (1990) The biological role of the carboxyl-terminal extension of human chorionic gonadotropin [corrected] beta-subunit. *Endocrinology*, **126**, 376–383.

57 Muyan, M. *et al.* (1998) Dissociation of early folding events from assembly of the human lutropin beta-subunit. *Molecular Endocrinology*, **12**, 1640–1649.

58 Bousfield, G.R. *et al.* (2008) Chromatofocusing fails to separate hFSH isoforms on the basis of glycan structure. *Biochemistry*, **47**, 1708–1720.

59 Ulloa-Aguirre, A. *et al.* (1995) Follicle-stimulating isohormones: characterization and physiological relevance. *Endocrine Reviews*, **16**, 765–787.

60 Ulloa-Aguirre, A. *et al.* (1999) Role of glycosylation in function of follicle-stimulating hormone. *Endocrine*, **11**, 205–215.

61 Robert, P. (1995) Contribution a l'étude des domaines dínteraction entre les hormones gonadotropes hypophysaires et leurs récepteurs, in *Pharmaceutical Sciences*, Université René Descartes de Paris, Paris, pp. 101.

62 Helenius, A. and Aebi, M. (2001) Intracellular functions of N-linked glycans. *Science*, **291**, 2364–2369.

63 Damian-Matsumura, P. *et al.* (1999) Oestrogens regulate pituitary alpha2,3-sialyltransferase messenger ribonucleic acid levels in the female rat. *Journal of Molecular Endocrinology*, **23**, 153–165.

64 Smith, P.L. and Baenziger, J.U. (1988) A pituitary *N*-acetylgalactosamine transferase that specifically recognizes glycoprotein hormones. *Science*, **242**, 930–933.

65 Smith, P.L. and Baenziger, J.U. (1990) Recognition by the glycoprotein hormone-specific *N*-acetylgalactosaminetransferase is independent of hormone native conformation. *Proceedings of the National Academy of Sciences of the United States of America* **87**, 7275–7279.

66 Smith, P.L. and Baenziger, J.U. (1992) Molecular basis of recognition by the glycoprotein hormone-specific *N*-acetylgalactosamine-transferase. *Proceedings of the National Academy of Sciences of the United States of America* **89**, 329–333.

67 Bousfield, G.R. *et al.* (2000) Carbohydrate analysis of glycoprotein hormones. *Methods*, **21**, 15–39.

68 Dharmesh, S.M. and Baenziger, J.U. (1993) Estrogen modulates expression of the glycosyltransferases that synthesize sulfated oligosaccharides on lutropin. *Proceedings of the National Academy of Sciences of the United States of America* **90**, 11127–11131.

69 Herbert, D.C. (1975) Localization of antisera to LHbeta and FSHbeta in the rat pituitary gland. *The American Journal of Anatomy*, **144**, 379–385.

70 Herbert, D.C. (1976) Immunocytochemical evidence that luteinizing hormone (LH) and follicle stimulating hormone (FSH) are present in the same cell type in the rhesus monkey pituitary gland. *Endocrinology*, **98**, 1554–1557.

71 Bielinska, M., Matzuk, M.M. and Boime, I. (1989) Site-specific processing of the N-linked oligosaccharides of the human chorionic gonadotropin alpha subunit. *The Journal of Biological Chemistry*, **264**, 17113–17118.

72 Corless, C.L. *et al.* (1987) Gonadotropin alpha subunit. Differential processing of free and combined forms in human trophoblast and transfected mouse cells. *The Journal of Biological Chemistry*, **262**, 14197–14203.

73 Olivares, A. *et al.* (2000) Reactivity of different LH and FSH standards and preparations in the World Health Organization matched reagents for enzyme-linked immunoassays of gonadotrophins. *Human Reproduction*, **15**, 2285–2291.

74 Stanton, P.G. *et al.* (1996) Structural and functional characterisation of hFSH and hLH isoforms. *Molecular and Cellular Endocrinology*, **125**, 133–141.

75 Wide, L. and Bakos, O. (1993) More basic forms of both human follicle-stimulating hormone and luteinizing hormone in serum at midcycle compared with the follicular or luteal phase. *The Journal of Clinical Endocrinology and Metabolism*, **76**, 885–889.

76 Baenziger, J.U. (2003) Glycoprotein hormone GalNAc-4-sulphotransferase. *Biochemical Society Transactions*, **31**, 326–330.

77 Schaaf, L. *et al.* (1997) Glycosylation variants of human TSH selectively activate signal transduction pathways. *Molecular and Cellular Endocrinology*, **132**, 185–194.

78 Nguyen, V.T. *et al.* (2003) Inositol phosphate stimulation by LH requires the entire alpha Asn56 oligosaccharide. *Molecular and Cellular Endocrinology*, **199**, 73–86.

79 Wide, L. (1985) Median charge and charge heterogeneity of human pituitary FSH, LH and TSH. I. Zone electrophoresis in agarose suspension. *Acta Endocrinologica*, **109**, 181–189.

80 Daya, S. (2002) Updated meta-analysis of recombinant follicle-stimulating hormone (FSH) versus urinary FSH for ovarian stimulation in assisted reproduction. *Fertility and Sterility*, **77**, 711–714.

81 Andersen, A.N., Devroey, P. and Arce, J.C. (2006) Clinical outcome following stimulation with highly purified hMG or recombinant FSH in patients undergoing IVF: a randomized assessor-blind controlled trial. *Human Reproduction*, **21**, 3217–3227.

82 Platteau, P. *et al.* (2006) Similar ovulation rates, but different follicular development with highly purified menotrophin compared with recombinant FSH in WHO Group II anovulatory infertility: a randomized controlled study. *Human Reproduction*, **21**, 1798–1804.

83 Andersen, C.Y., Westergaard, L.G. and van Wely, M. (2004) FSH isoform composition of commercial gonadotrophin preparations: a neglected aspect? *Reproductive Biomedicine Online*, **9**, 231–236.

84 Watanabe, S. *et al.* (2002) Sialylation of *N*-glycans on the recombinant proteins expressed by a baculovirus-insect cell system under beta-*N*-acetylglucosaminidase inhibition. *The Journal of Biological Chemistry*, **277**, 5090–5093.

85 Ulloa-Aguirre, A. *et al.* (2003) Impact of carbohydrate heterogeneity in function of follicle-stimulating hormone: studies derived from in vitro and in vivo models. *Biology of Reproduction*, **69**, 379–389.

86 Wide, L. (1985) Median charge and charge heterogeneity of human pituitary FSH, LH and TSH. II. Relationship to sex and age. *Acta Endocrinologica*, **109**, 190–197.

87 Padmanabhan, V. *et al.* (1988) Modulation of serum follicle-stimulating hormone bioactivity and isoform distribution by estrogenic steroids in normal women and in gonadal dysgenesis. *The Journal of Clinical Endocrinology and Metabolism*, **67**, 465–473.

88 Wide, L., Albertsson-Wikland, K. and Phillips, D.J. (1996) More basic isoforms of serum gonadotropins during gonadotropin-releasing hormone agonist therapy in pubertal children. *The Journal of Clinical Endocrinology and Metabolism*, **81**, 216–221.

89 Zambrano, E. *et al.* (1995) Dynamics of basal and gonadotropin-releasing hormone-releasable serum follicle-stimulating hormone charge isoform distribution throughout the human menstrual cycle. *The Journal of Clinical Endocrinology and Metabolism*, **80**, 1647–1656.

90 Ulloa-Aguirre, A. *et al.* (1995) On the nature of the follicle-stimulating signal delivered to the ovary during exogenously controlled follicular maturation. A search into the immunological and biological attributes and the molecular composition of two preparations of urofollitropin. *Archives of Medical Research*, **26 Spec No**, S219–S230.

91 Ulloa-Aguirre, A., Timossi, C. and Mendez, J.P. (2001) Is there any physiological role for gonadotrophin oligosaccharide heterogeneity in humans? I. Gondatrophins are synthesized and released in multiple molecular forms. A matter of fact. *Human Reproduction*, **16**, 599–604.

92 Ashwell, G. and Harford, J. (1982) Carbohydrate-specific receptors of the liver. *Annual Review of Biochemistry*, **51**, 531–554.

93 Chappel, S. *et al.* (1987) Production of bovine follicle stimulating hormone (FSH) by recombinant DNA technology. 69th annual Meeting of the Endocrine Society, Indianapolis, IN, USA,, Abstracts p. 130.

94 Bishop, L.A., Nguyen, T.V. and Schofield, P.R. (1995) Both of the beta-subunit carbohydrate residues of follicle-stimulating hormone determine the metabolic clearance rate and in vivo potency. *Endocrinology*, **136**, 2635–2640.

95 Wide, L. and Wide, M. (1984) Higher plasma disappearance rate in the mouse for pituitary follicle-stimulating hormone of young women compared to that of men and elderly women. *The Journal of Clinical Endocrinology and Metabolism*, **58**, 426–429.

96 Timossi, C. *et al.* (1998) A less acidic human follicle-stimulating hormone preparation induces tissue-type plasminogen activator enzyme activity earlier than a predominantly acidic analogue in phenobarbital-blocked pro-oestrous rats. *Molecular Human Reproduction*, **4**, 1032–1038.

97 Yding Andersen, C. *et al.* (1999) FSH-induced resumption of meiosis in mouse oocytes: effect of different isoforms. *Molecular Human Reproduction*, **5**, 726–731.

98 Berger, P. *et al.* (1993) Variants of human chorionic gonadotropin from pregnant women and tumor patients recognized by monoclonal antibodies. *The Journal of Clinical Endocrinology and Metabolism*, **77**, 347–351.

99 Elliott, M.M. *et al.* (1997) Carbohydrate and peptide structure of the alpha- and beta-subunits of human chorionic gonadotropin from normal and aberrant pregnancy and choriocarcinoma. *Endocrine*, **7**, 15–32.

100 Kovalevskaya, G. *et al.* (1999) Early pregnancy human chorionic gonadotropin (hCG) isoforms measured by an immunometric assay for choriocarcinoma-like hCG. *The Journal of Endocrinology*, **161**, 99–106.

101 Birken, S. *et al.* (1988) Structure of the human chorionic gonadotropin beta-subunit fragment from pregnancy urine. *Endocrinology*, **123**, 572–583.

102 Blithe, D.L. *et al.* (1988) Purification of beta-core fragment from pregnancy urine and demonstration that its carbohydrate moieties differ from those of native human chorionic gonadotropin-beta. *Endocrinology*, **122**, 173–180.

103 Blithe, D.L., Wehmann, R.E. and Nisula, B.C. (1989) Carbohydrate composition of beta-core. *Endocrinology*, **125**, 2267–2272.

104 Birken, S. *et al.* (1993) Separation of nicked human chorionic gonadotropin (hCG), intact hCG, and hCG beta fragment from standard reference preparations and raw urine samples. *Endocrinology*, **133**, 1390–1397.

105 Birken, S., Kovalevskaya, G. and O'Connor, J. (2001) Immunochemical measurement of early pregnancy isoforms of HCG: potential applications to fertility research, prenatal diagnosis, and cancer. *Archives of Medical Research*, **32**, 635–643.

106 Iles, R.K. *et al.* (1992) Immunoreactive beta-core-like material in normal postmenopausal urine: human chorionic gonadotrophin or LH origin? Evidence for

the existence of LH core. *The Journal of Endocrinology*, **133**, 459–466.
107 Birken, S. *et al.* (1993) Structure and significance of human luteinizing hormone-beta core fragment purified from human pituitary extracts. *Endocrinology*, **133**, 985–989.
108 Neven, P. *et al.* (1993) Substantial urinary concentrations of material resembling beta-core fragment of chorionic gonadotropin beta-subunit in mid-menstrual cycle. *Clinical Chemistry*, **39**, 1857–1860.
109 O'Connor, J.F. *et al.* (1998) The expression of the urinary forms of human luteinizing hormone beta fragment in various populations as assessed by a specific immunoradiometric assay. *Human Reproduction*, **13**, 826–835.
110 Thotakura, N.R. and Blithe, D.L. (1995) Glycoprotein hormones: glycobiology of gonadotrophins, thyrotrophin and free alpha subunit. *Glycobiology*, **5**, 3–10.
111 Diaz-Cueto, L. *et al.* (1996) More in-vitro bioactive, shorter-lived human chorionic gonadotrophin charge isoforms increase at the end of the first and during the third trimesters of gestation. *Molecular Human Reproduction*, **2**, 643–650.
112 Diaz-Cueto, L. *et al.* (1994) Amplitude regulation of episodic release, in vitro biological to immunological ratio, and median charge of human chorionic gonadotropin in pregnancy. *The Journal of Clinical Endocrinology and Metabolism*, **78**, 890–897.
113 Wide, L. and Hobson, B. (1987) Some qualitative differences of hCG in serum from early and late pregnancies and trophoblastic diseases. *Acta Endocrinologica*, **116**, 465–472.
114 Wide, L., Lee, J.Y. and Rasmussen, C. (1994) A change in the isoforms of human chorionic gonadotropin occurs around the 13th week of gestation. *The Journal of Clinical Endocrinology and Metabolism*, **78**, 1419–1423.
115 Skarulis, M.C. *et al.* (1992) Glycosylation changes in human chorionic gonadotropin and free alpha subunit as gestation progresses. *The Journal of Clinical Endocrinology and Metabolism*, **75**, 91–96.
116 Cole, L.A. (1987) *O*-Glycosylation of proteins in the normal and neoplastic trophoblast. *Trophoblast Research*, **2**, 139–148.
117 Cole, L.A. and Khanlian, S.A. (2007) Hyperglycosylated hCG: a variant with separate biological functions to regular hCG. *Molecular and Cellular Endocrinology*, **260–262**, 228–236.
118 Cole, L.A. *et al.* (2003) Hyperglycosylated hCG (invasive trophoblast antigen, ITA) a key antigen for early pregnancy detection. *Clinical Biochemistry*, **36**, 647–655.
119 Cole, L.A. *et al.* (2006) Hyperglycosylated hCG in gestational implantation and in choriocarcinoma and testicular germ cell malignancy tumorigenesis. *The Journal of Reproductive Medicine*, **51**, 919–929.
120 Toll, H. *et al.* (2006) Glycosylation patterns of human chorionic gonadotropin revealed by liquid chromatography-mass spectrometry and bioinformatics. *Electrophoresis*, **27**, 2734–2746.
121 Kobata, A. and Takeuchi, M. (1999) Structure, pathology and function of the N-linked sugar chains of human chorionic gonadotropin. *Biochimica et Biophysica Acta*, **1455**, 315–326.
122 Bishop, L.A. *et al.* (1994) Specific roles for the asparagine-linked carbohydrate residues of recombinant human follicle stimulating hormone in receptor binding and signal transduction. *Molecular Endocrinology*, **8**, 722–731.
123 Muyan, M. and Boime, I. (1998) The carboxyl-terminal region is a determinant for the intracellular behavior of the chorionic gonadotropin beta subunit: effects on the processing of the Asn-linked oligosaccharides. *Molecular Endocrinology*, **12**, 766–772.
124 Sairam, M.R. (1989) Role of carbohydrates in glycoprotein hormone signal transduction. *FASEB Journal*, **3**, 1915–1926.

125 Sairam, M.R. and Bhargavi, G.N. (1985) A role for glycosylation of the alpha subunit in transduction of biological signal in glycoprotein hormones. *Science*, **229**, 65–67.

126 Smith, P.L. *et al.* (1990) The sialylated oligosaccharides of recombinant bovine lutropin modulate hormone bioactivity. *The Journal of Biological Chemistry*, **265**, 874–881.

127 Dias, J.A. (2001) Is there any physiological role for gonadotrophin oligosaccharide heterogeneity in humans? II. A biochemical point of view. *Human Reproduction*, **16**, 825–830.

128 Merz, W.E. (1988) Evidence for impaired subunit interaction in chemically deglycosylated human choriogonadotropin. *Biochemical and Biophysical Research Communications*, **156**, 1271–1278.

129 Ryan, R.J. *et al.* (1987) Structure-function relationships of gonadotropins. *Recent Progress in Hormone Research*, **43**, 383–429.

130 Huth, J.R. *et al.* (1994) Bacterial expression and in vitro folding of the beta-subunit of human chorionic gonadotropin (hCG beta) and functional assembly of recombinant hCG beta with hCG alpha. *Endocrinology*, **135**, 911–918.

131 Ren, P., Sairam, M.R. and Yarney, T.A. (1995) Bacterial expression of human chorionic gonadotropin alpha subunit: studies on refolding, dimer assembly and interaction with two different beta subunits. *Molecular and Cellular Endocrinology*, **113**, 39–51.

132 Feng, W. *et al.* (1995) Asparagine-linked oligosaccharides facilitate human chorionic gonadotropin beta-subunit folding but not assembly of prefolded beta with alpha. *Endocrinology*, **136**, 52–61.

133 Feng, W. *et al.* (1995) The asparagine-linked oligosaccharides of the human chorionic gonadotropin beta subunit facilitate correct disulfide bond pairing. *The Journal of Biological Chemistry*, **270**, 11851–11859.

134 Feng, W. *et al.* (1996) Novel covalent chaperone complexes associated with human chorionic gonadotropin beta subunit folding intermediates. *The Journal of Biological Chemistry*, **271**, 18543–18548.

135 Flack, M.R. *et al.* (1994) Site-directed mutagenesis defines the individual roles of the glycosylation sites on follicle-stimulating hormone. *The Journal of Biological Chemistry*, **269**, 14015–14020.

136 Ulloa-Aguirre, A. *et al.* (1992) Effects of gonadotrophin-releasing hormone, recombinant human activin-A and sex steroid hormones upon the follicle-stimulating isohormones secreted by rat anterior pituitary cells in culture. *The Journal of Endocrinology*, **134**, 97–106.

137 Blithe, D.L. and Iles, R.K. (1995) The role of glycosylation in regulating the glycoprotein hormone free alpha-subunit and free beta-subunit combination in the extraembryonic coelomic fluid of early pregnancy. *Endocrinology*, **136**, 903–910.

138 Morell, A.G. *et al.* (1971) The role of sialic acid in determining the survival of glycoproteins in the circulation. *The Journal of Biological Chemistry*, **246**, 1461–1467.

139 Lefort, G.P., Stolk, J.M. and Nisula, B.C. (1984) Evidence that desialylation and uptake by hepatic receptors for galactose-terminated glycoproteins are immaterial to the metabolism of human choriogonadotropin in the rat. *Endocrinology*, **115**, 1551–1557.

140 Nisula, B.C. *et al.* (1989) Metabolic fate of human choriogonadotropin. *Journal of Steroid Biochemistry*, **33**, 733–737.

141 Barrios-De-Tomasi, J. *et al.* (2002) Assessment of the in vitro and in vivo biological activities of the human follicle-stimulating isohormones. *Molecular and Cellular Endocrinology*, **186**, 189–198.

142 Ulloa-Aguirre, A. *et al.* (1992) Biological characterization of the naturally occurring analogues of intrapituitary human follicle-stimulating hormone. *Human Reproduction*, **7**, 23–30.

143 Wide, L. (1986) The regulation of metabolic clearance rate of human FSH in mice by variation of the molecular structure of the hormone. *Acta Endocrinologica*, **112**, 336–344.

144 Fiete, D. and Baenziger, J.U. (1997) Isolation of the SO_4-4-GalNAcbeta1,4GlcNAcbeta1,2Manalpha-specific receptor from rat liver. *The Journal of Biological Chemistry*, **272**, 14629–14637.

145 Fiete, D. *et al.* (1991) A hepatic reticuloendothelial cell receptor specific for SO_4-4GalNAc beta 1,4GlcNAc beta 1,2Man alpha that mediates rapid clearance of lutropin. *Cell*, **67**, 1103–1110.

146 Klett, D. *et al.* (2003) Fast renal trapping of porcine luteinizing hormone (pLH) shown by 123I-scintigraphic imaging in rats explains its short circulatory half-life. *Reproductive Biology and Endocrinology*, **1**, 64.

147 Legardinier, S. *et al.* (2005) Biological activities of recombinant equine luteinizing hormone/chorionic gonadotropin (eLH/CG) expressed in Sf9 and Mimic insect cell lines. *Journal of Molecular Endocrinology*, **34**, 47–60.

148 Legardinier, S. *et al.* (2005) Mammalian-like nonsialyl complex-type *N*-glycosylation of equine gonadotropins in Mimic insect cells. *Glycobiology*, **15**, 776–790.

149 Duijkers, I.J. *et al.* (2002) Single dose pharmacokinetics and effects on follicular growth and serum hormones of a long-acting recombinant FSH preparation (FSH-CTP) in healthy pituitary-suppressed females. *Human Reproduction*, **17**, 1987–1993.

150 Bouloux, P.M. *et al.* (2001) First human exposure to FSH-CTP in hypogonadotrophic hypogonadal males. *Human Reproduction*, **16**, 1592–1597.

151 Perlman, S. *et al.* (2003) Glycosylation of an N-terminal extension prolongs the half-life and increases the in vivo activity of follicle stimulating hormone. *The Journal of Clinical Endocrinology and Metabolism*, **88**, 3227–3235.

152 Weenen, C. *et al.* (2004) Long-acting follicle-stimulating hormone analogs containing N-linked glycosylation exhibited increased bioactivity compared with o-linked analogs in female rats. *The Journal of Clinical Endocrinology and Metabolism*, **89**, 5204–5212.

153 Sharpless, J.L. *et al.* (1999) Disappearance of endogenous luteinizing hormone is prolonged in postmenopausal women. *The Journal of Clinical Endocrinology and Metabolism*, **84**, 688–694.

154 Wide, L. *et al.* (2007) Sulfonation and sialylation of gonadotropins in women during the menstrual cycle, after menopause, and with polycystic ovarian syndrome and in men. *The Journal of Clinical Endocrinology and Metabolism*, **92**, 4410–4417.

155 West, C.R. *et al.* (2002) Acidic mix of FSH isoforms are better facilitators of ovarian follicular maturation and E2 production than the less acidic. *Endocrinology*, **143**, 107–116.

156 Matzuk, M.M., Keene, J.L. and Boime, I. (1989) Site specificity of the chorionic gonadotropin N-linked oligosaccharides in signal transduction. *The Journal of Biological Chemistry*, **264**, 2409–2414.

157 Valove, F.M. *et al.* (1994) Receptor binding and signal transduction are dissociable functions requiring different sites on follicle-stimulating hormone. *Endocrinology*, **135**, 2657–2661.

158 Bousfield, G.R. *et al.* (2004) Differential effects of alpha subunit asparagine56 oligosaccharide structure on equine lutropin and follitropin hybrid conformation and receptor-binding activity. *Biochemistry*, **43**, 10817–10833.

159 Rcbois, R.V. and Liss, M.T. (1987) Antibody binding to the beta-subunit of deglycosylated chorionic gonadotropin converts the antagonist to an agonist. *The Journal of Biological Chemistry*, **262**, 3891–3896.

160 Butney, V.Y. *et al.* (1998) Hormone-specific inhibitory influence of alpha-subunit Asn56 oligosaccharide on in vitro

subunit association and follicle-stimulating hormone receptor binding of equine gonadotropins. *Biology of Reproduction*, **58**, 458–469.
161 Heikoop, J.C. *et al.* (1998) Partially deglycosylated human choriogonadotropin, stabilized by intersubunit disulfide bonds, shows full bioactivity. *European Journal of Biochemistry*, **253**, 354–356.
162 Butnev, V.Y. *et al.* (2002) Truncated equine LH beta and asparagine(56)-deglycosylated equine LH alpha combine to produce a potent FSH antagonist. *The Journal of Endocrinology*, **172**, 545–555.
163 Fan, Q.R. and Hendrickson, W.A. (2005) Structural biology of glycoprotein hormones and their receptors. *Endocrine*, **26**, 179–188.
164 Arey, B.J. *et al.* (1997) Induction of promiscuous G protein coupling of the follicle-stimulating hormone (FSH) receptor: a novel mechanism for transducing pleiotropic actions of FSH isoforms. *Molecular Endocrinology*, **11**, 517–526.
165 Andersen, C.Y. *et al.* (2001) Effect of different FSH isoforms on cyclic-AMP production by mouse cumulus-oocyte-complexes: a time course study. *Molecular Human Reproduction*, **7**, 129–135.
166 Barrios-de-Tomasi, J. *et al.* (2006) Effects of human pituitary FSH isoforms on mouse follicles in vitro. *Reproductive Biomedicine Online*, **12**, 428–441.
167 Timossi, C.M. *et al.* (1998) A naturally occurring basically charged human follicle-stimulating hormone (FSH) variant inhibits FSH-induced androgen aromatization and tissue-type plasminogen activator enzyme activity in vitro. *Neuroendocrinology*, **67**, 153–163.
168 Timossi, C.M. *et al.* (2000) Differential effects of the charge variants of human follicle-stimulating hormone. *The Journal of Endocrinology*, **165**, 193–205.
169 Creus, S. *et al.* (2001) Human FSH isoforms: carbohydrate complexity as determinant of in-vitro bioactivity. *Molecular and Cellular Endocrinology*, **174**, 41–49.
170 Stanton, P.G. *et al.* (1993) Isolation and characterization of human LH isoforms. *The Journal of Endocrinology*, **138**, 529–543.
171 Stanton, P.G. *et al.* (1992) Isolation and physicochemical characterization of human follicle-stimulating hormone isoforms. *Endocrinology*, **130**, 2820–2832.
172 Reddy, B.V., Bartoszewicz, Z. and Rebois, R.V. (1996) Modification of the sialic acid residues of choriogonadotropin affects signal transduction. *Cellular Signalling*, **8**, 35–41.
173 Kovalevskaya, G. *et al.* (2002) Trophoblast origin of hCG isoforms: cytotrophoblasts are the primary source of choriocarcinoma-like hCG. *Molecular and Cellular Endocrinology*, **194**, 147–155.
174 Kovalevskaya, G. *et al.* (2002) Differential expression of human chorionic gonadotropin (hCG) glycosylation isoforms in failing and continuing pregnancies: preliminary characterization of the hyperglycosylated hCG epitope. *The Journal of Endocrinology*, **172**, 497–506.
175 Cole, L.A. *et al.* (2006) Gestational trophoblastic diseases: 1. Pathophysiology of hyperglycosylated hCG. *Gynecologic Oncology*, **102**, 145–150.
176 Lei, Z.M. *et al.* (1999) Human chorionic gonadotropin promotes tumorigenesis of choriocarcinoma JAR cells. *Trophoblast Research*, **13**, 147–159.
177 Hamada, A.L. *et al.* (2005) Transfection of antisense chorionic gonadotropin beta gene into choriocarcinoma cells suppresses the cell proliferation and induces apoptosis. *The Journal of Clinical Endocrinology and Metabolism*, **90**, 4873–4879.
178 Cole, L.A. *et al.* (1988) Urinary human chorionic gonadotropin free beta-subunit and beta-core fragment: a new marker of gynecological cancers. *Cancer Research*, **48**, 1356–1360.
179 Lundin, M. *et al.* (2001) Tissue expression of human chorionic gonadotropin beta

predicts outcome in colorectal cancer: a comparison with serum expression. *International Journal of Cancer*, **95**, 18–22.

180 Mora, J. *et al.* (1996) Different hCG assays to measure ectopic hCG secretion in bladder carcinoma patients. *British Journal of Cancer*, **74**, 1081–1084.

181 Hotakainen, K. *et al.* (1999) Detection of messenger RNA for the beta-subunit of chorionic gonadotropin in urinary cells from patients with transitional cell carcinoma of the bladder by reverse transcription-polymerase chain reaction. *International Journal of Cancer*, **84**, 304–308.

182 Hotakainen, K. *et al.* (2007) Overexpression of human chorionic gonadotropin beta genes 3, 5 and 8 in tumor tissue and urinary cells of bladder cancer patients. *Tumour Biology: The Journal of the International Society, for Oncodevelopmental Biology and Medicine* **28**, 52–56.

183 Yokotani, T. *et al.* (1997) Expression of alpha and beta genes of human chorionic gonadotropin in lung cancer. *International Journal of Cancer*, **71**, 539–544.

184 Liu, T.C. and Jackson, G.L. (1990) 17-Beta-estradiol potentiates luteinizing hormone glycosylation and release induced by veratridine, diacylglycerol, and phospholipase C in rat anterior pituitary cells. *Neuroendocrinology*, **51**, 642–648.

185 Wide, L. and Naessen, T. (1994) 17 Beta-oestradiol counteracts the formation of the more acidic isoforms of follicle-stimulating hormone and luteinizing hormone after menopause. *Clinical Endocrinology*, **40**, 783–789.

186 Bergendah, M. and Veldhuis, J.D. (2001) Is there a physiological role for gonadotrophin oligosaccharide heterogeneity in humans? III. Luteinizing hormone heterogeneity: a medical physiologist's perspective. *Human Reproduction*, **16**, 1058–1064.

187 Wide, L. (1982) Male and female forms of human follicle-stimulating hormone in serum. *The Journal of Clinical Endocrinology and Metabolism*, **55**, 682–688.

188 Castro-Fernandez, C. *et al.* (2000) A preponderance of circulating basic isoforms is associated with decreased plasma half-life and biological to immunological ratio of gonadotropin-releasing hormone-releasable luteinizing hormone in obese men. *The Journal of Clinical Endocrinology and Metabolism*, **85**, 4603–4610.

189 Tsatsoulis, A., Shalet, S.M. and Robertson, W.R. (1988) Changes in the qualitative and quantitative secretion of luteinizing hormone (LH) following orchidectomy in man. *Clinical Endocrinology*, **29**, 189–194.

190 Campo, S. *et al.* (2007) Carbohydrate complexity and proportions of serum FSH isoforms in the male: lectin-based studies. *Molecular and Cellular Endocrinology*, **260–262**, 197–204.

191 Weintraub, B.D. *et al.* (1989) Effect of TRH on TSH glycosylation and biological action. *Annals of the New York Academy of Sciences*, **553**, 205–213.

192 Weintraub, B.D. *et al.* (1990) Pre-translational and post-translational regulation of TSH: relationship to bioactivity. *Hormone and Metabolic Research. Supplement*, **23**, 9–11.

193 Khan, S.A. *et al.* (1985) Influence of gonadectomy on isoelectrofocusing profiles of pituitary gonadotropins in rhesus monkeys. *Journal of Medical Primatology*, **14**, 177–194.

194 Solano, A.R. *et al.* (1980) Modulation of serum and pituitary luteinizing hormone bioactivity by androgen in the rat. *Endocrinology*, **106**, 1941–1948.

195 Christianson, S.L., Zalesky, D.D. and Grotjan, H.E. (1998) Ovine luteinizing hormone heterogeneity: androgens increase the percentage of less basic isohormones. *Domestic Animal Endocrinology*, **15**, 87–92.

196 Mitchell, R. *et al.* (1994) Less acidic forms of luteinizing hormone are associated with lower testosterone secretion in men on haemodialysis treatment. *Clinical Endocrinology*, **41**, 65–73.

197 Sharma, O.P. *et al.* (1990) Effects of androgens on bioactivity and immunoreactivity of pituitary FSH in GnRH antagonist-treated male rats. *Acta Endocrinologica*, **122**, 168–174.

198 Kennedy, J. and Chappel, S. (1985) Direct pituitary effects of testosterone and luteinizing hormone-releasing hormone upon follicle-stimulating hormone: analysis by radioimmuno- and radioreceptor assay. *Endocrinology*, **116**, 741–748.

199 Helton, T.E. and Magner, J.A. (1994) Beta-1,4-galactosyltransferase and alpha-mannosidase-II messenger ribonucleic acid levels increase with different kinetics in thyrotrophs of hypothyroid mice. *Endocrinology*, **135**, 1980–1985.

200 Helton, T.E. and Magner, J.A. (1994) Sialyltransferase messenger ribonucleic acid increases in thyrotrophs of hypothyroid mice: an in situ hybridization study. *Endocrinology*, **134**, 2347–2353.

201 Helton, T.E. and Magner, J.A. (1995) Beta-galactoside alpha-2,3-sialyltransferase messenger RNA increases in thyrotrophs of hypothyroid mice. *Thyroid*, **5**, 315–317.

202 Persani, L. *et al.* (1998) Changes in the degree of sialylation of carbohydrate chains modify the biological properties of circulating thyrotropin isoforms in various physiological and pathological states. *The Journal of Clinical Endocrinology and Metabolism*, **83**, 2486–2492.

203 Trojan, J. *et al.* (1998) Modulation of human thyrotropin oligosaccharide structures–enhanced proportion of sialylated and terminally galactosylated serum thyrotropin isoforms in subclinical and overt primary hypothyroidism. *The Journal of Endocrinology*, **158**, 359–365.

204 Persani, L. *et al.* (2000) Circulating thyrotropin bioactivity in sporadic central hypothyroidism. *The Journal of Clinical Endocrinology and Metabolism*, **85**, 3631–3635.

205 DeCherney, G.S. *et al.* (1989) Alterations in the sialylation and sulfation of secreted mouse thyrotropin in primary hypothyroidism. *Biochemical and Biophysical Research Communications*, **159**, 755–762.

206 Gyves, P.W. *et al.* (1990) Changes in the sialylation and sulfation of secreted thyrotropin in congenital hypothyroidism. *Proceedings of the National Academy of Sciences of the United States of America* **87**, 3792–3796.

207 Simoni, M. *et al.* (1996) Effects of gonadotropin-releasing hormone on bioactivity of follicle-stimulating hormone (FSH) and microstructure of FSH, luteinizing hormone and sex hormone-binding globulin in a testosterone-based contraceptive trial: evaluation of responders and non-responders. *European Journal of Endocrinology/European Federation of, Endocrine Societies* **135**, 433–439.

208 Zarinan, T. *et al.* (2001) Changes in the biological:immunological ratio of basal and GnRH-releasable FSH during the follicular, pre-ovulatory and luteal phases of the human menstrual cycle. *Human Reproduction*, **16**, 1611–1618.

209 Ebling, F.J. *et al.* (1990) Metabolic interfaces between growth and reproduction. III. Central mechanisms controlling pulsatile luteinizing hormone secretion in the nutritionally growth-limited female lamb. *Endocrinology*, **126**, 2719–2727.

210 Hassing, J.M. *et al.* (1993) Pulsatile administration of gonadotropin-releasing hormone does not alter the follicle-stimulating hormone (FSH) isoform distribution pattern of pituitary or circulating FSH in nutritionally growth-restricted ovariectomized lambs. *Endocrinology*, **132**, 1527–1536.

211 Liu, T.C., Jackson, G.L. and Gorski, J. (1976) Effects of synthetic gonadotropin-releasing hormone on incorporation of radioactive glucosamine and amino acids into luteinizing hormone and total protein by rat pituitaries in vitro. *Endocrinology*, **98**, 151–163.

212 Khar, A., Debeljuk, L., Jutisz, M. (1978) Biosynthesis of gonadotropins by rat pituitary cells in culture and in pituitary homogenates: effect of gonadotropin-releasing hormone. *Molecular and Cellular Endocrinology*, **12**, 53–65.

213 Baird, D.T. (2001) Is there a place for different isoforms of FSH in clinical medicine? IV. The clinician's point of view. *Human Reproduction*, **16**, 1316–1318.

214 Lunenfeld, B. (2004) Historical perspectives in gonadotrophin therapy. *Human Reproduction Update*, **10**, 453–467.

215 Talbot, J.A. *et al.* (1996) Recombinant human luteinizing hormone: a partial physicochemical, biological and immunological characterization. *Molecular Human Reproduction*, **2**, 799–806.

216 Gervais, A. *et al.* (2003) Glycosylation of human recombinant gonadotrophins: characterization and batch-to-batch consistency. *Glycobiology*, **13**, 179–189.

217 Horsman, G. *et al.* (2000) A biological, immunological and physico-chemical comparison of the current clinical batches of the recombinant FSH preparations Gonal-F and Puregon. *Human Reproduction*, **15**, 1898–1902.

218 van Wely, M. *et al.* (2003) Effectiveness of human menopausal gonadotropin versus recombinant follicle-stimulating hormone for controlled ovarian hyperstimulation in assisted reproductive cycles: a meta-analysis. *Fertility and Sterility*, **80**, 1086–1093.

219 Van Wely, M. *et al.* (2003) Human menopausal gonadotropin versus recombinant follicle stimulation hormone for ovarian stimulation in assisted reproductive cycles. *Cochrane Database of Systematic Reviews*, **1**, CD003973.

220 Ledger, W. (2005) Clinical pharmacology of gonadotrophin preparations. *Reproductive Biomedicine Online*, **10** (Suppl. 3), 19–24.

221 Coomarasamy, A. *et al.* (2008) Urinary hMG versus recombinant FSH for controlled ovarian hyperstimulation following an agonist long down-regulation protocol in IVF or ICSI treatment: a systematic review and meta-analysis. *Human Reproduction*, **23**, 310–315.

222 Lambert, A. *et al.* (1998) Gonadotrophin heterogeneity and biopotency: implications for assisted reproduction. *Molecular Human Reproduction*, **4**, 619–629.

223 Hard, K. *et al.* (1990) Isolation and structure determination of the intact sialylated N-linked carbohydrate chains of recombinant human follitropin expressed in Chinese hamster ovary cells. *European Journal of Biochemistry*, **193**, 263–271.

224 Rafferty, B. *et al.* (1995) Differences in carbohydrate composition of FSH preparations detected with lectin-ELISA systems. *The Journal of Endocrinology*, **145**, 527–533.

225 Amoresano, A. *et al.* (1996) Structural characterisation of human recombinant glycohormones follitropin, lutropin and choriogonadotropin expressed in Chinese hamster ovary cells. *European Journal of Biochemistry*, **242**, 608–618.

226 Flack, M.R. *et al.* (1994) Increased biological activity due to basic isoforms in recombinant human follicle-stimulating hormone produced in a human cell line. *The Journal of Clinical Endocrinology and Metabolism*, **79**, 756–760.

227 Phillip, M. *et al.* (1998) Male hypogonadism due to a mutation in the gene for the beta-subunit of follicle-stimulating hormone. *The New England Journal of Medicine*, **338**, 1729–1732.

228 Depenbusch, M. *et al.* (2002) Maintenance of spermatogenesis in hypogonadotropic hypogonadal men with human chorionic gonadotropin alone. *European Journal of Endocrinology/European Federation of, Endocrine Societies* **147**, 617–624.

229 Finkel, D.M., Phillips, J.L. and Snyder, P.J. (1985) Stimulation of spermatogenesis by gonadotropins in men with hypogonadotropic hypogonadism. *The New England Journal of Medicine*, **313**, 651–655.

230 Kliesch, S., Behre, H.M. and Nieschlag, E. (1995) Recombinant human follicle-stimulating hormone and human chorionic gonadotropin for induction of spermatogenesis in a hypogonadotropic male. *Fertility and Sterility*, **63**, 1326–1328.

231 Liu, P.Y. *et al.* (1999) Efficacy and safety of recombinant human follicle stimulating hormone (Gonal-F) with urinary human chorionic gonadotrophin for induction of spermatogenesis and fertility in gonadotrophin-deficient men. *Human Reproduction*, **14**, 1540–1545.

232 Jablonka-Shariff, A. *et al.* (2006) Single-chain, triple-domain gonadotropin analogs with disulfide bond mutations in the alpha-subunit elicit dual follitropin and lutropin activities in vivo. *Molecular Endocrinology*, **20**, 1437–1446.

233 Garone, L.M. *et al.* (2006) Biological properties of a novel follicle-stimulating hormone/human chorionic gonadotropin chimeric gonadotropin. *Endocrinology*, **147**, 4205–4212.

234 Agrawal, R. *et al.* (1997) Pregnancy after treatment with three recombinant gonadotropins. *Lancet*, **349**, 29–30.

235 Macklon, N.S. and Fauser, B.C. (2002) Gonadotropin therapy for the treatment of anovulation and for ovarian hyperstimulation for IVF. *Molecular and Cellular Endocrinology*, **186**, 159–161.

236 Macklon, N. (2005) Gonadotrophins in ovulation induction. *Reproductive Biomedicine Online*, **10** (Suppl. 3), 25–31.

237 Balen, A. *et al.* (2007) Highly purified FSH is as efficacious as recombinant FSH for ovulation induction in women with WHO Group II anovulatory infertility: a randomized controlled non-inferiority trial. *Human Reproduction*, **22**, 1816–1823.

238 Macklon, N.S. and Fauser, B.C. (2001) Follicle-stimulating hormone and advanced follicle development in the human. *Archives of Medical Research*, **32**, 595–600.

239 Fauser, B.C., Devroey, P. and Macklon, N.S. (2005) Multiple birth resulting from ovarian stimulation for subfertility treatment. *Lancet*, **365**, 1807–1816.

240 Macklon, N.S. *et al.* (2006) The science behind 25 years of ovarian stimulation for in vitro fertilization. *Endocrine Reviews*, **27**, 170–207.

241 Group, E.R.L.S. (2001) Human recombinant luteinizing hormone is as effective as, but safer than, urinary human chorionic gonadotropin in inducing final follicular maturation and ovulation in in vitro fertilization procedures: results of a multicenter double-blind study. *The Journal of Clinical Endocrinology and Metabolism*, **86**, 2607–2618.

242 Chang, P. *et al.* (2001) Recombinant human chorionic gonadotropin (rhCG) in assisted reproductive technology: results of a clinical trial comparing two doses of rhCG (Ovidrel) to urinary hCG (Profasi) for induction of final follicular maturation in in vitro fertilization-embryo transfer. *Fertility and Sterility*, **76**, 67–74.

243 Piani, D. and Malkowski, J.P. (2001) Recombinant hCG (OVIDREL) and recombinant interferon-beta1a (REBIF) (No. 13 in a series of articles to promote a better understanding of the use of genetic engineering). *Journal of Biotechnology*, **87**, 281–283.

244 The European Recombinant Human Chorionic Gonadotropin Study Group (2000) Induction of final follicular maturation and early luteinization in women undergoing ovulation induction for assisted reproduction treatment–recombinant HCG versus urinary HCG. *Human Reproduction*, **15**, 1446–1451.

245 Lathi, R.B. and Milki, A.A. (2001) Recombinant gonadotropins. *Current Women's Health Reports*, **1**, 157–163.

246 Lispi, M. *et al.* (2006) Comparative assessment of the consistency and quality of a highly purified FSH extracted from human urine (urofollitropin) and a recombinant human FSH (follitropin alpha). *Reproductive Biomedicine Online*, **13**, 179–193.

247 Durnerin, C.I. *et al.* (2008) Effects of recombinant LH treatment on folliculogenesis and responsiveness to FSH stimulation. *Human Reproduction*, **23**, 421–426.

248 Nyboeandersen, A. *et al.* (2008) Recombinant LH supplementation to recombinant FSH during the final days of controlled ovarian stimulation for in vitro fertilization. A multicentre, prospective, randomized, controlled trial. *Human Reproduction*, **23**, 427–434.

249 Mochtar, M.H. *et al.* (2007) Recombinant luteinizing hormone (rLH) for controlled ovarian hyperstimulation in assisted reproductive cycles. *Cochrane Database of Systematic Reviews*, **2**, CD005070.

250 Kaufmann, R. *et al.* (2007) Recombinant human luteinizing hormone, lutropin alfa, for the induction of follicular development and pregnancy in profoundly gonadotrophin-deficient women. *Clinical Endocrinology*, **67**, 563–569.

251 Beckers, N.G. *et al.* (2003) First live birth after ovarian stimulation using a chimeric long-acting human recombinant follicle-stimulating hormone (FSH) agonist (recFSH-CTP) for in vitro fertilization. *Fertility and Sterility*, **79**, 621–623.

252 Klein, J. *et al.* (2003) Development and characterization of a long-acting recombinant hFSH agonist. *Human Reproduction*, **18**, 50–56.

253 Bernard, M.P. *et al.* (2005) Crosslinked bifunctional gonadotropin analogs with reduced efficacy. *Molecular and Cellular Endocrinology*, **233**, 25–31.

6
Yeast Glycosylation and Engineering in the Context of Therapeutic Proteins

Terrance A. Stadheim and Natarajan Sethuraman

6.1
Introduction

The advent of recombinant technologies for the production of proteins in the 1980s revolutionized the use of proteins as therapeutic entities. By overcoming the laborious task of purifying biologically active proteins from animal sources, researchers for the first time had the ability to exercise control over the process by which a particular biologic could be produced. Recombinant protein expression first began with bacteria as the preferred host. Bacteria were attractive for several reasons, including ease of genetic manipulations, the short turnaround time from transformation to protein expression and the relatively good yields of recombinant protein that could be generated. Drawbacks to using bacteria include the need for extraction of protein from inclusion bodies and subsequent resolubilization in an oxidative environment to achieve disulfide bond formation. Alternately, proteins produced in bacteria could be engineered with a secretion signal for delivery to the periplasmic space where the environment is more oxidative [1]. It is well appreciated that substantial improvements have been made in the optimization of *Escherichia coli* as an expression host for therapeutic proteins and this has been reviewed elsewhere [2]. Despite this progress, a fundamental deficiency for the bacterial system remains the inability to express proteins that require glycosylation for therapeutic efficacy.

It was soon realized that the recombinant technologies that had worked so well with bacterial systems could be applied to eukaryotic systems, including yeast and mammalian cells. Although technically more difficult, these systems offered several significant advantages, including the ability to form disulfide bonds *in vivo* as well as the intrinsic ability to perform glycosylation. This latter modification has proved extremely valuable in the case of proteins that are glycosylated in their endogenous state. Glycoproteins made in their aglycosylated form in bacteria show abbreviated pharmacokinetic profiles and are potentially more immunogenic due to the unmasking of epitopes that would otherwise be shielded by the hydrodynamic properties of the glycan moiety.

Post-translational Modification of Protein Biopharmaceuticals. Edited by Gary Walsh

ISBN: 978-3-527-32074-5

N-linked glycosylation is a post-translational modification that is conserved across yeast and other eukaryotes, including the metazoans. *N*-Glycosylation begins in the endoplasmic reticulum (ER) with the transfer of a preassembled dolichol phosphate-linked 14-sugar oligosaccharide to asparagine residues in a protein (see References [3, 4] as well as Chapter 2 for more thorough reviews of this subject). The preferred site for *N*-glycosylation contains the three amino acid sequence Asn-Xaa-Ser/Thr, where the second position can be any amino acid except Pro [5]. This oligosaccharide contains three glucose residues on the non-reducing end of a $Man_9GlcNAc_2$ core structure attached to the protein and is trimmed by ER glucosidases and a resident ER α-1,2 mannosidase to a $Man_8GlcNAc_2$ structure prior to entry into the Golgi compartment [6]. Despite the highly conserved mechanisms and composition of N-linked glycosylation in the ER, divergence between yeast and mammalian cells occurs once the glycoprotein exits the ER and enters the Golgi apparatus. As discussed in further detail below, fungi generally add additional monosaccharides to the $Man_8GlcNAc_2$ core. While mannose is by far the most commonly used substrate by fungi, other monosaccharides, modified or not, can be used in the construction of a mature oligosaccharide. In mammalian cells, including CHO and human cell lines, a range of exoglycosidases and glycosyltransferases function in the Golgi to remodel the $Man_8GlcNAc_2$ core oligosaccharide into a heterogeneous mix of glycoforms, including high mannose ($Man_5GlcNAc_2$-$Man_7GlcNAc_2$), hybrid type, complex type (both fucosylated and afucosylated) and multiantennary structures (reviewed in Reference [7]) (see also Figure 6.3 below). Oligosaccharides with a terminal galactose can also receive sialic acid. Additionally, a subset of mammalian proteins, such as luteinizing hormone and thyroid-stimulating hormone, contain a unique glycan terminating in SO_4-4-GalNAc [8].

6.2
N-Glycosylation in Fungi

The budding yeast *Saccharomyces cerevisiae* is probably the most studied of the fungi. Thus, it should come as no surprise that most of our knowledge regarding eukaryotic glycosylation, and specifically fungal glycosylation, is based on this organism. In *S. cerevisiae*, N-linked glycan chains exist in two major forms, a relatively compact *N*-glycosylation structure ($Man_{8-14}GlcNAc_2$), or core-type oligosaccharide, and a large oligosaccharide containing from 50 to 200 mannoses or mannan-type glycan [9]. The $Man_8GlcNAc_2$ core oligosaccharide that enters the Golgi apparatus is typically extended by mannosyltransferases. Upon entry of a glycoprotein into the early Golgi, the addition of a single mannose in the α-1,6 linkage to the α-1,3 arm of the core N-linked oligosaccharide occurs (Figure 6.1). This reaction requires GDP-mannose and is catalyzed by a specific non-redundant α-1,6 mannosyltransferase called Och1p [10, 11]. The resulting structure provides for the formation of an α-1,6 linked backbone by a series of α-1,6 specific mannosyltransferases. In *S. cerevisiae*, two enzyme complexes termed Man Pol I and Man Pol II have been characterized. The former complex consists of Mnn9p and Van1p with the latter

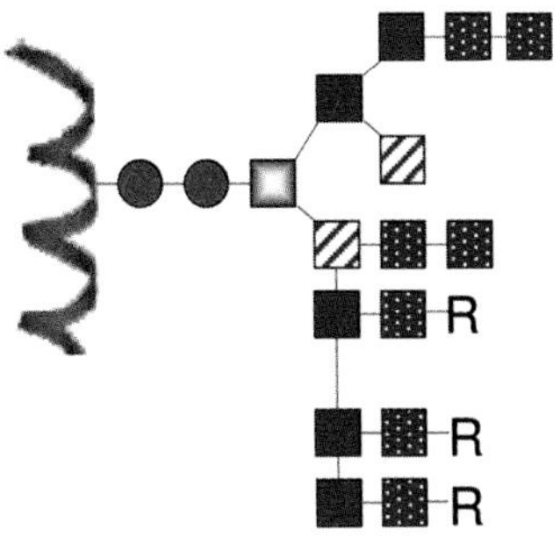

Species	R Groups	
Saccharomyces cerevisiae		α-1,2 linked mannose α-1,6 linked mannose-1-phosphate, α-1,3 linked mannose
Pichia pastoris		α-1,2 linked mannose α-1,6 linked mannose-1-phosphate β-1,2 linked mannose
Kluyveromyces lactis		α-1,2 linked mannose α-1,2 linked N-acetylglucosamine
Candida albicans		α-1,2 linked mannose α-1,3 linked mannose β-1,2 linked mannose mannose-1-phosphate
Schizosaccharomyces pombe		α-1,2 linked mannose α-1,2 linked galactose

Compound	Symbol
α-1,2-Galactose	
ß-1,2-Mannose	
ß-1,4-Mannose	
α-1,2-Mannose	
α-1,3-Mannose	
α-1,6-Mannose	
N-acetylglucosamine	
Mannosylphosphate	

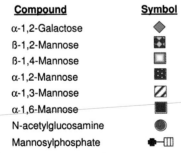

Figure 6.1 N-linked glycosylation in various yeasts. The composition of N-linked glycans from different yeasts is depicted. The R groups represent the common carbohydrate moieties found with each yeast species.

containing these proteins as well as Mnn10p, Mnn11p, Anp1p and Hoc1p [12]. Upon assembly of the α-1,6 backbone, Mnn2p and Mnn5p, α-1,2 specific mannosyltransferases, further extend the outer chain with branches that provide a substrate for further glycosylation [13].

For many yeast, the final "capping" on N-linked glycans consists primarily of mannose although they may differ with respect to linkage and anomericity (Figure 6.1). In *Candida albicans,* the outer chain capping residues consist of mannose in various linkages (α-1,2, α-1,3 and β-1,2) as well as phosphate, the latter determining the presence of either acid-labile or acid stable mannan fractions [14]. The exact composition of the outer chain of *C. albicans* varies with the serotype and can be quite large. *S. cerevisiae* contains α-1,2 with α-1,3 linked mannose catalyzed by the MNN1 gene product serving as the termination step [15]. The addition of α-1,3 linked mannose occurs in competition with the transfer of mannosylphosphate. Two genes, MNN4 and MNN6, have been shown to be important for at least some of the mannosylphosphate transferase activity in *S. cerevisiae* [15–17]. As mentioned previously, the outer chain mannose content for *S. cerevisiae* can reach up to 200 mannoses. *Pichia pastoris* shares elements of these two yeasts in that mannosylphosphate and β-linked mannose are found in the outer chain [18–20]. However, the size of the outer chain in *P. pastoris* is typically much smaller, displaying on average 15–30 mannoses [18].

Mannose is not the only sugar that fungi can use for incorporation into the outer chain or use to terminate glycosylation. For example, the yeasts *Kluyveromyces lactis* and *Schizosaccharomyces pombe* add α-1,2 linked GlcNAc [21] and α-1,2 linked galactose [22, 23], respectively (Figure 6.1), while filamentous fungi such as *Apergillus niger* terminate *N*-glycans with α-linked galactofuranose [24, 25]. Why fungi show such variation in the content of N-linked glycosylation is not well understood although it can be appreciated that such differences may confer certain advantages to the organism in its wild-type state.

6.3
O-Glycosylation in Fungi and Mammals

In fungi and yeast, O-linked glycosylation starts by the addition of a single mannose from dolichol-phosphate mannose to a Ser or Thr residue on a nascent glycoprotein in the ER [26, 27]. This reaction is catalyzed by the action of the PMT family of *O*-mannosyl transferases [28]. Subsequent elongation of the *O*-glycan chain occurs in the Golgi where subsequent mannose residues are transferred in the alpha configuration by GDP-Man. As is the case for N-linked glycosylation, there exist distinct differences across various yeast species with respect to carbohydrate content in *O*-glycans. *S. cerevisiae* and *P. pastoris* *O*-glycans contain mannosylphosphate, with the former containing α-1,3 linked mannose and the latter β-1,2 linked mannose [29–31]. *C. albicans,* in contrast, produces *O*-glycans predominately two to three mannoses long with an α-1,2 linkage [32]. *S. pombe* is unique in that it transfers terminal galactose to the O-linked mannose backbone. This galactose attachment occurs through an α-1,2 and/or α-1,3 linkage [33]. Sequence specificity for guiding *O*-glycosylation has not been determined although it is presumed that folding and solubility of the protein play important roles in the position and number of O-linked glycans that occur on a given protein. Support for this comes from a study where recombinant human granulocyte macrophage-colony stimulating factor (rhGM-CSF) expressed in *S. cerevisiae* and COS-1 mammalian cells produced *O*-glycosylation at the same Ser and Thr acceptor sites [34].

Mammalian *O*-glycosylation is complex due to a variety of enzymes involved in multiple pathways resulting in myriad glycoforms. The most common form of human *O*-glycosylation occurs in the Golgi compartment and involves the transfer of a GalNAc residue from UDP-GalNAc to a Ser or Thr though the action of a GalNAc transferase followed by the subsequent transfer of any number of different sugars in different linkages, including galactose, GalNAc and GlcNAc (Figure 6.2). Branching of the *O*-GalNAc chain as well as the addition of lactosamine also occurs, providing yet additional oligosaccharide diversity [35]. The O-linked glycan chains can also be capped with sialic acid [36].

Although less common, mammalian proteins bearing *O*-mannosylated glycans have been observed in brain, peripheral-nerve and muscle glycoproteins [37]. The most well-studied of these, α-dystroglycan, is a heavily glycosylated extracellular membrane glycoprotein that functions to regulate important interactions with extracellular matrix proteins [38]. *O*-Mannosylation on α-dystroglycan plays a critical

Mucin type*

(R -GalNAc-ol
where core (R) can be:
Gal(β1-3)
Gal (β1-3)[GlcNAc(β1-6)];
GlcNAc β-1,3);
GlcNAc β1-3)[GlcNAc(β1-6)];
GalNAc(α1,3);
GlcNAc β1-6);
GlcNAc α1,6);
Gal(α1,3)

*O-linked glycans are typically capped with NeuAc

O-linked mannose

NeuAc(α2-3)Gal(β1-4)GlcNAc(β1-2)Man-ol

Figure 6.2 O-linked glycosylation in mammalian species. Two major types of O-linked glycosylation are found in mammalian systems. Mucin-type *O*-glycosylation is initiated with the transfer of a GalNAc residue to a serine or threonine acceptor. This is followed by a series of carbohydrate attachments that are heterogeneous, depending on tissue and physiological context. O-linked mannosylation also occurs with the addition of a mannose to a serine or threonine acceptor by action of the enzyme POMT. This glycan can be further elaborated by serial addition of GlcNAc, Gal and sialic acid.

role in normal physiology, as illustrated clinically with the dystroglycanopathies [39]. Genetic diseases, including Walker–Warburg Syndrome (WWS) and Muscle-Eye-Brain disease (MEB) have been linked to genetic insufficiencies in POMT1, the enzyme responsible for the initial transfer of *O*-mannose to Ser or Thr in the ER [40], and POMGNT1, the enzyme required for the transfer of GlcNAc to the *O*-mannosyl substrate [41], respectively. Although most *O*-mannosyl glycans are Siaα2 $\rightarrow$ 3Galβ1 $\rightarrow$ 4GlcNAcβ1 $\rightarrow$ 2Man, terminally galactosylated *O*-mannosyl glycans have also been detected [42]. Early work on rat brain proteoglycan identified a heterogeneous mix of mannose-linked *O*-glycans, including a single mannitol containing oligosaccharide [43].

6.4
Remodeling Yeast Glycosylation for Therapeutic Protein Production

The use of yeast for the production of clinically relevant proteins is desirable in many circumstances, including the ease of handling and industrial scale-up of processes, the use of chemically defined media, decreased cost and reduced gene-to-product cycle times (reviewed in References [44, 45]). However, when considering yeast for the production of glycosylated proteins, the fungal glycosylation pattern is considered undesirable and in many circumstances precludes consideration of yeast as a host. Thus, for yeast to become a viable alternative to the currently preferred mammalian cell culture lines such as human embryonic kidney (HEK) 293 and Chinese hamster ovary (CHO), significant remodeling of the yeast glycosylation machinery must occur.

The first step in the effort to humanize the *N*-glycosylation pathway in yeast requires the elimination or disruption of *OCH1*. However, deletion of *OCH1* comes at a price. The loss of the outer chain of N-linked oligosaccharides on cell wall glycoproteins affects cell robustness and the ability to grow at 37 °C [46]. Strains with an *och1* phenotype produce glycoproteins with a $Man_8GlcNAc_2$ core oligosaccharide, an essential intermediate for the synthesis of human-like glycosylation in yeast [47]. While the genetic inhibition of *OCH1* is viewed as a critical first step, it is not sufficient. As a glycoprotein makes its way through the secretory pathway of an *och1* host, Golgi-resident mannosyltransferases attach additional mannoses to the Man_8-$GlcNAc_2$ core [47, 48]. Candidate enzymes responsible for this activity have been identified in *S. cerevisiae* and include Mnn1p [15], Mnn2p [49] and members of the KTR family [50–52]. These proteins belong to a large group of Type II membrane proteins and are characterized by an N-terminal transmembrane domain followed by a stem region and end with a C-terminal catalytic domain situated in the lumen of the Golgi apparatus [52, 53]. Additionally, there exist enzyme activities in the Golgi apparatus that contribute to the addition of mannosylphosphate. In *S. cerevisiae,* Mnn4p and Mnn6p have been reported to contribute to mannosylphosphate transfer to oligosaccharides [16, 17, 54]. It is likely that other gene products are involved as *mnn6* mutants still contain some mannosylphosphate [54, 55]. A homolog to MNN4 in *P. pastoris* called *PNO1* incompletely reduces mannosylphosphate content when eliminated [56]. These and other modifications that are non-human in nature thus need to be eliminated through either genetic methods or through enzyme competition. The latter strategy is particularly important as it pertains to α-1,2 mannosyltransferases where substantial redundancy exists in the yeast secretory pathway and genetic control through enzyme deletion is challenging. Therefore, the preferred path has been to stably introduce an α-mannosidase into the secretory pathway. The intended outcome would be trimming of the $Man_8GlcNAc_2$ core as well as any products of the mannosyltransferases, thereby generating the human high mannose glycoform $Man_5GlcNAc_2$.

Several groups have reported the creation of $Man_5GlcNAc_2$ in yeast with varying levels of success. A soluble α-1,2 mannosidase from *Aspergillus saitoi* expressed in *S. cerevisiae* showed the ability to hydrolyze a $Man_9GlcNAc_2$ standard into $Man_5GlcNAc_2$ [57, 58]. Subsequent work by Chiba and coworkers reported a strategy involving the attachment of a HDEL tag to the carboxyl terminus of *A. saitoi* mannosidase [59]. The HDEL tag represents a "retrieval" signal in the yeast secretory pathway that facilitates the retrograde transport of enzymes to the ER [60]. In this manner, the recombinant α-mannosidase-HDEL fusion is targeted to the late ER-early Golgi region of the secretory pathway. While their strategy was successful in that proteins in the medium contained $Man_5GlcNAc_2$, this structure made up only a small percentage of the N-linked oligosaccharide pool and it was unclear whether the mannosidase activity occurred *in vivo* or after the mannosidase enzyme was released from the cell through leakage, secretion or as a result of cell lysis [60].

This strategy has also been pursued in *P. pastoris* using *Trichoderma reesei* α-1,2 mannosidase with an HDEL tag, although in this case *OCH1* was not disrupted [61]. Thus, it is unclear how efficient the recombinant mannosidase was on the core N-linked oligosaccharide structure and, similar to the Chiba work described above, leakage of mannosidase into the medium was observed [61]. Further glycoengineering efforts focused on the introduction of recombinant α-1,2 mannosidase into the secretory pathway of an *och1* null *P. pastoris* host [48], an essential step towards assembly of human N-linked glycan structures. As a comprehensive alternative to the mannosidase-HDEL fusion, Choi and coworkers developed an α-mannosidase fusion library where the transmembrane domains of various fungal type II enzymes were fused in-frame and N-terminal to the catalytic domains of a range of α-mannosidases [48]. By screening hundreds of strains harboring unique combinations of mannosidase-leader fusion proteins, several clones were identified that secreted a reporter glycoprotein with a high degree of $Man_5GlcNAc_2$ content. To determine whether mannosidase activity in the medium was responsible for the highly efficient trimming of $Man_8GlcNAc_2$ to $Man_5GlcNAc_2$, a standard was incubated with media from several strains with highly uniform $Man_5GlcNAc_2$ *N*-glycan content. From these studies, strains with highly efficient $Man_5GlcNAc_2$ production were identified for which there was no measurable mannosidase activity in the media.

The production of $Man_5GlcNAc_2$ is an early step in the human secretory pathway and subsequent engineering steps are necessary for the complete humanization of the yeast glycosylation pathway. However, a glycoprotein with a $Man_5GlcNAc_2$ structure may also be therapeutically relevant. *S. cerevisiae* was engineered to secrete fibroblast growth factor-1 (FGF-1) with a $Man_5GlcNAc_2$ oligosaccharide profile [62]. This material was injected into mice and tissue distribution of the FGF-1 Man_5-$GlcNAc_2$ material was measured and compared with FGF-1 derived from wild-type yeast (high mannose), *och1* yeast ($Man_8GlcNAc_2$) and CHO cells. Both $Man_8GlcNAc_2$ and $Man_5GlcNAc_2$ forms of FGF-1 showed a high concentration in the kidney where the authors suggest that it may be possible to preferentially target glycoproteins to specific tissues based on *N*-glycan composition. Indeed, this means of tissue targeting is already in commercial use; recombinant human glucocerebrosidase is produced in CHO and *in vitro* converted into a pauci-mannosidic ($Man_3GlcNAc_2$) glycoform that selectively targets liver and macrophages via mannose binding receptors [63].

The *in vivo* fate of $Man_5GlcNAc_2$ has also been studied in the context of monoclonal antibodies. Antibodies with $Man_5GlcNAc_2$ glycosylation were purified from an antibody pool produced by mouse NS0 cells and administered to mice [64]. The authors reported no difference in pharmacokinetics between the purified monoclonal antibody containing predominately $Man_5GlcNAc_2$ glycosylation and the control monoclonal antibody (predominately complex fucosylated glycosylation). A separate team used a different approach where a CHO cell line expressing a monoclonal antibody was treated with the α-mannosidase inhibitor kufunensine [65]. This treatment resulted in a monoclonal antibody with glycosylation that consisted primarily of $Man_{8-9}GlcNAc_2$. When this antibody was compared with a monoclonal antibody produced under normal conditions, hence containing complex fucosylated glycans, no difference in pharmacokinetics was observed. Taken together, these studies suggest that, in contrast to glucocerebrosidase, the mannose receptor does not appear to play a significant role in pharmacokinetic properties of antibodies with $Man_{5-9}GlcNAc_2$ glycans. In contrast, chimeric antibodies expressed in Lec 1 CHO cells show a significant reduction in half-life, mainly due to an accelerated alpha phase [66]. Since Lec 1 mutants are deficient in β-1,2 GlcNAc transferase I (GNT I) activity, proteins secreted in this host display a mannose terminated N-linked glycosylation profile. In these studies, IgG clearance via mannose binding receptor was implicated [66, 67]. It is apparent that further studies need to be performed to fully elucidate the utility of IgGs with mannose terminated glycosylation.

With the production of a fungal host capable of synthesizing $Man_5GlcNAc_2$ established, the next step towards humanized *N*-glycosylation requires UDP-GlcNAc and GnT I activity. While initial attempts to introduce GnT I into yeast were successful, obtaining *in vivo* GnT I activity proved to be more elusive [68]. GnT I activity, UDP-GlcNAc and the $Man_5GlcNAc_2$ substrate must be placed spatially and temporally in the secretory pathway for maximal conversion into the mammalian hybrid glycoform $GlcNAcMan_5GlcNAc_2$. First demonstrated by Choi *et al.* [48], targeted expression of GnT I into a *P. pastoris* genetic background engineered to express $Man_5GlcNAc_2$ resulted in the secretion of glycoproteins with an N-linked glycosylation profile containing $GlcNAcMan_5GlcNAc_2$ as the predominant glycoform. The production of a human hybrid glycoform in *P. pastoris* was later confirmed by Vervecken and coworkers [69].

To achieve de novo synthesis of a human complex glycoform in a fungal host, several additional steps remain. These involve the conversion from a human glycan intermediate into the final sialylated complex biantennary form (Figure 6.3). While the creation of this structure in any amount represents a significant breakthrough in glycoengineering, a high degree of glycan uniformity is required for such yeast to serve as a host for the production of therapeutic glycoproteins. Specifically, Hamilton and coworkers expressed α-mannosidase II (MNSII) and GlcNAc transferase II (GnT II) fused to various fungal transmembrane domains in a manner analogous to Choi [48] into a yeast host capable of hybrid-type glycosylation [70]. MNS II specifically recognizes the $GlcNAcMan_5GlcNAc_2$ substrate and hydrolyses the non-reducing α-1,3 and α-1,6 linked mannoses, resulting in the formation of a $GlcNAcMan_3GlcNAc_2$ hybrid structure. To avoid significant glycan heterogeneity during the engineering of the human glycosylation pathway, it is essential that a near total conversion of N-linked glycans into the hybrid form be complete during the time of exposure to MNSII in the secretory pathway. This structure is a specific substrate for GnT II, which catalyzes the transfer of a β-1,2 linked GlcNAc to the α-1,6 linked mannose on the core, thus creating the basic element of the human biantennary complex glycan $GlcNAc_2Man_3GlcNAc_2$. By screening a large number of MNSII and GNTII fusion proteins, the selection of a strain that secreted protein with essentially uniform $GlcNAc_2Man_3GlcNAc_2$ was isolated [70].

With the development of a yeast host capable of biantennary complex glycosylation complete, the final steps for the full elaboration of the human biantennary glycoform involved the introduction of machinery to allow the transfer of galactose followed by terminal capping with sialic acid. The *in vivo* transfer of galactose to a glycoprotein substrate in *P. pastoris* was demonstrated by Vervecken and others by the introduction of a human β-1,4 galactosyltransferase anchored to the secretory pathway with a fungal transmembrane sequence [69]. This resulted in the partial transfer of galactose to a hybrid-type glycan substrate [69]. Bobrowicz and coworkers, working in a *Pichia* host capable of GlcNAc-terminated complex biantennary glycan synthesis, introduced a multifunctional chimeric protein that consists of an epimerase capable of transforming UDP-glucose to UDP-galactose and human β-1,4 galactosyltransferase [71]. This fusion protein was shown to catalyze quantitative transfer of galactose to the GlcNAc2Man3GlcNAc2 substrate.

The use of an organism capable of complex biantennary glycosylation without sialic acid transfer is generally undesirable as a production host for therapeutic proteins due to real concerns regarding increased clearance potentially leading to loss of efficacy. The production of recombinant antibodies provides an exception. Monoclonal antibodies have emerged as one of the most important classes of therapeutic proteins with respect to both their clinical efficacy and market potential [72]. As such, the effector function capability of an antibody requires not only high-affinity antigen binding but the presence and composition of glycosylation at Asn297. Specifically, antigen-dependent cellular cytotoxicity (ADCC) is an important effector function for oncology directed antibodies [73]. Furthermore, the presence of a core α-1,6 linked fucose attached to the GlcNAc proximal to Asn modulates ADCC, with absence of fucose conferring enhanced ADCC [74]. A central mechanism

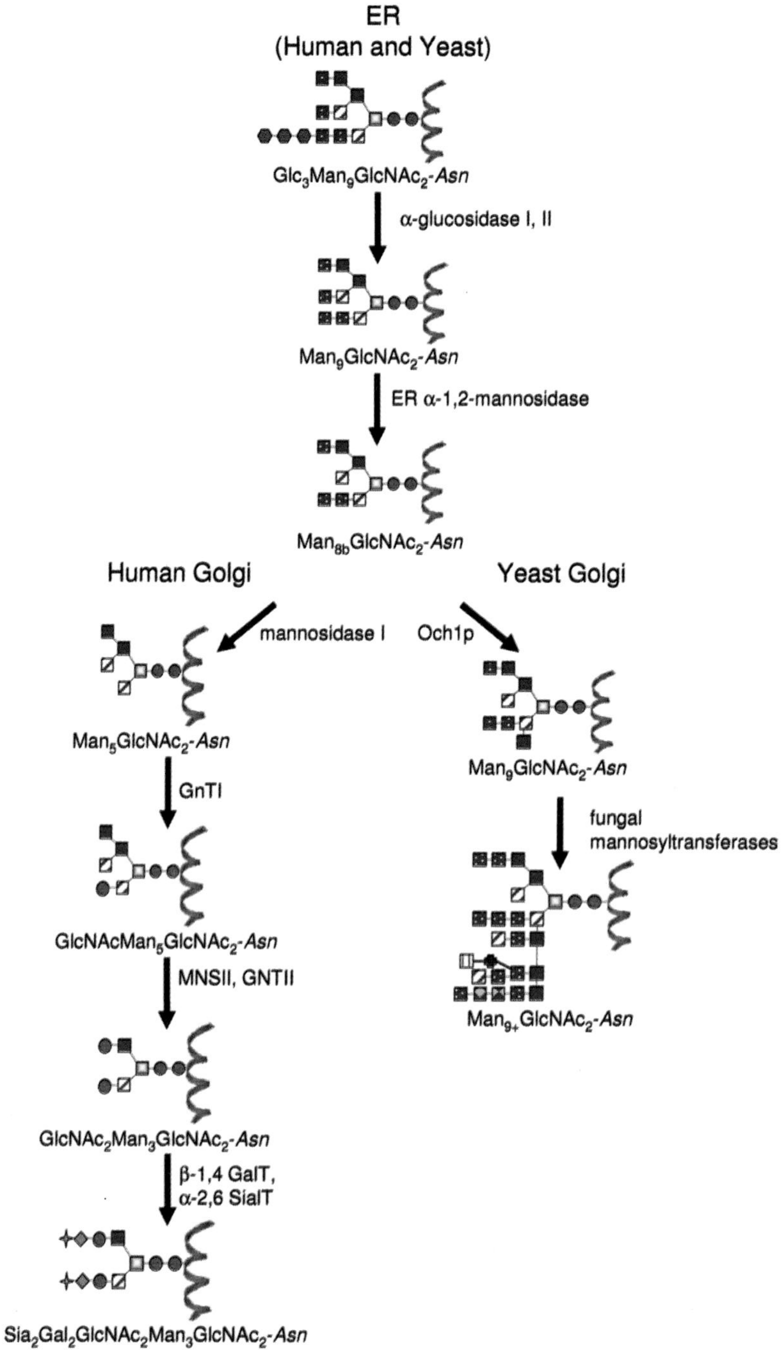
ER
(Human and Yeast)
Glc3Man9GlcNAc2-Asn
α-glucosidase I, II
Man9GlcNAc2-Asn
ER α-1,2-mannosidase
Man8bGlcNAc2-Asn
Human Golgi
Yeast Golgi
mannosidase I
Och1p
Man5GlcNAc2-Asn
GnTI
GlcNAcMan5GlcNAc2-Asn
MNSII, GNTII
GlcNAc2Man3GlcNAc2-Asn
β-1,4 GalT,
α-2,6 SialT
Sia2Gal2GlcNAc2Man3GlcNAc2-Asn
Man9GlcNAc2-Asn
fungal
mannosyltransferases
Man9+GlcNAc2-Asn

for this is presumed to be mediated through FcγRIIIa found on natural killer cells. Fungi neither have the genetic blueprint to express the machinery necessary for the production of GDP-fucose nor do they have the fucosyl transferase to attach fucose to the oligosaccharide. As such, the synthesis of complex biantennary glycoforms in a host without fucose transfer is an attractive host for the design and production of effector function dependent antibodies. Li and coworkers expressed an anti-CD20 monoclonal antibody in a *P. pastoris* strain engineered to secrete antibody with afucosylated complex biantennary glycosylation and found that it possessed greater FcγRIIIa receptor binding and caused a higher degree of B-cell depletion than a fucosylated counterpart produced in CHO [75].

The final step in the construction of the human biantennary complex glycoform in yeast involves the successful anchoring of a sialyltransferase in the secretory pathway such that the $Gal_2GlcNAc_2Man_3GlcNAc_2$ substrate is capped with sialic acid. Adding to the complexity of this final step is the fact that yeast such as *P. pastoris* do not synthesize the necessary CMP-sialic acid substrate. To address this, Hamilton and coworkers first introduced the necessary enzymes for the *in vivo* synthesis of CMP-sialic acid, its transport into the Golgi compartment, and finally the transfer of sialic acid to a galactose acceptor in a single transformation step [76]. The result of these studies yielded the secretion of a rat recombinant erythropoietin (EPO) with a high degree of terminal sialylation. When administered in a murine hematocrit model, EPO with terminal sialylation showed a superior increase in hematocrit than the same EPO sequence expressed in a wild-type *P. pastoris* host.

By engineering yeast to perform N-linked glycosylation with a composition similar to humans, therapeutic glycoproteins can now be considered for production. The global market for protein based biopharmaceuticals is growing at a remarkable rate, with experts predicting annual growth between 10% and 30%. In addition, the approval rate for biopharmaceuticals is predicted to grow at 16–30%, compared with about a 4% growth of traditional small-molecule pharmaceuticals [77]. Of particular importance in this fastest growing sector of the pharmaceutical industry are production of monoclonal antibodies and Fc-fusion proteins. Currently, mammalian cell culture is the production system of choice for the manufacture of therapeutic glycoproteins with bacteria and yeast being the major hosts for proteins where glycosylation is not required.

Yeast have generally not been used for the production of therapeutic glycoproteins. This is because yeast secrete proteins with a characteristic high-mannose type glycosylation that confers a short *in vivo* half-life to the protein and may render it less efficacious. However, yeast and fungal protein expression systems have been useful for production of industrially relevant enzymes, production of proteins that

Figure 6.3 Comparison of N-linked glycosylation between mammals and yeast. N-linked glycosylation is to a large degree conserved in the endoplasmic reticulum of yeast and mammalian systems. Divergence occurs in the Golgi, where mannosidases in mammals trim the $Man_8GlcNAc_2$ core to a $Man_5GlcNAc_2$ base structure followed by subsequent addition of non-mannose sugars and further trimming of mannose residues. In contrast, yeast build on the $Man_8GlcNAc$ core with additional mannoses and, depending on the yeast species, non-mannose sugars are used for capping. *Pichia pastoris* is shown for illustrative purposes.

cannot be actively expressed in *E. coli* and production of proteins that do not require glycosylation for proper folding and biological activity.

The ability to engineer yeast with the capability of producing glycoproteins with human glycosylation creates the opportunity to systematically investigate glycosylation-dependent structure–activity relationships and thus create specific glycoproteins with improved therapeutic properties.

Acknowledgments

The authors would like to thank Karen Page for valuable editorial assistance in the preparation of the manuscript.

References

1 Le, H.V. and Trotta, P.P. (1991) Purification of secreted recombinant proteins from Escherichia coli. *Bioprocess Technology*, **12**, 163–181.

2 Swartz, J.R. (2001) Advances in Escherichia coli production of therapeutic proteins. *Current Opinion in Biotechnology*, **12**, 195–201.

3 Burda, P. and Aebi, M. (1999) The dolichol pathway of N-linked glycosylation. *Biochimica et Biophysica Acta*, **1426**, 239–257.

4 Knauer, R. and Lehle, L. (1999) The oligosaccharyltransferase complex from yeast. *Biochimica et Biophysica Acta*, **1426**, 259–273.

5 Marshall, R.D. (1972) Glycoproteins. *Annual Review of Biochemistry*, **41**, 673–702.

6 Herscovics, A. (1999) Processing glycosidases of Saccharomyces cerevisiae. *Biochimica et Biophysica Acta*, **1426**, 275–285.

7 Kornfeld, R. and Kornfeld, S. (1985) Assembly of asparagine-linked oligosaccharides. *Annual Review of Biochemistry*, **54**, 631–664.

8 Baenziger, J.U. and Green, E.D. (1988) Pituitary glycoprotein hormone oligosaccharides: structure, synthesis and function of the asparagine-linked oligosaccharides on lutropin, follitropin and thyrotropin. *Biochimica et Biophysica Acta*, **947**, 287–306.

9 Dean, N. (1999) Asparagine-linked glycosylation in the yeast Golgi. *Biochimica et Biophysica Acta*, **1426**, 309–322.

10 Romero, P.A. and Herscovics, A. (1989) Glycoprotein biosynthesis in Saccharomyces cerevisiae. Characterization of alpha-1,6-mannosyltransferase which initiates outer chain formation. *The Journal of Biological Chemistry*, **264**, 1946–1950.

11 Nakayama, K., Nagasu, T., Shimma, Y., Kuromitsu, J. and Jigami, Y. (1992) OCH1 encodes a novel membrane bound mannosyltransferase: outer chain elongation of asparagine-linked oligosaccharides. *The EMBO Journal*, **11**, 2511–2519.

12 Jungmann, J., Rayner, J.C. and Munro, S. (1999) The Saccharomyces cerevisiae protein Mnn10p/Bed1p is a subunit of a Golgi mannosyltransferase complex. *The Journal of Biological Chemistry*, **274**, 6579–6585.

13 Rayner, J.C. and Munro, S. (1998) Identification of the MNN2 and MNN5 mannosyltransferases required for forming and extending the mannose branches of the outer chain mannans of Saccharomyces cerevisiae.

The Journal of Biological Chemistry, **273**, 26836–26843.

14 Hazen, K.C., Singleton, D.R. and Masuoka, J. (2007) Influence of outer region mannosylphosphorylation on *N*-glycan formation by Candida albicans: normal acid-stable N-glycan formation requires acid-labile mannosylphosphate addition. *Glycobiology*, **17**, 1052–1060.

15 Yip, C.L. *et al.* (1994) Cloning and analysis of the Saccharomyces cerevisiae MNN9 and MNN1 genes required for complex glycosylation of secreted proteins. *Proceedings of the National Academy of Sciences of the United States of America*, **91**, 2723–2727.

16 Odani, T., Shimma, Y., Wang, X.H. and Jigami, Y. (1997) Mannosylphosphate transfer to cell wall mannan is regulated by the transcriptional level of the MNN4 gene in Saccharomyces cerevisiae. *FEBS Letters*, **420**, 186–190.

17 Odani, T., Shimma, Y., Tanaka, A. and Jigami, Y. (1996) Cloning and analysis of the MNN4 gene required for phosphorylation of N-linked oligosaccharides in Saccharomyces cerevisiae. *Glycobiology*, **6**, 805–810.

18 Grinna, L.S. and Tschopp, J.F. (1989) Size distribution and general structural features of N-linked oligosaccharides from the methylotrophic yeast, Pichia pastoris. *Yeast (Chichester, England)*, **5**, 107–115.

19 Kobayashi, H., Shibata, N. and Suzuki, S. (1986) Acetolysis of Pichia pastoris IFO 0948 strain mannan containing alpha-1,2 and beta-1,2 linkages using acetolysis medium of low sulfuric acid concentration. *Archives of Biochemistry and Biophysics*, **245**, 494–503.

20 Miele, R.G., Castellino, F.J. and Bretthauer, R.K. (1997) Characterization of the acidic oligosaccharides assembled on the Pichia pastoris-expressed recombinant kringle 2 domain of human tissue-type plasminogen activator. *Biotechnology and Applied Biochemistry*, **26** (Pt 2), 79–83.

21 Raschke, W.C. and Ballou, C.E. (1972) Characterization of a yeast mannan containing *N*-acetyl-D-glucosamine as an immunochemical determinant. *Biochemistry*, **11**, 3807–3816.

22 Moreno, S., Ruiz, T., Sanchez, Y., Villanueva, J.R. and Rodriguez, L. (1985) Subcellular localization and glycoprotein nature of the invertase from the fission yeast Schizosaccharomyces pombe. *Archives of Microbiology*, **142**, 370–374.

23 Chappell, T.G. and Warren, G. (1989) A galactosyltransferase from the fission yeast Schizosaccharomyces pombe. *The Journal of Cell Biology*, **109**, 2693–2702.

24 Takayanagi, T., Kushida, K., Idonuma, K. and Ajisaka, K. (1992) Novel N-linked oligo-mannose type oligosaccharides containing an alpha-D-galactofuranosyl linkage found in alpha-D-galactosidase from Aspergillus niger. *Glycoconjugate Journal*, **9**, 229–234.

25 Wallis, G.L., Easton, R.L., Jolly, K., Hemming, F.W. and Peberdy, J.F. (2001) Galactofuranoic-oligomannose N-linked glycans of alpha-galactosidase A from Aspergillus niger. *European Journal of Biochemistry/FEBS*, **268**, 4134–4143.

26 Goto, M. (2007) Protein *O*-glycosylation in fungi: diverse structures and multiple functions. *Bioscience, Biotechnology, and Biochemistry*, **71**, 1415–1427.

27 Haselbeck, A. and Tanner, W. (1983) *O*-Glycosylation in Saccharomyces cerevisiae is initiated at the endoplasmic reticulum. *FEBS Letters*, **158**, 335–338.

28 Willer, T., Valero, M.C., Tanner, W., Cruces, J. and Strahl, S. (2003) *O*-Mannosyl glycans: from yeast to novel associations with human disease. *Current Opinion in Structural Biology*, **13**, 621–630.

29 Nakayama, K., Feng, Y., Tanaka, A. and Jigami, Y. (1998) The involvement of mnn4 and mnn6 mutations in mannosylphosphorylation of O-linked oligosaccharide in yeast Saccharomyces cerevisiae. *Biochimica et Biophysica Acta*, **1425**, 255–262.

30 Trimble, R.B. *et al.* (2004) Characterization of N- and O-linked glycosylation of recombinant human bile salt-stimulated lipase secreted by Pichia pastoris. *Glycobiology*, **14**, 265–274.

31 Romero, P.A. *et al.* (1999) Mnt2p and Mnt3p of Saccharomyces cerevisiae are members of the Mnn1p family of alpha-1,3-mannosyltransferases responsible for adding the terminal mannose residues of O-linked oligosaccharides. *Glycobiology*, **9**, 1045–1051.

32 Hayette, M.P. *et al.* (1992) Presence of human antibodies reacting with Candida albicans O-linked oligomannosides revealed by using an enzyme-linked immunosorbent assay and neoglycolipids. *Journal of Clinical Microbiology*, **30**, 411–417.

33 Gemmill, T.R. and Trimble, R.B. (1999) Schizosaccharomyces pombe produces novel Gal0-2Man1-3 O-linked oligosaccharides. *Glycobiology*, **9**, 507–515.

34 Ernst, J.F., Mermod, J.J. and Richman, L.H. (1992) Site-specific *O*-glycosylation of human granulocyte/macrophage colony-stimulating factor secreted by yeast and animal cells. *European Journal of Biochemistry/FEBS*, **203**, 663–667.

35 Muller, S. and Hanisch, F.G. (2002) Recombinant MUC1 probe authentically reflects cell-specific *O*-glycosylation profiles of endogenous breast cancer mucin High density and prevalent core 2-based glycosylation. *The Journal of Biological Chemistry*, **277**, 26103–26112.

36 Sewell, R. *et al.* (2006) The ST6GalNAc-I sialyltransferase localizes throughout the Golgi and is responsible for the synthesis of the tumor-associated sialyl-Tn *O*-glycan in human breast cancer. *The Journal of Biological Chemistry*, **281**, 3586–3594.

37 Endo, T. (1999) *O*-Mannosyl glycans in mammals. *Biochimica et Biophysica Acta*, **1473**, 237–246.

38 Barresi, R. and Campbell, K.P. (2006) Dystroglycan: from biosynthesis to pathogenesis of human disease. *Journal of Cell Science*, **119**, 199–207.

39 Muntoni, F., Brockington, M., Blake, D.J., Torelli, S. and Brown, S.C. (2002) Defective glycosylation in muscular dystrophy. *Lancet*, **360**, 1419–1421.

40 Beltran-Valero de Bernabe, D. *et al.* (2002) Mutations in the *O*-mannosyltransferase gene POMT1 give rise to the severe neuronal migration disorder Walker-Warburg syndrome. *American Journal of Human Genetics*, **71**, 1033–1043.

41 Yoshida, A. *et al.* (2001) Muscular dystrophy and neuronal migration disorder caused by mutations in a glycosyltransferase. *POMGnT1. Developmental Cell*, **1**, 717–724.

42 Smalheiser, N.R., Haslam, S.M., Sutton-Smith, M., Morris, H.R. and Dell, A. (1998) Structural analysis of sequences O-linked to mannose reveals a novel Lewis X structure in cranin (dystroglycan) purified from sheep brain. *The Journal of Biological Chemistry*, **273**, 23698–23703.

43 Finne, J., Krusius, T., Margolis, R.K. and Margolis, R.U. (1979) Novel mannitol-containing oligosaccharides obtained by mild alkaline borohydride treatment of a chondroitin sulfate proteoglycan from brain. *The Journal of Biological Chemistry*, **254**, 10295–10300.

44 Wildt, S. and Gerngross, T.U. (2005) The humanization of *N*-glycosylation pathways in yeast. *Nature Reviews Microbiology*, **3**, 119–128.

45 Gerngross, T.U. (2004) Advances in the production of human therapeutic proteins in yeasts and filamentous fungi. *Nature Biotechnology*, **22**, 1409–1414.

46 Nagasu, T. *et al.* (1992) Isolation of new temperature-sensitive mutants of Saccharomyces cerevisiae deficient in mannose outer chain elongation. *Yeast (Chichester, England)*, **8**, 535–547.

47 Nakanishi-Shindo, Y., Nakayama, K., Tanaka, A., Toda, Y. and Jigami, Y. (1993) Structure of the N-linked oligosaccharides that show the complete loss of alpha-1,6-polymannose outer chain from och1, och1 mnn1, and och1 mnn1 alg3 mutants of Saccharomyces cerevisiae. *The Journal of Biological Chemistry*, **268**, 26338–26345.

48 Choi, B.K. *et al.* (2003) Use of combinatorial genetic libraries to humanize N-linked glycosylation in the yeast Pichia pastoris. *Proceedings of the National Academy of Sciences of the United States of America*, **100**, 5022–5027.

49 Devlin, C. and Ballou, C.E. (1990) Identification and characterization of a gene and protein required for glycosylation in the yeast Golgi. *Molecular Microbiology*, **4**, 1993–2001.

50 Lussier, M. *et al.* (1997) Completion of the Saccharomyces cerevisiae genome sequence allows identification of KTR5, KTR6 and KTR7 and definition of the nine-membered KRE2/MNT1 mannosyltransferase gene family in this organism. *Yeast (Chichester, England)*, **13**, 267–274.

51 Lussier, M., Sdicu, A.M., Camirand, A. and Bussey, H. (1996) Functional characterization of the YUR1, KTR1, and KTR2 genes as members of the yeast KRE2/MNT1 mannosyltransferase gene family. *The Journal of Biological Chemistry*, **271**, 11001–11008.

52 Lussier, M., Sdicu, A.M., and Bussey, H. (1999) The KTR and MNN1 mannosyltransferase families of Saccharomyces cerevisiae. *Biochimica et Biophysica Acta*, **1426**, 323–334.

53 Paulson, J.C. and Colley, K.J. (1989) Glycosyltransferases. Structure, localization, and control of cell type-specific glycosylation. *The Journal of Biological Chemistry*, **264**, 17615–17618.

54 Jigami, Y. and Odani, T. (1999) Mannosylphosphate transfer to yeast mannan. *Biochimica et Biophysica Acta*, **1426**, 335–345.

55 Wang, X.H., Nakayama, K., Shimma, Y., Tanaka, A. and Jigami, Y. (1997) MNN6 a member of the KRE2/MNT1 family, is the gene for mannosylphosphate transfer in Saccharomyces cerevisiae. *The Journal of Biological Chemistry*, **272**, 18117–18124.

56 Miura, M., Hirose, M., Miwa, T., Kuwae, S. and Ohi, H. (2004) Cloning and characterization in Pichia pastoris of PNO1 gene required for phosphomannosylation of N-linked oligosaccharides. *Gene*, **324**, 129–137.

57 Fujita, A., Yoshida, T. and Ichishima, E. (1997) Five crucial carboxyl residues of 1,2-alpha-mannosidase from Aspergillus saitoi (A. phoenicis), a food microorganism, are identified by site-directed mutagenesis. *Biochemical and Biophysical Research Communications*, **238**, 779–783.

58 Inoue, T., Yoshida, T. and Ichishima, E. (1995) Molecular cloning and nucleotide sequence of the 1,2-alpha-D-mannosidase gene, msdS, from Aspergillus saitoi and expression of the gene in yeast cells. *Biochimica et Biophysica Acta*, **1253**, 141–145.

59 Chiba, Y. *et al.* (1998) Production of human compatible high mannose-type (Man5GlcNAc2) sugar chains in Saccharomyces cerevisiae. *The Journal of Biological Chemistry*, **273**, 26298–26304.

60 Pelham, H.R. (1988) Evidence that luminal ER proteins are sorted from secreted proteins in a post-ER compartment. *The EMBO Journal*, **7**, 913–918.

61 Callewaert, N. *et al.* (2001) Use of HDEL-tagged Trichoderma reesei mannosyl oligosaccharide 1,2-alpha-D-mannosidase for *N*-glycan engineering in Pichia pastoris. *FEBS Letters*, **503**, 173–178.

62 Takamatsu, S. *et al.* (2004) Monitoring of the tissue distribution of fibroblast growth factor containing a high mannose-type sugar chain produced in mutant yeast. *Glycoconjugate Journal*, **20**, 385–397.

63 Mistry, P.K., Wraight, E.P. and Cox, T.M. (1996) Therapeutic delivery of proteins to macrophages: implications for treatment of Gaucher's disease. *Lancet*, **348**, 1555–1559.

64 Millward, T.A. *et al.* (2008) Effect of constant and variable domain glycosylation on pharmacokinetics of therapeutic antibodies in mice. *Biologicals*, **36**, 41–47.

65 Zhou, Q. *et al.* (2008) Development of a simple and rapid method for producing non-fucosylated oligomannose containing antibodies with increased effector function. *Biotechnology and Bioengineering*, **99**, 652–665.

66 Wright, A. and Morrison, S.L. (1994) Effect of altered CH2-associated carbohydrate structure on the functional properties and in vivo fate of chimeric mouse-human immunoglobulin G1. *The Journal of Experimental Medicine*, **180**, 1087–1096.

67 Wright, A. and Morrison, S.L. (1998) Effect of C2-associated carbohydrate structure on Ig effector function: studies with chimeric mouse-human IgG1 antibodies in glycosylation mutants of Chinese hamster ovary cells. *Journal of Immunology*, **160**, 3393–3402.

68 Kalsner, I., Hintz, W., Reid, L.S. and Schachter, H. (1995) Insertion into Aspergillus nidulans of functional UDP-GlcNAc: alpha 3-D-mannoside beta-1,2-*N*-acetylglucosaminyl-transferase I, the enzyme catalysing the first committed step from oligomannose to hybrid and complex *N*-glycans. *Glycoconjugate Journal*, **12**, 360–370.

69 Vervecken, W. *et al.* (2004) In vivo synthesis of mammalian-like, hybrid-type *N*-glycans in Pichia pastoris. *Applied and Environmental Microbiology*, **70**, 2639–2646.

70 Hamilton, S.R. *et al.* (2003) Production of complex human glycoproteins in yeast. *Science*, **301**, 1244–1246.

71 Bobrowicz, P. *et al.* (2004) Engineering of an artificial glycosylation pathway blocked in core oligosaccharide assembly in the yeast Pichia pastoris: production of complex humanized glycoproteins with terminal galactose. *Glycobiology*, **14**, 757–766.

72 Reichert, J.M. and Valge-Archer, V.E. (2007) Development trends for monoclonal antibody cancer therapeutics. *Nature Reviews*, **6**, 349–356.

73 Iannello, A. and Ahmad, A. (2005) Role of antibody-dependent cell-mediated cytotoxicity in the efficacy of therapeutic anti-cancer monoclonal antibodies. *Cancer Metastasis Reviews*, **24**, 487–499.

74 Rothman, R.J., Perussia, B., Herlyn, D. and Warren, L. (1989) Antibody-dependent cytotoxicity mediated by natural killer cells is enhanced by castanospermine-induced alterations of IgG glycosylation. *Molecular Immunology*, **26**, 1113–1123.

75 Li, H. *et al.* (2006) Optimization of humanized IgGs in glycoengineered Pichia pastoris. *Nature Biotechnology*, **24**, 210–215.

76 Hamilton, S.R. *et al.* (2006) Humanization of yeast to produce complex terminally sialylated glycoproteins. *Science*, **313**, 1441–1443.

77 Nagle, P.C., Nicita, C.A., Gerdes, L.A. and Schmeichel, C.J. (2008) Characteristics of and trends in the late-stage biopharmaceutical pipeline. *The American Journal of Managed Care*, **14**, 226–229.

7
Insect Cell Glycosylation Patterns in the Context of Biopharmaceuticals

Christoph Geisler and Don Jarvis

7.1
Introduction

The baculovirus expression vector system (BEVS) was originally described in 1983 [1]. This binary expression system consists of a lepidopteran insect cell or insect host and a recombinant baculoviral vector encoding one or more proteins of interest (for reviews see [2–7]).

The insect cell lines used in the BEVS are derived from pupae, larvae or eggs of a wide variety of lepidopteran insect species. The most commonly used cell lines are derived from *Trichoplusia ni* (BTI-TN-5B1-4, commercially available as High Five [8]) and *Spodoptera frugiperda* (IPLB-Sf21AE [9] and Sf9 [6]). These immortalized cell lines can be grown as adherent cultures in T-flasks or as suspension cultures in Fernbach flasks, spinner flasks, or in airlift, stirred-tank or WAVE bioreactors.

Baculovirus vectors are recombinant viruses that can productively infect insect, but not mammalian cells. Recombinant baculoviruses can be produced using several different approaches involving either homologous recombination [10, 11] or DNA transposition [12]. Genes encoding the protein of interest in baculovirus expression vectors are typically placed under the control of strong transcriptional promoters derived from the very late baculoviral *polyhedrin* or *p10* genes. Alternatively, they can be placed under the control of weaker promoters derived from baculoviral late (e.g., *p6.9*) [13] or immediate early (e.g., *ie1*) promoters [14].

Recombinant baculovirus vectors are used to infect insects or insect cells, thus inducing production of the recombinant protein as a by-product of the infection process. Recombinant protein production is transient because baculovirus infections are cytolytic. Nonetheless, the BEVS can produce recombinant proteins at very high levels due to amplification of the vector and the extraordinary strength of the very late gene promoters typically used to drive foreign gene expression. Furthermore, the BEVS is a eukaryotic system that can produce soluble, biologically active recombinant proteins with co- and post-translational modifications, including glycosylation.

Post-translational Modification of Protein Biopharmaceuticals. Edited by Gary Walsh

ISBN: 978-3-527-32074-5

Hence, the ability to provide high expression levels and eukaryotic protein processing are two major advantages of the BEVS. Other advantages of this system include biosafety and the ability to isolate recombinant baculoviruses much more quickly than transformed cell lines, thus hastening product development. These features have contributed to the widespread use of the BEVS for recombinant protein production in academic and industrial laboratories all over the world.

An important issue concerning the use of the BEVS for the production of recombinant therapeutic glycoproteins is that insect protein glycosylation patterns differ from those of mammals, and it is increasingly clear that this difference can have an impact on biomedically relevant glycoproteins. For example, the absence of terminal sialic acids on recombinant glycoproteins produced in the BEVS negatively impacts their pharmacokinetic behavior and constrains their therapeutic efficacy [15]. Similarly, the presence of core α1,3-linked fucose residues on recombinant glycoproteins produced in the BEVS contributes to their allergenicity and reduces their safety. In contrast, there is growing evidence to suggest that the presence of mannose-terminated glycans on recombinant glycoprotein vaccines produced in the BEVS can positively impact their efficacy by mediating more efficient interactions with dendritic cells [16]. In this chapter we discuss insect cell protein glycosylation, its impact on the use of the BEVS for the production of biopharmaceuticals, and efforts to overcome the negative impacts in further detail, with a focus on protein *N*-glycosylation.

7.2
Recombinant *N*-Glycoproteins in the BEVS Product Pipeline

To date, there is no BEVS-derived *N*-glycoprotein product licensed for human clinical use in the USA. However, several products have been examined extensively in phase I, II and III human clinical trials. Interestingly, each of these products was a vaccine consiststing of one or more recombinant glycoproteins and each obviously had insect cell *N*-glycosylation patterns that could have a negative or positive impact on their efficacy. Table 7.1 lists these BEVS-produced recombinant *N*-glycoproteins, and their principal features are described in the following sub-sections.

Table 7.1 Selected BEVS-produced glycoproteins currently in human clinical trials.

Product name	Company	Product type
Chimigen™	ViRexx	Chimeric vaccines
FluBlok™	Protein Sciences	Influenza subunit vaccine
Influenza virus-like particles	Novavax	Influenza VLP vaccine
Provenge®	Dendreon	Prostate cancer immunotherapy
Specifid™ (formerly FavId™)	Favrille	Lymphoma immunotherapy

7.2.1 Chimigen™ Vaccines

Chimigen™ technology involves the use of the BEVS to produce subunit vaccines consisting of chimeric proteins in which an antigen derived from a specific pathogen is fused to a murine IgG Fc fragment. One example is a chimeric recombinant protein in which the antigen is the E2 glycoprotein of Western Equine Encephalitis virus. The purpose in fusing specific antigens to the Fc fragment is to overcome immune tolerance to pathogens and to enhance the immune response in humans by targeting these antigens to dendritic cells.

7.2.2 FluBlok™

FluBlok™ is an influenza viral subunit vaccine consisting of three recombinant influenza hemagglutinins, which is currently in phase III human clinical trials. The sequence of the influenza hemagglutinins changes from year to year, and the three proteins in FluBlok™ are derived from three different influenza strains circulating in humans. All three hemagglutinins are produced as recombinant, membrane bound *N*-glycoprotein precursors in a commercial cell line known as expresSF+ [17].

7.2.3 Influenza Virus-Like Particles

Novavax technology involves the use of the BEVS to produce virus-like particles composed of various recombinant proteins. One prominent example is an influenza virus-like particle vaccine candidate consisting of three influenza units, including the two envelope *N*-glycoproteins, hemagglutinin and neuraminidase, and the M1 matrix protein. A single recombinant baculovirus vector encodes all three proteins. Thus, infection of insect cells with this virus leads to the production of these proteins and the assembly of influenza virus-like particles, which are secreted into the culture medium [18, 19].

7.2.4 Provenge®

Provenge® is a prostate cancer vaccine candidate currently undergoing phase III human clinical trials. It is a fusion protein consisting of the full-length human prostatic acid phosphatase, which is a prostate-specific cancer antigen, and the full length human granulocyte-macrophage colony-stimulating factor, which targets the fusion protein to dendritic cells [20]. After the BEVS is used to produce Provenge®, the product is purified and used to stimulate peripheral blood mononuclear cells, which include dendritic cells, isolated from an individual patient's blood. These activated cells are then returned to the same patient, in which they can attack prostate cancer cells [21, 22].

7.2.5 Specifid™

Specifid™ was a patient-specific immunotherapeutic directed against non-Hodgkin's B-cell lymphomas [23]. B-cell lymphomas are clonal B-cell populations that express a unique, tumor-specific immunoglobulin variable region antigen known as an idiotypic determinant. Specifid™ consisted of a patient-specific, recombinant IgG protein that includes that patient's B-cell lymphoma idiotypic determinant. Recombinant baculovirus expression vectors encoding each individual, patient-specific form of Specifid™ were produced using the portion of the gene encoding the idiotypic determinant from an individual patient's B-cell lymphoma. Following expression in the BEVS, the secreted full-length IgG protein was purified and chemically conjugated to an immunogenic carrier protein. The conjugated recombinant IgG was then used to immunize the patient against their own lymphoma [24]. Unfortunately, Specifid™ is no longer in the BEVS product pipeline because a recent, large phase III clinical trial failed to demonstrate its efficacy.

7.3 Insect Glycoprotein *N*-Glycan Structure

The BEVS produces recombinant forms of higher eukaryotic *N*-glycoproteins that are highly similar to the native, mammalian proteins with respect to folding, post-translational processing and modifications, except for the structures of their *N*-glycans (Figure 7.1). Both mammalian and insect cell-derived glycoproteins can have oligomannose-type *N*-glycans in cases in which those *N*-glycans remain in this form and are not subjected to further processing [25]. However, in those cases in which oligomannose *N*-glycans are transient intermediates that are further processed, mammalian and insect cells produce different *N*-glycan structures. While mammalian cells can produce multiply branched, terminally sialylated, complex *N*-glycans (Chapter 2), insect cells typically produce much simpler mono- or biantennary, paucimannose or hybrid structures lacking sialic acid. In addition, insect cell *N*-glycans can include allergenic epitopes in the form of fucose residues α1,3-linked to the *N*-glycan core.

7.3.1 Typical *N*-Glycan Structures

Plasma membrane glycoproteins include the full range of *N*-glycan structures produced by processing reactions in the secretory pathway. As such, the structures of these glycans reflect the *N*-glycan processing capacity of a given cell type. Studies on the structures of lepidopteran insect cell plasma membrane *N*-glycans revealed a range of oligomannosidic *N*-glycans with or without core fucose, with a paucity of terminal *N*-acetylglucosamine and no detectable galactose [26]. Similar *N*-glycan structures have been observed on recombinant glycoproteins produced by the BEVS.

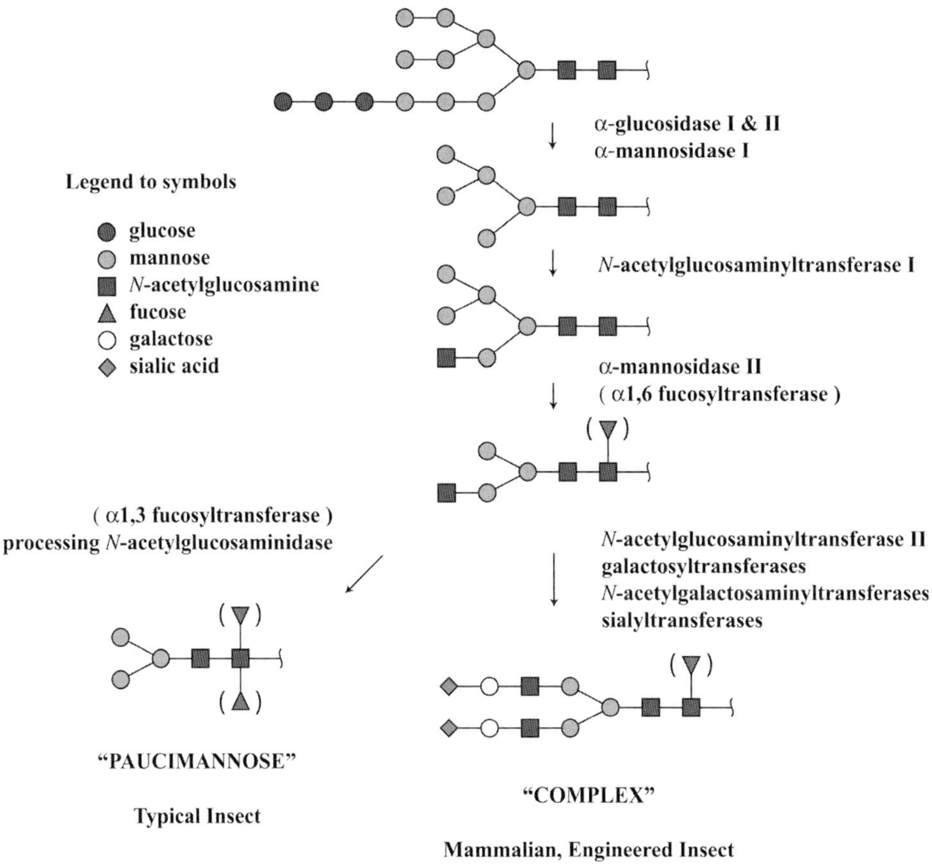

Figure 7.1 Protein *N*-glycosylation pathways in insect and mammalian cells. Monosaccharides are indicated by their standard symbolic representations, as defined in the key on the left-hand side of the figure. The insect and mammalian *N*-glycosylation pathways each begin with the transfer of a preassembled glycan to a nascent protein. This is followed by a series of common processing reactions that effectively trim the precursor and produce a common intermediate. Insect cells have a processing enzyme that further trims this intermediate and produces paucimannosidic *N*-glycans. In contrast, mammalian cells have processing enzymes that elongate the intermediate to produce complex *N*-glycans. Transgenic insect cells transformed with genes encoding mammalian elongation functions also can produce complex, "mammalianized" *N*-glycans. The core fucose residues are in parentheses because they are found on some, but not all, *N*-glycans.

Recombinant interferon-γ secreted from baculovirus-infected *Sf*9 cells was found to have only paucimannose *N*-glycans, in contrast to the complex, sialylated *N*-glycans found on native interferon-γ secreted by CHO cells [27, 28]. Likewise, recombinant human transferrin produced by baculovirus-infected *T. ni* cells had mostly paucimannose *N*-glycans, but also a minority of *N*-acetylglucosamine terminated hybrid *N*-glycans [29]. A similar result was obtained with transferrin from *Lymantria dispar* cells [30]. Human interferon-β secreted by baculovirus-infected

Bombyx mori larvae also had mostly paucimannose structures with or without core fucose residues [31]. In a direct comparison of secreted alkaline phosphatase produced by four different species of insect larvae and cell lines, *N*-glycosylation patterns were found to be very similar, consisting of almost exclusively oligo- and paucimannose structures [32]. Thus, both insect larvae and cell lines produce recombinant *N*-glycoproteins with mostly paucimannose *N*-glycans, minor amounts of hybrid structures and a range of oligomannose glycans. The latter arise at least in part due to lysis of infected cells, which would release intracellular forms of the proteins expected to have only partially processed *N*-glycan intermediates.

7.3.2
Hybrid/Complex *N*-Glycans

Several reports have detailed the presence of galactose-terminated *N*-glycans on glycoproteins produced in the BEVS. Analysis of IgG secreted by *T. ni* cells showed that about 65% of its *N*-glycans had terminal *N*-acetylglucosamine residues and about 20% were terminally galactosylated [33]. Other reports also revealed that recombinant glycoproteins derived from *T. ni* had terminally galactosylated *N*-glycans [34, 35]. Similarly, interferon-γ produced by the cell line Ea4 (derived from *Estigmene acrea*) had about 20% *N*-glycans with terminal galactose residues [36]. Thus, some insect cell lines, notably High Five and Ea4, appear to have a low level endogenous capacity to produce terminally galactosylated *N*-glycans.

7.3.3
Sialylated *N*-Glycans

The vast majority of searches for sialic acid in insect cells and BEVS-produced recombinant *N*-glycoproteins have failed to reveal any evidence for the presence of sialic acid or sialylated glycoconjugates (reviewed in Reference [37]). However, there are reports of the synthesis of complex-type, sialylated *N*-glycans on recombinant human plasminogen [38–40] and secreted alkaline phosphatase [41–43] produced in non-engineered insect cells.

7.3.4
Summary

The BEVS typically yields recombinant glycoproteins with either oligomannosidic or paucimannosidic *N*-glycans. Reports of BEVS-derived glycoproteins carrying hybrid or complex *N*-glycans are relatively limited and indicate that these structures make up only a minor fraction of the total *N*-glycans on these proteins. Finally, there are even fewer reports of sialylated recombinant *N*-glycoproteins produced in the BEVS. Thus, for all practical purposes, investigators using the BEVS for recombinant glycoprotein production should not expect to obtain products with complex-type, sialylated *N*-glycans.

The production of human thyrotropin in the BEVS provides an example of the pharmacokinetic problems resulting from the inability of insect cells to sialylate recombinant glycoproteins [44]. Although human thyrotropin produced in the BEVS had higher levels of *in vitro* activity than the same protein produced in CHO cells, the *in vivo* activity was much lower due to its rapid clearance from the circulation. Various human tissues express mannose binding proteins, which are responsible for the rapid clearance of glycoproteins with terminally mannosylated *N*-glycans. Thus, the insect *N*-glycosylation patterns provided by the BEVS are not suitable for therapeutic glycoproteins, which must remain in the circulation for extended time periods.

7.4
N-Glycan Processing Enzymes in the BEVS

Despite the differences in *N*-glycan structures described above, insect cells have a protein *N*-glycan processing pathway that is highly similar to that of mammalian cells. In fact, the steps involved in the assembly, transfer and processing of *N*-glycans leading to the $GlcNAcMan_3GlcNAc_2$ intermediate (Figure 7.1) are highly similar or identical in insect and mammalian cells. However, while the mammalian pathway typically includes multiple additional processing steps involving the transfer of new monosaccharides to this trimmed *N*-glycan, insect cells usually fail to add any new residues and, in fact, remove the terminal *N*-acetylglucosamine residue instead. This section focuses on the enzymes responsible for this and other differences between the mammalian and insect *N*-glycan processing pathways.

7.4.1
Processing β-*N*-Acetylglucosaminidase

The paucimannose *N*-glycan structures on insect-cell derived glycoproteins are produced by a processing β-*N*-acetylglucosaminidase (GlcNAcase). This enzyme cleaves the *N*-acetylglucosamine residue from the α1,3 branch mannose and distinguishes higher from lower eukaryotic *N*-glycan processing pathways, as it is not expressed in higher eukaryotic cells. The presence of a processing GlcNAcase activity in insect cells was first demonstrated by Altmann *et al.* [45]. Whereas typical degradative GlcNAcases can cleave a broad range of substrates, the insect processing GlcNAcase is highly specific, as it cleaves only the *N*-acetylglucosamine residue on the α1,3 branch mannose, not the one on the α1,6 branch mannose. The gene encoding this highly specific processing enzyme was first cloned from *Drosophila* [46]. As a mutation in this enzyme caused fusion of the mushroom body lobes of the *Drosophila* brain, the gene was referred to as fused lobes, or *fdl* [47]. A recombinant form of the *Drosophila* FDL protein had the expected substrate specificity, cleaving only the terminal *N*-acetylglucosamine residue from the α1,3 branch mannose on *N*-glycans [46]. An *fdl* cDNA was recently isolated from Sf9 cells and, like the *Drosophila* FDL protein and the endogenous Sf9 cell processing activity, the product of this gene cleaved only *N*-acetylglucosamine from the α1,3 branch mannose [48].

7.4.2 Core α1,3 Fucosyltransferase

Another feature that sometimes distinguishes insect from mammalian glycoprotein *N*-glycans is the presence of an α1,3-linked fucose residue on the proximal core *N*-acetylglucosamine residue. This is unlike α1,6-linked core fucose, which is commonly found on both mammalian and insect *N*-glycans. Core α1,3 fucosylation is not limited to insects as it can also be found on nematode and plant *N*-glycans. As mammals do not produce core α1,3 fucosylated *N*-glycans, core α1,3 fucose can be recognized as foreign by the mammalian immune system. Antibodies raised against the plant *N*-glycoprotein horseradish peroxidase (HRP) recognize core α1,3 fucose; consequently, this glycan modification is known as the HRP epitope. Bee venom lipase contains a sizeable fraction of α1,3 core fucosylated *N*-glycans and is a major allergen in bee venom [49]. Antibodies against bee venom have been shown to be directed to the HRP epitope [50–53]. In plants, nematodes and insects core α1,3 fucose is introduced by a core α1,3-fucosyltransferase (core FT3) [50, 54, 55]. The *Drosophila* core FT3 has been shown to be the sole fucosyltransferase responsible for the generation of the anti-HRP epitope in this insect. A lepidopteran insect core FT3 has recently been cloned, indicating that Lepidoptera also have a core FT3 involved in the generation of the anti-HRP epitope (X. Shi and D.L. Jarvis, unpublished data). With respect to the two lepidopteran insect cell lines used most commonly in the BEVS, Sf9 cells produce a relatively small fraction while High Five cells produce a much larger fraction of *N*-glycans with α1,3-linked fucose [33, 56].

7.4.3 Lack of Mannose-6-Phosphate

The mannose-6-phosphate modification (Man-6-P) is a mammalian lysosomal targeting signal, responsible for the transport of newly synthesized *N*-glycoproteins to lysosomes. Man-6-P is synthesized by two sequentially acting enzymes. An *N*-acetylglucosamine-phosphotransferase adds an *N*-acetylglucosamine residue in phosphodiester linkage to a terminal mannose residue on an *N*-glycan [57]. Subsequently, the *N*-acetylglucosamine residue is removed by the action of a specific phosphodiester glycosidase, leaving behind a phosphorylated mannose residue [58].

Lysosomal storage diseases are a group of over 40 human genetic disorders that are caused by mutations in lysosomal enzymes [59]. Recombinant glycoproteins carrying the Man-6-P modification are currently used for the treatment of these disorders. One example is Gaucher's disease, which is treated with recombinant glucocerebrosidase. Cell surface receptors recognize the Man-6-P modification and allow the recombinant enzyme to be taken up and trafficked to lysosomes. Insect cells have a different targeting mechanism for lysosomal enzymes, which does not rely upon a specific *N*-glycan modification [60], and insect genomes do not encode identifiable homologues of the phosphotransferase subunits [61–63] or phosphodiester glycosidase [64]. Therefore, insect glycoprotein *N*-glycans are not decorated with mannose-6-phosphate [65] and the BEVS is generally unsuitable for the production of therapeutic glycoproteins for enzyme replacement therapy. One interesting exception is recombinant glucocerebrosidase, as the presence of terminal mannose residues on the insect-cell derived glycoprotein enhances its uptake by macrophages, which is one of the desired targets of this therapeutic product [66].

7.5 Lack of Glycosyltransferases

A key difference between insect and mammalian cells is that they have different levels of glycosyltransferase activities involved in *N*-glycan elongation. Whereas mammalian cells have high levels of each of these sequentially acting glycosyltransferases, including sialyltransferases, insect cells have only *N*-acetylglucosaminyltransferase I (GNT-I) at levels comparable to mammalian cells [67]. Although the downstream glycosyltransferases are predicted to be encoded by insect genomes, their products have only been detected using the most sensitive analyses [68]. Notably, *Drosophila* mutants lacking enzymes required for the synthesis of complex *N*-glycans have reduced longevity, fertility and viability, indicating that despite their very limited expression these enzymes have important developmental functions in insects [69–71].

7.5.1 N-Acetylglucosaminyltransferases II–IV

N-Acetylglucosaminyltransferase II (GNT-II) transfers an *N*-acetylglucosamine residue to the α6-linked mannose of the glycan core. Low levels of GNT-II activity have been reported in lepidopteran insect cell lysates [67], and a putative GNT-II gene has been cloned from *Drosophila melanogaster* [72, 73]. Low levels of *N*-acetylglucosamine-substituted α1,6-linked mannose have been reported on *N*-glycans from several lepidopteran insect cell lines [26]. Enzymes encoding *N*-acetylglucosaminyltransferases III and IV (GNT-III, -IV) are predicted to be encoded by the *D. melanogaster* genome, but have not yet been cloned. The products of these enzymes have been detected at very low levels in the *D. melanogaster* glycome [68].

7.5.2 N-acetylgalactosaminyl-Galactosyl- and Sialyltransferase

Genes encoding β1,4-*N*-acetylgalactosaminyltransferases (GalNAcTs) have been cloned and their products characterized from *T. ni* [74] and *D. melanogaster* [71]. GalNAcT activity has been detected in lysates of several lepidopteran insect cell lines [75], and low but significant levels of galactosyltransferase activity have been detected in *T. ni* cells [76]. A *D. melanogaster* sialyltransferase has been cloned and characterized [77, 78], but no sialyltransferase activity has been detected in lysates of any insect cells. Other genes involved in biosynthesis of CMP-sialic acid, the donor substrate for sialyltransferases, have also been cloned from *D. melanogaster* [79, 80], indicating that this insect has the enzymatic machinery required for the synthesis of sialylated *N*-glycans.

7.5.3
Glycosyltransferase Donor Substrates

In contrast to the glycosyltransferases themselves, the donor substrates used by *N*-acetylglucosaminyl-, galactosyl- and *N*-acetylgalactosaminyltransferase enzymes are readily detectable in lepidopteran insect cells [81]. However, there is no evidence for the presence of the donor substrate for sialyltransferases, CMP-sialic acid [27, 81].

7.6
Use of Baculoviruses to Extend BEVS *N*-Glycosylation

Although some insect genomes encode the enzymes required for the synthesis of complex type *N*-glycans, these functions are *not* expressed in lepidopteran insect cell lines at levels sufficient to produce complex, sialylated *N*-glycans. Exogenous genes encoding the required glycosyltransferases and other activities must be introduced into lepidopteran insect cells to obtain recombinant glycoproteins with mammalian-type *N*-glycans. One strategy devised for this purpose has been to co-infect the cells with a baculovirus encoding the glycoprotein of interest, together with one or more virus(es) encoding the requisite enzymatic activities.

7.6.1
Promoter Choice

The recombinant glycoprotein of interest (GOI) is typically expressed under the transcriptional control of a very late promoter such as the *polyhedrin* promoter. Although it is possible to express both the GOI and *N*-glycan processing genes under control of the same promoter, this approach could reduce expression of the GOI. Moreover, expression of the GOI would ensue before the mammalian *N*-glycan processing machinery is in place. Thus, it is preferable to express the genes encoding *N*-glycan processing functions under the control of promoters that are active in the early stages of infection. The identification of baculovirus genes expressed during the early stages of infection allowed the subsequent identification of their promoter sequences. Baculovirus immediate early (*ie*) gene promoters were found to be transcribed by host RNA polymerase II [82]. Also, transcriptional enhancers were discovered in the baculovirus genome in the form of several highly repetitive and conserved sequences [83, 84]. When placed in *cis*, these enhancers increased transcription levels of genes under the control of *ie* promoters several-fold. A combination of a baculovirus enhancer with a baculovirus *ie* promoter was demonstrated to be an effective means of expressing foreign genes during the early phases of infection [14].

7.6.2 Baculovirus Encoded Glycosyltransferases

7.6.2.1 *N*-Acetylglucosaminyltransferase I

In a pivotal experiment, Wagner *et al.* infected Sf9 cells with baculoviruses encoding human GNT-I and the fowl plague virus hemagglutinin under the control of the *polyhedrin* promoter [85]. The infected cells functionally expressed mammalian GNT-I, as indicated by elevated cellular GNT-I activity, and this resulted in the production of recombinant hemagglutinin with *N*-glycans containing terminal *N*-acetylglucosamine residues.

7.6.2.2 Galactosyltransferase

Jarvis and Finn [86] showed that Sf9 cells infected with a virus encoding a bovine β1-4 galactosyltransferase (β1,4GalT) under *ie* promoter control expressed high levels of galactosyltransferase activity. Moreover, the virion envelope glycoprotein, gp64, of progeny isolated from cells infected with this virus was shown to have galactose-terminated *N*-glycans. Subsequently, *T. ni* cells co-infected with viruses encoding β1,4GalT and human transferrin were shown to produce recombinant transferrin with monoantennary, terminally galactosylated *N*-glycans [29].

These reports demonstrated that mammalian glycosyltransferases localize in the secretory pathway of insect cells in such a way that they can functionally extend the endogenous *N*-glycan processing pathway. They also showed that the *N*-glycosylation patterns of recombinant glycoproteins expressed at high levels under the *polyhedrin* promoter can be changed using virally encoded glycosyltransferases.

7.6.2.3 Sialyltransferase

In subsequent experiments, transgenic Sf9 cells constitutively expressing a bovine β1,4GalT gene were infected with a recombinant baculovirus encoding a rat α2,6 sialyltransferase (ST6GalI) under *ie* promoter control [87]. The infected cells had high levels of sialyltransferase activity and produced viral progeny with terminally sialylated gp64 [87]. Although insect cells had been shown to contain low levels of sialic acid that were increased when fetal bovine serum was included in the growth medium [88], they contained no detectable CMP-sialic acid [27, 81]. Thus, the ability of insect cells to produce sialylated glycoproteins was surprising, as a galactosyl- and sialyltransferase, but no enzymes required for CMP-sialic acid synthesis had been introduced. A cell line constitutively expressing both glycosyltransferases yielded similar results [89]. Hollister *et al.* [90] provided an explanation for these findings by showing that insect cells expressing a mammalian galactosyltransferase and sialyltransferase require an exogenous source of sialic acid for the *de novo* synthesis of recombinant, sialylated *N*-glycoproteins. Alternatively, supplementation of the sialic acid precursor *N*-acetylmannosamine also resulted in low levels of sialylation. These results showed that lepidopteran insect cells have an endogenous capacity to utilize an exogenous source of sialic acid or *N*-acetylmannosamine for the production of sialylated *N*-glycans.

As a follow-up to these studies, Jarvis *et al.* constructed a single virus encoding both the bovine β1,4GalT and the rat ST6GalI under *ie* promoter control [91]. Both genes were found to be expressed and provided high levels of glycosyltransferase activity by 12 hours post infection. Moreover, viral progeny contained gp64 with terminally sialylated *N*-glycans. Thus, this virus induced synthesis of monoantennary sialylated *N*-glycans in lepidopteran insect cells and also functioned as a BEV for the expression of a recombinant protein under *polyhedrin* control.

7.6.2.4 *N*-acetylglucosaminyltransferase II

The expression of GNT-II is required for the synthesis of biantennary *N*-glycans. Lepidopteran insect cells typically have very low levels of GNT-II activity and, accordingly, *N*-glycans produced by these cells typically lack terminal *N*-acetylglucosamine residues on their α1,6 branch [67]. Tomiya *et al.* infected transgenic insect cells constitutively expressing the bovine β1,4GalT with a virus encoding the human GNT-II under *ie* promoter control as well as human transferrin under *polyhedrin* promoter control [92]. GNT-II activities were increased over 100-fold compared to wild-type virus infected cells by 12 hours post infection. Moreover, a detailed analysis revealed a 50% abundance of digalactosylated, biantennary *N*-glycans on recombinant transferrin. Similarly, Chang *et al.* [93] used a single baculovirus to introduce the human β1,4GalT, ST6GalI, GNT-II and human anti-trypsin as a model glycoprotein under control of the *p10* and *polyhedrin* promoters, resulting in the synthesis of sialylated anti-trypsin.

7.6.2.5 Trans-Sialidase

The protozoan parasite *Trypanosoma cruzi* expresses a developmentally-regulated *trans*-sialidase (TS), which allows it to sialylate glycoproteins on its cell surface [94]. The TS enzyme transfers α2,3-linked sialic acids to a galactose-terminated *N*-glycan using host glycoconjugates as the donor substrate [95]. Marchal *et al.* expressed a chimeric, membrane bound TS in insect cells using a baculovirus vector and found that it was able to sialylate galactose-terminated *N*-glycans using fetuin or sialyllactose as donor substrates [96]. The expression of membrane-bound TS in insect cells expressing a galactosyltransferase could allow for the production of recombinant, sialylated glycoproteins in serum-supplemented culture medium.

7.6.3 Baculoviruses Encoded Sugar Processing Genes

7.6.3.1 UDP-GlcNAc 2-Epimerase/*N*-Acetylmannosamine Kinase

In addition to glycosyltransferases, enzymes involved in the production of precursors required for the biosynthesis of complex *N*-glycans have also been expressed in insect cells using baculovirus vectors. The first dedicated precursor of the sialic acid pathway is *N*-acetylmannosamine. In mammals, *N*-acetylmannosamine is produced from UDP-*N*-acetylglucosamine (UDP-GlcNAc) by the bifunctional enzyme UDP-GlcNAc 2-epimerase/*N*-acetylmannosamine kinase (epimerase/kinase). Sialic acid is then synthesized from *N*-acetylmannosamine-6-P by the enzyme sialic acid-9-phosphate synthase (SAS). *N*-Acetylmannosamine is usually added to the culture medium as insect cells have UDP-GlcNAc epimerase levels that are about 30-fold lower than found in rat liver [97]. Insect cell *N*-acetylmannosamine kinase activity, in contrast, is 50-fold higher than the UDP-GlcNAc epimerase activity [97], and is not limiting in the production of sialic acid. Viswanathan *et al.* showed that insect cells infected with viruses encoding SAS and epimerase/kinase had high levels of sialic acid, which could be increased further with *N*-acetylglucosamine supplementation [98].

7.6.3.2 Sialic Acid Synthetase

Lawrence *et al.* showed that sialic acid levels in Sf9 cells supplemented with *N*-acetylmannosamine increased 200-fold following infection with a virus encoding the human SAS [88]. Under these circumstances, sialic acid content was sixfold higher than observed in CHO cells. However, the human SAS can also effectively utilize mannose-6-phosphate as a substrate to synthesize 2-keto-3-deoxy-D-*glycero*-D-*galacto*-nononic acid (KDN). As the occurrence of KDN is rare and almost undetectable in normal adult human tissues [99], and potentially immunogenic, the presence of this sugar on recombinant therapeutic glycoproteins is undesirable. The subsequently cloned murine SAS was shown to use only *N*-acetylmannosamine-6-phosphate as substrate to yield sialic acid-9-phosphate [100]. Thus, the murine, not the human, SAS is the better choice for the engineering of insect cells for sialic acid production.

7.6.3.3 CMP-Sialic Acid Synthetase

Sialic acid is activated by the CMP-sialic acid synthetase (CSAS) enzyme. Lawrence *et al.* showed that CMP-sialic acid levels in Sf9 cells supplemented with *N*-acetylmannosamine increased over 100-fold following infection with viruses encoding the human CSAS and SAS under *polyhedrin* promoter control, as compared to infection with a SAS-encoding virus alone [101]. Viswanathan *et al.* achieved similar levels of CMP-sialic acid without sugar supplementation by co-infecting Sf9 cells with three separate BEVs encoding the mammalian epimerase/kinase, SAS and CSAS under *polyhedrin* promoter control [102]. Genes encoding murine CSAS [103] and SAS have also been placed under dual *ie* promoter control in a single baculovirus to permit endogenous production of sialic acid and CMP-sialic acid [104]. SfSWT-1 cells [105] cultured in the presence of *N*-acetylmannosamine had high sialic acid and CMP-sialic acid levels and were able to produce recombinant glycoproteins with sialylated *N*-glycans following infection with this virus.

7.6.4 Summary

Mammalian glycosyltransferases and sugar processing enzymes have been functionally expressed in lepidopteran insect cells using baculovirus vectors, allowing for the synthesis of recombinant glycoproteins with hybrid and complex, instead of paucimannosidic *N*-glycans. As more than one baculovirus encoding one or more functions can be used to simultaneously infect lepidopteran insect cells, the advantage of this approach lies in its flexibility. These viruses can be used in a modular fashion to produce recombinant glycoproteins with the desired *N*-glycan structure from any cell line susceptible to infection. Also, there is no need to continuously maintain transgenic cell lines. A disadvantage is the continued need for titered virus stocks, which are expended with each infection. As high multiplicities of infection are needed for each separate virus to ensure infection of each cell, virus stock consumption can be quite high. This also raises the issue of which multiplicity of infection is appropriate for which virus(es): should infections be balanced and if so, how? These questions can only be answered by thorough investigations on a case-by case basis. Also, co-infection strategies could reduce the yield of glycoproteins that are expressed under control of the same promoter as the processing genes. However, this can be avoided by expressing the processing genes under control of an early promoter and the glycoprotein of interest under control of a late or very late promoter.

7.7
Transgenic Insect Cell Lines

As mentioned before, the baculovirus *ie* promoters recruit host cell machinery to initiate transcription. Thus, such promoters can be used to constitutively express genes in uninfected lepidopteran insect cells if the transgene has been stably incorporated in the host cell genome. To achieve this, plasmid(s) encoding the gene(s) of interest as well as a resistance marker are co-transfected into insect cells. It is critical to use cells that have been passaged in the absence of standard cell culture antibiotics, as cells grown in the presence of these might not be sensitive to the antibiotic used for selection. Following antibiotic selection, a polyclonal population is generated, from which clonal cells lines are obtained by limiting dilution or other methods. Subsequently, the clonal cell lines are screened for expression of the unselected transgenes. While initial screening can be performed at the transcriptional level, expression of functional transgene product(s) must ultimately be examined.

7.7.1
Proof of Concept

Jarvis *et al.* first showed that a baculovirus *ie* promoter could be used to generate transformed insect cell lines that constitutively express a transgene. In this experiment, Sf9 cells were co-transfected with plasmids encoding human tissue plasminogen activator (t-PA) or *Escherichia coli* β-galactosidase as well as neomycin phosphotransferase, all under the transcriptional control of the *ie* promoter [106]. After an antibiotic selection period, several clones were identified in which the transgene was transcribed and the transgene product was expressed. These transgenes were expressed over a large number of consecutive cell passages, indicating that they were stably integrated into the host cell genome. Interestingly, the glycoprotein t-PA was expressed at levels comparable to those observed after infection with a BEV encoding t-PA, but without as much proteolytic degradation. Further characterization of these cell lines revealed that baculovirus infection transiently stimulated, then repressed transcription of transgenes integrated into the cellular genome [107].

7.7.2 Transgenic Glycosyltransferases

7.7.2.1 Galactosyltransferase

Hollister *et al.* isolated a transgenic insect cell line constitutively expressing a mammalian glycosyltransferase by co-transfecting Sf 9 cells with plasmids encoding bovine β1,4GalT and neomycin phosphotransferase, with each gene under the control of an *ie* promoter [108]. A clone with the highest galactosyltransferase activity was designated Sfβ4GalT and further characterized. The levels of galactosyltransferase activity in these cells peaked early after baculovirus infection and then declined to starting levels, confirming and extending previous observations of the influence of virus infection on transgene expression, as mentioned above [107]. Shutdown of transgene transcription following viral infection did not present a problem, however, as the glycosyltransferases resident in the secretory pathway have long half-lives. Thus, upon infection with a baculovirus encoding human t-PA, Sfβ4GalT cells produced recombinant t-PA with terminally galactosylated, monoantennary *N*-glycans, whereas no galactosylated t-PA was detected when the same BEV was used to infect the parental Sf9 cells.

7.7.2.2 Sialyltransferase

The *N*-glycosylation pathway of the Sfβ4GalT cell line was further extended by transforming these cells with plasmids encoding ST6GalI and hygromycin phosphotransferase under *ie* promoter control [105]. A clone with galactosyltransferase levels comparable to the parental cell line plus high levels of sialyltransferase activity was identified and designated Sfβ4GalT/ST6. This cell line was able to produce recombinant glycoproteins with sialylated *N*-glycans following BEV infection. Again, this finding was surprising because this cell line had not been engineered to produce endogenous sialic acid or CMP-sialic acid, which is present at extremely low or undetectable levels in Sf9 cells. Subsequently, a potential explanation was provided with evidence indicating that Sf9 cells can salvage sialic acids from glycoproteins present in the culture medium [90].

The Sfβ4GalT/ST6 cell line was isolated by successively inserting transgenes into the host cell genome. This approach limits the final number of transgenes that can be incorporated because only a limited number of antibiotic markers are available. Moreover, the serial cloning steps required by this approach are laborious and time-consuming. On the other hand, one can simultaneously insert two or more transgenes into the host genome by either using a single plasmid encoding several genes or several plasmids encoding single genes in one transformation and selection step.

The simultaneous insertion of two transgenes into a lepidopteran insect cell genome was first accomplished by Breitbach and Jarvis [109], who used a plasmid encoding mammalian ST6GalI and β1,4GalT genes separated by an enhancer element in a back-to-back configuration. *T. ni* cells were co-transfected with this plasmid and a hygromycin-resistance marker, followed by antibiotic selection and cell cloning. This resulted in the isolation of a clone expressing high levels of both sialyltransferase and galactosyltransferase activities and capable of producing recombinant glycoproteins with sialylated *N*-glycans. This report also revealed that there are some important technical differences in transformation of different insect cell lines, as *T. ni* cells were apparently resistant to neomycin and susceptible to the loss of transgene expression in the absence of antibiotic selection pressure.

7.7.2.3 **N-Acetylglucosaminyltransferase II**

Stable expression of GNT-II to yield biantennary *N*-glycans was first demonstrated by Hollister *et al.* [105]. A transgenic cell line encoding five mammalian glycosyltransferases was obtained by transforming the Sfβ4GalT cell line with plasmids encoding GNT-I, GNT-II, ST6GalI, α2,3 sialyltransferase IV (ST3GalIV) and a hygromycin resistance marker. A cell line that expressed all five glycosyltransferases was designated SfSWT-1. The *N*-glycosylation potential of SfSWT-1 cells was subsequently compared to that of Sf9, Sfβ4GalT and Sfβ4GalT/ST6 cells. This report confirmed the absence of hybrid and complex *N*-glycans on glycoproteins from Sf9 cells and the presence of galactosylated, monoantennary *N*-glycans on glycoproteins derived from Sfβ4GalT cells. Also, sialylated, monoantennary *N*-glycans were found on glycoproteins derived from Sfβ4GalT/ST6 cells, corroborating and extending the earlier report [105]. In addition, SfSWT-1 cells were found to be able to produce glycoproteins with monosialylated, biantennary *N*-glycans.

Unlike previously characterized transgenic cell lines, SfSWT-1 cells had significantly different growth characteristics than the parental Sf9 cell line. SfSWT-1 cells had a longer initial lag phase, but were able to grow to higher densities than Sf9 cells [110]. Based upon lectin blotting analyses, Legardinier *et al.* later reported that SfSWT-1 (MIMIC) cells were unable to sialylate the *N*-glycans on recombinant equine gonadotropin [111]. Thus, it is possible that there are recombinant *N*-glycoproteins that cannot be sialylated by transgenic insect cell lines expressing mammalian glycosyltransferases, such as SfSWT-1 cells.

7.7.3
Transgenic Sugar Processing Genes

As the SfSWT-1 cell line was unable to produce its own sialic acid and had very low levels of intracellular CMP-sialic acid, it was further engineered to express murine SAS and CSAS. An expression plasmid encoding both genes as well as a zeocin resistance marker was used to produce a cell line that constitutively expressed SAS, CSAS and the SfSWT-1 suite of glycosyltransferases. This cell line, called SfSWT-3, was able to synthesize sialic acid and CMP-sialic acid when grown in a serum-free medium supplemented with *N*-acetylmannosamine. Unlike the parental cell line SfSWT-1, which required serum supplementation, the SfSWT-3 cell line was able to produce recombinant glycoproteins with sialylated *N*-glycans when grown in a serum-free medium supplemented with *N*-acetylmannosamine. Surprisingly, SfSWT-3 cells had growth characteristics similar to the Sf9 cell line, but unlike the parental SfSWT-1 cell line. However, our recent in-house experience with this cell line has not been highly satisfactory, as it has been difficult to maintain consistently under our usual culture conditions (J.J. Aumiller and D.L. Jarvis, unpublished data).

It was later discovered that the SfSWT-1 and SfSWT-3 cell lines are unable to α2,3-sialylate *N*-glycoproteins and that extracts of these cells have no α2,3-sialyltransferase activity (B. Harrison and D.L. Jarvis, unpublished data). These cell lines were transformed with a murine gene encoding the ST3GalIV isozyme. Hence, this enzyme is the first example to our knowledge of a mammalian glycosyltransferase gene that failed to induce the appropriate enzyme activity when expressed in native form in lepidopteran insect cells. In contrast, cell extracts from Sf9 cells transiently transfected with a murine gene encoding the ST3GalIII isozyme had α2,3-sialyltransferase activity [112], indicating that this gene might be useful for the isolation of a transgenic insect cell line capable of producing α2,3-sialylated *N*-glycoproteins.

7.7.4 Use of Transposon-Based Systems

The *piggyBac* system is derived from a *T. ni* transposon called the *piggyBac* element. The *piggyBac* element encodes a functional transposase, which is required for its transposition. The *piggyBac* element has terminal inverted repeats that function as recognition markers for the *piggyBac* transposase [113, 114]. In the *piggyBac* system, the transposase is encoded by a helper plasmid and mobilizes a separate DNA element flanked by the inverted repeats [115, 116]. Novel tools for the generation of transgenic insect cell lines using the *piggyBac* system have been described by Shi *et al.* [112]. These investigators positioned a cassette encoding two genes encoding *N*-glycan processing functions together with an eye-specific marker between the *piggyBac* terminal repeat sequences. This arrangement was designed to permit the *piggyBac* transposase to move the gene cassette into the host cell genome. The presence of the eye-specific marker in this gene cassette allows this approach to be used for the isolation of not only transgenic insect cell lines but also transgenic insects.

7.7.5 Summary

Baculovirus immediate early promoters are transcribed by host transcriptional machinery and can be used to express transgenes constitutively in lepidopteran insect cells. Several transgenic insect cell lines have been engineered to express glycosyltransferases and other enzymes involved in carbohydrate metabolism using these promoters. Importantly, the glycosyltransferases functionally extend the endogenous protein *N*-glycosylation pathway, as indicated by the ability of the transgenic insect cell lines to produce recombinant glycoproteins with hybrid or complex *N*-glycans. Moreover, the growth properties, infectivity and expression levels of the transgenic cell lines are typically similar to those of the parental cell lines. To allow expression of recombinant glycoproteins with sialylated, complex *N*-glycans in the absence of serum supplementation, insect cell lines have also been engineered to synthesize their own sialic acid and CMP-sialic acid from the sialic acid precursor, *N*-acetylmannosamine.

7.8 Sugar Supplementation

Although lepidopteran insect cells are able to salvage sialic acid from exogenous glycoconjugates [90], they have very low or undetectable levels of sialic acids or sialic acid precursors. Supplementation of insect cell growth media with sialic acid precursors can dramatically increase intracellular levels of CMP-sialic acid, although it is currently not clear what levels are required to result in optimum sialylation.

Insect genomes do not appear to encode a UDP-GlcNAc 2-epimerase, which is the mammalian enzyme that produces *N*-acetylmannosamine, the first dedicated intermediate in the sialic acid pathway. However, very low levels of UDP-GlcNAc 2-epimerase activity have been reported in lepidopteran insect cells [97]. As the endogenous levels of *N*-acetylmannosamine are too low to support sialylation, this sugar is typically added to the culture medium. Insect cell culture media can be supplemented with either *N*-acetylmannosamine or its peracetylated derivative. Supplementation with 10 mM *N*-acetylmannosamine supports sialylation of recombinant *N*-glycoproteins [104, 110]. However, peracetylated *N*-acetylmannosamine is more hydrophobic and can readily diffuse though the insect cell plasma membrane, where cytosolic esterases will remove the acetyl groups. Thus, utilization of peracetylated sugar analogues is substantially more efficient than use of the unsubstituted forms [117]. Insect cells also lack an *N*-acetylmannosamine kinase, but *N*-acetylmannosamine can be phosphorylated by the non-specific action of an *N*-acetylglucosamine kinase. As this enzyme has a pronounced preference for *N*-acetylglucosamine, supplementation of *N*-acetylglucosamine together with *N*-acetylmannosamine abolishes the production of *N*-acetylmannosamine-6-P and hence sialic acids [98]. An alternative to *N*-acetylmannosamine supplementation is the expression of the mammalian epimerase/kinase, which was shown to yield comparable levels of sialic acid [98, 102]. Even higher sialic acid levels can be achieved by expressing the mammalian epimerase/kinase enzyme in insect cells cultured in a growth medium supplemented with *N*-acetylglucosamine [98].

7.9 Future Directions

Although the BEVS has been successfully engineered to produce recombinant glycoproteins with terminally sialylated *N*-glycans, several challenges remain in mammalianizing the lepidopteran insect cell *N*-glycosylation pathway. Most notably, enzymes required for the synthesis of complex *N*-glycans with more than two antennae and with terminal sialic acids with different linkages remain to be engineered into the system.

7.9.1
Completing the *N*-Glycosylation Pathway

As described previously, insect cells have been engineered to functionally express GNT-I and GNT-II. However, it should be considered that lepidopteran insect cell lines typically have GNT-I levels similar to mammalian cells and that endogenous GNT-II activity is also present, albeit at low levels. Besides GNT-I and II, insect genomes are also predicted to encode GNT-III and GNT-IV, but these enzyme activities have not been detected in lepidopteran insect cell lines. Thus, to create insect cell lines that have the ability to introduce bi- and trisecting *N*-acetylglucosamine residues, GNT-III and GNT-IV need to be introduced. The BEVS has been used to produce enzymatically active forms of recombinant GNT-III [118], GNT-V [119] and GNT-VI [120], indicating that lepidopteran insect cells can synthesize and correctly fold these mammalian enzymes. However, the ability to functionally extend the endogenous protein *N*-glycosylation pathway using these enzymes remains to be established.

Lepidopteran insect cells have been successfully engineered to synthesize *N*-glycans with terminal α2,6-linked sialic acids, but not yet with terminal sialic acids in other linkages. Recent studies have shown that lepidopteran insect cells can produce catalytically active tagged [121] and native forms of ST3GalIII [112]. Thus, efforts should be pursued to produce a transgenic insect cell lines expressing this isozyme, rather than the ST3GalIV isozyme, as a cell line expressing a functional α2,3-sialyltransferase would be a valuable tool for the production of recombinant glycoproteins such as human erythropoietin, which is virtually exclusively α2,3-sialylated [122].

7.9.2
Reducing Deleterious Activities

RNA-mediated interference can be used to reduce the levels of endogenously expressed enzymes that yield allergenic epitopes or that divert *N*-glycan substrates from enzymes involved in their elongation [123]. The constructs used for this purpose may encode an inverted repeat derived from a portion of the target gene placed under the control of either a constitutive or an inducible promoter [124]. Expression of the inverted repeat would give rise to a double-stranded RNA (dsRNA) molecule that could trigger cellular RNA interference pathways. In addition, an intron, such as the *Drosophila* white intron, can be positioned between the inverted repeats to stabilize the construct in bacterial cells. Excision of this intron by normal cellular splicing mechanisms would give rise to the desired dsRNA molecule, which would consist of the two single-stranded RNAs encoded by the inverted repeat [125].

The use of this type of dsRNA-encoding construct to create a transgenic lepidopteran insect cell line was first reported as a way to suppress the expression of a baculovirus-encoded gene [126]. In this study, a transgenic cell line was shown to be able to suppress the expression of a BEV-encoded firefly luciferase gene for at least 20 passages. To be able to use this same type of approach to suppress the expression of an endogenous gene, that gene must first be molecularly cloned. This was first demonstrated by the isolation of a transgenic insect cell line expressing a dsRNA construct targeting the endogenous processing GlcNAcase of Sf9 cells [48]. This cell line had <50% of the processing GlcNAcase activity of the parental cell line. A similar approach could be used to target the insect cell core α1,3 fucosyltransferase and reduce the levels of HRP epitope on BEVS-produced recombinant glycoproteins. However, the long-term stability of transgenic constructs encoding a double-stranded RNA targeting an endogenously expressed protein has not been assessed.

References

1 Smith, G.E., Fraser, M.J. and Summers, M.D. (1983) Molecular engineering of the *Autographa californica* nuclear polyhedrosis virus genome: deletion mutations within the polyhedrin gene. *Journal of Virology*, **46**, 584–593.

2 O'Reilly, D.R., Miller, L.K. and Luckow, V.A. (1992) *Baculovirus Expression Vectors: A Laboratory Manual*, W.H. Freeman and Company, New York.

3 Summers, M.D. (2006) Milestones leading to the genetic engineering of baculoviruses as expression vector systems and viral pesticides. *Advances in Virus Research*, **68**, 3–73.

4 Kost, T.A., Condreay, J.P. and Jarvis, D.L. (2005) Baculovirus as versatile vectors for protein expression in insect and mammalian cells. *Nature Biotechnology*, **23**, 567–575.

5 Jarvis, D.L. (1997) Baculovirus expression vectors, in *The Baculoviruses* (ed. L.K. Miller), Plenum Press, New York, pp. 389–431.

6 Summers, M.D. and Smith, G.E. (1987) A manual of methods for baculovirus vectors and insect cell culture procedures, *Texas Agricultural Experiment Station Bulletin* No 1555 College Station, TX.

7 Murhammer, D.W. (2007) Baculovirus and insect cell expression protocols, in *Methods in Molecular Biology*, 2nd edn, Vol. **388**, Humana Press.

8 Wickham, T.J. *et al.* (1992) Screening of insect cell lines for the production of recombinant proteins and infectious virus in the baculovirus expression system. *Biotechnology Progress*, **8**, 391–396.

9 Vaughn, J.L. *et al.* (1977) The establishment of two cell lines from the insect *Spodoptera frugiperda* (Lepidoptera; Noctuidae), *In Vitro*, **13**, 213–217.

10 Pennock, G.D., Shoemaker, C. and Miller, L.K. (1984) Strong and regulated expression of *Escherichia coli* β-galactosidase in insect cells with a baculovirus vector. *Molecular and Cellular Biology*, **4**, 399–406.

11 Smith, G.E., Summers, M.D. and Fraser, M.J. (1983) Production of human beta interferon in insect cells infected with a baculovirus expression vector. *Molecular and Cellular Biology*, **3**, 2156–2165.

12 Luckow, V.A. *et al.* (1993) Efficient generation of infectious recombinant baculoviruses by site-specific transposon-

mediated insertion of foreign genes into a baculovirus genome propagated in *Escherichia coli*. *Journal of Virology*, **67**, 4566–4579.

13 Hill-Perkins, M.S. and Possee, R.D. (1990) A baculovirus expression vector derived from the basic protein promoter of *Autographa californica* nuclear polyhedrosis virus. *The Journal of General Virology*, **71**, 971–976.

14 Jarvis, D.L., Weinkauf, C. and Guarino, L.A. (1996) Immediate-early baculovirus vectors for foreign gene expression in transformed or infected insect cells. *Protein Expression and Purification*, **8**, 191–203.

15 Morell, A.G. *et al.* (1971) The role of sialic acid in determining the survival of glycoproteins in the circulation. *The Journal of Biological Chemistry*, **246**, 1461–1467.

16 Hervas-Stubbs, S. *et al.* (2007) Insect baculoviruses strongly potentiate adaptive immune responses by inducing type I IFN. *The Journal of Immunology*, **178**, 2361–2369.

17 Huber, V.C. and McCullers, J.A. (2008) FluBlok, a recombinant influenza vaccine. *Current Opinion in Molecular Therapeutics*, **10**, 75–85.

18 Pushko, P. *et al.* (2005) Influenza virus-like particles comprised of the HA, NA, and M1 proteins of H9N2 influenza virus induce protective immune responses in BALB/c mice. *Vaccine*, **23**, 5751–5759.

19 Bright, R.A. *et al.* (2007) Influenza virus-like particles elicit broader immune responses than whole virion inactivated influenza virus or recombinant hemagglutinin. *Vaccine*, **25**, 3871–3878.

20 Burch, P.A. *et al.* (2000) Priming tissue-specific cellular immunity in a phase I trial of autologous dendritic cells for prostate cancer. *Clinical Cancer Research*, **6**, 2175–2182.

21 Harzstark, A.L. and Small, E.J. (2007) Immunotherapy for prostate cancer using antigen-loaded antigen-presenting cells: APC8015 (Provenge). *Expert Opinion on Biological Therapy*, **7**, 1275–1280.

22 Patel, P.H. and Kockler, D.R. (2008) Sipuleucel-T: a vaccine for metastatic, asymptomatic, androgen-independent prostate cancer. *Annals of Pharmacotherapy*, **42**, 91–98.

23 Reinis, M. (2007) Drug evaluation: FavId, a patient-specific idiotypic vaccine for non-Hodgkin's lymphoma. *Current Opinion in Molecular Therapeutics*, **9**, 291–298.

24 Redfern, C.H. *et al.* (2006) Phase II trial of idiotype vaccination in previously treated patients with indolent non-Hodgkin's lymphoma resulting in durable clinical responses. *Journal of Clinical Oncology*, **24**, 3107–3112.

25 Kornfeld, R. and Kornfeld, S. (1985) Assembly of asparagine-linked oligosaccharides. *Annual Review of Biochemistry*, **54**, 631–664.

26 Kubelka, V. *et al.* (1994) Structures of the N-linked oligosaccharides of the membrane glycoproteins from three lepidopteran cell lines (Sf-21, IZD-Mb-0503, Bm-N). *Archives of Biochemistry and Biophysics*, **308**, 148–157.

27 Hooker, A.D. *et al.* (1999) Constraints on the transport and glycosylation of recombinant IFN-γ in Chinese hamster ovary and insect cells. *Biotechnology and Bioengineering*, **63**, 559–572.

28 James, D.C. *et al.* (1995) *N*-Glycosylation of recombinant human interferon-γ produced in different animal expression systems. *Nature Biotechnology*, **13**, 592–596.

29 Ailor, E. *et al.* (2000) *N*-Glycan patterns of human transferrin produced in *Trichoplusia ni* insect cells: effects of mammalian galactosyltransferase. *Glycobiology*, **10**, 837–847.

30 Choi, O. *et al.* (2003) *N*-Glycan structures of human transferrin produced by *Lymantria dispar* (gypsy moth) cells using the LdMNPV expression system. *Glycobiology*, **13**, 539–548.

31 Misaki, R. *et al.* (2003) N-linked glycan structures of mouse interferon-β produced by *Bombyx mori* larvae. *Biochemical and Biophysical Research Communications*, **311**, 979–986.

32 Kulakosky, P.C., Hughes, P.R. and Wood, H.A. (1998) *N*-Linked glycosylation of a baculovirus-expressed recombinant glycoprotein in insect larvae and tissue culture cells. *Glycobiology*, **8**, 741–745.

33 Hsu, T.A. *et al.* (1997) Differential *N*-glycan patterns of secreted and intracellular IgG produced in *Trichoplusia ni* cells. *The Journal of Biological Chemistry*, **272**, 9062–9070.

34 Joshi, L. *et al.* (2000) Influence of baculovirus-host cell interactions on complex N-linked glycosylation of a recombinant human protein. *Biotechnology Progress*, **16**, 650–656.

35 Rudd, P.M. *et al.* (2000) Hybrid and complex glycans are linked to the conserved *N*-glycosylation site of the third eight-cysteine domain of LTBP-1 in insect cells. *Biochemistry*, **39**, 1596–1603.

36 Ogonah, O.W. *et al.* (1996) Isolation and characterization of an insect cell line able to perform complex *N*-linked glycosylation on recombinant proteins. *Nature Biotechnology*, **14**, 197–202.

37 Marchal, I. *et al.* (2001) Glycoproteins from insect cells: sialylated or not? *Biological Chemistry*, **382**, 151–159.

38 Davidson, D.J., Fraser, M.J. and Castellino, F.J. (1990) Oligosaccharide processing in the expression of human plasminogen cDNA by lepidopteran insect (*Spodoptera frugiperda*) cells. *Biochemistry*, **29**, 5584–5590.

39 Davidson, D.J. and Castellino, F.J. (1991) Asparagine-linked oligosaccharide processing in lepidopteran insect cells. Temporal dependence of the nature of the oligosaccharides assembled on asparagine-289 of recombinant human plasminogen produced in baculovirus vector infected *Spodoptera frugiperda* (IPLB-SF-21AE) cells. *Biochemistry*, **30**, 6167–6174.

40 Davidson, D.J. and Castellino, F.J. (1991) Structures of the asparagine-289-linked oligosaccharides assembled on recombinant human plasminogen expressed in a *Mamestra brassicae* cell line (IZD-MBO503). *Biochemistry*, **30**, 6689–6696.

41 Joosten, C.E. and Shuler, M.L. (2003) Production of a sialylated N-linked glycoprotein in insect cells: Role of glycosidases and effect of harvest time on glycosylation. *Biotechnology Progress*, **19**, 193–201.

42 Joosten, C.E. and Shuler, M.L. (2003) Effect of culture conditions on the degree of sialylation of a recombinant glycoprotein expressed in insect cells. *Biotechnology Progress*, **19**, 739–749.

43 Palomares, L.A. *et al.* (2003) Novel insect cell line capable of complex *N*-glycosylation and sialylation of recombinant proteins. *Biotechnology Progress*, **19**, 185–192.

44 Grossmann, M. *et al.* (1997) Expression of biologically active human thyrotropin (hTSH) in a baculovirus system: effect of insect cell glycosylation on hTSH activity *in vitro* and *in vivo*. *Endocrinology*, **138**, 92–100.

45 Altmann, F. *et al.* (1995) Insect cells contain an unusual, membrane-bound β-*N*-acetylglucosaminidase probably involved in the processing of protein *N*-glycans. *The Journal of Biological Chemistry*, **270**, 17344–17349.

46 Leonard, R. *et al.* (2006) The *Drosophila* fused lobes gene encodes an *N*-acetylglucosaminidase involved in *N*-glycan processing. *The Journal of Biological Chemistry*, **281**, 4867–4875.

47 Boquet, I. *et al.* (2000) Central brain postembryonic development in *Drosophila*: implication of genes expressed at the interhemispheric junction. *Journal of Neurobiology*, **42**, 33–48.

48 Geisler, C., Aumiller, J.J. and Jarvis, D.L. (2008) A fused lobes gene encodes the processing β-*N*-acetylglucosaminidase in

Sf9 cells. *The Journal of Biological Chemistry*, **283**, 11330–11339.

49 Kubelka, V. *et al.* (1993) Primary structures of the N-linked carbohydrate chains from honeybee venom phospholipase A_2. *European Journal of Biochemistry*, **213**, 1193–1204.

50 Fabini, G. *et al.* (2001) Identification of core α1,3-fucosylated glycans and cloning of the requisite fucosyltransferase cDNA from *Drosophila melanogaster.* Potential basis of the neural anti-horseradish peroxidase epitope. *The Journal of Biological Chemistry*, **276**, 28058–28067.

51 Bencurova, M. *et al.* (2004) Specificity of IgG and IgE antibodies against plant and insect glycoprotein glycans determined with artificial glycoforms of human transferrin. *Glycobiology*, **14**, 457–466.

52 Tretter, V. *et al.* (1993) Fucose α1,3-linked to the core region of glycoprotein *N*-glycans creates an important epitope for IgE from honeybee venom allergic individuals. *International Archives of Allergy and Immunology*, **102**, 259–266.

53 Faye, L. *et al.* (1993) Affinity purification of antibodies specific for Asn-linked glycans containing α1 → 3 fucose or β1 → 2 xylose. *Analytical Biochemistry*, **209**, 104–108.

54 Leiter, H. *et al.* (1999) Purification, cDNA cloning, and expression of GDP-L-Fuc:Asn-linked GlcNAc α1,3-fucosyltransferase from mung beans. *The Journal of Biological Chemistry*, **274**, 21830–21839.

55 Paschinger, K. *et al.* (2004) Molecular basis of anti-horseradish peroxidase staining in *Caenorhabditis elegans. The Journal of Biological Chemistry*, **279**, 49588–49598.

56 Staudacher, E., Kubelka, V. and März, L. (1992) Distinct *N*-glycan fucosylation potentials of three lepidopteran cell lines. *European Journal of Biochemistry*, **207**, 987–993.

57 Reitman, M.L. and Kornfeld, S. (1981) Lysosomal enzyme targeting. *N*-acetylglucosaminylphosphotransferase selectively phosphorylates native lysosomal enzymes. *The Journal of Biological Chemistry* **256**, 11977–11980.

58 Varki, A. and Kornfeld, S. (1981) Purification and characterization of rat liver α-*N*-acetylglucosaminyl phosphodiesterase. *The Journal of Biological Chemistry*, **256**, 9937–9943.

59 Winchester, B., Vellodi, A. and Young, E. (2000) The molecular basis of lysosomal storage diseases and their treatment. *Biochemical Society Transactions*, **28**, 150–154.

60 Dennes, A. *et al.* (2005) The novel *Drosophila* lysosomal enzyme receptor protein mediates lysosomal sorting in mammalian cells and binds mammalian and *Drosophila* GGA adaptors. *The Journal of Biological Chemistry*, **280**, 12849–12857.

61 Kudo, M. *et al.* (2005) The α- and β-subunits of the human UDP-*N*-acetylglucosamine:lysosomal enzyme *N*-acetylglucosamine-1-phosphotransferase are encoded by a single cDNA. *The Journal of Biological Chemistry*, **280**, 36141–36149.

62 Tiede, S. *et al.* (2005) Mucolipidosis II is caused by mutations in GNPTA encoding the α/β GlcNAc-1-phosphotransferase. *Nature Medicine*, **11**, 1109–1112.

63 Raas-Rothschild, A. *et al.* (2000) Molecular basis of variant pseudo-hurler polydystrophy (mucolipidosis IIIC). *The Journal of Clinical Investigation*, **105**, 673–681.

64 Kornfeld, R. *et al.* (1999) Molecular cloning and functional expression of two splice forms of human *N*-acetylglucosamine-1-phosphodiester α-*N*-acetylglucosaminidase. *The Journal of Biological Chemistry*, **274**, 32778–32785.

65 Aeed, P.A. and Elhammer, A.P. (1994) Glycosylation of recombinant prorenin in insect cells: the insect cell line Sf9 does not express the mannose 6-phosphate recognition signal. *Biochemistry*, **33**, 8793–8797.

66 Van Patten, S.M. *et al.* (2007) Effect of mannose chain length on targeting of glucocerebrosidase for enzyme replacement therapy of Gaucher disease. *Glycobiology*, **17**, 467–478.

67 Altmann, F. *et al.* (1993) Processing of asparagine-linked oligosaccharides in insect cells. *N*-Acetylglucosaminyl-transferase I and II activities in cultured lepidopteran cells. *Glycobiology*, **3**, 619–625.

68 Aoki, K. *et al.* (2007) Dynamic developmental elaboration of N-linked glycan complexity in the *Drosophila melanogaster* embryo. *The Journal of Biological Chemistry*, **282**, 9127–9142.

69 Sarkar, M. *et al.* (2006) Null mutations in *Drosophila N*-acetylglucosaminyl-transferase I produce defects in locomotion and a reduced life span. *The Journal of Biological Chemistry*, **281**, 12776–12785.

70 Pitts, J.D. (2006) *Effects of Sialyltransferase Mutation on Drosophila melanogaster Viability, Fertility, and Longevity*, Texas A&M University College Station, TX.

71 Haines, N. and Irvine, K.D. (2005) Functional analysis of *Drosophila* β1,4-*N*-acetlygalactosaminyltransferases. *Glycobiology*, **15**, 335–346.

72 Tsitilou, S.G. and Grammenoudi, S. (2003) Evidence for alternative splicing and developmental regulation of the *Drosophila melanogaster* Mgat2 (*N*-acetylglucosaminyltransferase II) gene. *Biochemical and Biophysical Research Communications*, **312**, 1372–1376.

73 Fabini, G. and Wilson, I.B. (2001) Glycosyltransferases in *Drosophila melanogaster. Drosophila Information Service*, **84**, 122–129.

74 Vadaie, N. and Jarvis, D.L. (2004) Molecular cloning and functional characterization of a Lepidopteran insect β4-*N*-acetylgalactosaminyltransferase with broad substrate specificity, a functional role in glycoprotein biosynthesis, and a potential functional role in glycolipid biosynthesis. *The Journal of Biological Chemistry*, **279**, 33501–33518.

75 van Die, I. *et al.* (1996) Glycosylation in lepidopteran insect cells: identification of a β1 → 4-*N*-acetylgalactosaminyltrans-ferase involved in the synthesis of complex-type oligosaccharide chains. *Glycobiology*, **6**, 157–164.

76 Abdul-Rahman, B. *et al.* (2002) β-(1 → 4)-galactosyltransferase activity in native and engineered insect cells measured with time-resolved europium fluorescence. *Carbohydrate Research*, **337**, 2181–2186.

77 Koles, K., Irvine, K.D. and Panin, V.M. (2004) Functional characterization of *Drosophila* sialyltransferase. *The Journal of Biological Chemistry*, **279**, 4346–4357.

78 Fabini, G. and Wilson, I.B. (2002) Stage and tissue specific expression of a sialyltransferase-like gene (CG4871) in *Drosophila melanogaster. Drosophila Information Service*, **85**, 45–49.

79 Kim, K. *et al.* (2002) Expression of a functional *Drosophila melanogaster N*-acetylneuraminic acid (Neu5Ac) phosphate synthase gene: evidence for endogenous sialic acid biosynthetic ability in insects. *Glycobiology*, **12**, 73–83.

80 Viswanathan, K. *et al.* (2006) Expression of a functional *Drosophila melanogaster* CMP-sialic acid synthetase: differential localization of the *Drosophila* and human enzymes. *The Journal of Biological Chemistry*, **281**, 15929–15940.

81 Tomiya, N. *et al.* (2001) Determination of nucleotides and sugar nucleotides involved in protein glycosylation by high-performance anion-exchange chromatography: sugar nucleotide contents in cultured insect cells and mammalian cells. *Analytical Biochemistry*, **293**, 129–137.

82 Friesen, P.D. (1997) Regulation of baculovirus early gene expression, in *The Baculoviruses* (ed. L.K. Miller), Plenum Press, New York, pp. 141–170.

83 Guarino, L.A., Gonzalez, M.A. and Summers, M.D. (1986) Complete sequence and enhancer function of the homologous DNA regions of *Autographa californica* nuclear polyhedrosis virus. *Journal of Virology*, **60**, 224–229.

84 Guarino, L.A. and Summers, M.D. (1986) Functional mapping of a trans-activating gene required for expression of a baculovirus delayed-early gene. *Journal of Virology*, **57**, 563–571.

85 Wagner, R. *et al.* (1996) Elongation of the *N*-glycans of fowl plague virus hemagglutinin expressed in *Spodoptera frugiperda* (Sf9) cells by coexpression of human β1,2-*N*-acetylglucosaminyltransferase I. *Glycobiology*, **6**, 165–175.

86 Jarvis, D.L. and Finn, E.E. (1996) Modifying the insect cell *N*-glycosylation pathway with immediate early baculovirus expression vectors. *Nature Biotechnology*, **14**, 1288–1292.

87 Seo, N.S., Hollister, J.R. and Jarvis, D.L. (2001) Mammalian glycosyltransferase expression allows sialoglycoprotein production by baculovirus-infected insect cells. *Protein Expression and Purification*, **22**, 234–241.

88 Lawrence, S.M. *et al.* (2000) Cloning and expression of the human *N*-acetylneuraminic acid phosphate synthase gene with 2-keto-3-deoxy-D-*glycero*-D-*galacto*-nononic acid biosynthetic ability. *The Journal of Biological Chemistry*, **275**, 17869–17877.

89 Hollister, J.R. and Jarvis, D.L. (2001) Engineering lepidopteran insect cells for sialoglycoprotein production by genetic transformation with mammalian β1,4-galactosyltransferase and α2,6-sialyltransferase genes. *Glycobiology*, **11**, 1–9.

90 Hollister, J., Conradt, H. and Jarvis, D.L. (2003) Evidence for a sialic acid salvaging pathway in lepidopteran insect cells. *Glycobiology*, **13**, 487–495.

91 Jarvis, D.L., Howe, D. and Aumiller, J.J. (2001) Novel baculovirus expression vectors that provide sialylation of recombinant glycoproteins in lepidopteran insect cells. *Journal of Virology*, **75**, 6223–6227.

92 Tomiya, N. *et al.* (2003) Complex-type biantennary *N*-glycans of recombinant human transferrin from *Trichoplusia ni* insect cells expressing mammalian β1,4-galactosyltransferase and β1,2-*N*-acetylglucosaminyltransferase II. *Glycobiology*, **13**, 23–34.

93 Chang, G.D. *et al.* (2003) Improvement of glycosylation in insect cells with mammalian glycosyltransferases. *Journal of Biotechnology*, **102**, 61–71.

94 Schenkman, S. *et al.* (1994) Structural and functional properties of Trypanosoma *trans*-sialidase. *Annual Review of Microbiology*, **48**, 499–523.

95 Scudder, P. *et al.* (1993) Enzymatic characterization of [β]-D-galactoside [α]2,3-*trans*-sialidase from *Trypanosoma cruzi*. *The Journal of Biological Chemistry*, **268**, 9886–9891.

96 Marchal, I. *et al.* (2001) Expression of a membrane-bound form of *Trypanosoma cruzi trans*-sialidase in baculovirus-infected insect cells: a potential tool for sialylation of glycoproteins produced in the baculovirus-insect cells system. *Glycobiology*, **11**, 593–603.

97 Effertz, K., Hinderlich, S. and Reutter, W. (1999) Selective loss of either the epimerase or kinase activity of UDP-*N*-acetylglucosamine 2-epimerase/*N*-acetylmannosamine kinase due to site-directed mutagenesis based on sequence alignments. *The Journal of Biological Chemistry*, **274**, 28771–28778.

98 Viswanathan, K. *et al.* (2003) Engineering sialic acid synthetic ability into insect cells: identifying metabolic bottlenecks and devising strategies to overcome them. *Biochemistry*, **42**, 15215–15225.

99 Inoue, S. and Kitajima, K., (2006) KDN (deaminated neuraminic acid): dreamful past and exciting future of the newest member of the sialic acid family. *Glycoconjugate Journal*, **23**, 277–290.

100 Nakata, D. *et al.* (2000) Molecular cloning and expression of the mouse *N*-acetylneuraminic acid 9-phosphate synthase which does not have

deaminoneuraminic acid (KDN) 9-phosphate synthase activity. *Biochemical and Biophysical Research Communications*, **273**, 642–648.
101 Lawrence, S.M. *et al.* (2001) Cloning and expression of human sialic acid pathway genes to generate CMP-sialic acids in insect cells. *Glycoconjugate Journal*, **18**, 205–213.
102 Viswanathan, K. *et al.* (2005) Engineering intracellular CMP-sialic acid metabolism into insect cells and methods to enhance its generation. *Biochemistry*, **44**, 7526–7534.
103 Münster, A.K. *et al.* (1998) Mammalian cytidine 5′-monophosphate *N*-acetylneuraminic acid synthetase: a nuclear protein with evolutionarily conserved structural motifs. *Proceedings of the National Academy of Sciences of the United States of America*, **95**, 9140–9145.
104 Hill, D.R. *et al.* (2006) Isolation and analysis of a baculovirus vector that supports recombinant glycoprotein sialylation by SfSWT-1 cells cultured in serum-free medium. *Biotechnology and Bioengineering*, **95**, 37–47.
105 Hollister, J. *et al.* (2002) Engineering the protein *N*-glycosylation pathway in insect cells for production of biantennary, complex *N*-glycans. *Biochemistry*, **41**, 15093–11504.
106 Jarvis, D.L. *et al.* (1990) Use of early baculovirus promoters for continuous expression and efficient processing of foreign gene products in stably transformed lepidopteran cells. *Nature Biotechnology*, **8**, 950–955.
107 Jarvis, D.L. (1993) Effects of baculovirus infection on IE1-mediated foreign gene expression in stably transformed insect cells. *Journal of Virology*, **67**, 2583–2591.
108 Hollister, J.R., Shaper, J.H. and Jarvis, D.L. (1998) Stable expression of mammalian β1,4-galactosyltransferase extends the *N*-glycosylation pathway in insect cells. *Glycobiology*, **8**, 473–480.
109 Breitbach, K. and Jarvis, D.L. (2001) Improved glycosylation of a foreign protein by Tn-5B1-4 cells engineered to express mammalian glycosyltransferases. *Biotechnology and Bioengineering*, **74**, 230–239.
110 Aumiller, J.J., Hollister, J.R. and Jarvis, D.L. (2003) A transgenic insect cell line engineered to produce CMP-sialic acid and sialylated glycoproteins. *Glycobiology*, **13**, 497–507.
111 Legardinier, S. *et al.* (2005) Mammalian-like nonsialyl complex-type *N*-glycosylation of equine gonadotropins in Mimic™ insect cells. *Glycobiology*, **15**, 776–790.
112 Shi, X. *et al.* (2007) Construction and characterization of new *piggyBac* vectors for constitutive or inducible expression of heterologous gene pairs and the identification of a previously unrecognized activator sequence in *piggyBac*. *BMC Biotechnology*, **7**, 5.
113 Cary, L.C. *et al.* (1989) Transposon mutagenesis of baculoviruses: analysis of *Trichoplusia ni* transposon IFP2 insertions within the FP-locus of nuclear polyhedrosis viruses. *Virology*, **172**, 156–169.
114 Elick, T.A., Lobo, N. and Fraser, M.J. Jr (1997) Analysis of the cis-acting DNA elements required for *piggyBac* transposable element excision. *Molecular and General Genetics*, **255**, 605–610.
115 Fraser, M.J. *et al.* (1995) Assay for movement of lepidopteran transposon IFP2 in insect cells using a baculovirus genome as a target DNA. *Virology*, **211**, 397–407.
116 Elick, T.A., Bauser, C.A. and Fraser, M.J. (1996) Excision of the *piggyBac* transposable element *in vitro* is a precise event that is enhanced by the expression of its encoded transposase. *Genetica*, **98**, 33–41.
117 Jones, M.B. *et al.* (2004) Characterization of the cellular uptake and metabolic conversion of acetylated *N*-acetylmannosamine (ManNAc) analogues to sialic acids. *Biotechnology and Bioengineering*, **85**, 394–405.

118 Ikeda, Y. *et al.* (2000) Kinetic basis for the donor nucleotide-sugar specificity of β1,4-*N*-acetylglucosaminyltransferase III. *Journal of Biochemistry*, **128**, 609–619.

119 Sasai, K. *et al.* (2002) UDP-GlcNAc concentration is an important factor in the biosynthesis of β1,6-branched oligosaccharides: regulation based on the kinetic properties of *N*-acetylglucosaminyltransferase V. *Glycobiology*, **12**, 119–127.

120 Watanabe, T. *et al.* (2006) A specific detection of GlcNAcβ1-6Manα1 branches in N-linked glycoproteins based on the specificity of *N*-acetylglucosaminyltransferase VI. *Glycobiology*, **16**, 431–439.

121 Ivannikova, T. *et al.* (2003) Recombinant (2 → 3)-α-sialyltransferase immobilized on nickel-agarose for preparative synthesis of sialyl Lewis^x and Lewis^a precursor oligosaccharides. *Carbohydrate Research*, **338**, 1153–1161.

122 Sasaki, H. *et al.* (1987) Carbohydrate structure of erythropoietin expressed in Chinese hamster ovary cells by a human erythropoietin cDNA. *The Journal of Biological Chemistry*, **262**, 12059–12076.

123 Hammond, S.M. *et al.* (2000) An RNA-directed nuclease mediates post-transcriptional gene silencing in *Drosophila* cells. *Nature*, **404**, 293–296.

124 Kennerdell, J.R. and Carthew, R.W. (2000) Heritable gene silencing in *Drosophila* using double-stranded RNA. *Nature Biotechnology*, **18**, 896–898.

125 Lee, Y.S. and Carthew, R.W. (2003) Making a better RNAi vector for *Drosophila*: use of intron spacers. *Methods (San Diego, CA)*, **30**, 322–329.

126 Lin, C.C. *et al.* (2006) Stable RNA interference in *Spodoptera frugiperda* cells by a DNA vector-based method. *Biotechnology Letters*, **28**, 271–277.

8
Getting Bacteria to Glycosylate

Michael Kowarik and Mario F. Feldman

8.1
Introduction

8.1.1
Overview and Background

Bacteria encode an enormous diversity of pathways that synthesize and degrade carbohydrates [1]. Those poly-or oligosaccharides occur as conjugates linked to either proteins or lipids, but they can also be found in form of genuine oligo- or polysaccharides either freely soluble or polymerized to high molecular weight capsular structures. The products of carbohydrate metabolism in prokaryotes are often displayed on the surface of cells [2, 3]. Thus, they are important in pathogenesis as they interact with the immune system [4].

Different developments in recent years have led to the perspective that the enormous biochemical diversity contained in bacteria could be efficiently exploited to produce glycoproteins of industrial importance. First, the growing number of sequenced bacterial genomes increases the information about bacterial polysaccharide metabolizing enzymes on an almost daily basis. Second, three protein glycosylation enzymes – one for *N*- and two for *O*-glycosylation – were discovered, allowing the *in vivo* conjugation of poly- and oligosaccharides to polypeptides [5, 6]. Third, it was shown that these conjugation machineries are active in *Escherichia coli* [7, 8], the most frequently used and genetically best characterized bacterium in the world. Although it is clear that the combination of the knowledge carries an enormous potential, many details of the different processes are still poorly understood, and thus the product range accessible by bacterial glycosylation is still limited.

Two classes of commercially relevant products are predicted to be within reach of the bacterial glycosylation technologies: first, glycoproteins containing defined eukaryotic sugar structures for therapeutic use in humans, and, second, glycoproteins containing bacterial sugar structures for use as polysaccharide conjugate-like vaccines, herein called "bioconjugates" [9]. In this chapter we overview the bacterial

Post-translational Modification of Protein Biopharmaceuticals. Edited by Gary Walsh

ISBN: 978-3-527-32074-5

N- and O-linked protein glycosylation systems, and evaluate the challenges for the synthesis of the mentioned product classes.

8.1.2
Bacterial Protein Glycosylation

For a long time the dogma of glycobiology was that bacteria and archeae lack the ability to glycosylate proteins [10]. The first glycoproteins found in prokaryotic cells were surface layer glycoproteins, the major components of the cell wall, in the Gram-negative halophile archaea *Halobacterium salinarium* [11]. After this initial report of glycosylated proteins in procaryotic cells, various glycoproteins have been discovered in the eubacterial domain. The glycoproteins described were mostly O-linked and integrated into cell-surface appendages, such as pili and flagella [10, 12]. The first N-linked protein glycosylation system was discovered in the Gram-negative, food borne pathogen *Campylobacter jejuni* [13], showing that also bacteria exhibit the ability for *N*-glycosylation.

8.2
N-Glycosylation

8.2.1
Introduction

Today about two dozen organisms from different subgroups of the proteobacteria are assumed to encode an *N*-glycosylation system in their genomes. They all represent potential sources for genes with different desirable activities. Among them are many Campylobacterales like *C. coli*, *C. lari* and *C. upsaliensis*, but also *Wollinella succinogenes* and members of the delta subgroup of proteobacteria like *Desulfovibrio desulfuricans* and *D. gigas* [14, 15].

The functions of glycosylation in bacteria are still under debate. Pioneering studies showed that protein *N*-glycosylation is important for adhesion and invasion of host cells by *C. jejuni in vivo* and *in vitro* [16, 17]. Although important for survival and colonization in the natural habitat, *C. jejuni in vitro* growth is unaffected in glycosylation defective mutants [18]. This is in striking contrast to the eukaryotic system, where most *N*-glycosylation knockouts are lethal. Thus, bacterial cells represent a suitable and promising host cell system for glycosylation engineering because their survival is independent of the glycosylation machinery. This fact encodes an advantage towards all eukaryotic glycosylation systems, where engineering always interferes with an essential endogenous process.

Because *N*-glycosylation was first discovered in *C. jejuni*, this system represents a prototype for *N*-glycosylation in bacteria. The proteins required for the *C. jejuni* *N*-glycosylation are encoded in a single genetic locus, called the *pgl* (protein glycosylation) cluster. It consists of eleven genes [19] and its recombinant expression functionally reconstitutes the pathway in *E. coli* [13]. This allowed a rapid

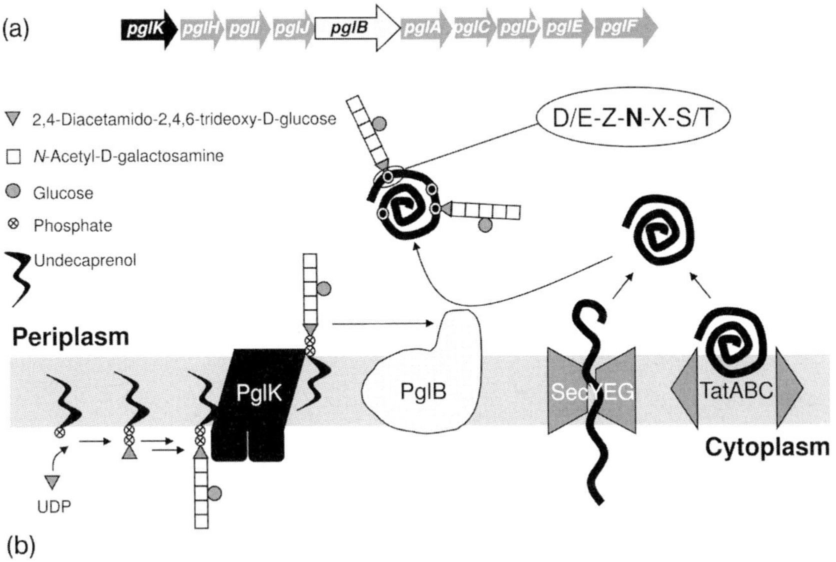

Figure 8.1 Bacterial *N*-glycosylation pathway from *Campylobacter jejuni*. (a) Organization of the *pgl* cluster in the *C. jejuni* genome. Genes encoding proteins for biosynthesis of nucleotide activated sugars are indicated in grey. Black is the flippase, white the OTase. Gene names are indicated. (b) Cartoon of the *N*-glycosylation machinery of *C. jejuni*. Assembly of the oligosaccharide starts on the cytoplasmic side of the membrane on the polyisoprenyl lipid undecaprenol-phosphate. First, a glycosylphosphate transferase, then different glycosyltransferases elongate the non-reducing end of the oligosaccharide, forming the lipid-linked glycan substrate, resulting in a heptasaccharide linked to the pyrophosphate-lipid (for the precise structure see Figure 8.3). After flipping to the extracytoplasmic side of the membrane by the enzyme PglK, condensation of glycan and protein occurs in the bacterial periplasm. Like the eukaryotic system, the *C. jejuni* glycosylation functions in a general manner. This means that a wide variety of proteins can be modified at asparagine residues embedded in a defined consensus sequon as indicated. The consensus sequence around the modified asparagine attachment site for the glycans is conserved. The pyrophosphate (PP) moiety linking the OS substrate to a lipid provides the energy for the condensation reaction.

genetic and biochemical analysis of the different elements of the *C. jejuni* pathway (Figure 8.1).

To understand the usability of the bacterial *N*-glycosylation system for the production of *N*-glycoproteins we present the details and intrinsic characteristics of the *C. jejuni pgl* system below.

8.2.2
The Acceptor Protein

Here we summarize relevant data that describe the chemical and conformational conditions required in a protein to make it an acceptor substrate for *N*-glycosylation by the *C. jejuni pgl* machinery.

8.2.2.1 Primary Acceptor Consensus

A first important characteristic in *C. jejuni N*-glycosylation is the primary consensus sequence on an acceptor protein required for glycan attachment. The *C. jejuni* PglB OST recognizes a pentapeptide sequence motif, D/E-X-N-Z-S/T, where X and Z cannot be proline [20]. In contrast, the eukaryotic acceptor motif lacks the two N-terminal amino acids, resulting in N-X-S/T, where X cannot be P. Thus, the consensus sequences overlap significantly and also share a conformational characteristic shown by the exclusion of P from position +1. However, the negatively charged residue in the −2 position, which is essential for glycosylation in *C. jejuni*, is rarely found in eukaryotic sites. Further statistical analysis of bacterial *N*-glycosylation sites is difficult due to the limited number of bacterial *N*-glycoproteins identified to date. Twenty two proteins containing 32 active acceptor sites were unequivocally identified from *C. jejuni* glycoproteins. With the striking exception of the D/E at position –2, residues found around *C. jejuni* glycosylation sites are generally similar to those from analyses of eukaryotic *N*-glycosylation sites [21–24].

An optimal consensus sequon would be desirable to insert into proteins for *N*-glycoprotein production in bacteria. One such sequence (DQNAT) was identified using a *C. jejuni in vitro* glycosylation system with short peptides as acceptors and chemically synthesized lipid-linked oligosaccharides (LLOs) [25]. In this study, position −1 was found to prefer amido and large hydrophobic side chain functionalities, whereas charged amino acids were disfavored at that position. The +1 position favored the positively charged functionalities of lysine and arginine, along with alanine and serine, while disfavoring large hydrophobic functionalities. Interestingly, the alignment of naturally occurring consensus sequons differs from the DQNAT [20]. In the context of a full-length folded protein, the differences between variations of the glycosylation sequences were found to be consistent with the trends observed from the peptides, though less dramatic because of additional conformational influences. Taken together, these data emphasize the importance of the conformational context of a glycosylation site for efficient glycosylation.

8.2.2.2 Conformational Requirements for *N*-Glycosylation

That conformation and also folding are important characteristics for glycosylation is known from the eukaryotic system [26]. In eukaryotes, *N*-glycosylation is functionally coupled to translocation and folding of the acceptor site consensus sequence [27]. Folding generally inhibits glycosylation [28]. The acceptor motif has to be flexible to become glycosylated and that folding restrains the flexibility. This condition was also observed in the *C. jejuni* system when using a modified version of the protein bovine ribonuclease A (RNaseA) as a model acceptor substrate [29]. This RNase variant (RNaseA*) contained a single point mutation to encode a bacterial glycosylation site at the position of the natural one. Using an *in vitro* glycosylation system, chemical unfolding of RNaseA* led to a higher glycosylation efficiency. When RNase A was present in a random coil conformation, the glycosylation efficiency was similar to the situation in a molten globule or unnaturally oxidized, compact state of the acceptor protein. In contrast, the enzymatically active, fully folded protein was a poor acceptor.

This data suggest that, similar to the situation in eukaryotes, a flexible state of the acceptor site supports glycosylation in the bacterial system.

8.2.2.3 Crystal Structures of Bacterial *N*-Glycoproteins

Two crystal structures of unglycosylated forms of *C. jejuni* *N*-glycoproteins allow further analysis of the structural context of bacterial *N*-glycosylation sites [30, 31]. These proteins, Peb3 and CjaA, contain sequons resembling each other in their side-chain arrangement, the exposure on the protein surface and their positions between two secondary structural elements. The side chains of the amino acids −2, 0 and +2 within the consensus sequence point outwards into the solvent, whereas the side chains of residues −1 and +1 are buried in the body of the protein. It is not known if these folded, crystallized structures represent glycosylation competent acceptor proteins for *N*-glycosylation. However, experimental evidence obtained with the *C. jejuni* glycoprotein AcrA suggests that a purified and folded protein can accept *N*-glycans for glycosylation *in vitro* and *in vivo* on its natural sites [29]. This suggests that, in a folded protein, the glycosylation sequon needs to be present in a glycosylation competent conformation to be active. Thus, the conformations of the glycosylation sites represented in the crystal structures of Peb3 and CjaA may be useful templates for the 3D design of bacterial *N*-glycosylation sites into arbitrary proteins of interest.

To directly compare the differences in site selection by the glycosylation machineries from *C. jejuni* (recombinantly expressed in *E. coli*) and eukaryotes, the *C. jejuni* *N*-glycoprotein AcrA was expressed in the yeast *Saccharomyces cerevisiae* [29]. The yeast *N*-glycosylation machinery selected three out of the five eukaryotic tripeptide consensus sequences for glycosylation. Only one of them was of the bacteria specific pentapeptide type, of which there are two in this protein. Of course, both native bacterial *N*-glycosylation sites encoded in the polypeptide sequence of AcrA were modified in the bacterial system. However, glycosylation of artificially inserted consensus sequons in AcrA was dependent on where the consensus sequence was introduced. Thus, glycosylation is conformation specific as well as primary sequence specific. This implies that eukaryotes and bacteria differ and overlap in site usage and thus represent individually specific systems.

8.2.3
The LLO Substrate and the N-OTase, PglB

Evidentially, the oligosaccharyl transferase PglB is a key factor in the glycosylation process. PglB specifically recognizes the acceptor sequon and the lipid-linked oligosaccharide (LLO) and transfers the carbohydrate moiety to the amide nitrogen atom of the modified asparagine residue. It is active as a single polypeptide [9] and contains the conserved and catalytically essential motifs WWDYGYand DxxK present in bacterial, archeal and eukaryotic Stt3 homologs [13, 32].

An advantage for the commercial use of the bacterial glycosylation system is the relaxed sugar substrate specificity of the PglB protein [7, 9]. As shown in Figure 8.2,

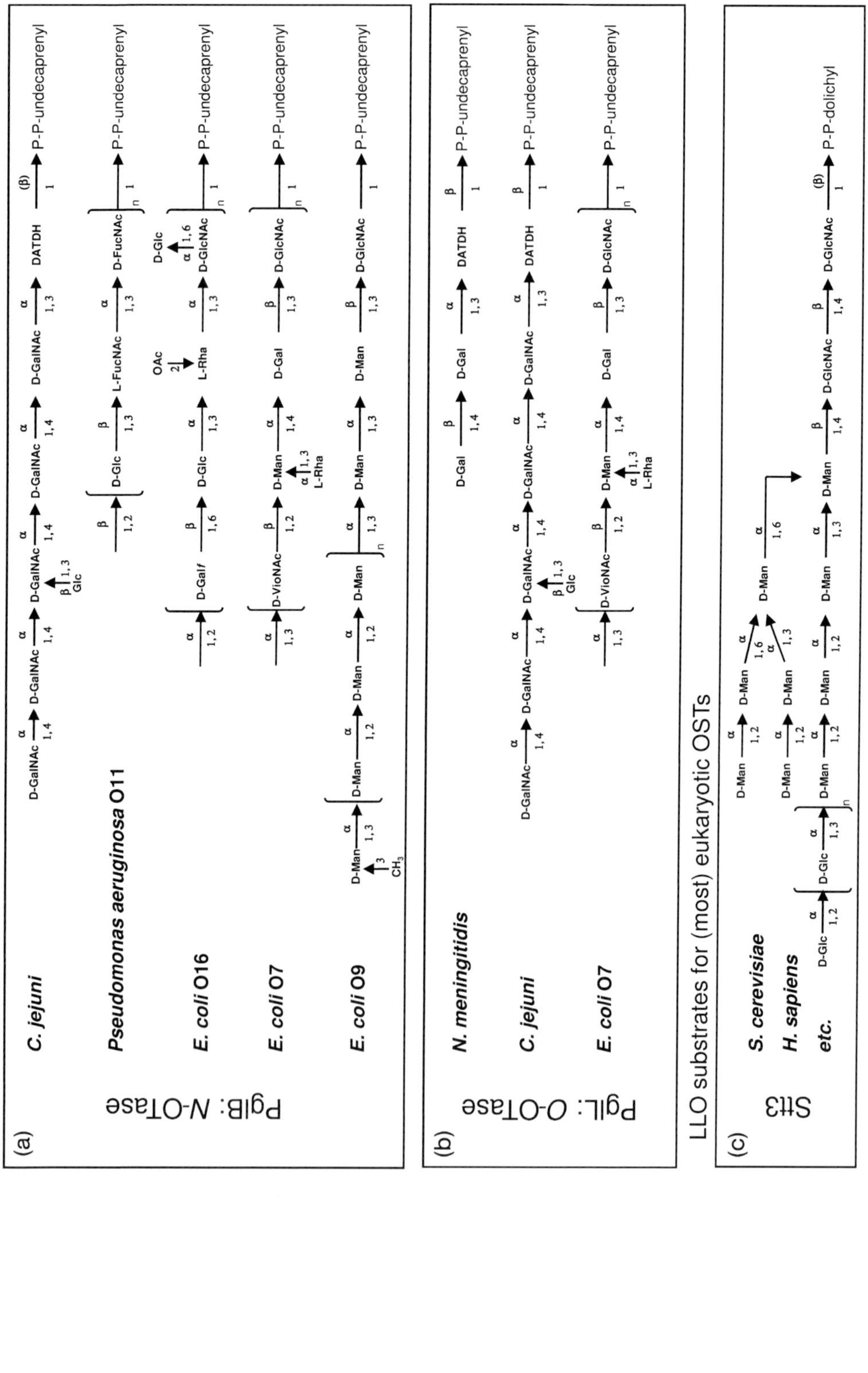
(a)
PglB: N-OTase
C. jejuni
Pseudomonas aeruginosa O11
E. coli O16
E. coli O7
E. coli O9
P-P-undecaprenyl
(b)
PglL: O-OTase
N. meningitidis
C. jejuni
E. coli O7
LLO substrates for (most) eukaryotic OSTs
(c)
Stt3
S. cerevisiae
H. sapiens
etc.
P-P-dolichyl

the natural glycan substrate of PglB is a heptasaccharide consisting of six linearly arranged acetamido sugars and a branching glucose [33]. However, also, other oligo- and polysaccharides are transferred by PglB (Figure 8.2). They share discrete structural characteristics: (i) an acetamido group at position C2 of the monosaccharide at the non-reducing end of the LLO, probably due to a transition state stabilization effect during catalysis; (ii) an alpha or beta 1,3-linkage to the second sugar; and (iii) an undecaprenol-pyrophosphate (Und-PP) lipid as the glycan carrier. *In vitro* analysis with different, chemically synthesized lipid anchors showed specificity of PglB towards the *cis*-double bond geometry and α-unsaturation of the lipid, whereas the precise polyisoprene length was less critical [34]. Polysaccharides containing different D-form monosaccharides at the non-reducing end were shown to act as substrates for PglB (GlcNAc, FucNAc or 2,4-diacetamido-2,4,6-trideoxyglucose). Furthermore, the length of the sugar can range from 1 to over 100 monosaccharides. This probably means that any polymeric sugar can act as a glycosylation substrate as long as the glycan is linked to Und-PP, is present on the periplasmic side of the cytoplasmic membrane and contains a C2 acetamido group on the monosaccharide at the non-reducing end [13].

The lipid Und-PP acts as a generic biosynthetic intermediate for the assembly of higher order carbohydrate polymers (i.e., for the O-antigen of lipopolysaccharide, for capsular polysaccharides of classes 1 and 4, peptidoglycan and enterobacterial common antigen). Some of those structures contain C2 acetamido sugars at the reducing end of the LLO and thus represent potential substrates for PglB, because it might be possible to attach such unrelated Und-PP linked polysaccharide structures (for glycan donor production) to recombinantly expressed *C. jejuni* acceptor proteins by the use of PglB [9]. This opens up the perspective that many different sugar pathways could be associated to OTase-dependent glycosylation in *E. coli* for the production of tailor made *N*-glycoproteins.

8.3 *O*-Glycosylation

O-Glycosylation can be found in several prokaryotic organisms. Two mechanisms have been described for *O*-glycosylation. In most cases, sugars are attached to the

Figure 8.2 Oligosaccharide substrates transferred to proteins by PglB and PilL/PilO. Oligosaccharide structures from different bacterial oligo-/polysaccharide biosynthesis pathways (indicated on the left) are shown. Brackets indicate the repeating units of O-antigen polymer structures. The lipid anchor is shown to the right, and the polymer extends from right to left (reducing to non-reducing end) as in the order of biosynthesis. (a) Glycans shown to be transferred to protein by the *N*-OTase PglB of *C. jejuni*. (b) Glycans shown to be transferred by the *O*-OTase of *N. meningitidis* or *P. aeruginosa*. (c) The *N*-glycan of eukaryotes for comparison. Note that all sugars transferred by bacterial OTases exhibit at the reducing terminus a hexosamine residue containing an C2-*N*-acetyl group and a 1,3-linkage to the second monosaccharide. *f*: = Furanose.

proteins by glycosyltransferases directly from their nucleotide-activated donors. This is the model for *O*-glycosylation of flagellins in *Listeria, Helicobacter, Campylobacter,* and several adhesins in some *E. coli* strains [14]. Only short glycan chains, generally monosaccharides, are attached to proteins using this pathway. The glycosyltransferases have high specificity for both the target protein and the sugars attached, limiting the applications of these systems for glycoengineering. In a second pathway, oligosaccharides are sequentially assembled by glycosyltransferases onto a lipid carrier, usually Und-PP, in a pathway similar to the one just described for *N*-glycosylation. OTases are responsible for the subsequent transfer en bloc of the glycans from the lipid to select serine or threonine residues in the target proteins. This pathway has been described for pilin *O*-glycosylation in *Neisseria* [8] and *Pseudomonas* [35], and it has been proposed for S-layer formation in Gram-positive organisms [36]. This second mechanism, which has never been reported to occur in eukaryotes, has great potential for glycoengineering and it is therefore the focus of the next subsections.

8.3.1
O-Glycosylation in *Pseudomonas aeruginosa*

8.3.1.1 Introduction

Pilin glycosylation was first described in a single strain of *Pseudomonas aeruginosa,* strain 1244 [6], and more recently it has been found to occur in several clinical isolates [37]. Two lines of evidence suggested that PilO activity is located to the periplasm. First, translocation of the Und-PP-linked into the periplasm is required for glycosylation [8]. Second, topology studies showed that PilO contains regions required for enzymatic activity that are located in the periplasm [38].

8.3.1.2 The Acceptor Protein for *O*-Glycosylation

Pilin appears to be the only protein that is *O*-glycosylated by PilO. Unlike in the *C. jejuni* system, there is no obvious consensus sequence to which the glycans are attached. Instead, the glycosylation site is the C-terminal serine residue of the protein. PilO is able to transfer truncated O-antigen subunits to its acceptor protein, but it cannot accept long glycan chains [8, 39]. Likely, PilO has evolved to prevent transfer of long O-antigen chains to pilin, as they could interfere with pilus assembly. A *P. aeruginosa* 1244 PilO mutant strain was able to produce pili normal in appearance and quantity [6]. It formed normal biofilms but exhibited reduced twitching motility. Competition index analysis using a mouse respiratory model comparing wild type and PilO deficient 1244 strains showed that the presence of the pilin glycan improves the survival in the lung environment [40]. These results collectively suggested that the pilin glycan is a significant virulence factor and may aid in the establishment of infection. Interestingly, strains capable of expressing the type of pilin that can be glycosylated are more prevalent in cystic fibrosis patients than in environmental strains, suggesting the pilin glycosylation could be advantageous for the bacteria in colonization of these patients [37]. However, the specific role for the pilin glycan remains unknown.

8.3.1.3 Glycan Structures in *P. aeruginosa* *O*-Glycosylation

In the 1244 strain, pilin is glycosylated by the PilO OTase at its C-terminal serine residue [41]. The glycan moiety consists of a trisaccharide, which derives from the LPS biosynthetic pathway [42]. Indeed, the glycan attached to pilin consists of a single O-antigen subunit. DiGiandomenico *et al.* expressed PilO heterologously in other *Pseudomonas* strains, where it was able to transfer various O-antigen subunits to pilin [35]. PilO has also been functionally expressed in *E. coli.* By co-expressing PilO with pilin and the genes included in the *pgl* locus of *C. jejuni*, Faridmoayer *et al.* showed that PilO can also transfer the *C. jejuni* heptasaccharide to pilin [8]. These two experiments demonstrated that PilO, as the PglB transferase of the *C. jejuni* *N*-glycosylation system, possesses relaxed sugar specificity. Recently, the composition of the glycan attached to pilin the *P. aeruginosa* strain 5196 was determined [43]. In this strain, the sugar moiety does not originate from the O-antigen pathway. Instead, the pilin glycan is composed of an unusual homo-oligomer of alpha-1,5-linked d-arabinofuranose (d-Araf). This sugar is uncommon in prokaryotes, but is one of the main components of the arabinogalactan and lipoarabinomannan (LAM) polymers found in mycobacteria cell walls. There is no obvious PilO homologous in the 5196 strain, and therefore the OTase has not yet been identified.

8.3.2
O-Glycosylation in *Neisseria*

8.3.2.1 Introduction

In *Neisseria gonorrhoea* there is only a single *O*-glycosylation substrate protein known. It is, as was just described for *P. aeruginosa*, the pilus-forming protein pilin. Removal of the pilin glycan by mutagenesis of the glycan attachment site did not produce any significant change in piliation or subsequent pilus-mediated adhesion [44]. However, it was shown that the glycan increases the solubilization of pilin monomers and/or individual pilus fibers. Thus, the actual role of pilin glycosylation remains unclear.

8.3.2.2 PglL, the *O*-OTase of *N. meningitidis*

Glycosylation of pilin in *Neisseria meningitidis* and *N. gonorrhoea* has been for several years, as it was first described by Virji *et al.* in 1993 [45]. Power *et al.* showed that a mutation in a gene named *pglL* resulted in a shift in pilin electrophoretic mobility compatible with the loss of the pilin glycan [46]. These authors also found that PglL exhibits a domain with homology to the PilO OTase involved in *P. aeruginosa* pilin glycosylation, and the O-antigen ligases implicated in the last step of LPS biosynthesis (transfer of O-antigen from its Und-PP carrier to the LipidA-core). Based on these observations, the authors hypothesized that PglL could be the OTase involved in pilin glycosylation in *Neisseria*. This hypothesis was supported by the work of Aas *et al.* showing that a mutation in PglL in *N. gonorrhoeae* (they re-named the enzyme as PglO) actually abolished pilin glycosylation [47]. Final confirmation that PglL is an OTase was provided by Faridmoayer *et al.* (Figure 8.3). These authors reconstituted the *O*-pilin glycosylation process in *E. coli,* by co-expressing PglL and PilE, the

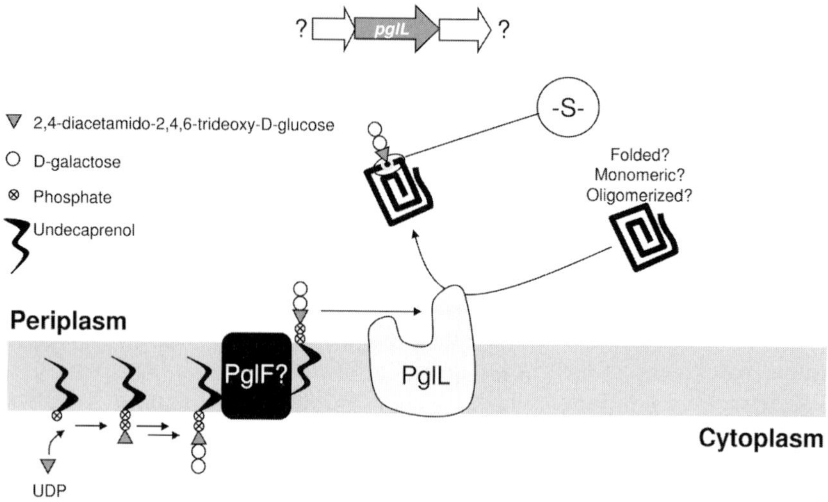

Figure 8.3 *O*-Glycosylation machinery of *Neisseria meningitides*. Compared to the *N*-glycosylation pathway, the details of *O*-glycosylation are much less understood. Glycan biosynthesis occurs on the cytoplasmic side of the membrane, and the glycosylation reaction is dependent on the flipping of the sugar substrate to the periplasmic space. Neither the enzymes for sugar biosynthesis nor the flippase (black) are identified. The acceptor site does not exhibit an obvious primary consensus sequence and only a single protein acceptor (pilin) is known up to date.

Neisseria pilin protein, in the presence of the genes required for the synthesis of several Und-PP-linked glycan substrates [8].

8.3.2.3 Glycan Substrates for *N. meningitidis O*-Glycosylation

In *N. gonorrhoea* the glycan is a disaccharide composed of galactose and *N*-acetylglucosamine at the reducing end [48]. In *N. meningitidis*, the glycan attached to pilin is composed of a terminal 1-4-linked digalactose moiety covalently linked to a 2,4-diacetamido-2,4,6-trideoxyhexose sugar which is directly attached to pilin. Interestingly, the *C. jejuni N*-glycan and the neisserial *O*-glycans contain the same sugar at the reducing end. Until recently it was thought that these sugars were sequentially added to pilin by glycosyltransferases [14]. However, it is now clear that this is not the pathway used. By their genetically reconstituted PilL *O*-glycosylation system in *E. coli*, Faridmoayer *et al.* showed that PglL can transfer a wide variety of Und-PP-linked glycans to the *N. meningitidis* pilin. PglL was sufficient for pilin glycosylation, as no other *Neisseria* protein was present in the *E. coli* cells. The glycans transferred were the *C. jejuni* heptasaccharide and different versions of the *E. coli* O7 polysaccharide (Figure 8.2). This demonstrated that PglL, as PglB and PilO, exhibits low glycan specificity. However, only PglL and PglB are able to transfer polysaccharides. Faridmoayer *et al.* noticed that, in opposition to PilO, PglL and PglB act in organisms that do not produce a typical O-antigen in their LPS, and therefore there is no need for these enzymes to discriminate between short and long glycan chains. The consequences for their putative applications in biotechnology are discussed in the

Table 8.1 Characteristics of *N*- and *O*-glycosylation in bacteria.

Characteristic	*Campylobacter jejuni*	*Neisseria meningitidis*
General		
Process knock-out	Loss of pathogenicity	Pathogenicity marginally affected
Organelle	Periplasm	Periplasm
Protein acceptors/targets	Secretory and membrane proteins	Only one known: pilin
Enzymology		
Elongation of glycan	Cytoplasmic only	Cytoplasmic only
OST	Single protein: PglB	Single protein: PglL
Flippase	ABC transporter	Putative flippase
Mode of glycosylation	*En bloc* transfer	*En bloc* transfer
Substrates		
Lipid carrier for oligosaccharide	Undecaprenol-PP	*En bloc* transfer
Glycan transferred	See Figures 8.1/8.2	See Figures 8.3/8.2
Acceptor site consensus	Pentapeptide: D/E-Z-N-X-S/T	S, consensus unknown

next section. Table 8.1 presents selected characteristics of N- and *O*-linked glycosylation in *C. jejuni* and *N. meningitides.*

8.4 Exploitation of N- and O-Linked Glycosylation

8.4.1 Therapeutic (Human) Proteins

Production of recombinant proteins for clinical applications constitutes a multibillion dollar industry. Glycosylation potentially affects pharmacokinetic and other properties of the recombinant proteins [49]. In some cases, specific glycans are essential for activity. However, different eukaryotic cell lines glycosylate their proteins differentially, potentially limiting the use of mammalian cells for production of recombinant glycoproteins for biotechnological purposes. Glycosylation is an essential process in mammalian cells, and therefore manipulation of glycosylation pathways is extremely difficult. Instead, it has been shown that yeast can be modified to produce glycoproteins containing defined and homogeneous humanized *N*-glycans [50]. This is a significant achievement that could potentially revolutionize the production of recombinant glycoproteins. However, although it has been shown that *N*-glycosylation pathways can be modified in yeast, it has not been demonstrated that the *O*-glycosylation status of the glycoproteins can be also modified.

Bacteria, in particular *E. coli* cells, constitute a perfect toolbox for glycoengineering. As protein glycosylation is normally absent in these bacteria, *E. coli* cells can tolerate the incorporation and manipulation of foreign bacterial glycosylation pathways. In principle, it should be possible to engineer *E. coli* cells to produce glycoproteins

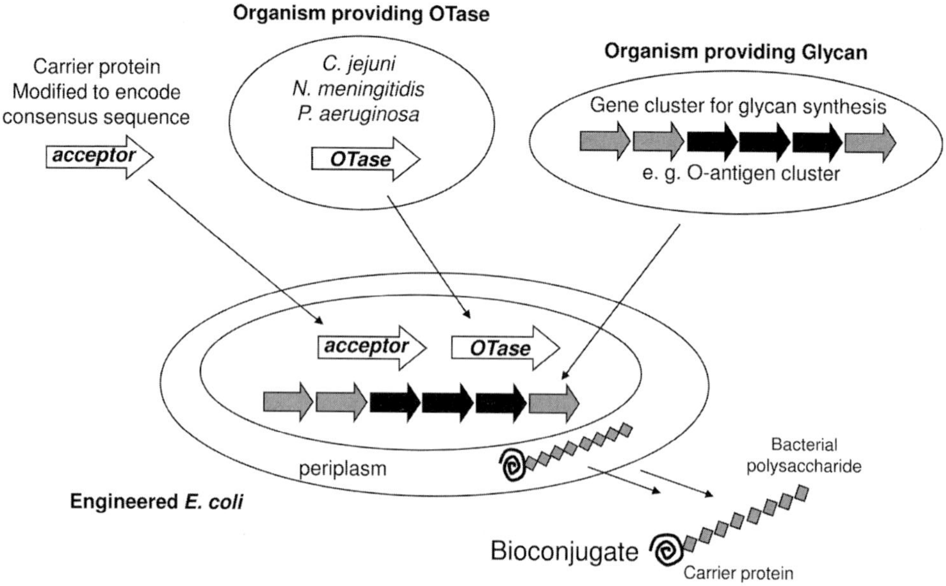

Figure 8.4 Bioconjugate production in engineered *Escherichia coli*. Three components are needed for the synthesis of a glycoprotein: the acceptor protein, the Und-PP-linked glycan and the OTase. The acceptor proteins can be endogenous or artificially designed (glycol) proteins that need to be secreted into the periplasm to become glycosylated. The O-antigen or other glycans attached to the undecaprenyl-PP carrier will be transferred by PglB, creating an N-linkage, or by PglL or PilO creating an O-linkage. PilO activity is restricted to short glycan chains, whereas PglB and PglL can transfer polysaccharides. The bioconjugate can be directly purified from the periplasm of engineered *E. coli*.

containing humanized glycans (Figure 8.4). A major challenge will be the coordinate expression of different glycosyltransferases that would assemble the glycans usually found in human glycoproteins onto the Und-PP carrier. Assuming that this is possible, the bacterial flippases should be able to translocate the Und-PP-linked glycan into the periplasm, as it has been shown that the activity of the flippases is independent of the glycan structure [51, 52]. The relaxed specificity of PglB would allow the transfer of these glycans onto human proteins, previously translocated to the periplasm [9, 13]. These human proteins should be modified to contain the bacterial consensus acceptor sequence [20]. Although, in theory all this is achievable, the assembly of humanized homogenous glycans onto human proteins in bacteria with good yields will be an arduous and lengthy task (Figure 8.4).

8.4.2 Bioconjugate Vaccines

More realistically, bacterial glycoproteins will constitute a new generation of conjugate vaccines in the near future. Glycoconjugate vaccines composed of a cell surface carbohydrate from a microorganism chemically linked to an appropriate carrier protein are effective means to generate protective immune responses to prevent a wide range of diseases [53]. The best example is the conjugate vaccines against *Haemophilus* influenza type b [54]. Other conjugate vaccines have been licensed and several others are currently in clinical trials.

Although these chemically conjugated vaccines have been successfully introduced, the current technology presents some drawbacks. The carbohydrates employed for the formulation of these vaccines are usually obtained from pathogenic organisms, which may constitute a hazardous risk for the patients and may result in higher levels of biosafety for the production. Some of the glycans containing acid-labile sugars do not resist the chemical treatment required for their purification and crosslinking to proteins. Furthermore, the chemical crosslinking is not controllable and the site of linkage between proteins and sugars is not reproducible, resulting in unreliable products that may vary from batch to batch. The use of recombinant bacterial glycoproteins as vaccines can overcome some of these issues. Introducing the genes necessary for the synthesis of capsules or O-antigens in *E. coli*, along with an OTase and a suitable protein carrier, the culture of big amounts of pathogenic or slow-growing bacteria can be avoided. As the length of the polysaccharides and the glycosylation sites is genetically determined, the variability from sample to sample is minimized.

8.4.3 Glycoengineering

Current work in our laboratories focuses on the demonstration that bacterial glycoproteins containing bacterial polysaccharides can elicit protective immune responses. In principle, glycoproteins obtained exploiting both PglB and PglL can be used. Both enzymes exhibit low sugar specificity and can transfer polysaccharides to defined amino acids in their target proteins. PglB presents the advantage that different carrier proteins can be easily engineered [20, 29]. Well-characterized carriers such as the Cholera toxin can be used as carrier by simply including the *C. jejuni* *N*-glycan consensus sequence into a glycosylation competent location. It has not yet been studied whether PglL or PglO glycosylate pilin before, during or after folding of the pilus subunits. It can be speculated that some sort of folding state must be acquired before *O*-glycosylation occurs, as it seems that *O*-OTases recognize a particular motif rather than a sequon on the target protein. More work is still needed to define the minimal glycosylation domain recognized by PglL and PglO during the *O*-glycosylation process. This would enable the glycosylation of other proteins or the construction of "glycotags" (i.e., sequences that can be added to other proteins to incorporate glycans onto carrier proteins). The use of PilO was a priori limited by the inability of this enzyme to transfer long-chain glycans. However, Horzempa *et al.* recently demonstrated that glycosylated *P. aeruginosa* 1244 pilin protects mice from a challenge with a pilus-null isogenic mutant, suggesting that the glycan-specific immune response was responsible for the protection [55]. The importance of the length of the glycans as well as the type of for the efficacy of the vaccines needs to be evaluated.

Taken together it is feasible that, in the future, it will be possible to synthesize desired glycans *in vivo* and *in vitro* by combining different glycosyltransferases [56] and transferring them to suitable endogenous or artificially designed acceptor proteins by the use of bacterial OTases. The exploitation of these different technologies will have an impact on the design of novel vaccines and therapeutics and will enable the production of novel molecules exhibiting new activities.

References

1 Guo, H. *et al.* (2008) Current understanding on biosynthesis of microbial polysaccharides. *Current Topics in Medicinal Chemistry*, **8**, 141–151.

2 Raetz, C.R. and Whitfield, C. (2002) Lipopolysaccharide endotoxins. *Annual Review of Biochemistry*, **71**, 635–700.

3 Whitfield, C. (2006) Biosynthesis assembly of capsular polysaccharides in *Escherichia coli*. *Annual Review of Biochemistry*, **75**, 39–68.

4 Gioannini, T.L. and Weiss, J.P. (2007) Regulation of interactions of Gram-negative bacterial endotoxins with mammalian cells. *Immunologic Research*, **39**, 249–260.

5 Szymanski, C.M. *et al.* (1999) Evidence for a system of general protein glycosylation in *Campylobacter jejuni*. *Molecular Microbiology*, **32**, 1022–1030.

6 Castric, P. (1995) pilO, a gene required for glycosylation of Pseudomonas aeruginosa 1244 pilin. *Microbiology*, **141**(Pt 5), 1247–1254.

7 Wacker, M. *et al.* (2006) Substrate specificity of bacterial oligosaccharyltransferase suggests a common transfer mechanism for the bacterial and eukaryotic systems. *Proceedings of the National Academy of Sciences of the United States of America*, **103**, 7088–7093.

8 Faridmoayer, A. *et al.* (2007) Functional characterization of bacterial oligosaccharyltransferases involved in O-linked protein glycosylation. *Journal of Bacteriology*, **189**, 8088–8098.

9 Feldman, M.F. *et al.* (2005) Engineering N-linked protein glycosylation with diverse O antigen lipopolysaccharide structures in *Escherichia coli*. *Proceedings of the National Academy of Sciences of the United States of America*, **102**, 3016–3021.

10 Messner, P. (2004) Prokaryotic glycoproteins: unexplored but important. *Journal of Bacteriology*, **186**, 2517–2519.

11 Mescher, M.F., Hansen, U. and Strominger, J.L. (1976) Formation of lipid-linked sugar compounds in Halobacterium salinarium. Presumed intermediates in glycoprotein synthesis. *The Journal of Biological Chemistry*, **251**, 7289–7294.

12 Weerapana, E. and Imperiali, B. (2006) Asparagine-linked protein glycosylation: from eukaryotic to prokaryotic systems. *Glycobiology*, **16**, 91R–101R.

13 Wacker, M. *et al.* (2002) N-linked glycosylation in Campylobacter jejuni and its functional transfer into *E. coli*. *Science*, **298**, 1790–1793.

14 Szymanski, C.M. and Wren, B.W. (2005) Protein glycosylation in bacterial mucosal pathogens. *Nature Reviews Microbiology*, **3**, 225–237.

15 Santos-Silva, T. *et al.* (2007) Crystal structure of the 16 heme cytochrome from Desulfovibrio gigas: a glycosylated protein in a sulphate-reducing bacterium. *Journal of Molecular Biology*, **370**, 659–673.

16 Karlyshev, A.V. *et al.* (2004) The Campylobacter jejuni general glycosylation system is important for attachment to human epithelial cells and

in the colonization of chicks. *Microbiology*, **150**, 1957–1964.

17 Szymanski, C.M., Burr, D.H. and Guerry, P. (2002) Campylobacter protein glycosylation affects host cell interactions. *Infection and Immunity*, **70**, 2242–2244.

18 Hendrixson, D.R. and DiRita, V.J. (2004) Identification of *Campylobacter jejuni* genes involved in commensal colonization of the chick gastrointestinal tract. *Molecular Microbiology*, **52**, 471–484.

19 Linton, D. *et al.* (2005) Functional analysis of the *Campylobacter jejuni* N-linked protein glycosylation pathway. *Molecular Microbiology*, **55**, 1695–1703.

20 Kowarik, M. *et al.* (2006) Definition of the bacterial *N*-glycosylation site consensus sequence. *The EMBO Journal*, **25**, 1957–1966.

21 Ben-Dor, S. *et al.* (2004) Biases and complex patterns in the residues flanking protein *N*-glycosylation sites. *Glycobiology*, **14**, 95–101.

22 Petrescu, A.J. *et al.* (2004) Statistical analysis of the protein environment of *N*-glycosylation sites: implications for occupancy, structure, and folding. *Glycobiology*, **14**, 103–114.

23 Shakin-Eshleman, S.H., Spitalnik, S.L. and Kasturi, L. (1996) The amino acid at the X position of an Asn-X-Ser sequon is an important determinant of N-linked core-glycosylation efficiency. *The Journal of Biological Chemistry*, **271**, 6363–6366.

24 Mellquist, J.L. *et al.* (1998) The amino acid following an asn-X-Ser/Thr sequon is an important determinant of N-linked core glycosylation efficiency. *Biochemistry*, **37**, 6833–6837.

25 Chen, M.M., Glover, K.J. and Imperiali, B. (2007) From peptide to protein: comparative analysis of the substrate specificity of N-linked glycosylation in *C. jejuni*. *Biochemistry*, **46**, 5579–5585.

26 Chen, W. and Helenius, A. (2000) Role of ribosome and translocon complex during folding of influenza hemagglutinin in the endoplasmic reticulum of living cells. *Molecular Biology of the Cell*, **11**, 765–772.

27 Mingarro, I. *et al.* (2000) Different conformations of nascent polypeptides during translocation across the ER membrane. *BMC Cell Biology*, **1**, 3.

28 Bulleid, N.J. *et al.* (1992) Cell-free synthesis of enzymically active tissue-type plasminogen activator. Protein folding determines the extent of N-linked glycosylation. *The Biochemical Journal*, **286** (Pt 1), 275–280.

29 Kowarik, M. *et al.* (2006) N-linked glycosylation of folded proteins by the bacterial oligosaccharyltransferase. *Science*, **314**, 1148–1150.

30 Muller, A. *et al.* (2005) An ATP-binding cassette-type cysteine transporter in Campylobacter jejuni inferred from the structure of an extracytoplasmic solute receptor protein. *Molecular Microbiology*, **57**, 143–155.

31 Rangarajan, E.S. *et al.* (2007) Structural context for protein *N*-glycosylation in bacteria: The structure of PEB3, an adhesin from *Campylobacter jejuni*. *Protein Science*, **16**, 990–995.

32 Igura, M. *et al.* (2008) Structure-guided identification of a new catalytic motif of oligosaccharyltransferase. *The EMBO Journal*, **27**, 234–243.

33 Young, N.M. *et al.* (2002) Structure of the N-linked glycan present on multiple glycoproteins in the Gram-negative bacterium, *Campylobacter jejuni*. *The Journal of Biological Chemistry*, **277**, 42530–42539.

34 Chen, M.M. *et al.* (2007) Polyisoprenol specificity in the Campylobacter jejuni N-linked glycosylation pathway. *Biochemistry*, **46**, 14342–14348.

35 DiGiandomenico, A. *et al.* (2002) Glycosylation of Pseudomonas aeruginosa 1244 pilin: glycan substrate specificity. *Molecular Microbiology*, **46**, 519–530.

36 Steiner, K. *et al.* (2007) Functional characterization of the initiation enzyme of S-layer glycoprotein glycan biosynthesis in Geobacillus stearothermophilus NRS 2004/3a. *Journal of Bacteriology*, **189**, 2590–2598.

37 Kus, J.V. *et al.* (2004) Significant differences in type IV pilin allele distribution among Pseudomonas aeruginosa isolates from cystic fibrosis (CF) versus non-CF patients. *Microbiology*, **150**, 1315–1326.

38 Qutyan, M., Paliotti, M. and Castric, P. (2007) PilO of Pseudomonas aeruginosa 1244: subcellular location and domain assignment. *Molecular Microbiology*, **66**, 1444–1458.

39 Horzempa, J. *et al.* (2006) Glycosylation substrate specificity of Pseudomonas aeruginosa 1244 pilin. *The Journal of Biological Chemistry*, **281**, 1128–1136.

40 Smedley, J.G. 3rd. *et al.* (2005) Influence of pilin glycosylation on Pseudomonas aeruginosa 1244 pilus function. *Infection and Immunity*, **73**, 7922–7931.

41 Comer, J.E. *et al.* (2002) Identification of the Pseudomonas aeruginosa 1244 pilin glycosylation site. *Infection and Immunity*, **70**, 2837–2845.

42 Castric, P., Cassels, F.J. and Carlson, R.W. (2001) Structural characterization of the Pseudomonas aeruginosa 1244 pilin glycan. *The Journal of Biological Chemistry*, **276**, 26479–26485.

43 Voisin, S. *et al.* (2007) Glycosylation of Pseudomonas aeruginosa strain Pa5196 type IV pilins with mycobacterium-like alpha-1,5-linked d-Araf oligosaccharides. *Journal of Bacteriology*, **189**, 151–159.

44 Marceau, M. *et al.* (1998) Consequences of the loss of O-linked glycosylation of meningococcal type IV pilin on piliation and pilus-mediated adhesion. *Molecular Microbiology*, **27**, 705–715.

45 Virji, M. *et al.* (1993) Pilus-facilitated adherence of Neisseria meningitidis to human epithelial and endothelial cells: modulation of adherence phenotype occurs concurrently with changes in primary amino acid sequence and the glycosylation status of pilin. *Molecular Microbiology*, **10**, 1013–1028.

46 Power, P.M., Seib, K.L. and Jennings, M.P. (2006) Pilin glycosylation in Neisseria meningitidis occurs by a similar pathway to wzy-dependent O-antigen biosynthesis in *Escherichia coli. Biochemical and Biophysical Research Communications*, **347**, 904–908.

47 Aas, F.E. *et al.* (2007) Neisseria gonorrhoeae O-linked pilin glycosylation: functional analyses define both the biosynthetic pathway and glycan structure. *Molecular Microbiology*, **65**, 607–624.

48 Parge, H.E. *et al.* (1995) Structure of the fibre-forming protein pilin at 2.6 A resolution. *Nature*, **378**, 32–38.

49 Sinclair, A.M. and Elliott, S. (2005) Glycoengineering: the effect of glycosylation on the properties of therapeutic proteins. *Journal of Pharmaceutical Sciences*, **94**, 1626–1635.

50 Hamilton, S.R. *et al.* (2006) Humanization of yeast to produce complex terminally sialylated glycoproteins. *Science*, **313**, 1441–1443.

51 Feldman, M.F. *et al.* (1999) The activity of a putative polyisoprenol-linked sugar translocase (Wzx) involved in *Escherichia coli* O antigen assembly is independent of the chemical structure of the O repeat. *The Journal of Biological Chemistry*, **274**, 35129–35138.

52 Alaimo, C. *et al.* (2006) Two distinct but interchangeable mechanisms for flipping of lipid-linked oligosaccharides. *The EMBO Journal*, **25**, 967–976.

53 Jones, C. (2005) Vaccines based on the cell surface carbohydrates of pathogenic bacteria. *Anais da Academia Brasileira de Ciencias*, **77**, 293–324.

54 Williams, C. and Masterton, R. (2008) Pneumococcal immunisation in the 21st century. *The Journal of Infection*, **56**, 13–19.

55 Horzempa, J. *et al.* (2008) Immunization with Pseudomonas aeruginosa 1244 pilin provides O-antigen-specific protection. *Clinical and Vaccine Immunology*, **15**, 590–597.

56 Antoine, T. *et al.* (2003) Large-scale *in vivo* synthesis of the carbohydrate moieties of gangliosides GM1 and GM2 by metabolically engineered *Escherichia coli. Chembiochem: A European Journal of Chemical Biology*, **4**, 406–412.

Part Two
Other Modifications

Post-translational Modification of Protein Biopharmaceuticals. Edited by Gary Walsh

ISBN: 978-3-527-32074-5

9
Biopharmaceuticals: Post-Translational Modification Carboxylation and Hydroxylation

Mark A. Brown and Leisa M. Stenberg

9.1
Introduction

The binding of calcium ions is critical for the structural integrity and biological activity of many proteins. In some cases, specific residues in the polypeptide chain must be post-translationally modified to form functional Ca^{2+}-binding sites that contribute to conformational stability and protein–protein and/or protein–membrane interactions. Two post-translational processes that modify amino acids involved in Ca^{2+} binding are γ-carboxylation of glutamyl (Glu) residues to form γ-carboxyglutamate (Gla), and β-hydroxylation of aspartyl (Asp) or asparaginyl (Asn) residues to form *erythro*-β-hydroxyaspartate (Hya) or *erythro*-β-hydroxyasparagine (Hyn). Hydroxylation of other amino acids also occurs for different purposes in several proteins. So far, only a few γ-carboxylated or hydroxylated proteins have been developed for use as biopharmaceuticals, though it can be argued that their high cost consumes a disproportionate amount of healthcare funding [1]. These two post-translational processes and their importance for the structure and function of biopharmaceutical proteins are discussed, with particular reference to factor VIIa, factor IX, activated protein C, prothrombin and pipeline neuropeptides (conotoxins).

9.2
γ-Carboxylation

9.2.1
Biological Function of γ-Carboxylation

The post-translational conversion of a glutamyl residue (Glu) into γ-carboxy glutamate (Gla) is a vitamin K-dependent reaction that confers an important

Post-translational Modification of Protein Biopharmaceuticals. Edited by Gary Walsh

ISBN: 978-3-527-32074-5

Ca^{2+}-binding functionality to the modified polypeptide. The dicarboxylate side chain of a Gla residue, which carries a double negative charge at physiological pH, allows it to chelate Ca^{2+} – a property central to the structure and function of Gla-containing proteins (for reviews, see References [2–5]). γ-Carboxylated proteins have roles in a sweep of biological processes, including hemostasis, apoptosis, signal transduction and tissue mineralization. If the γ-carboxylation process is impaired serious physiological consequences or death may result [6, 7]. Moreover, mutations affecting the Gla residues of vitamin K-dependent proteins underlie various disorders of hemostasis [5, 6, 8].

Gla was first discovered in 1974 in a blood clotting protein, pro thrombin [9, 10], and the essential role played by γ-carboxylation in hemostasis soon became apparent. In a seminal series of papers culminating in the discovery of Gla, Stenflo *et al.* demonstrated the role of this unusual amino acid in binding Ca^{2+} and the requirement for vitamin K for its biosynthesis [10–16]. Thus, the mechanism whereby coumarins such as warfarin (3-α-phenyl-β-acetylethyl-4-hydroxycoumarin) inhibit blood clotting was elucidated. Important contributions during the same period also came from other research groups (e.g., References [9, 17–19]). Gla was subsequently found in several more hemostatic proteins and in proteins involved in other processes, such as tissue mineralization. Thus, the role played by γ-carboxylation in disorders such as hemophilia, thrombosis, fetal warfarin syndrome and Keutel syndrome was clarified. The development of γ-carboxylated proteins as therapeutics has followed. Plasma-derived or recombinant forms of vitamin K-dependent proteins are currently used for treatment of hemophilia B [20, 21], hemorrhage [22] and acute sepsis [23].

To date, 14 proteins that are known or suspected to undergo γ-carboxylation have been identified in humans, and homologs are widely distributed amongst other vertebrates. Several of these proteins circulate in the blood and play important roles in hemostasis as serine proteases or cofactors. Factors VII, IX, X, prothrombin and Gas6 have procoagulant functions, whereas proteins C, S and Z have anticoagulant roles [5, 24]. Gas6 is a ligand for the Axl/Tyro3 group of transmembrane receptors [25]. In addition, four transmembrane receptors with unknown functions have been cloned: the proline-rich Gla proteins, PRGP1 and PRGP2, and the transmembrane Gla proteins, TMG3 and TMG4 [26, 27]. Two other γ-carboxylated proteins, bone Gla protein (osteocalcin) and matrix Gla protein, are associated with bone and the extracellular matrix and play roles in mineralization [28–30]. All of these γ-carboxylated proteins contain multiple Gla residues that, with the exception of the bone and matrix Gla proteins, reside within a stretch of ~45 amino acids known as the Gla domain located at the N-terminus of the mature polypeptides (Figure 9.1) [31]. Depending on the protein, from 9 to 13 Glu residues are present in the Gla domain and all are normally γ-carboxylated. The bone and matrix Gla proteins contain fewer Gla residues (three and five, respectively) and do not possess a Gla domain as such. In all cases, γ-carboxylation is involved in Ca^{2+} binding, which induces a conformational rearrangement in the proteins [32, 33]. In Gla domain-containing proteins this structural change is requisite for their interaction with biological membranes and other proteins [34]. In the bone and matrix Gla proteins

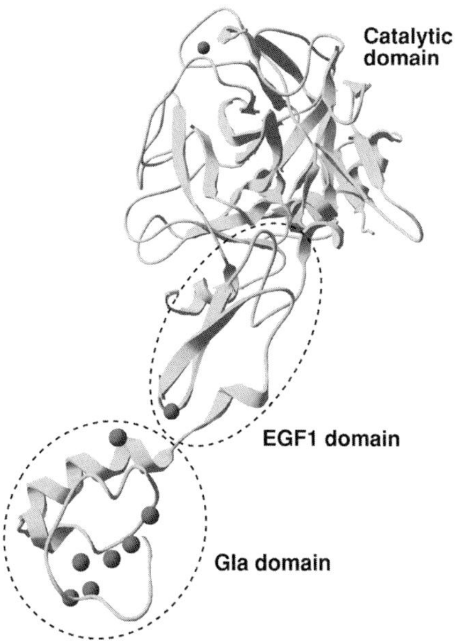

Figure 9.1 Model of the Ca^{2+}-loaded structure of mature human factor VIIa [80]. Ca^{2+} ions are depicted as spheres, with seven bound within the N-terminal Gla domain, one within the adjacent epidermal growth factor-like domain (EGF1) and one within the catalytic (serine protease) domain.

γ-carboxylation is required to mediate mineral binding [6, 29]. In some instances extensive γ-carboxylation of the precursor is required for its efficient secretion and to prevent degradation by the proteosomal pathway [35].

In addition to the vertebrate proteins, several γ-carboxylated peptides have been isolated from the venom of predatory marine snails belonging to the genus *Conus* [36–41]. These Gla-containing peptides bear little resemblance to the vertebrate γ-carboxylated proteins and function as neurotoxins for paralyzing or killing the snail's prey. They form a subset within a large group of non-carboxylated neuroactive peptides termed conotoxins (conopeptides) that have considerable pharmacological potential. In several cases the γ-carboxylated conotoxins undergo a pronounced Gla-dependent conformational rearrangement upon metal ion binding [42–44]. Though in some instances the Gla residues in conotoxins are critical for biological activity, in others they are not. Indeed, for some conotoxins the main function of γ-carboxylation seems to be to facilitate proper folding of the peptide in the endoplasmic reticulum [45]. Aside from the conotoxins, one other γ-carboxylated protein has been identified in cone snails: a ∼40-kDa putative protease termed GlaCrisp, which contains a single Gla residue [46]. Cone snails are so far the only invertebrates shown to contain γ-carboxylate proteins. However, the genes encoding the enzymatic machinery for γ-carboxylation are widely present in both vertebrate and invertebrate taxa, which indicates that γ-carboxylation is an evolutionarily ancient process with a broad phylogenetic distribution [47–54].

9.2.2
The Gla Domain

The Gla domain found at the N-terminus of most mature γ-carboxylated vertebrate proteins has a highly conserved sequence with a characteristic spacing of Gla and Cys residues (Figure 9.2). The motif appears early in the lineage leading to vertebrates, being found in tunicate proteins, for example at the N-terminus of membrane-bound receptor tyrosine kinases [55, 56], but it has not been found in invertebrates. The conserved intron/exon organization and the nature of the splice junctions observed in the genetic regions encoding Gla domains from different proteins and species indicate that the domains have been recruited by duplication of a common ancestral gene sequence [5, 57]. The positions of most of the Gla residues are conserved, as are those of the only two Cys residues in the domain, which bond to form a crucial loop. Mutations disrupting this disulfide linkage severely affect the

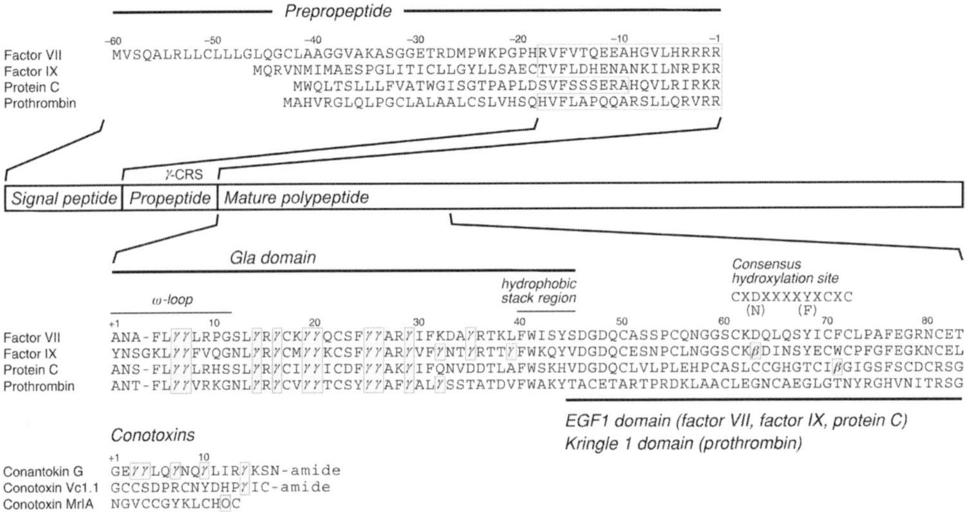

Figure 9.2 Partial amino acid sequences of post-translationally γ-carboxylated and/or hydroxylated polypeptides currently in use or in clinical development as biopharmaceuticals. The locations of Gla (γ), Hya (β) and hydroxyprolyl (O) residues are indicated (boxed and shaded). The prepro-region of the human proteins is depicted with the propeptide region containing the γ-carboxylation recognition site (γ-CRS) boxed (note that the length of the propeptide has not been confirmed experimentally for factor VII). A β-hydroxylation consensus sequence (Cys-Xxx-Asp/Asn-Xxx_4-Tyr/Phe-Xxx-Cys-Xxx-Cys), which is necessary but not definitive for specifying β-hydroxylation, is located in the EGF1 domain of factors VII and IX and protein C. The consensus is shown aligned with the factor VII and IX sequences. Prothrombin contains kringle domains instead of EGF-like domains and only a C-terminal derivative of the protein (thrombin; not shown) is used as a therapeutic.

function of the proteins [58]. Though γ-carboxylation of the Glu residues is essential for Ca^{2+} binding and biological activity, systematic mutation of the residues by recombinant methods has revealed that the importance of individual Gla residues is variable. While some substitutions may greatly impair Ca^{2+} binding and biological activity, others have little effect [59–62].

Whereas the affinity of a single Gla residue for Ca^{2+} is very low (K_D ~30 mM), the multiple Gla residues interspersed throughout the Gla domain contribute to form several sites that bind the metal with positive cooperativity [14, 31]. Thus, the average K_D for Ca^{2+} binding at these sites is ~0.3–0.7 mM and they are presumed to be saturated at the concentration of free Ca^{2+} in extracellular fluids (i.e., ~1.2 mM) [4,33]. Binding of Ca^{2+} induces the domain to fold into a conformation that is required for mediating binding of the proteins to biological membranes [63–69]. This structural rearrangement sequesters several of the Gla residues and Ca^{2+} ions inside the domain and forces the side chains of hydrophobic residues located in the "ω-loop" (residues 1–12) near the N-terminus into the solvent [63, 64, 67–69] (Figure 9.3). Exposure of the hydrophobic residues appears to be essential for the interaction with membranes. Binding occurs specifically to surfaces upon which anionic phospholipids (i.e., phosphatidylserine) are exposed, such as damaged endothelium or the surface of activated blood platelets [70–72]. These surfaces can be mimicked *in vitro* by negatively charged phospholipid vesicles [73]. The Gla domain also contributes to certain protein–protein interactions. Examples include

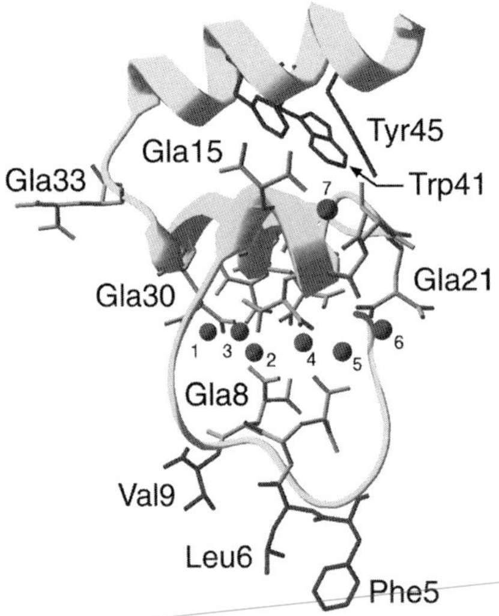

Figure 9.3 Model of the Ca^{2+}-loaded Gla domain of bovine prothrombin [67]. The seven bound Ca^{2+} ions are shown as spheres and the side chains of the Gla residues depicted. The exposed hydrophobic side chains of Phe5, Leu6 and Val9 in the ω-loop are proposed to form part of the membrane-binding site. Measurements of the intrinsic fluorescence of Trp41 have often been used to monitor Ca^{2+}-induced folding of Gla domains.

the interactions between protein C and the endothelial cell protein C receptor [74, 75], factor IX and factor VIIIa [76, 77], factor IX and collagen IV [78], prothrombin and factor Va [79], and factor VIIa and tissue factor [80, 81].

Three-dimensional structures of the metal-free forms of Gla domain-containing proteins or their fragments have been determined by X-ray crystallography [82–85]. In these studies the Gla domain was largely disordered, implying a considerable degree of flexibility in the polypeptide backbone in the apo-form. The only defined structural element observed was an α-helix encompassing the hydrophobic stack region. By contrast, the Ca^{2+}-loaded domain produced a highly ordered crystal structure, with a similar fold observed for Gla domains from various proteins [66, 67, 74, 80, 86–88]. In the case of the prothrombin Gla domain seven Ca^{2+} ions are bound almost exclusively by the side chains of the Gla residues, which point towards the interior of the domain so that several of the metal ions are buried away from the solvent (Figure 9.3). The N-terminal ω-loop implicated in membrane binding is folded so that Ala1 is buried within the domain and the side chains of Phe5, Leu6 and Val9 project into the solvent. Bonding interactions between Ala1 and several Gla residues contribute to stabilizing the ω-loop.

Solution NMR structures of metal-free [64, 68, 69], Ca^{2+}-bound [63, 66] and Mg^{2+}-bound Gla domains [65] have clearly revealed the pronounced conformational restructuring that occurs upon metal binding. The Ca^{2+}-bound NMR structures largely concur with those determined by X-ray crystallography, showing sequestration of several Gla residues within the folded domain and stable formation of the ω-loop structure [63, 89]. By contrast, Mg^{2+} ions, which do not support membrane binding, elicited a similar structure for much of the Gla domain, but residues 1–11 failed to assume the ω-loop conformation [65]. Remarkably, in the metal-free structures the side chains of the Gla residues face the solvent, while those of the hydrophobic residues in the ω-loop are clustered and face the interior of the structure. Also, the N-terminus is mobile and does not form contacts with other parts of the Gla domain [64, 68, 69]. Thus, Ca^{2+} binding induces a very specific fold in the Gla domain, connects Gla residues that are remote in the primary structure, and provides the energy required to force hydrophobic side chains into the solvent to unmask the membrane binding site(s).

Determination of the three-dimensional structure of the prothrombin Gla domain bound to a synthetic phospholipid (lyso-PS) has revealed the structural basis for the unique specificity exhibited by the domain for membranes upon which phosphatidylserine is exposed [66]. Multiple interactions between the Gla domain and lyso-PS were observed, with the headgroup of the phospholipid forming contacts with protein-bound Ca^{2+} ions and two Gla residues (Gla17, Gla21) and the glycerophosphate backbone interacting with basic residues (Lys3, Arg10, Arg16) as well as with Leu6 and Gla7. This preference for phosphatidylserine is a crucial regulatory mechanism for directing the vitamin K-dependent hemostatic proteins to the wound site where phosphatidylserine is exposed to the blood, while limiting their interaction with inappropriate membrane surfaces.

Crystal structure studies have also provided evidence for the interaction of the Gla domain with other proteins. For example, the structure of human factor VIIa in complex with tissue factor indicates that the Gla domain, which had seven Ca^{2+} ions bound to it, forms several contacts with tissue factor [80]. Likewise, the ω-loop of the human protein C Gla domain (also with seven Ca^{2+} ions bound) was found to form important contacts with the endothelial cell protein C receptor, suggesting a further function for the loop in addition to its role in membrane binding [74].

9.2.3
Biosynthesis of Gla

Vitamin K-dependent polypeptides are synthesized as single-chain precursors. They typically have a defined prepropeptide structure consisting of an N-terminal signal peptide, an intervening propeptide and an adjacent region containing the Glu residues slated for modification (Figure 9.2) [5]. The signal peptide directs transport to the secretory pathway of the cell and is cleaved off prior to the synthesis of Gla. The PRGP1 and TMG3 precursors are exceptions in that they lack a conventional signal peptide [26, 27]. The propeptide, commonly 10–28 amino acids long, contains a γ-carboxylation recognition site (γ-CRS) that is both necessary and sufficient to direct γ-carboxylation of the polypeptide [90–92]. In rare instances the γ-CRS is incorporated elsewhere within the polypeptide, as in the case of matrix Gla protein [93], or even within a "postpeptide" at the extreme C-terminus, as occurs in some conotoxins [36]. The γ-CRS does not constitute a clearly defined sequence motif but rather appears to be more a topological determinant. The propeptide is removed in the Golgi by a subtilisin-like serine protease (PACE/furin) after γ-carboxylation has been completed [94–96]. This is an essential step in maturation of the proteins. Mutations affecting residues within the propeptide can prevent its cleavage or result in the presence of extra amino acids at the N-terminus of the secreted protein. This prohibits the Gla domain from adopting a proper conformation and severely reduces the biological activity of the protein [97–101]. Depending on the protein affected, this can produce complications such as a bleeding tendency or a hypercoagulable state [100, 102].

The propeptide mediates binding of the polypeptide substrate to γ-glutamyl carboxylase, a 758-amino acid transmembrane protein found in the endoplasmic reticulum of most cells. The enzyme catalyzes the conversion of particular Glu residues in the polypeptide into Gla in a processive fashion [103, 104]. The number of Gla residues modified ranges from as few as one in some conotoxins to as many as 13 in protein Z. The propeptide serves not only to tether the substrate but also to directly activate the enzyme. The carboxylase is a bifunctional enzyme and catalyzes two linked reactions: γ-glutamyl carboxylation and vitamin K epoxidation [3, 5, 24, 105, 106]. In a reaction that requires CO_2, O_2 and the reduced (dihydroquinone) form of vitamin K as cofactors, the γ-carboxylase replaces a proton on the γ-carbon of a Glu residue with a CO_2 molecule to form the dicarboxylate side chain of Gla (Figure 9.4). The mechanism for proton abstraction is unclear but it is assumed that the carboxylase converts vitamin K dihydroquinone into a peroxide intermediate that reacts further to form a basic dialkoxide capable of abstracting the proton. The dialkoxide then collapses to a 2,3-epoxide [107, 108]. An endogenous inhibitor of the carboxylase, the chaperone protein calumenin, has been identified [109]. Though the mechanism of inhibition is unclear, silencing expression of calumenin enhances the rate of γ-carboxylation in cultured cells (Section 9.2.5.2).

The biosynthesis of Gla is coupled to a vitamin K redox cycle that regenerates the inactive epoxide by reducing it to the active dihydroquinone [3, 106, 110, 111] (Figure 9.4). Recycling of the epoxide is crucial, as only low concentrations of vitamin K are present in cells. Its complete reduction is effected in two steps, both of which are catalyzed by vitamin K epoxide reductase, a 163-amino acid membrane-bound enzyme found in the endoplasmic reticulum [52, 53, 112]. It appears that, during reduction of the vitamin, thiols in the active site of the reductase are oxidized to form a disulfide bridge, thereby inactivating the enzyme [110, 113, 114]. It is re-activated by an unidentified redox protein, possibly protein disulfide isomerase [115]. Dithiothreitol and other dithiols can suffice to activate the enzyme *in vitro* [116].

Coumarin-based vitamin K antagonists selectively inhibit the first reductive step involved in recycling the epoxide, an effect exploited in anticoagulant therapy with warfarin [5, 117, 118]. The precise mechanism whereby coumarins exert their effect is poorly understood but inhibition appears to occur when the enzyme's active site sulfhydryl groups are in the oxidized (inactive) state [113, 119]. Inhibition of the reductase indirectly inhibits the γ-carboxylation process so that partially and/or non-carboxylated forms of the vitamin K-dependent proteins

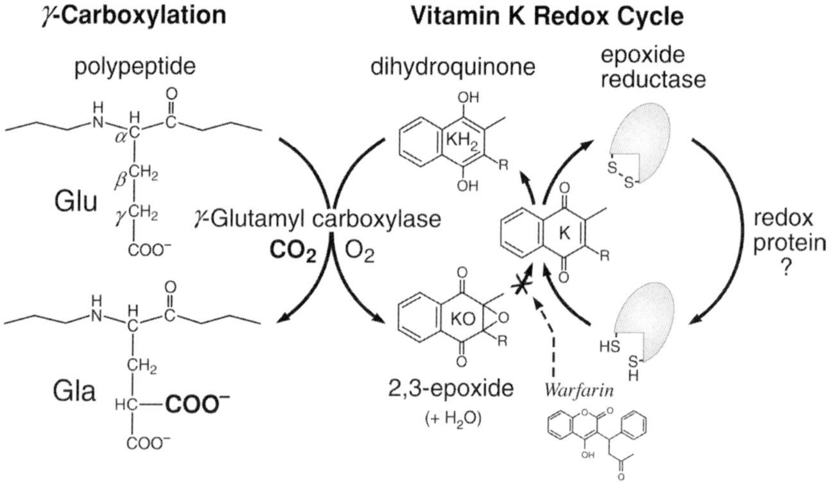

Figure 9.4 Biosynthesis of Gla and the linked vitamin K redox cycle. The enzyme γ-glutamyl carboxylase utilizes CO_2, O_2 and vitamin K dihydroquinone to catalyze the replacement of a proton on the side chain of a glutamyl residue with a carboxyl group, thus forming γ-carboxyglutamate (Gla). Vitamin K is converted into an epoxide during the reaction and is subsequently recycled in two steps to its active (reduced) form by the enzyme vitamin K epoxide reductase, which is sensitive to coumarin compounds such as warfarin. Reduction of the epoxide causes thiols in the catalytic center of the reductase to become oxidized, thereby inactivating the enzyme. It is presumably re-activated by an undefined redox protein.

accumulate in the endoplasmic reticulum and their plasma concentrations decrease [5, 120]. Many of the secreted molecules are deficient in Gla and therefore biologically inactive. Because humans cannot synthesize vitamin K, which is obtained primarily from green leafy vegetables, a dietary deficiency of the vitamin produces a similar effect to coumarins [121]. Neonates are born in a natural state of vitamin K deficiency and are routinely administered injections of the vitamin shortly after birth.

The liver is the primary site of synthesis for several of the vitamin K-dependent proteins and they are expressed throughout fetal development, childhood and adulthood. Factors VII, IX, X, prothrombin, protein C and protein Z are expressed mainly in hepatocytes, though some synthesis occurs in extra-hepatic tissues [122–124]. Protein S, Gas6, the PRGP/TMG proteins and matrix Gla protein are expressed in a wide variety of tissues [26, 27, 125, 126]. In contrast, expression of osteocalcin occurs exclusively in osteoblasts and odontoblasts [127]. Conotoxins are expressed in a specialized venom gland, as are the γ-carboxylated factor Xa homologs found in the venom of certain snakes [40, 128].

9.2.4 γ-Carboxylated Biopharmaceuticals

Currently, three γ-carboxylated proteins have been approved for therapeutic use in humans: factor IX, factor VIIa and activated protein C (Table 9.1). In addition, thrombin derived from its γ-carboxylated plasma-derived zymogen has been approved as a topical agent to control bleeding. Certain synthetic γ-carboxylated peptides based on conotoxins are also in the pharmaceutical pipeline. These proteins are reviewed briefly below.

9.2.4.1 Factor IX

Factor IX (FIX) is an integral component of the blood coagulation cascade. In its activated form (FIXa) it associates with a cofactor (factor VIIIa) on the surface of platelets to form the "tenase" complex. Its principle function is to activate factor X, which then associates with membrane-bound factor Va to form the prothrombinase complex that converts prothrombin into thrombin. FIX is synthesized in the liver as a single polypeptide chain of 454–461 amino acids [129]. After cellular processing, including removal of the signal peptide and an 18-residue propeptide, it is secreted into the blood as a single-chain zymogen of a serine protease (plasma concentration $\sim 5\,\mu g\,mL^{-1}$). The circulating zymogen ($\sim$55 kDa; 415 amino acids) consists of an N-terminal Gla domain followed by two epidermal growth factor-like domains (EGF1 and EGF2), an activation peptide and the serine protease domain [33, 130].

γ-Carboxylation of multiple Glu residues in the Gla domain of FIX is required for membrane binding and biological activity. Normally all 12 Glu residues in the domain undergo γ-carboxylation prior to secretion (Figure 9.2). However, the functional importance of individual Gla residues varies. Thus, while natural mutations affecting Gla7 or Gla27 cause hemophilia B, the Gla residues at positions 36 and

Table 9.1 Some properties of post-translationally carboxylated and/or hydroxylated proteins with biopharmaceutical potential[a].

	Gla residues	β/γ-OH residues	Carbohydrate (%)	Mass (Da)	ε(280 nm) ($mg\,mL^{-1}\,cm^{-1}$)	Physiological function	Therapeutic application	Main target	Reference
Human factor IX	12	1(Hya)	17	55 000	1.3	Coagulant	Hemophilia B	FX	[130]
Human factor VIIa	10	0	13	50 000	1.4	Coagulant	Acute bleeding	FIX, FX	[144]
Human activated protein C	9	1(Hya)	23	56 200	1.5	Anticoagulant	Sepsis	FVa, FVIIIa	[166]
Human prothrombin[b]	10	0	8	72 000	1.4	Coagulant	Acute bleeding	Fibrinogen	[168]
Bovine prothrombin[b]	10	0	10	72 000	1.4	Coagulant	Acute bleeding	Fibrinogen	[168]
Conantokin-G	5	0	0	2 264	~0	Venom toxin	Analgesic	NMDA receptor	[39]
Conotoxin Vc1.1	1	0	0	1 855	0.9	Venom toxin	Analgesic	nAch receptor	[183]
Conotoxin MrIA	0	1(Hyp)	0	1 411	1.2	Venom toxin	Analgesic	NET reporter	[171]

[a] Abbreviations: Hya, *erythro*-β-hydroxyaspartic acid; Hyp, 4-hydroxyproline; NMDA, *N*-methyl-D-aspartate; nAch, nicotinic acetylcholine; NET, norepinephrine.
[b] The therapeutic product is thrombin, a non-carboxylated serine protease derived from the γ-carboxylated prothrombin zymogen.

40 are not required for biological activity [59, 61]. Other post-translational modifications to FIX include β-hydroxylation of Asp64 in the EGF1 domain, *N*- and *O*-glycosylation, sulfation of Tyr155 and *O*-phosphorylation of Ser158. FIX is proteolytically activated when the clotting cascade is triggered, through the action of either the factor VIIa–tissue factor complex or factor XIa in the presence of Ca^{2+} [131, 132]. Activation occurs in two steps. In the first, the Arg145–Ala146 peptide bond is hydrolyzed to produce a two-chain inactive intermediate, FIXα, consisting of a light chain (amino acids 1–145) disulfide-bonded to a heavy chain (residues 146–415). This cleavage is thought to expose a binding site for factor VIIIa [133]. Second, the heavy chain of FIXα is cleaved at the Arg180–Val181 bond, releasing a 35-residue activation peptide to generate the active protease, FIXaβ [134, 135].

Administration of factor IX (FIX) has proven effective in the treatment of hemophilia B, an X-linked recessive bleeding disorder affecting about 1 in 30 000 males, which is caused by a deficiency of functional FIX protein [20, 21]. The diathesis arises from any of hundreds of mutations identified in the factor IX gene, including some that affect Glu residues normally converted into Gla; for instance, factor IX Seattle 3 (Gla27Lys) and factor IX Oxford b2 (Gla7Ala) [136, 137]. Symptoms include spontaneous and prolonged bleeding in the muscle and organs. Patients with a severe FIX deficiency (levels <1% of normal) often bleed into the joints, leading to arthropathy. The disorder is most commonly treated by infusion of plasma-derived FIX concentrates or recombinant FIX (BeneFIX; Wyeth, Madison, NJ). The recombinant protein is produced by culturing a Chinese hamster ovary (CHO) cell line engineered to co-express FIX and a soluble form of the propeptide-processing enzyme PACE/furin [138, 139]. The latter modification to the cell line was required to enhance removal of the propeptide, as its retention prevents the Gla domain from adopting the proper Ca^{2+}-induced conformation required membrane binding. After removal of poorly γ-carboxylated forms by chromatography, a fair amount of the purified recombinant protein (~40%) exhibits partial γ-carboxylation of one or both of the Glu residues at positions 36 and 40, but this does not significantly affect the activity of the protein [59, 61, 140].

9.2.4.2 **Factor VIIa**

Factor VII (FVII) plays an important role early in the blood clotting cascade. A small portion (~1%) of the circulating protein resides in the blood in the activated form (FVIIa). Damage to the vascular endothelium exposes a membrane-bound cofactor, tissue factor, which binds and allosterically stimulates the activity of FVIIa. The FVIIa–tissue factor complex then cleaves and activates factors IX and X, thus amplifying the clotting stimulus and promoting formation of a blood clot [141–143].

FVII is synthesized in the liver as a 466-amino acid precursor that includes a 60-residue prepropeptide region. It undergoes extensive intracellular processing, including removal of the signal peptide and propeptide, *N*- and *O*-glycosylation and γ-carboxylation of all ten Glu residues in the Gla domain, but is not modified by β-hydroxylation (Figure 9.2) [144, 145]. It is secreted into the blood as a ~50-kDa

(406 amino acids) single-chain zymogen of a serine protease (plasma concentration ~0.5 μg mL^{-1}). The circulating zymogen consists of an N-terminal Gla domain followed by two EGF-like domains and a serine protease domain (Figure 9.1). A small amount of the zymogen becomes cleaved to form the active serine protease under basal conditions. This occurs by hydrolysis of the Arg152–Ile153 peptide bond, producing a disulfide-bonded dimer. Larger amounts of the zymogen are activated when the clotting cascade is triggered; probably mainly through the action of factor Xa, though activation can also be effected by thrombin, FIXa or FXIIa [146].

Recombinant human FVIIa (NovoSeven; Novo Nordisk, Copenhagen, Denmark) was originally employed for treatment of hemophilia patients who had developed antibody inhibitors against factors VIII or IX [145]. Infusions were found to be effective for controlling spontaneous bleeding episodes as well as during surgical procedures in these patients. It has since been approved for use in FVII-deficient patients and those with Glanzmann thrombasthenia, and shows promise as a general therapeutic for treatment of acute bleeding. The recombinant protein is manufactured using a baby hamster kidney (BHK) cell line transfected with the FVII gene [145]. Efficient activation of the secreted zymogen is achieved by autolysis and the protein is purified by several steps of ion-exchange and immunoaffinity chromatography.

9.2.4.3 **Protein C/Activated Protein C**

Protein C (PC) is the central component of a natural anticoagulant pathway. When bound to its cofactor, protein S, the activated form of PC (APC) regulates blood coagulation by proteolytically inactivating two blood clotting cofactors, factors Va and VIIIa [147–149]. Factor V also serves as a cofactor for APC in the degradation of factor VIIIa [147]. Individuals with a homozygous or heterozygous PC deficiency are prone to thrombotic episodes that can be life-threatening. In addition to being a potent anticoagulant, APC possesses anti-inflammatory, anti-apoptotic and fibrinolytic activity [150–152]. It has proved beneficial as a therapeutic agent for treating patients with severe sepsis.

PC is synthesized in the liver as a single polypeptide chain of 461 amino acids with an 18-residue signal peptide and a 24-residue propeptide (Figure 9.2) [90]. It undergoes several post-translational modifications, including cleavage of the signal and propeptides, γ-carboxylation of nine Glu residues in the Gla domain, β-hydroxylation of Asp71, *N*-glycosylation and specific endoproteolysis, before being secreted into the blood as a ~62-kDa zymogen of a serine protease (plasma concentration ~4 μg mL^{-1}) [153–155]. A portion (5–15%) of PC circulates as a single-chain molecule but the majority is converted into a two-chain form prior to secretion by excision of a dipeptide (Lys156–Arg157) [156, 157]. The dimer form consists of a light chain (~21 kDa; 155 amino acids), consisting of the Gla domain and two EGF-like domains, that is linked by a single cystine to a heavy chain (~41 kDa; 262 amino acids), which constitutes the catalytic domain [158]. The zymogen is activated to a serine protease (APC) by the thrombin–thrombomodulin complex that forms on the surface of endothelial cells when the blood vessel wall is damaged. Thrombin activates dimeric PC by cleaving a single peptide bond (Arg12–Leu13) near the amino-terminus of the heavy chain to release a 12-residue activation peptide [159]. Thrombin can also cleave single-chain PC to yield a two-chain molecule that retains the activation peptide on the light chain but has full enzymatic activity [156, 160, 161]. Extensive γ-carboxylation of the Gla domain of PC is required for biological activity and molecules containing six or fewer of the full complement of nine Gla residues exhibit greatly reduced anticoagulant activity [162]. However, mutagenesis studies have indicated that the functional importance of individual Gla residues varies, depending on their location in the Gla domain. For instance, the Gla residues at positions 7, 16, 20 and 26 are indispensable for biological activity, whereas those at positions 6, 14 and 19 seem to be relatively unimportant [62].

Infusion of PC- or APC-containing plasma concentrates or purified APC (plasma-derived or recombinant protein) has demonstrated utility in thrombolytic therapy and in the treatment of acute infections [23, 147, 163–165]. In cases of severe septic shock the plasma pool of PC becomes depleted and replenishment with exogenous APC helps to mitigate the associated effects of microvascular thrombosis and disseminated intravascular coagulation. An approved recombinant form of human APC [drotrecogin alfa (activated), Xigris; Eli Lilly and Co., Indianapolis, IN] is produced in engineered human embryonic kidney (HEK293) cells. The molecules that have undergone full γ-carboxylation are selected by immunoaffinity chromatography [166].

9.2.4.4 **Prothrombin**

Prothrombin is not used as a therapeutic protein *per se*, but thrombin derived from it is employed as a topical treatment to control bleeding. Prothrombin circulates as a single-chain zymogen (~72 kDa; 579 amino acids) consisting of an N-terminal Gla domain containing ten Gla residues (Figures 9.2 and 9.3) followed by two kringle domains and the thrombin catalytic domain. The zymogen is proteolytically activated by the factor Xa–factor Va (prothrombinase) complex when blood clotting is triggered. Activation results in the cleavage of two peptide bonds (Arg271–Thr272 and Arg320–Ile321), thereby excising the Gla and kringle domains and releasing thrombin, a dimer joined by a single cystine linkage [8, 167, 168]. Thrombin is the terminal serine protease of the coagulation cascade and serves to cleave fibrinogen to form insoluble monomers that stabilize the blood clot. The central role of thrombin in hemostasis is attested to by the fact that most of the vitamin K-dependent proteins involved in hemostasis function to either up-regulate (factors VII, IX, X) or down-regulate (proteins C, S and Z) the conversion of prothrombin to thrombin [5, 24].

Plasma-derived thrombin is currently available in two forms: a bovine thrombin product (Thrombin-JMI; King Pharmaceuticals, Bristol, TN) and human thrombin (Evithrom; Omrix Pharmaceuticals, New York, NY). The first recombinant form of human thrombin (Recothrom) was approved by the FDA in January 2008. Recothrom was developed by Zymogenetics (Seattle, WA) and is produced from a Gla domain-less precursor expressed in mammalian cells. Its biosynthesis therefore does not require a γ-carboxylation processing step. Recothrom exhibits a comparable efficacy to bovine thrombin when used for surgical hemostasis and is less immunogenic than its bovine counterpart [169].

9.2.4.5 Conotoxins

The venom of cone snails contains a plethora of structurally diverse peptides termed conotoxins or conopeptides that exert their toxic effects by binding to receptors and ion channels in the neuromuscular system of the snail's prey [40]. Many of the peptides undergo extensive post-translational modifications, including γ-carboxylation and hydroxylation [39, 170]. Several γ-carboxylated conotoxins exhibit specificity for mammalian targets such as voltage- and ligand-gated ion channels, G-protein-coupled receptors and neurotransmitter reporters [171–174]. They are widely employed as tools in neuroscience research and are being intensively investigated as pharmacological leads in the development of analgesics and other therapeutics [175–177].

The mature conotoxins are typically 10–30 amino acids long and are generated from larger precursors by proteolytic removal of amino acids from the N-terminus and in some cases also the C-terminus of the molecule [178, 179]. In the case of the γ-carboxylated conotoxins the excised regions include the signal peptide and the pro- or post-peptide that specifies γ-carboxylation. One or more Glu residues in the peptides are modified by a γ-carboxylase that exhibits many mechanistic similarities to its counterpart in vertebrates [36, 49, 180]. γ-Carboxylation is important for binding Ca^{2+}, which imparts structural stability to several conotoxins [42–44, 181]. The short length of the mature peptides has facilitated their characterization and the chemical synthesis of analogs. Synthetic peptides based on the mature forms of at least two γ-carboxylated conotoxins are being investigated as potential analgesics (Figure 9.2 and Table 9.1) [175, 176]. One, a conantokin-G derivative (CGX-1007; Cognetix Inc., Salt Lake City, UT) blocks agonist binding to the NMDA receptor. It is a 17-amino acid peptide containing five Gla residues, some of which are required for receptor binding [44]. Another, Vc1.1 (16 amino acids) is a nicotinic acetylcholine receptor agonist containing a single Gla residue, but the Gla is not incorporated into the synthetic analog (*ACV1*; Metabolic Pharmaceuticals Ltd., Melbourne, Australia) [182,183].

9.2.5 Enhancement of Cellular Carboxylation Capacity

The therapeutic use of γ-carboxylated proteins sourced from plasma is attended by major concerns, most notably the possibility for transmission of pathogens. This has driven successful efforts to bring to market recombinant forms of some of the vitamin K-dependent proteins, particularly FIX, FVIIa and APC (Section 9.2.4). However, obtaining high yields of biologically active recombinant protein from heterologous expression systems has proved problematic because of inefficiencies in the γ-carboxylation process. Only at low expression rates has it been possible to obtain a high proportion of molecules containing their full complement of Gla residues.

Thus, at high expression levels a large proportion of non- or poorly carboxylated protein as well as molecules retaining the propeptide is found in the secreted protein pool. This has been observed both for recombinant cell lines [60, 184–188] and transgenic animals [189]. So far, attempts to increase the γ-carboxylation capacity of cultured cells via selective pressure or genetic engineering have met with only partial success, though promising advances have been made. Brief coverage of current developments in this area is given below.

9.2.5.1 Enhanced Expression of the γ-Carboxylation Machinery

The identification and cloning of the gene encoding the γ-glutamyl carboxylase [190, 191] raised hopes that it would be possible to increase the yield of fully carboxylated recombinant proteins expressed in cultured cells by simply co-expressing the γ-carboxylase. In fact, overexpression of the γ-carboxylase either by transient transfection with a vector containing the γ-carboxylase cDNA or by stable integration of the cDNA into the cell's genome failed to substantially improve (if at all) the efficiency of γ-carboxylation, or even inhibited it [114, 190, 192–194]. This was observed for several cell lines, including HEK293, BHK and CHO cells. For FIX, the secretion of partially carboxylated forms suggested that a build up of high concentrations of the precursor in the secretory pathway somehow perturbed the carboxylase–substrate interaction, leading to an increased rate of dissociation from the enzyme [195]. Moreover, the observation that much higher γ-carboxylase activity could be introduced into the cells without any improvement in the efficiency of carboxylation suggested that another step in the γ-carboxylation process was rate-limiting, such as recycling of the vitamin K cofactor (Figure 9.4) [192].

The cloning of the gene encoding the vitamin K epoxide reductase [52, 53] presented an opportunity to test whether enhancing the expression level of the reductase in cells would improve the γ-carboxylation capacity by increasing the amount of reduced vitamin K available to the carboxylase. If the reduced vitamin is added to the culture medium it quickly becomes oxidized to the quinone, so the action of the epoxide reductase is required to generate significant amounts of the active cofactor in cells. Stable co-expression of recombinant FIX and the epoxide reductase in BHK cells produced a 2–3-fold increase in the proportion of fully carboxylated protein secreted into the culture medium [194, 196, 197]. Similarly, a 2–3-fold improvement in carboxylation of recombinant FX expressed in CHO cells and recombinant FVII expressed in HEK293 cells was achieved by co-expressing the reductase [193, 198]. However, as ascertained for production of FIX the magnitude of the increase was much less than the 14-fold increase in epoxide reductase activity measured in the cells [196]. It has, thus, been postulated that some other step such as recycling of the reductase may be saturated in the engineered cell lines [194, 196].

Importantly, co-expression of both the γ-carboxylase and epoxide reductase has failed to enhance γ-carboxylation of recombinant FIX or FX above the level obtained by overexpressing only the reductase [193, 194]. Together, the above studies have revealed that achieving a sufficient increase in the biosynthetic rate to effect full carboxylation of overexpressed proteins will probably require enhanced expression of not only the γ-carboxylase and epoxide reductase but also a redox component capable of recycling the reductase to its active (reduced) state. The nature of the endogenous electron donor in cells is currently uncertain but it seems its identification will be necessary to allow the complete γ-carboxylation system in cells to be reconstituted and enhanced by recombinant methods.

9.2.5.2 Inhibition of Calumenin Expression

Calumenin, an ~47-kDa chaperone protein of the endoplasmic reticulum, has been identified as an endogenous inhibitor of the γ-carboxylase [109, 199]. It appears to associate closely with the carboxylase based on immunoprecipitation experiments [109]. Though the mechanism of inhibition is unclear, transient inhibition of calumenin expression via siRNA silencing in BHK cells overexpressing the vitamin K epoxide reductase markedly improved the efficiency of γ-carboxylation of recombinant FIX. An almost twofold increase in the yield of functional FIX was obtained compared to control cells [197]. Later experiments employing shRNA technology to stably inhibit calumenin expression also yielded an improvement in the carboxylation of recombinant FVII expressed in HEK293 cells [198].

9.2.5.3 Propeptide/Propeptidase Engineering

The propeptides of vitamin K-dependent precursors mediate their interaction with the γ-carboxylase [90–92, 200]. They vary in length and amino acid sequence depending on the protein (Figure 9.2) and display different affinities for the γ-carboxylase that affect the rate at which the protein is bound and released from the enzyme [201]. One strategy to improve the efficiency of γ-carboxylation has been to alter the propeptide to modify its interaction with the carboxylase. For example, substitution of the native prepropeptide of FX with that from prothrombin (to reduce its affinity for the carboxylase) improved the yield of fully γ-carboxylated FX produced by HEK293 cells by almost threefold [202]. Other efforts have sought to increase the yield of functional γ-carboxylated protein by enhancing proteolysis of the propeptide. For instance, substitution of Thr-2 for Arg in the propeptide cleavage site of FX facilitates excision of the propeptide by HEK293 cells [203]. At high expression levels the endogenous proteolytic system becomes saturated, leading to secretion of a large proportion of molecules that retain the propeptide and are therefore biologically inactive [190]. Co-expression of the propeptide-processing enzyme PACE (paired basic amino acid cleavage enzyme)/furin has proved effective for improving the yield of functional recombinant protein. This strategy increased the amount of functional FIX produced by CHO cells by circa threefold [190] and has been incorporated in the manufacture of recombinant FIX [138, 139].

9.2.6 Purification of γ-Carboxylated Proteins

Many methods have been developed for purifying γ-carboxylated proteins. Several exploit physicochemical properties imparted by the Gla modification itself. For instance, the purification of Gla-containing proteins from blood plasma has traditionally incorporated an early precipitation step in which the Gla domain is adsorbed to an insoluble salt such as barium citrate (formed by the addition of barium chloride to citrated plasma). This is often followed by anion-exchange and affinity chromatography (e.g., on heparin- or benzamidine-conjugated resins) [204, 205]. Chromatographic methods for purifying recombinant Gla-containing proteins from cell culture medium often preferentially recover the fully γ-carboxylated protein fraction by exploiting differences in binding affinity between completely and incompletely processed molecules. Such methods include tailored ion-exchange protocols, some of which utilize Ca^{2+}-induced elution of the bound protein, chromatography on hydroxyapatite matrices, and immunoaffinity chromatography with immobilized conformation-dependent antibodies [60, 77, 139, 194, 206]. Monoclonal antibodies that bind specifically to the Gla residues have also proved suitable for purifying a range of γ-carboxylated proteins [46, 207].

9.2.7 Analytical Characterization of γ-Carboxylated Proteins

9.2.7.1 Methods for Detecting Gla

The detection of Gla can prove problematic by biochemical methods, particularly if dealing with crude protein preparations. This is in part due to the labile nature of the Gla side chain, which becomes decarboxylated under acidic conditions such as those commonly used in determining amino acid compositions. Also, the Gla–phenylthiohydantoin derivative is not extracted during conventional amino acid sequencing, so that a blank cycle or a small amount of decarboxylated Gla (Glu) is observed. Alternative procedures to detect the Gla modification have been developed, including methyl esterification of the Gla residues prior to sequencing [208], the use of specific stains [209], isotopic labeling [210], immunochemical detection [207], and analysis of alkaline hydrolysates using capillary electrophoresis or anion-exchange, reversed-phase, gas or thin layer methodologies [16, 211, 212]. A detailed protocol for quantitative detection of Gla in base hydrolyates by HPLC is available [213]. Several mass spectrometry-based methods have been applied very effectively to detect and localize the positions of Gla residues, including electron capture dissociation, isotopic labeling, derivatization of the Gla residues and monitoring decarboxylation events [36, 46, 214–216].

9.2.7.2 Metal Content

Gla residues can chelate several divalent and trivalent metal ions, though the interaction with Ca^{2+} is the most physiologically significant. Stoichiometric studies of metal binding to γ-carboxylated proteins have most often utilized dialysis methods [14, 217–219]. These have produced variable results depending on buffer composition, ionic strength, pH, protein concentration, and so on [220]. Ca^{2+}-selective electrode titrations have also been employed [221]. In the Ca^{2+}-loaded crystal structures determined for FVIIa, FIX, APC and prothrombin, between five and seven Ca^{2+} ions are observed to be bound to the Gla domain (Figures 9.1 and 9.3) [66, 67, 74, 80, 86–88]. These stoichiometries are comparable to those obtained from some equilibrium dialysis experiments [217, 219]. However, discrepancies occur when compared to other experimentally determined binding stoichiometries. For example, while the crystal structure of the factor VIIa–tissue factor complex indicates that seven Ca^{2+} ions are bound within the Gla domain [80], membrane filtration experiments and other studies suggest that only around four Ca^{2+}-binding sites are present [218]. Assumptions as to the metal content of the proteins *in vivo* are at best a guess.

9.2.7.3 Metal Binding-Induced Structural Changes

The pronounced conformational transition induced in most Gla-containing proteins by metal ion binding can been monitored with various techniques. These have included rocket electrophoresis [11, 222], circular dichroism [223, 224], intrinsic fluorescence quenching assays [63, 67, 225–227], ESR spectroscopy [228], NMR spectroscopy [68, 224, 229], differential scanning calorimetry [230, 231], Fourier-transform infrared spectroscopy [232] and immunoassays using conformation-specific antibodies [206]. Fluorescence measurements have been commonly employed for proteins containing a Gla domain. The metal-induced transition alters the interaction of the disulfide loop region of the Gla domain with the hydrophobic stack region, thereby burying the indole ring of Trp41/42 and quenching its intrinsic fluorescence (Figure 9.3) [31].

9.2.7.4 Phospholipid Membrane Binding Assays

The interaction between the Gla domain and phospholipid-containing membranes is critical for the function of several vitamin K-dependent proteins. The nature and affinity of the interaction, which varies significantly among the proteins, has been probed by several methods, including fluorescence energy transfer measurements [233, 234], differential scanning calorimetry [230], light scattering assays [219, 235] and Fourier-transform infrared spectroscopy [232].

9.2.7.5 γ-Carboxylase Enzyme Assays

Kinetic parameters for the post-translational conversion of Glu into Gla can be measured *in vitro* by assaying the incorporation of $^{14}CO_2$ into a peptide substrate containing Glu residues [236–238]. A frequently employed substrate has been the pentapeptide FLEEL (Phe-Leu-Glu-Glu-Leu), though the γ-carboxylase will process a wide variety of Glu-containing peptides of varying length and sequence. Incorporating an N-terminal sequence containing a γ-carboxylation recognition sequence (γ-CRS) into the peptide reduces the K_m for the reaction by orders of magnitude [103, 239]. Microsomal preparations made from a suitable tissue (e.g., liver) or cultured cells may serve as a source of the γ-carboxylase, though the purified recombinant enzyme has also been utilized [103]. The reactions are carried out in the presence of exogenous reduced vitamin K and $NaH^{14}CO_3$. Unincorporated radiolabel is removed by heating and the samples are counted in a scintillation counter. Alternatively, the samples can be resolved by SDS-PAGE and quantitated by PhosporImager analysis [104].

9.3
Post-Translational Hydroxylation

9.3.1
Biological Function of Hydroxylation

Post-translational hydroxylation of several amino acids has been demonstrated. Perhaps the most widely known examples occur in collagen, where β- and γ-hydroxylation of Pro residues and 5-hydroxylation of Lys residues occur. These modifications are catalyzed by well-characterized prolyl and lysyl hydroxylases and they function to augment the structural stability of the collagen triple helix by providing sites for hydrogen bond formation [240]. The post-translational hydroxylation of Leu has been noted in unstable forms of hemoglobin from patients with hemolytic anemia [241] and hydroxylated forms of Pro, Lys and D-Val have been found in conotoxins isolated from the venom of marine cone snails [242–246]. Little is known about the function of the conotoxin modifications. β-Hydroxylation of Asp or Asn to form *erythro*-β-hydroxyaspartic acid (Hya) or *erythro*-β-hydroxyasparagine (Hyn), respectively, occurs in the epidermal growth factor (EGF) precursor and in some of the Ca^{2+}-binding EGF-like domains found in other proteins, for example, certain receptors and their ligands, structural proteins of the extracellular matrix, complement proteins and vitamin K-dependent hemostatic proteins [247–250]. β-Hydroxylation occurs in two important biopharmaceutical proteins, factor IX and protein C (Table 9.1).

Hya was first discovered in a protein by Stenflo *et al.* in the early 1980s, within a heptapeptide isolated from bovine protein C (PC) [251]. Detailed analysis revealed that the peptide possessed Hya in its third position, corresponding to position 71 in the light chain of PC, that is, within the first EGF-like domain (Figure 9.2) [251]. Hya was soon found in several other vitamin K-dependent proteins: factor IX (FIX), factor X (FX), protein S and protein Z [252–255]. Each of these proteins contains two or four (protein S) EGF-like domains arranged in tandem but, at most, only a single Hya residue is present per molecule. In each case it is located within the EGF-like domain nearest to the N-terminus (EGF1). Protein S also contains a Hyn residue in each of its other three EGF-like domains [254]. Neither Hya nor Hyn is found in plasma-derived prothrombin or human FVII [144, 256], though Hya has been detected in recombinant FVII [185].

The EGF domain, or module, is an evolutionarily ancient structural unit found in both animals and plants, usually at multiple adjacent locations in the protein [33, 257]. The fold of this ~45-amino acid domain is dominated by a two-stranded β-hairpin and the structure is typically constrained by three disulfide bridges, with the Cys residues linked in the pattern 1–3, 2–4, 5–6 (there are four disulfides in the EGF1 domain of protein C) [257–260]. EGF-like domains can serve to provide rigidity to a protein, such as in fibrillin-1, or to provide a surface for interprotein interactions. For instance, the N-terminal (Ca^{2+}-binding) EGF-like domain of FVIIa forms an important interface for the association with tissue factor, which in part explains the Ca^{2+} dependence observed for the interaction [80]. EGF-like domains also function as "spacers" in some coagulation proteases (e.g., FVII, FIX, APC), where they contribute to positioning the catalytic center at an appropriate distance from the biological membrane for contact with the cofactor and substrate [257].

A subset of EGF-like domains are β-hydroxylated, containing either a Hya or Hyn residue. The modified residue participates in the formation of a functionally important, high-affinity Ca^{2+}-binding site that chelates a single metal ion (Figure 9.5). Whereas an isolated EGF-like domain binds Ca^{2+} with a K_D of around 1 mM, linkage to adjacent domains can increase the affinity 10–100-fold, so that the site is presumed to be saturated at the physiological concentration of extracellular free Ca^{2+} [257, 261, 262]. Seven oxygen atoms (one from a water molecule) serve as ligands for the metal ion and are arranged in a pentagonal bipyramidal geometry [33, 80, 258]. The Hya/Hyn residue is not required for chelating Ca^{2+}, as was once thought. Equilibrium dialysis measurements of FIX and NMR studies of chemically synthesized EGF-like domains from FIX and FX have revealed that Ca^{2+} is bound at the site with similar affinity whether Asp or Hya is present [260, 263, 264].

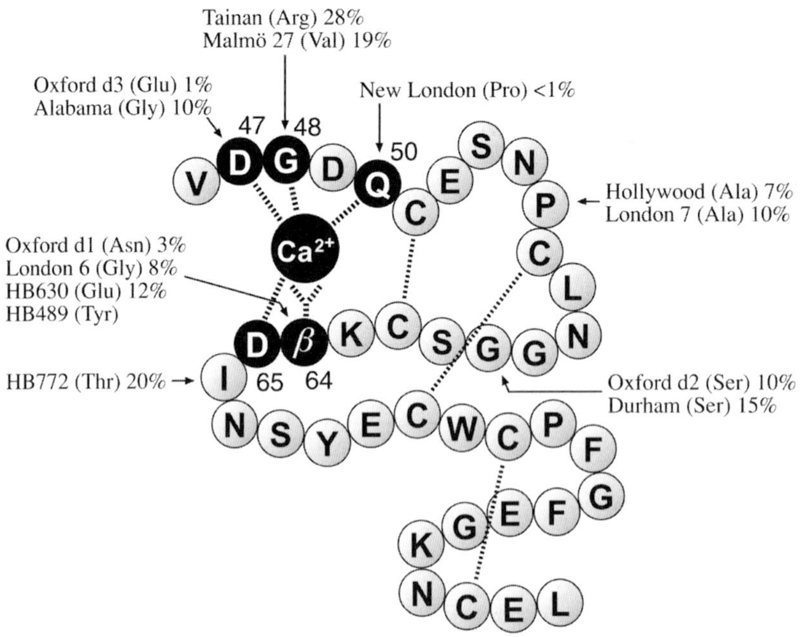

Figure 9.5 Schematic of the Ca^{2+}-binding site in the EGF1 domain of human FIX. The positions of residues that provide ligands for the Ca^{2+} ion are depicted as black spheres and numbered according to the amino acid sequence of mature FIX. Seven oxygen atoms serve as ligands for the metal ion. The Hya residue located at position 64 (labeled "β") forms a bidentate interaction with Ca^{2+} via the two oxygen atoms of its side chain carboxyl group. The five other ligands are provided by Asp47 (side chain carboxyl), Gly48 (backbone carbonyl group), Gln50 (side chain carboxamide), Asp65 (backbone carbonyl) and an H_2O molecule [258, 260]. Some naturally occurring point substitutions in FIX are indicated along with the percent clotting activity of the variants relative to the wild-type protein [265–267, 269–273, 302, 303]. The three disulfide linkages in the domain are indicated by dashed lines.

This suggested that the side chain hydroxyl group of Hya is not a direct ligand for Ca^{2+}; a notion confirmed by determination of the X-ray crystal structure of the domain [258]. In fact, it is the side chain carboxyl group of the Asp/Hya residue (or carboxamide group in the case of Asn/Hyn) that serves as the ligand and the β-hydroxyl group faces away from the Ca^{2+} ion.

The functional importance of the EGF-type Ca^{2+}-binding site is well established and is underscored by naturally occurring mutations affecting amino acids that serve as Ca^{2+} ligands. Point substitutions of these residues, including those that are normally β-hydroxylated to form Hya or Hyn, affect the biological activity of the proteins by reducing the affinity of the site for Ca^{2+}. They are associated with disorders such as hemophilia B in the case of FIX, and Marfan's syndrome in the case of fibrillin-1 [265–273]. Chelation of Ca^{2+} at the site stabilizes the N-terminal region of the EGF-like domain [257]. In the vitamin K-dependent proteins this reduces the flexibility of the hinge region between EGF1 and the adjacent Gla domain and locks them in a conformation that contributes to maximal biological activity [69, 261]. However, unlike the Gla domain, Ca^{2+} binding to the EGF1 domain is associated with only minor structural changes.

A specific role for β-hydroxylation of EGF-like domains has not been clarified. No effect on biological activity has been observed if β-hydroxylation of recombinant FIX is blocked [274]. Likewise, inhibiting β-hydroxylation of recombinant protein S does not affect its biological activity as a cofactor in APC-catalyzed inactivation of factor Va [275]. It has been suggested that β-hydroxylation may be involved in negatively regulating glycosylation of protein C [276] and in modulating signaling in the Notch pathway [277]. Interestingly, expression of the Asp/Asn β-hydroxylase is elevated in human carcinoma cells [278], while knocking out the enzyme's cognate gene in a mice tumor model results in tumorigenesis and developmental defects [277].

9.3.2 Biosynthesis of Hydroxylated Amino Acids

9.3.2.1 Biosynthesis of Hya/Hyn

The post-translational conversion of Asp into Hya or Asn into Hyn is catalyzed by a Fe^{2+}/2-oxoglutarate-dependent dioxygenase, aspartyl(asparaginyl) β-hydroxylase (EC 1.14.11.16). The enzyme modifies only a single site in a subset of the Ca^{2+}-binding EGF or EGF-like domains found in proteins. A characteristic consensus sequence associated with β-hydroxylation has been identified between the third and fourth Cys residues in EGF-like domains that contain Hya/Hyn: Cys-Xxx-Asp*/Asn*-Xxx_4-Tyr/Phe-Xxx-Cys-Xxx-Cys, where an asterisk denotes the hydroxylated residue (Figures 9.2 and 9.5) [249, 254]. For hydroxylation to occur the consensus sequence must form part of the major β-hairpin of the domain, with the Asp/Asn residue spatially located adjacent to the Tyr/Phe residue. Even so, the motif is merely necessary but not definitive for specifying β-hydroxylation. In addition to the above consensus, there is considerable conservation among the Ca^{2+} ligands preceding the first Cys in the domain [257, 279].

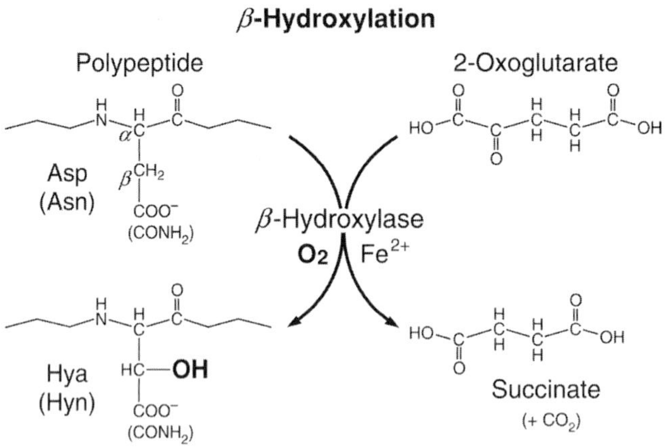

Figure 9.6 Biosynthesis of Hya/Hyn. The enzyme aspartyl(asparaginyl) β-hydroxylase utilizes O_2, Fe^{2+} and 2-oxoglutarate as co-substrates to catalyze stereospecific hydroxylation of the β-carbon on the side chain of an aspartyl or asparaginyl residue. *erythro*-β-Hydroxyaspartic acid (Hya) or *erythro*-β-hydroxyasparagine (Hyn) are formed, concomitant with the stoichiometric decarboxylation of 2-oxoglutarate to generate succinate and CO_2.

The β-hydroxylase utilizes O_2, Fe^{2+} and 2-oxoglutarate as co-substrates to catalyze the formation of an hydroxyl β–OH) group on the β-carbon of the targeted Asp or Asn residue (Figure 9.6) [280, 281]. The enzyme is stereospecific and only produces the *erythro* form of the modified amino acids. Concomitant with β-hydroxylation of the substrate the 2-oxoglutarate molecule is decarboxylation to release CO_2, thus generating succinate. Hence, one oxygen atom is incorporated into the new hydroxyl group and the other into CO_2. The Asp/Asn β-hydroxylase has been purified and its cDNA cloned [274, 280–283]. The human enzyme is translated from an alternative splice variant of its cognate gene. It is localized in the endoplasmic reticulum and is predicted to assume an anchored transmembrane topology with its C-terminal catalytic domain in the lumen [278, 282]. Several compounds are effective inhibitors of the enzyme, including *o*-phenanthroline, 2,9-dimethylphenanthroline and pyridine-2,4-dicarboxylic acid [274, 282]. A separate cytosolic hydroxylase catalyzes formation of the *threo* isomer of β-hydroxylated Asn, which has been found in the transcription factor hypoxia-inducible factor 1a [284].

9.3.2.2 **Biosynthesis of Hydroxyproline**

Though best known as a component of collagen and elastin in mammals, 4-*trans*-hydroxyPro (Hyp) is also found in several conotoxins, including MrIA [171] – a synthetic form of which is being developed as an analgesic (Section 9.3.3.3). Hyp differs from its non-hydroxylated precursor (Pro) in having an hydroxyl group attached to the γ carbon atom. The prolyl hydroxylase responsible for modifying the natural cone snail peptides has not been identified. In mammals, hydroxylation is specified by an –Xxx-Pro-Gly- motif in the substrate and is catalyzed by a multi-subunit dioxygenase, prolyl 4-hydroxylase (EC 1.14.11.2) [285]. The enzyme requires Fe^{2+}, O_2 and 2-oxoglutarate to oxidize the Pro residue. The reaction is accompanied both by the stoichiometric decarboxylation of 2-oxoglutarate to release CO_2 (as depicted in Figure 9.6 for hydroxylation of Asp/Asn) and by oxidation of the enzyme-bound iron atom to Fe^{3+}. Ascorbic acid (vitamin C) is used as a source of reducing equivalents to return the iron to its reduced state. A deficiency of the vitamin leads to impairment of β-hydroxylation and presents as the disease scurvy.

9.3.3 Hydroxylated Biopharmaceuticals

9.3.3.1 Factor IX

Factor IX (FIX) is an important biopharmaceutical used for treatment of hemophilia B. Aspects of the structure, function, activation and therapeutic applications of FIX have been covered above (see especially Section 9.2.4.1). β-Hydroxylation of human FIX occurs at Asp64 within the EGF1 domain located directly adjacent to the Gla domain. Asp64 forms part of a canonical β-hydroxylation consensus sequence (Cys62-Xxx-Asp64-Xxx_4-Tyr69-Xxx-Cys71-Xxx-Cys73) and occupies a crucial position in the high affinity ($K_D \sim 0.1$ mM) Ca^{2+}-binding site in EGF1 (Figures 9.2 and 9.5) [249, 254, 286, 287]. The side chain carboxyl oxygen atoms of the amino acid provide two of the ligands for chelating the Ca^{2+} ion, binding of which stabilizes the hinge region between the EGF1 and Gla domains [33, 258]. The integrity of this Ca^{2+}-binding site is essential for full biological activity of FIX and naturally occurring mutations in Asp64 or other residues that disrupt Ca^{2+} binding give rise to the hemophilia B phenotype (Figure 9.5) [265–267, 269–273]. Likewise, substitution of Asp64 in recombinant FIX with various charged or neutral residues produces molecules with greatly reduced procoagulant activity, at least in part by affecting their ability to interact with FVIIIa [288, 289].

Despite the importance of position 64, the function of β-hydroxylation at this site is unknown. Recombinant FIX molecules lacking Hya have been produced by treating FIX-secreting cell lines with inhibitors that block the modification [274]. The recombinant protein, despite containing Asp rather than Hya at position 64, exhibits normal binding to endothelial cells and normal activity in one-stage clotting assays. Moreover, β-hydroxylation of FIX is naturally a variable event *in vivo* and only ~40% of the circulating protein contains Hya, the remainder possessing an unmodified Asp [140, 263]. Ca^{2+} binds to the EGF1 site with similar affinity whether Asp or Hya is present [263, 264]. Recombinant preparations of FIX used in therapy contain a slightly higher proportion of molecules that are β-hydroxylated than plasma-derived FIX (~46% vs. 37%) [140, 256].

9.3.3.2 Protein C/Activated Protein C

Protein C (PC) or its activated form (APC) is used in thrombolytic therapy and in the treatment of severe infections. Aspects of the structure, function, activation and therapeutic applications of PC/APC have been covered above (see especially Section 9.2.4.3). Protein C normally undergoes β-hydroxylation at Asp71 in the EGF1 domain (Figure 9.2) [251]. Asp71 forms part of a canonical β-hydroxylation consensus sequence in the domain (Cys69-Xxx-Asp71-Xxx_4-Tyr76-Xxx-Cys78-Xxx-Cys80). The EGF1 Ca^{2+}-binding site chelates one Ca^{2+} ion with high-affinity ($K_D \sim 0.1$ mM) [290]. The residues forming the site have not been fully defined but, by comparison to other Ca^{2+}-binding EGF-like domains, prospective ligands are presumed to include Hya71, Asp46, Gly47 and Gln49. The EGF1 domain in PC is unusual in containing a seven-residue insert relative to the EGF-like domains of other vitamin K-dependent proteins, and in possessing four (rather than three) disulfide bonds linked in the pattern 1–5, 2–4, 3–6 and 7–8 [291].

The functional importance of Asp/Hya at position 71 has been investigated by site-directed mutagenesis of recombinant PC. Substitution of the Hya71 residue in PC for Glu [292, 293], Gly [293] or Ala [294, 295] greatly reduces its anticoagulant activity in plasma-based assays and reduces or abolishes its activity in factor Va and factor VIIIa inactivation assays. Though an Asp/Hya at position 71 seems to be critical for Ca^{2+} binding and anticoagulant activity, the significance of β-hydroxylation at the site is unknown. The integrity of the EGF1 Ca^{2+}-binding site does not appear to be important for the interaction of PC with the thrombin–thrombomodulin complex or for the observed Ca^{2+}-dependent inhibition of thrombin-catalyzed activation of PC [294]. It has been suggested that β-hydroxylation of Asp71 may be involved in negatively regulating glycosylation of protein C [276]. Analysis of the FDA-approved recombinant form of human APC [drotrecogin alfa (activated), Xigris; Eli Lilly and Co., Indianapolis, IN], which is expressed in human embryonic kidney (HEK293) cells, indicates that the protein is properly β-hydroxylated during the manufacturing process [166].

9.3.3.3 **Conotoxins**

Aspects of the structure, function and biopharmaceutical potential of conotoxins have been covered above (see especially Section 9.2.4.5). Several conotoxins have been found to undergo post-translational hydroxylation of specific residues. The hydroxylated amino acids identified so far are 4-hydroxyPro (Hyp), 5-hydroxyLys and D-4-hydroxyVal; the latter being formed after L-valine has been first converted into the D-isomer by an epimerase [242–246]. Within the current crop of biopharmaceutical proteins, Hyp is present as the penultimate residue in conotoxin MrIA from *Conus marmoreus* [171]. A synthetic form of MrIA (Xen-2174; Xenome Ltd., Brisbane, Australia) is being tested in clinical trials for relieving neuropathic pain [296–298]. The mature 13-residue MrIA peptide (Figure 9.2) contains two disulfide bonds, with the Cys residues connected in the pattern 1–4, 2–3. It acts as a non-competitive inhibitor of the norepinephrine transporter. The importance of the Hyp residue at position 12 is unclear but its replacement with Ala has no effect on binding to the transporter [299]. The solution NMR structure of the peptide suggests that noncovalent interactions between Hyp12 and Val3 contribute to stabilizing the conformation of the peptide [300].

9.3.4
Analytical Characterization of β-Hydroxylated Proteins

9.3.4.1 Methods for Detecting Hya/Hyn

Several methods have been employed to detect the β-hydroxylated forms of Asp (Hya) and Asn (Hyn) in proteins. The original identification of Hya in protein C employed a combination of amino acid analysis, Edman degradation, NMR spectroscopy and mass spectrometry [251]. These techniques are still used in various formats to detect β-hydroxy-amino acids in proteins. Analysis of amino acid hydrolysates has been routinely used to detect Hya and Hyn [248, 252, 254, 277]. A detailed protocol for quantitative detection of Hya in acid hydrolyates by HPLC is available [213]. Hyn spontaneously undergoes conversion into Hya during acid hydrolysis and its suspected presence must therefore be confirmed by other analytical methods or by analysis of the cDNA sequence encoding the consensus sequence for β-hydroxylation [250]. In addition, a small proportion of *erythro*-Hya is converted into the *threo* form during acid hydrolysis (and vice versa) but the two isomers can be distinguished by their different elution positions during HPLC [251]. A blank cycle is encountered at positions occupied by Hya/Hyn during conventional protein sequencing via Edman degradation. However, modification of the HPLC conditions used to resolve their phenylthiohydantion derivatives permits Hya and Hyn to be identified by this method [254]. Hya and Hyn can be assigned in peptides from their resonances observed by NMR spectroscopy [251, 254]. They may also be identified from their molecular masses as determined by mass spectrometry. The β-hydroxyl group is relatively stable during ionization, allowing Hya/Hyn to be identified in peptide preparations by several mass spectrometric methods [140, 248, 264].

9.3.4.2 β-Hydroxylase Enzyme Assays

The hydroxylating activity of the Asp/Asn β-hydroxylase can be measured in crude microsomal extracts or purified enzyme preparations using CO_2 capture assays [277, 280, 281, 301]. An EGF domain-like sequence present in an intact protein, proteolytic fragment, or synthetic peptide can serve as the substrate. Isotopically labeled 2-oxoglutarate (2-oxo[1-^{14}C]-glutarate) is supplied as a co-substrate. Substrate-dependent conversion of the labeled 2-oxoglutarate into succinate form releases $^{14}CO_2$, which is trapped by an adsorbent such as calcium hydroxide or hyamine hydroxide and measured in a scintillation counter. The reaction produces 1 mole of succinate and 1 mole of CO_2 per mole of β-hydroxylated product (Figure 9.6).

9.4
Conclusions

Post-translational carboxylation and/or hydroxylation reactions serve to modify amino acids in several proteins or peptides with important biopharmaceutical applications. Significant progress has been made in understanding the mechanisms underlying these processes. In particular, the identification and cloning of the genes encoding the γ-carboxylase, vitamin K epoxide reductase and calumenin has allowed a good deal of progress to be made in developing cell systems capable of delivering larger amounts of fully γ-carboxylated and biologically active recombinant vitamin K-dependent proteins. Though some obstacles remain, the advances of late suggest that there is good reason to believe that an era of more cheaply produced γ-carboxylated therapeutics may not be far off. Indeed, someday the hope of affordable worldwide access to these complex biopharmaceuticals may be realized.

References

1 Aledort, L.M. (1999) Economic aspects of haemophilia care in the United States. *Haemophilia*, **5**, 282–285.

2 Furie, B., Bouchard, B.A. and Furie, B.C. (1999) Vitamin K-dependent biosynthesis of γ-carboxyglutamic acid. *Blood*, **93**, 1798–1808.

3 Stafford, D.W. (2005) The vitamin K cycle. *Journal of Thrombosis and Haemostasis*, **3**, 1873–1878.

4 Stenberg, L.M., Brown, M.A. and Stenflo, J. (2002) γ-Carboxyglutamic acid (Gla) domains, in *Encyclopedia of Molecular Medicine* (ed. T. Creighton), John Wiley & Sons, Inc, New York, pp. 1367–1370.

5 Stenflo, J. and Dahlbäck, B. (2001) Vitamin K-dependent proteins in blood coagulation, in *The Molecular Basis of Blood Diseases* (eds G. Stamatoyannopoulos, P.W. Majerus, R.M. Perlmutter and H. Varmus), W.B. Saunders Co, Philadelphia, PA, pp. 579–613.

6 Murshed, M. *et al.* (2004) Extracellular matrix mineralization is regulated locally; different roles of two Gla-containing proteins. *The Journal of Cell Biology*, **165**, 625–630.

7 Price, P.A., Faus, S.A. and Williamson, M.K. (1998) Warfarin causes rapid calcification of the elastic lamellae in rat arteries and heart valves. *Arteriosclerosis, Thrombosis, and Vascular Biology*, **18**, 1400–1407.

8 Brown, M.A., Stenberg, L.M. and Stenflo, J. (2004) Coagulation factor X, in *Handbook of Proteolytic Enzymes* (eds A.J. Barrett, N.D. Rawlings and J.F. Woessner), Elsevier Ltd., London, pp. 1662–1666.

9 Nelsestuen, G.L., Zytokovicz, T.H. and Howard, J.B. (1974) Mode of action of vitamin K: identification of γ-carboxyglutamic acid as a component of prothrombin. *The Journal of Biological Chemistry*, **249**, 6347–6350.

10 Stenflo, J. *et al.* (1974) Vitamin K dependent modifications of glutamic acid residues in prothrombin. *Proceedings of the National Academy of Sciences of the United States of America*, **71**, 2730–2733.

11 Stenflo, J. and Ganrot, P.-O. (1972) Vitamin K and the biosynthesis of prothrombin. I. Identification and purification of a dicoumarol-induced abnormal prothrombin from bovine plasma. *The Journal of Biological Chemistry*, **247**, 8160–8166.

12 Stenflo, J. (1972) Vitamin K and the biosynthesis of prothrombin. II. Structural comparison of normal and dicoumarol-induced bovine prothrombin. *The Journal of Biological Chemistry*, **247**, 8167–8175.

13 Stenflo, J. (1973) Vitamin K and the biosynthesis of prothrombin. III. Structural comparison of an NH_2-terminal fragment from normal and from dicoumarol-induced

bovine prothrombin. *The Journal of Biological Chemistry*, **248**, 6325–6332.

14 Stenflo, J. and Ganrot, P.-O. (1973) Binding of Ca^{2+} to normal and dicoumarol-induced prothrombin. *Biochemical and Biophysical Research Communications*, **50**, 98–104.

15 Stenflo, J. (1974) Vitamin K and the biosynthesis of prothrombin. IV. Isolation of peptides containing prosthetic groups from normal prothrombin and the corresponding peptides from dicoumarol-induced prothrombin. *The Journal of Biological Chemistry*, **249**, 5527–5535.

16 Fernlund, P. *et al.* (1975) Vitamin K and the biosynthesis of prothrombin. V. γ-Carboxyglutamic acids, the vitamin K-dependent structures in prothrombin. *The Journal of Biological Chemistry*, **250**, 6125–6133.

17 Nelsestuen, G.L. and Suttie, J.W. (1972) Mode of action of vitamin K: calcium binding properties of bovine prothrombin. *Biochemistry*, **11**, 4961–4964.

18 Nelsestuen, G.L. and Suttie, J.W. (1972) The purification and properties of an abnormal prothrombin protein produced by dicumarol-treated cows: a comparison to normal prothrombin. *The Journal of Biological Chemistry*, **247**, 8176–8182.

19 Magnusson, S. *et al.* (1974) Primary structure of the vitamin K-dependent part of prothrombin. *FEBS Letters*, **44**, 189–193.

20 Roth, D.A. *et al.* (2001) Human recombinant factor IX: safety and efficacy studies in hemophilia B patients previously treated with plasma-derived factor IX concentrates. *Blood*, **98**, 3600–3606.

21 Shapiro, A.D. *et al.* (2005) The safety and efficacy of recombinant human blood coagulation factor IX in previously untreated patients with severe or moderately severe hemophilia B. *Blood*, **105**, 518–525.

22 Hedner, U. (2001) Recombinant factor VIIa (NovoSeven) as a hemostatic agent. *Seminars in Hematology*, **38**, 43–47.

23 Esmon, C.T. (2002) Protein C pathway in sepsis. *Annals of Medicine*, **34**, 598–605.

24 Furie, B. and Furie, B.C. (2000) Molecular basis of blood coagulation, in *Hematology: Basic Principles and Practice* (eds R. Hoffman, E.J.J. Benz, S.J. Shattil, B. Furie, H.J. Cohen, L.E. Silberstein and P. McGlave), Churchill Livingstone, New York, pp. 1783–1804.

25 Manfioletti, G. *et al.* (1993) The protein encoded by a growth arrest-specific gene (*gas6*) is a new member of the vitamin K-dependent proteins related to protein S, a negative coregulator in the blood coagulation cascade. *Molecular and Cellular Biology*, **13**, 4976–4985.

26 Kulman, J.D. *et al.* (1997) Primary structure and tissue distribution of two novel proline-rich γ-carboxyglutamic acid proteins. *Proceedings of the National Academy of Sciences of the United States of America*, **94**, 9058–9062.

27 Kulman, J.D. *et al.* (2001) Identification of two novel transmembrane γ-carboxyglutamic acid proteins expressed broadly in fetal and adult tissues. *Proceedings of the National Academy of Sciences of the United States of America*, **98**, 1370–1375.

28 Price, P.A., Poser, J.W. and Raman, N. (1976) Primary structure of the γ-carboxyglutamic acid-containing protein from bovine bone. *Proceedings of the National Academy of Sciences of the United States of America*, **73**, 3374–3375.

29 Price, J.A. *et al.* (1980) Structure and function of the vitamin K-dependent protein of bone, in *Vitamin K Metabolism and Vitamin K-dependent Proteins (ed* J.W. Suttie), University Park Press, Baltimore, Maryland, pp. 219–226.

30 Laizé, V. *et al.* (2005) Evolution of matrix and bone γ-carboxyglutamic acid proteins in vertebrates. *The Journal of Biological Chemistry*, **280**, 26659–26668.

31 Brown, M.A. and Stenflo, J. (2004) Gla-domain, in *Handbook of Metalloproteins* (eds A. Messerschmidt, W. Bode and M. Cygler), John Wiley & Sons, Ltd., Chichester, England, pp. 573–583.

32 Hauschka, P.V. and Carr, S.A. (1982) Calcium-dependent α-helical structure in osteocalcin. *Biochemistry*, **21**, 2538–2547.

33 Stenflo, J. (1999) Contributions of Gla and EGF-like domains to the function of vitamin K-dependent coagulation factors. *Critical Reviews in Eukaryotic Gene Expression*, **9**, 59–88.

34 Stenflo, J. and Dahlbäck, B. (1994) Vitamin K-dependent proteins, in *The Molecular Basis of Blood Diseases* (eds G. Stamatoyannopoulos, A.W. Nienhuis, P.W. Majerus and H. Varmus), W.B. Saunders Co., Philadelphia, pp. 565–598.

35 Bolt, G., Steenstrup, T.D. and Kristensen, C. (2007) All post-translational modifications except propeptide cleavage are required for optimal secretion of coagulation factor VII. *Thrombosis and Haemostasis*, **98**, 988–997.

36 Brown, M.A. *et al.* (2005) Precursors of novel Gla-containing conotoxins contain a carboxy-terminal recognition site that directs γ-carboxylation. *Biochemistry*, **44**, 9150–9159.

37 Fainzilber, M. *et al.* (1998) γ-Conotoxin-PnVIIA, a γ-carboxyglutamate-containing peptide agonist of neuronal pacemaker cation currents. *Biochemistry*, **37**, 1470–1477.

38 Hansson, K. *et al.* (2004) The first γ-carboxyglutamic acid-containing contryphan: a selective L-type calcium ion channel blocker isolated from the venom of *Conus marmoreus*. *The Journal of Biological Chemistry*, **279**, 32453–32463.

39 McIntosh, J.M. *et al.* (1984) γ-Carboxyglutamate in a neuroactive toxin. *The Journal of Biological Chemistry*, **259**, 14343–14346.

40 Olivera, B.M. *et al.* (1990) Diversity of *Conus* neuropeptides. *Science*, **249**, 257–263.

41 Rigby, A.C. *et al.* (1999) A conotoxin from *Conus textile* with unusual posttranslational modifications reduces presynaptic Ca^{2+} influx. *Proceedings of the National Academy of Sciences of the United States of America*, **96**, 5758–5763.

42 Chen, Z. *et al.* (1998) Conformational changes in conantokin-G induced upon binding of calcium and magnesium as revealed by NMR structural analysis. *The Journal of Biological Chemistry*, **273**, 16248–16258.

43 Grant, M.A. *et al.* (2004) The metal-free and calcium-bound structures of a γ-carboxyglutamic acid-containing contryphan from *Conus marmoreus*, Glacontryphan-M. *The Journal of Biological Chemistry*, **279**, 32464–32473.

44 Zhou, L.M. *et al.* (1996) Synthetic analogues of conantokin-G: NMDA antagonists acting through a novel polyamine-coupled site. *Journal of Neurochemistry*, **66**, 620–628.

45 Bulaj, G. *et al.* (2003) Efficient oxidative folding of conotoxins and the radiation of venomous cone snails. *Proceedings of the National Academy of Sciences of the United States of America*, **100**, 14562–14568.

46 Hansson, K. *et al.* (2006) A single γ-carboxyglutamic acid residue in a novel cysteine-rich secretory protein without propeptide. *Biochemistry*, **45**, 12828–12839.

47 Bandyopadhyay, P.K. *et al.* (2002) γ-Glutamyl carboxylation: an extracellular posttranslational modification that antedates the divergence of molluscs, arthropods, and chordates. *Proceedings of the National Academy of Sciences of the United States of America*, **99**, 1264–1269.

48 Begley, G.S. *et al.* (2000) A conserved motif within the vitamin K-dependent carboxylase gene is widely distributed across animal phyla. *The Journal of Biological Chemistry*, **275**, 36245–36249.

49 Czerwiec, E. *et al.* (2002) Expression and characterization of recombinant vitamin K-dependent γ-glutamyl carboxylase from an invertebrate, *Conus textile*. *European Journal of Biochemistry*, **269**, 6162–6172.

50 Goodstadt, L. and Ponting, C.P. (2004) Vitamin K epoxide reductase: homology, active site and catalytic mechanism. *Trends in Biochemical Sciences*, **29**, 289–292.

51 Li, T. *et al.* (2000) Identification of a *Drosophila* vitamin K-dependent γ-glutamyl carboxylase. *The Journal of Biological Chemistry*, **275**, 18291–18296.

52 Li, T. *et al.* (2004) Identification of the gene for vitamin K epoxide reductase. *Nature*, **427**, 541–544.

53 Rost, S. *et al.* (2004) Mutations in *VKORC1* cause warfarin resistance and multiple coagulation factor deficiency type 2. *Nature*, **427**, 537–541.

54 Walker, C.S. *et al.* (2001) On a potential global role for vitamin K-dependent γ-carboxylation in animal systems: evidence for a γ-glutamyl carboxylase in *Drosophila*. *The Journal of Biological Chemistry*, **276**, 7769–7774.

55 Kulman, J.D. *et al.* (2006) Vitamin K-dependent proteins in *Ciona intestinalis*, a basal chordate lacking a blood coagulation cascade. *Proceedings of the National Academy of Sciences of the United States of America*, **103**, 15794–15799.

56 Wang, C.-P. *et al.* (2003) Identification of a gene encoding a typical γ-arboxyglutamic acid domain in the tunicate *Halocynthia roretzi*. *Journal of Thrombosis and Haemostasis*, **1**, 118–123.

57 MacGillivray, R.T.A. *et al.* (1993) Structure of the genes encoding proteins involved in blood clotting, in *Genetic Engineering: Principles and Methods (ed* J.K. Setlow), Plenum Press, New York, pp. 265–330.

58 Wojcik, E.G. *et al.* (1996) Mutations which introduce free cysteine residues in the Gla-domain of vitamin K dependent proteins result in the formation of complexes with α_1-microglobulin. *Thrombosis and Haemostasis*, **75**, 70–75.

59 Gillis, S. *et al.* (1997) γ-Carboxyglutamic acids 36 and 40 do not contribute to human factor IX function. *Protein Science*, **6**, 185–196.

60 Larson, P.J. *et al.* (1998) Structure/function analyses of recombinant variants of human factor Xa: factor Xa incorporation into prothrombinase on the thrombin-activated platelet surface is not mimicked by synthetic phospholipid vesicles. *Biochemistry*, **37**, 5029–5038.

61 White, G.C. II., Beebe, A. and Nielsen, B. (1997) Recombinant factor IX. *Thrombosis and Haemostasis*, **78**, 261–265.

62 Zhang, L., Jhingan, A. and Castellino, F.J. (1992) Role of individual γ-carboxyglutamic acid residues of activated human protein C in defining its *in vitro* anticoagulant activity. *Blood*, **80**, 942–952.

63 Freedman, S.J. *et al.* (1995) Structure of the calcium ion-bound γ-carboxyglutamic acid-rich domain of factor IX. *Biochemistry*, **34**, 12126–12137.

64 Freedman, S.J. *et al.* (1995) Structure of the metal-free γ-carboxyglutamic acid-rich membrane binding region of factor IX by two-dimensional NMR spectroscopy. *The Journal of Biological Chemistry*, **270**, 7980–7987.

65 Freedman, S.J. *et al.* (1996) Identification of the phospholipid binding site in the vitamin K-dependent blood coagulation protein factor IX. *The Journal of Biological Chemistry*, **271**, 16227–16236.

66 Huang, M. *et al.* (2003) Structural basis of membrane binding by Gla domains of vitamin K-dependent proteins. *Nature Structural Biology*, **10**, 751–756.

67 Soriano-Garcia, M. *et al.* (1992) The Ca^{2+} ion and membrane binding structure of the Gla domain of Ca-prothrombin fragment 1. *Biochemistry*, **31**, 2554–2566.

68 Sunnerhagen, M. *et al.* (1995) Structure of the Ca^{2+}-free Gla domain sheds light on membrane binding of blood coagulation proteins. *Nature Structural Biology*, **2**, 504–509.

69 Sunnerhagen, M. *et al.* (1996) The relative orientation of Gla and EGF domains in coagulation factor X is altered by Ca^{2+} binding to the first EGF domain: a combined NMR–small angle X-ray scattering study. *Biochemistry*, **35**, 11547–11559.

70 Gitel, S.N. *et al.* (1973) A polypeptide region of bovine prothrombin specific for binding to phospholipids. *Proceedings of the National Academy of Sciences of the United States of America*, **70**, 1344–1348.

71 Nelsestuen, G.L., Broderius, M. and Martin, G. (1976) Role of γ-carboxyglutamic acid: cation specificity of prothrombin and factor X-phospholipid binding. *The Journal of Biological Chemistry*, **251**, 6886–6893.

72 Stenflo, J. and Suttie, J.W. (1977) Vitamin K-dependent formation of γ-carboxyglutamic acid. *Annual Review of Biochemistry*, **46**, 157–172.

73 London, F. and Walsh, P.N. (1996) The role of electrostatic interactions in the assembly of the factor X activating complex on both activated platelets and negatively-charged phospholipid vesicles. *Biochemistry*, **35**, 12146–12154.

74 Oganesyan, V. *et al.* (2002) The crystal structure of the endothelial protein C receptor and a bound phospholipid. *The Journal of Biological Chemistry*, **277**, 24851–24854.

75 Regan, L.M. *et al.* (1997) The interaction between the endothelial cell protein C receptor and protein C is dictated by the γ-carboxyglutamic acid domain of protein C. *The Journal of Biological Chemistry*, **272**, 26279–26284.

76 Blostein, M.D. *et al.* (2003) The Gla domain of factor IXa binds to factor VIIIa in the tenase complex. *The Journal of Biological Chemistry*, **278**, 31297–31302.

77 Larson, P.J. *et al.* (1996) Structural integrity of the γ-carboxyglutamic acid domain of human blood coagulation factor IXa is required for its binding to cofactor VIIIa. *The Journal of Biological Chemistry*, **271**, 3869–3876.

78 Wolberg, A.S., Stafford, D.W. and Erie, D.A. (1997) Human factor IX binds to specific sites on the collagenous domain of collagen IV. *The Journal of Biological Chemistry*, **272**, 16717–16720.

79 Blostein, M.D. *et al.* (2000) The Gla domain of human prothrombin has a binding site for factor Va. *The Journal of Biological Chemistry*, **275**, 38120–38126.

80 Banner, D.W. *et al.* (1996) The crystal structure of the complex of blood coagulation factor VIIa with soluble tissue factor. *Nature*, **380**, 41–46.

81 Zhang, E., St Charles, R. and Tulinsky, A. (1999) Structure of extracellular tissue factor complexed with factor VIIa inhibited with a BPTI mutant. *Journal of Molecular Biology*, **285**, 2089–2104.

82 Park, C.H. and Tulinsky, A. (1986) Three-dimensional structure of the kringle sequence: structure of prothrombin fragment 1. *Biochemistry*, **25**, 3977–3982.

83 Tulinsky, A. and Park, C.H. (1988) Structure of prothrombin fragment 1 and its relation to calcium binding, in *Current Advances in Vitamin K Research*, (ed J.W. Suttie), Elsevier Science Publishing Co., Inc., New York, pp. 295–304.

84 Seshadri, T.P.J. *et al.* (1991) Structure of bovine prothrombin fragment 1 refined at 2.25 Å resolution. *Journal of Molecular Biology*, **220**, 481–494.

85 Brandstetter, H. *et al.* (1995) X-ray structure of clotting factor IXa: active site and module structure related to Xase activity and hemophilia B. *Proceedings of the National Academy of Sciences of the United States of America*, **92**, 9796–800.

86 Huang, M., Furie, B.C. and Furie, B. (2004) Crystal structure of the calcium-stabilized human factor IX Gla domain bound to a conformation-specific anti-factor IX antibody. *The Journal of Biological Chemistry*, **279**, 14338–14346.

87 Mizuno, H. *et al.* (1999) Crystal structure of coagulation factor IX-binding protein from Habu snake venom at 2.6 Å: implication of central loop swapping

based on deletion in the linker region. *Journal of Molecular Biology*, **289**, 103–112.

88 Shikamoto, Y. *et al.* (2003) Crystal structure of Mg^{2+}- and Ca^{2+}-bound Gla domain of factor IX complexed with binding protein. *The Journal of Biological Chemistry*, **278**, 24090–24094.

89 Li, L. *et al.* (1997) Refinement of the NMR solution structure of the γ-carboxyglutamic acid domain of coagulation factor IX using molecular dynamics simulation with initial Ca^{2+} positions determined by a genetic algorithm. *Biochemistry*, **36**, 2132–2138.

90 Foster, D.C. *et al.* (1987) Propeptide of human protein C is necessary for γ-carboxylation. *Biochemistry*, **26**, 7003–7011.

91 Huber, P. *et al.* (1990) Identification of amino acids in the γ-carboxylation recognition site on the propeptide of prothrombin. *The Journal of Biological Chemistry*, **265**, 12467–12473.

92 Jorgensen, M.J. *et al.* (1987) Recognition site directing vitamin K-dependent γ-carboxylation resides on the propeptide of factor IX. *Cell*, **48**, 185–191.

93 Price, P.A., Fraser, J.D. and Metz-Virca, G. (1987) Molecular cloning of matrix Gla protein: implications for substrate recognition by the vitamin K-dependent γ-carboxylase. *Proceedings of the National Academy of Sciences of the United States of America*, **84**, 8335–8339.

94 Wise, R.J. *et al.* (1990) Expression of a human proprotein processing enzyme: correct cleavage of the von Willebrand factor precursor at a paired basic amino acid site. *Proceedings of the National Academy of Sciences of the United States of America*, **87**, 9378–9382.

95 Bresnahan, P.A. *et al.* (1990) Human *fur* gene encodes a yeast KEX2-like endoprotease that cleaves pro-β-NGF *in vivo*. *The Journal of Cell Biology*, **111**, 2851–2859.

96 Stanton, C., Taylor, R. and Wallin, R. (1991) Processing of prothrombin in the secretory pathway. *The Biochemical Journal*, **277**, 59–65.

97 Bentley, A.K. *et al.* (1986) Defective propeptide processing of blood clotting factor IX caused by mutation of arginine to glutamine at position –4. *Cell*, **45**, 343–348.

98 Diuguid, D.L. *et al.* (1986) Molecular basis of hemophilia B: a defective enzyme due to an unprocessed propeptide is caused by a point mutation in the factor IX precursor. *Proceedings of the National Academy of Sciences of the United States of America*, **83**, 5803–5807.

99 Gandrille, S. *et al.* (1995) Identification of 15 different candidate causal point mutations and three polymorphisms in 19 patients with protein S deficiency using a scanning method for the analysis of the protein S active gene. *Blood*, **85**, 130–138.

100 Lind, B., Johnsen, A.H. and Thorsen, S. (1997) Naturally occurring Arg^{-1} to His mutation in human protein C leads to aberrant propeptide processing and secretion of dysfunctional protein C. *Blood*, **89**, 2807–2816.

101 Reitsma, P.H. *et al.* (1995) Protein C deficiency: a database of mutations, 1995 update. On behalf of the Subcommittee on Plasma Coagulation Inhibitors of the Scientific and Standardization Committee of the ISTH. *Thrombosis and Haemostasis*, **73**, 876–889.

102 Ware, J. *et al.* (1989) Factor IX San Dimas: substitution of glutamine for Arg-4 in the propeptide leads to incomplete γ-carboxylation and altered phospholipid binding properties. *The Journal of Biological Chemistry*, **264**, 11401–11406.

103 Stanley, T.B. *et al.* (1998) Role of the propeptide and γ-glutamic acid domain of factor IX for *in vitro* carboxylation by the vitamin K-dependent carboxylase. *Biochemistry*, **37**, 13262–13268.

104 Stenina, O. *et al.* (2001) Tethered processivity of the vitamin K-dependent carboxylase: factor IX is efficiently modified in a mechanism which distinguishes Gla's from Glu's and which accounts for comprehensive carboxylation *in vivo*. *Biochemistry*, **40**, 10301–10309.

105 Berkner, K.L. (2008) Vitamin K-dependent carboxylation. *Vitamins and Hormones*, **78**, 131–156.

106 Suttie, J.W. (1993) Synthesis of vitamin K-dependent proteins. *The FASEB Journal*, **7**, 445–452.

107 Dowd, P. *et al.* (1995) The mechanism of action of vitamin K. *Annual Review of Nutrition*, **15**, 419–440.

108 Davis, C.H. *et al.* (2007) A quantum chemical study of the mechanism of action of Vitamin K carboxylase (VKC) III. Intermediates and transition states. *Journal of Molecular Graphics & Modelling*, **26**, 409–414.

109 Wajih, N. *et al.* (2004) The inhibitory effect of calumenin on the vitamin K-dependent γ-carboxylation system: characterization of the system in normal and warfarin-resistant rats. *The Journal of Biological Chemistry*, **279**, 25276–25283.

110 Jin, D.-Y., Tie, J.-K. and Stafford, D.W. (2007) The conversion of vitamin K epoxide to vitamin K quinone and vitamin K quinone to vitamin K hydroquinone uses the same active site cysteines. *Biochemistry*, **46**, 7279–7283.

111 Wallin, R., Wajih, N. and Hutson, S.M. (2008) VKORC1: a warfarin-sensitive enzyme in vitamin K metabolism and biosynthesis of vitamin K-dependent blood coagulation factors. *Vitamins and Hormones*, **78**, 227–246.

112 Zimmermann, A. and Matschiner, J.T. (1974) Biochemical basis of hereditary resistance to warfarin in the rat. *Biochemical Pharmacology*, **23**, 1033–1040.

113 Fasco, M.J. *et al.* (1983) Warfarin inhibition of vitamin K 2,3-epoxide reductase in rat liver microsomes. *Biochemistry*, **22**, 5655–5660.

114 Wajih, N. *et al.* (2005) Engineering of a recombinant vitamin K-dependent γ-carboxylation system with enhanced γ-carboxyglutamic acid forming capacity: evidence for a functional CXXC redox center in the system. *The Journal of Biological Chemistry*, **280**, 10540–10547.

115 Wajih, N., Hutson, S.M. and Wallin, R. (2007) Disulfide dependent protein folding is linked to operation of the vitamin K cycle in the endoplasmic reticulum: a protein disulfide isomerase-VKORC1 redox enzyme complex appears to be responsible for vitamin k1 2,3-epoxide reduction. *The Journal of Biological Chemistry*, **282**, 2626–2635.

116 Thijssen, H.H.W., Baars, L.G.M. and Vervoort-Peters, H.T.M. (1988) Vitamin K 2,3-epoxide reductase: the basis for stereoselectivity of 4-hydroxycoumarin anticoagulant activity. *British Journal of Pharmacology*, **95**, 675–682.

117 Furie, B. (2000) Oral anticoagulant therapy, in *Hematology: Basic Principles and Practice* (eds R. Hoffman, E.J.J. Benz, S.J. Shattil, B. Furie, H.J. Cohen, L.E. Silberstein and P. McGlave), Churchill Livingstone, New York, pp. 2040–2046.

118 Suttie, J.W. (1987) The biochemical basis of warfarin therapy, in *The New Dimensions of Warfarin Prophylaxis* (eds S. Wessler, C.G. Becker and Y. Nemerson), Plenum Press, New York, pp. 3–16.

119 Silverman, R.B., Mukharji, I. and Nandi, D.L. (1988) Solubilization, partial purification, mechanism, and inactivation of vitamin K epoxide reductase, in *Current Advances in Vitamin K Research* (eds J.W. Suttie), Elsevier Science Publishing Co., Inc., New York, pp. 65–74.

120 Stanton, C. and Wallin, R. (1992) Processing and trafficking of clotting factor X in the secretory pathway: effects of warfarin. *The Biochemical Journal*, **284**, 25–31.

121 Shearer, M.J., Bach, A. and Kohlmeier, M. (1996) Chemistry, nutritional sources, tissue distribution and metabolism of vitamin K with special reference to bone health. *The Journal of Nutrition*, **126** (Suppl. 4), 1181S–1186.

122 Ferland, G. (1998) The vitamin K-dependent proteins: an update. *Nutrition Reviews*, **56**, 223–230.

123 Stenberg, L.M. *et al.* (2001) A functional prothrombin gene product is synthesized by human kidney cells. *Biochemical and Biophysical Research Communications*, **280**, 1036–1041.

124 Stenberg, L.M. *et al.* (2001) Synthesis of γ-carboxylated polypeptides by α-cells of the pancreatic islets. *Biochemical and Biophysical Research Communications*, **283**, 454–459.

125 Malm, J. *et al.* (1994) Vitamin K-dependent protein S in Leydig cells of human testis. *The Biochemical Journal*, **302**, 845–850.

126 Ishimoto, Y. and Nakano, T. (2000) Release of a product of growth arrest-specific gene 6 from rat platelets. *FEBS Letters*, **466**, 197–199.

127 Ducy, P. *et al.* (1996) Increased bone formation in osteolcalcin-deficient mice. *Nature*, **382**, 448–452.

128 Reza, M.A. *et al.* (2006) Molecular evolution caught in action: gene duplication and evolution of molecular isoforms of prothrombin activators in *Pseudonaja textilis* (brown snake). *Journal of Thrombosis and Haemostasis*, **4**, 1346–1353.

129 Kurachi, K. and Davie, E.W. (1982) Isolation and characterization of a cDNA coding for human factor IX. *Proceedings of the National Academy of Sciences of the United States of America*, **79**, 6461–6464.

130 High, K.A. and Roberts, H.R. (1995) Factor IX, in *Molecular Basis of Thrombosis and Hemostasis* (eds K.A. High and H.R. Roberts), Marcel Dekker, Inc., New York, pp. 215–237.

131 Bajaj, S.P. and Birktoft, J.J. (1993) Human factor IX and factor IXa. *Methods in Enzymology*, **222**, 96–128.

132 Gailani, D. *et al.* (2001) Model for a factor IX activation complex on blood platelets: dimeric conformation of factor XIa is essential. *Blood*, **97**, 3117–3122.

133 Lenting, P.J. *et al.* (1995) Cleavage at arginine 145 in human blood coagulation factor IX converts the zymogen into a factor VIII binding enzyme. *The Journal of Biological Chemistry*, **270**, 14884–14890.

134 Agarwala, K.L. *et al.* (1994) Activation peptide of human factor IX has oligosaccharides *O*-glycosidically linked to threonine residues at 159 and 169. *Biochemistry*, **33**, 5167–5171.

135 Di Scipio, R.G., Kurachi, K. and Davie, E.W. (1978) Activation of human factor IX (Christmas factor). *The Journal of Clinical Investigation*, **61**, 1528–1538.

136 Giannelli, F. *et al.* (1998) Haemophilia B: database of point mutations and short additions and deletions — eighth edition. *Nucleic Acids Research*, **26**, 265–268.

137 Sommer, S.S., Scaringe, W.A. and Hill, K.A. (2001) Human germline mutation in the factor IX gene. *Mutation Research*, **487**, 1–17.

138 Edwards, J. and Kirkby, N. (1999) Recombinant coagulation factor IX (BeneFIX®), in *Biopharmaceuticals, an Industrial Perspective* (eds G. Walsh and B. Murphy), Kluwer Academic Publishers, The Netherlands, pp. 73–108.

139 Harrison, S. *et al.* (1998) The manufacturing process for recombinant factor IX. *Seminars in Hematology*, **35**, 4–10.

140 Bond, M. *et al.* (1998) Biochemical characterization of recombinant factor IX. *Seminars in Hematology*, **35** (Suppl. 2), 11–17.

141 Broze, G.J. Jr (1992) The role of tissue factor pathway inhibitor in a revised coagulation cascade. *Seminars in Hematology*, **29**, 159–169.

142 Østerud, B. (1997) Tissue factor: a complex biological role. *Thrombosis and Haemostasis*, **78**, 755–758.

143 Giesen, P.L. *et al.* (1999) Blood-borne tissue factor: another view of thrombosis. *Proceedings of the National Academy of Sciences of the United States of America*, **96**, 2311–2315.

144 Thim, L. *et al.* (1988) Amino acid sequence and posttranslational modifications of human factor VIIa from plasma and transfected baby hamster kidney cells. *Biochemistry*, **27**, 7785–7793.

145 Erhardtsen, E. *et al.* (2006) Recombinant factor VIIa, in *Directory of Therapeutic Enzymes* (eds B.M. McGrath and G. Walsh), CRC Press, Boca Raton, Florida, pp. 189–207.

146 Butenas, S. and Mann, K.G. (1996) Kinetics of human factor VII activation. *Biochemistry*, **35**, 1904–1910.

147 Dahlbäck, B. and Villoutreix, B.O. (2005) The anticoagulant protein C pathway. *FEBS Letters*, **579**, 3310–3316.

148 Fay, P.J., Smudzin, T.M. and Walker, F.J. (1991) Activated protein C-catalyzed inactivation of human factor VIII and factor VIIIa: identification of cleavage sites and correlation of proteolysis with cofactor activity. *The Journal of Biological Chemistry*, **266**, 20139–20145.

149 Kalafatis, M., Rand, M.D. and Mann, K.G. (1994) The mechanism of inactivation of human factor V and human factor Va by activated protein C. *The Journal of Biological Chemistry*, **269**, 31869–31880.

150 Comp, P.C. and Esmon, C.T. (1981) Generation of fibrinolytic activity by infusion of activated protein C into dogs. *The Journal of Clinical Investigation*, **68**, 1221–1228.

151 Esmon, C.T., Taylor, F.B. Jr and Snow, T.R. (1991) Inflammation and coagulation: linked processes potentially regulated through a common pathway mediated by protein C. *Thrombosis and Haemostasis*, **66**, 160–165.

152 Snow, T.R. *et al.* (1991) Protein C activation following coronary artery occlusion in the *in situ* porcine heart. *Circulation*, **84**, 293–299.

153 Foster, D. and Davie, E.W. (1984) Characterization of a cDNA coding for human protein C. *Proceedings of the National Academy of Sciences of the United States of America*, **81**, 4766–47670.

154 Plutzky, J. *et al.* (1986) Evolution and organization of the human protein C gene. *Proceedings of the National Academy of Sciences of the United States of America*, **83**, 546–550.

155 Stenflo, J. (1976) A new vitamin K-dependent protein: purification from bovine plasma and preliminary characterization. *The Journal of Biological Chemistry*, **251**, 355–363.

156 Marlar, R.A. (1985) Plasma single chain protein C is functionally similar to the two chain form of plasma protein C. *Thrombosis and Haemostasis*, **54**, 216.

157 Greffe, B.S., Manco-Johnson, M.J. and Marlar, R.A. (1989) Molecular forms of human protein C: comparison and distribution in human adult plasma. *Thrombosis and Haemostasis*, **62**, 902–905.

158 Esmon, C.T. (1984) Protein C. *Progress in Hemostasis and Thrombosis*, **7**, 25–54.

159 Kisiel, W. (1979) Human plasma protein C: isolation, characterization, and mechanism of activation by α-thrombin. *The Journal of Clinical Investigation*, **64**, 761–769.

160 Oppenheimer, C. and Wydro, R. (1988) Cellular processing of vitamin K-dependent proteins, in *Current Advances in Vitamin K Research*, (ed J.W. Suttie), Elsevier Science Publishing Co., Inc., New York, pp. 165–171.

161 Foster, D.C. *et al.* (1990) Endoproteolytic processing of the dibasic cleavage site in the human protein C precursor in transfected mammalian cells: effects of sequence alterations on efficiency of cleavage. *Biochemistry*, **29**, 347–354.

162 Yan, S.C.B., Grinnell, B.W. and Wold, F. (1989) Post-translational modifications of proteins: some problems left to solve. *Trends in Biochemical Sciences*, **14**, 264–268.

163 Laterre, P.F. (2007) Clinical trials in severe sepsis with drotrecogin alfa (activated). *Critical Care (London, England)*, **11**, S5.

164 O'Brien, L.A., Gupta, A. and Grinnell, B.W. (2006) Activated protein C and sepsis. *Frontiers in Bioscience: a Journal and Virtual Library*, **11**, 676–698.

165 Rivard, G.E. *et al.* (1995) Treatment of purpura fulminans in meningococcemia with protein C concentrate. *The Journal of Pediatrics*, **126**, 646–652.

166 Grinnell, B.W., Yan, S.B. and Macias, W.L. (2006) Activated protein C, in *Directory of Therapeutic Enzymes*, (eds B.M. McGrath and G. Walsh), CRC Press, Boca Raton, Florida, pp. 69–95.

167 Mann, K.G. *et al.* (1990) Surface-dependent reactions of the vitamin K-dependent enzyme complexes. *Blood*, **76**, 1–16.

168 Jackson, C.M. (1994) Physiology and biochemistry of prothrombin, in *Haemostasis and Thrombosis* (eds A.L. Bloom, C.D. Forbes, D.P. Thomas and E.G.D. Tuddenham), Churchill Livingstone, Edinburgh, pp. 397–438.

169 Chapman, W.C. *et al.* (2007) A phase 3, randomized, double-blind comparative study of the efficacy and safety of topical recombinant human thrombin and bovine thrombin in surgical hemostasis. *Journal of the American College of Surgeons*, **205**, 256–265.

170 Craig, A.G. (2000) The characterization of conotoxins. *Journal of Toxicology. Toxin Reviews*, **19**, 53–93.

171 Sharpe, I.A. *et al.* (2001) Two new classes of conopeptides inhibit the α1-adrenoceptor and noradrenaline transporter. *Nature Neuroscience*, **4**, 902–907.

172 Santos, A.D. *et al.* (2004) The A-superfamily of conotoxins: structural and functional divergence. *The Journal of Biological Chemistry*, **279**, 17596–17606.

173 Terlau, H. and Olivera, B.M. (2004) *Conus* venoms: a rich source of novel ion channel-targeted peptides. *Physiological Reviews*, **84**, 41–68.

174 Armishaw, C.J. and Alewood, P.F. (2005) Conotoxins as research tools and drug leads. *Current Protein & Peptide Science*, **6**, 221–240.

175 Grant, M.A., Shanmugasundaram, K. and Rigby, A.C. (2007) Conotoxin therapeutics: a pipeline for success? *Expert Opinion Drug Discovery*, **2**, 453–468.

176 Olivera, B.M. and Teichert, R.W. (2007) Diversity of the neurotoxic Conus peptides: a model for concerted pharmacological discovery. *Molecular Interventions*, **7**, 251–260.

177 Sharp, D. (2005) Novel pain relief via marine snails. *Lancet*, **366**, 439–440.

178 Woodward, S.R. *et al.* (1990) Constant and hypervariable regions in conotoxin propeptides. *The EMBO Journal*, **9**, 1015–1020.

179 Kauferstein, S. *et al.* (2003) A novel conotoxin inhibiting vertebrate voltage-sensitive potassium channels. *Toxicon*, **42**, 43–52.

180 Stanley, T.B. *et al.* (1997) Identification of a vitamin K-dependent carboxylase in the venom duct of a *Conus* snail. *FEBS Letters*, **407**, 85–88.

181 Prorok, M. *et al.* (1996) Calcium binding properties of synthetic γ-carboxyglutamic acid-containing marine cone snail "sleeper" peptides, conantokin-G and conantokin-T. *Biochemistry*, **35**, 16528–16534.

182 Lang, P.M. *et al.* (2005) A *Conus* peptide blocks nicotinic receptors of unmyelinated axons in human nerves. *Neuroreport*, **16**, 479–483.

183 Sandall, D.W. *et al.* (2003) A novel α-conotoxin identified by gene sequencing is active in suppressing the vascular response to selective stimulation of sensory nerves *in vivo*. *Biochemistry*, **42**, 6904–6911.

184 Berkner, K.L. (1993) Expression of recombinant vitamin K-dependent proteins in mammalian cells: factors IX and VII. *Methods in Enzymology*, **222**, 450–477.

185 Chaing, S. *et al.* (1994) Severe factor VII deficiency caused by mutations abolishing the cleavage site for activation and altering binding to tissue factor. *Blood*, **83**, 3524–3535.

186 Kaufman, R.J. *et al.* (1986) Expression, purification, and characterization of recombinant γ-carboxylated factor IX synthesized in Chinese hamster ovary cells. *The Journal of Biological Chemistry*, **261**, 9622–9628.

187 Toomey, J.R., Smith, K.J. and Stafford, D.W. (1991) Localization of the human tissue factor recognition determinant of human factor VIIa. *The Journal of Biological Chemistry*, **266**, 19198–19202.

188 Yan, S.C.B. *et al.* (1990) Characterization and novel purification of recombinant human protein C from three mammalian cell lines. *Bio/Technology*, **8**, 655–661.

189 Van Cott, K.E. *et al.* (1999) Transgenic pigs as bioreactors: a comparison of gamma-carboxylation of glutamic acid in recombinant human protein C and factor IX by the mammary gland. *Genet Analysis*, **15**, 155–160.

190 Rehemtulla, A. *et al.* (1993) *In vitro* and *in vivo* functional characterization of bovine vitamin K-dependent γ-carboxylase expressed in Chinese hamster ovary cells. *Proceedings of the National Academy of Sciences of the United States of America*, **90**, 4611–4615.

191 Wu, S.-M. *et al.* (1991) Cloning and expression of the cDNA for human γ-glutamyl carboxylase. *Science*, **254**, 1634–1636.

192 Hallgren, K.W. *et al.* (2002) Carboxylase overexpression effects full carboxylation but poor release and secretion of factor IX: implications for the release of vitamin K-dependent proteins. *Biochemistry*, **41**, 15045–15055.

193 Sun, Y.-M. *et al.* (2005) Vitamin K epoxide reductase significantly improves carboxylation in a cell line overexpressing factor X. *Blood*, **106**, 3811–3815.

194 Wajih, N. *et al.* (2005) Increased production of functional recombinant human clotting factor IX by baby hamster kidney cells engineered to overexpress VKORC1, the vitamin K 2,3-epoxide-reducing enzyme of the vitamin K cycle. *The Journal of Biological Chemistry*, **280**, 31603–31607.

195 Berkner, K.L. (2005) The vitamin K-dependent carboxylase. *Annual Review of Nutrition*, **25**, 127–149.

196 Hallgren, K.W. *et al.* (2006) r-VKORC1 expression in factor IX BHK cells increases the extent of factor IX carboxylation but is limited by saturation of another carboxylation component or by a shift in the rate-limiting step. *Biochemistry*, **45**, 5587–5598.

197 Wajih, N., Hutson, S.M. and Wallin, R. (2006) siRNA silencing of calumenin enhances functional factor IX production. *Blood*, **108**, 3757–3760.

198 Wajih, N., Owen, J. and Wallin, R. (2008) Enhanced functional recombinant factor VII production by HEK 293 cells stably transfected with VKORC1 where the gamma-carboxylase inhibitor calumenin is stably suppressed by shRNA transfection. *Thrombosis Research*, **122**, 405–410.

199 Wallin, R. *et al.* (2001) A molecular mechanism for genetic warfarin resistance in the rat. *The FASEB Journal*, **15**, 2542–2544.

200 Handford, P.A., Winship, P.R. and Brownlee, G.G. (1991) Protein engineering of the propeptide of human factor IX. *Protein Engineering*, **4**, 319–323.

201 Stanley, T.B. *et al.* (1999) The propeptides of the vitamin K-dependent proteins possess different affinities for the vitamin K-dependent carboxylase. *The Journal of Biological Chemistry*, **274**, 16940–16944.

202 Camire, R.M. *et al.* (2000) Enhanced γ-carboxylation of recombinant factor X using a chimeric construct containing the prothrombin propeptide. *Biochemistry*, **39**, 14322–14329.

203 Rudolph, A.E. *et al.* (1997) Expression, purification, and characterization of recombinant human factor X. *Protein Expression and Purification*, **10**, 373–378.

204 Jackson, C.M., Johnson, T.F. and Hanahan, D.J. (1968) Studies on bovine factor X. I. Large-scale purification of the bovine plasma protein possessing factor X activity. *Biochemistry*, **7**, 4492–4505.

205 Di Scipio, R.G. *et al.* (1977) A comparison of human prothrombin, factor IX (Christmas factor), factor X (Stuart factor), and protein S. *Biochemistry*, **16**, 698–706.

206 Yan, S.B. (1996) Review of conformation-specific affinity purification methods for plasma vitamin K-dependent proteins. *Journal of Molecular Recognition*, **9**, 211–218.

207 Brown, M.A. *et al.* (2000) Identification and purification of vitamin K-dependent proteins and peptides with monoclonal antibodies specific for γ-carboxyglutamyl (Gla) residues. *The Journal of Biological Chemistry*, **275**, 19795–19802.

208 Cairns, J.R., Williamson, M.K. and Price, P.A. (1991) Direct identification of γ-carboxyglutamic acid in the sequencing of vitamin K-dependent proteins. *Analytical Biochemistry*, **199**, 93–97.

209 Jie, K.-S.G., Gijsbers, B.L.M.G. and Vermeer, C. (1995) A specific colorimetric staining method for γ-carboxyglutamic acid-containing proteins in polyacrylamide gels. *Analytical Biochemistry*, **224**, 163–165.

210 Hauschka, P.V. (1979) Specific tritium labeling of γ-carboxyglutamic acid in proteins. *Biochemistry*, **18**, 4992–4999.

211 Kuwada, M. and Katayama, K. (1983) An improved method for the determination of γ-carboxyglutamic acid in proteins, bone, and urine. *Analytical Biochemistry*, **131**, 173–179.

212 Vo, H.C. *et al.* (1999) Undercarboxylation of recombinant prothrombin revealed by analysis of γ-carboxyglutamic acid using capillary electrophoresis and laser-induced fluorescence. *FEBS Letters*, **445**, 256–260.

213 Castellino, F.J., Ploplis, V.A. and Zhang, L. (2008) γ-Glutamate and β-hyydroxyaspartate in proteins. *Methods in Molecular Biology (Clifton, NJ)*, **446**, 85–94.

214 Carr, S.A., Hauschka, P.V. and Biemann, K. (1981) Gas chromatographic mass spectrometric sequence determination of osteocalcin, a γ-carboxyglutamic acid-containing protein from chicken bone. *The Journal of Biological Chemistry*, **256**, 9944–9950.

215 Kalume, D.E. *et al.* (2000) Structure determination of two conotoxins from *Conus textile* by a combination of matrix-assisted laser desorption/ionization time-of-flight and electrospray ionization mass spectrometry and biochemical methods. *Journal of Mass Spectrometry*, **35**, 145–156.

216 Kelleher, N.L. *et al.* (1999) Localization of labile posttranslational modifications by electron capture dissociation: the case of γ-carboxyglutamic acid. *Analytical Chemistry*, **71**, 4250–4253.

217 Deerfield, D.W. II *et al.* (1987) Mg(II) binding by bovine prothrombin fragment 1 via equilibrium dialysis and the relative roles of Mg(II) and Ca(II) in blood coagulation. *The Journal of Biological Chemistry*, **262**, 4017–4023.

218 Sabharwal, A.K. *et al.* (1995) High affinity Ca^{2+}-binding site in the serine protease domain of human factor VIIa and its role in tissue factor binding and development of catalytic activity. *The Journal of Biological Chemistry*, **270**, 15523–15530.

219 Zapata, G.A. et al. (1988) Chemical modification of bovine prothrombin fragment 1 in the presence of Tb^{3+} ions: sequence studies on 3-γ-MGlu-fragment 1. *The Journal of Biological Chemistry*, **263**, 8150–8156.

220 Jackson, C.M. (1988) Calcium ion binding to γ-carboxyglutamic acid-containing proteins from the blood clotting system: what we still don't understand, in *Current Advances in Vitamin K Research*, (ed J.W. Suttie), Elsevier Science Publishing Co., Inc., New York, pp. 305–324.

221 Colpitts, T.L. and Castellino, F.J. (1994) Calcium and phospholipid binding properties of synthetic γ-carboxyglutamic acid-containing peptides with sequence counterparts in human protein C. *Biochemistry*, **33**, 3501–3508.

222 Stenflo, J. (1970) Dicumarol-induced prothrombin in bovine plasma. *Acta Chemica Scandinavica*, **24**, 3762–3763.

223 Björk, I. and Stenflo, J. (1973) A conformational study of normal and dicoumarol-induced prothrombin. *FEBS Letters*, **32**, 343–346.

224 Marsh, H.C. *et al.* (1979) Magnesium and calcium ion binding to bovine prothrombin fragment 1: a circular dichroism, fluorescence, and $^{43}Ca^{2+}$ and $^{25}Mg^{2+}$ nuclear magnetic resonance study. *The Journal of Biological Chemistry*, **254**, 10268–10275.

225 Häfner, A. *et al.* (2000) Calcium-induced conformational change in fragment 1-86 of factor X. *Biopolymers*, **57**, 226–234.

226 Hof, M. (1998) Picosecond tryptophan fluorescence of membrane-bound prothrombin fragment 1. *Biochimica et Biophysica Acta*, **1388**, 143–153.

227 Nelsestuen, G.L. (1976) Role of γ-carboxyglutamic acid: an unusual protein transition required for the calcium-dependent binding of prothrombin to phospholipid. *The Journal of Biological Chemistry*, **251**, 5648–5656.

228 Bajaj, S.P., Nowak, T. and Castellino, F.J. (1976) Interaction of manganese with bovine prothrombin and its thrombin-mediated cleavage products. *The Journal of Biological Chemistry*, **251**, 6294–6299.

229 Furie, B.C., Blumenstein, M. and Furie, B. (1979) Metal binding sites of a γ-carboxyglutamic acid-rich fragment of bovine prothrombin. *The Journal of Biological Chemistry*, **254**, 12521–12530.

230 Lentz, B.R., Zhou, C.-M. and Wu, J.R. (1994) Phosphatidylserine-containing membranes alter the thermal stability of prothrombin's catalytic domain: a differential scanning calorimetric study. *Biochemistry*, **33**, 5460–5468.

231 Vysotchin, A., Medved, L.V. and Ingham, K.C. (1993) Domain structure and domain–domain interactions in human coagulation factor IX. *The Journal of Biological Chemistry*, **268**, 8436–8446.

232 Wu, J.R. and Lentz, B.R. (1991) Fourier transform infrared spectroscopic study of Ca^{2+} and membrane-induced secondary structural changes in bovine prothrombin and prothrombin fragment 1. *Biophysical Journal*, **60**, 70–80.

233 Jacobs, M. *et al.* (1994) Membrane binding properties of the factor IX γ-carboxyglutamic acid-rich domain prepared by chemical synthesis. *The Journal of Biological Chemistry*, **269**, 25494–25501.

234 Falls, L.A. *et al.* (2001) The ω-loop region of the human prothrombin γ-carboxyglutamic acid domain penetrates anionic phospholipid membranes. *The Journal of Biological Chemistry*, **276**, 23895–23902.

235 Zhang, L. and Castellino, F.J. (1993) The contributions of individual γ-carboxyglutamic acid residues in the calcium-dependent binding of recombinant human protein C to acidic phospholipid vesicles. *The Journal of Biological Chemistry*, **268**, 12040–12045.

236 Suttie, J.W., Canfield, L.M. and Shah, D.V. (1980) Microsomal vitamin K-dependent carboxylase. *Methods in Enzymolgy*, **67**, 180–185.

237 Traverso, H.P., Hauschka, P.V. and Gallop, P.M. (1980) Vitamin K-dependent carboxylation in microsomal preparations derived from cultured kidney cells, chick embryo fibroblasts, and pancreas, in *Vitamin K Metabolism and Vitamin K-dependent Proteins* (ed J.W. Suttie), University Park Press, Baltimore, Maryland, pp. 311–314.

238 Vermeer, C. (1990) γ-Carboxyglutamate-containing proteins and the vitamin K-dependent carboxylase. *The Biochemical Journal*, **266**, 625–636.

239 Ulrich, M.M.W. *et al.* (1988) Vitamin K-dependent carboxylation: a synthetic peptide based upon the γ-carboxylation recognition site sequence of the prothrombin propeptide is an active substrate for the carboxylase *in vitro*. *The Journal of Biological Chemistry*, **263**, 9697–9702.

240 Brodsky, B. and Shah, N.K. (1995) Protein motifs. 8. The triple-helix motif in proteins. *The FASEB Journal*, **9**, 1537–1546.

241 Brennan, S.O. *et al.* (1993) Posttranslational modification of β141 Leu associated with the β75(E19) Leu → Pro mutation in Hb Atlanta. *Hemoglobin*, **17**, 1–7.

242 Aguilar, M.B. *et al.* (2005) A novel conotoxin from *Conus delessertii* with posttranslationally modified lysine residues. *Biochemistry*, **44**, 11130–11136.

243 Buczek, O., Bulaj, G. and Olivera, B.M. (2005) Conotoxins and the posttranslational modification of secreted gene products. *Cellular and Molecular Life Sciences*, **62**, 3067–3079.

244 Franco, A. *et al.* (2006) Hyperhydroxylation: a new strategy for neuronal targeting by venomous marine molluscs, in *Progress in Molecular and Subcellular Biology, Subseries Marine Molecular Biotechnology: Molluscs*, (eds G. Cimino and M. Gavagnin), Springer-Verlag, Berlin, pp. 83–103.

245 Pisarewicz, K. *et al.* (2005) Polypeptide chains containing D-γ-hydroxyvaline. *Journal of the American Chemical Society*, **127**, 6207–6215.

246 Stone, B.L. and Gray, W.R. (1982) Occurrence of hydroxyproline in a toxin from the marine snail *Conus geographus*. *Archives of Biochemistry and Biophysics*, **216**, 765–767.

247 Luo, C. *et al.* (1992) Recombinant human complement subcomponent C1s lacking β-hydroxyasparagine, sialic acid, and one of its two carbohydrate chains still reassembles with C1q and C1r to form a functional C1 complex. *Biochemistry*, **31**, 4254–4262.

248 Przysiecki, C.T. *et al.* (1987) Occurrence of β-hydroxylated asparagine residues in non-vitamin K-dependent proteins containing epidermal growth factor-like domains. *Proceedings of the National Academy of Sciences of the United States of America*, **84**, 7856–7860.

249 Stenflo, J. *et al.* (1988) β-Hydroxyaspartic acid or β-hydroxyasparagine in bovine low density lipoprotein receptor and in bovine thrombomodulin. *The Journal of Biological Chemistry*, **263**, 21–24.

250 Valcarce, C., Björk, I. and Stenflo, J. (1999) The epidermal growth factor precursor: a calcium-binding, β-hydroxyasparagine containing modular protein present on the surface of platelets. *European Journal of Biochemistry*, **260**, 200–207.

251 Drakenberg, T. *et al.* (1983) β-hydroxyaspartic acid in vitamin K-dependent protein C. *Proceedings of the National Academy of Sciences of the United States of America*, **80**, 1802–1806.

252 Fernlund, P. and Stenflo, J. (1983) β-hydroxyaspartic acid in vitamin K-dependent proteins. *The Journal of Biological Chemistry*, **258**, 12509–12512.

253 McMullen, B.A. *et al.* (1983) Complete amino acid sequence of the light chain of human blood coagulation factor X: evidence for identification of residue 63 as β-hydroxyaspartic acid. *Biochemistry*, **22**, 2875–2884.

254 Stenflo, J., Lundwall, Å. and Dahlbäck, B. (1987) β-Hydroxyasparagine in domains homologous to the epidermal growth factor precursor in vitamin K-dependent protein S. *Proceedings of the National Academy of Sciences of the United States of America*, **84**, 368–372.

255 Sugo, T., Fernlund, P. and Stenflo, J. (1984) *erythro*-β-Hydroxyaspartic acid in bovine factor IX and factor X. *FEBS Letters*, **165**, 102–106.

256 White, G.C. II *et al.* (1998) Mammalian recombinant coagulation proteins: structure and function. *Transfusion Science*, **19**, 177–189.

257 Boswell, E.J., Kurinawan, N.D. and Downing, A.K. (2004) Calcium-binding EGF-like domains, in *Handbook of Metalloproteins*, (eds A. Messerschmidt,

W. Bode and M. Cygler), John Wiley & Sons, Ltd., Chichester, England, pp. 1–18.

258 Rao, Z. *et al.* (1995) The structure of a Ca^{2+}-binding epidermal growth factor-like domain: its role in protein–protein interactions. *Cell*, **82**, 131–141.

259 Selander, M. *et al.* (1990) ^{1}H NMR assignment and secondary structure of the Ca^{2+}-free form of the amino-terminal epidermal growth factor like domain in coagulation factor X. *Biochemistry*, **29**, 8111–8118.

260 Selander-Sunnerhagen, M. *et al.* (1992) How an epidermal growth factor (EGF)-like domain binds calcium: high resolution NMR structure of the calcium form of the NH_2-terminal EGF-like domain in coagulation factor X. *The Journal of Biological Chemistry*, **267**, 19642–19649.

261 Muranyi, A. *et al.* (1998) Solution structure of the N-terminal EGF-like domain from human factor VII. *Biochemistry*, **37**, 10605–10615.

262 Stenflo, J., Stenberg, Y. and Muranyi, A. (2000) Calcium-binding EGF-like modules in coagulation proteinases: function of the calcium ion in module interactions. *Biochimica et Biophysica Acta (Protein Structure and Molecular Enzymology)*, **1477**, 51–63.

263 Morita, T. and Kisiel, W. (1985) Calcium binding to a human factor IXa derivative lacking γ-carboxyglutamic acid: evidence for two high-affinity sites that do not involve β-hydroxyaspartic acid. *Biochemical and Biophysical Research Communications*, **130**, 841–847.

264 Selander-Sunnerhagen, M. *et al.* (1993) The effect of aspartate hydroxylation on calcium binding to epidermal growth factor-like modules in coagulation factors IX and X. *The Journal of Biological Chemistry*, **268**, 23339–23344.

265 Davis, L.M. *et al.* (1987) Factor $IX_{Alabama}$: a point mutation in a clotting protein results in hemophilia B. *Blood*, **69**, 140–143.

266 Green, P.M. *et al.* (1989) Molecular pathology of haemophilia B. *The EMBO Journal*, **8**, 1067–1072.

267 Green, P.M. *et al.* (1991) Haemophilia B mutations in a complete Swedish population sample: a test of new strategy for the genetic counselling of diseases with high mutational heterogeneity. *British Journal of Haematology*, **78**, 390–397.

268 Kainulainen, K. *et al.* (1994) Mutations in the fibrillin gene responsible for dominant ectopia lentis and neonatal Marfan syndrome. *Nature Genetics*, **6**, 64–69.

269 Lin, S.W. and Shen, M.C. (1991) Characterization of genetic defects of hemophilia B of Chinese origin. *Thrombosis and Haemostasis*, **66**, 459–463.

270 Liu, J.-Z. *et al.* (2000) The human factor IX gene as germline mutagen test: samples from Mainland China have the putatively endogenous pattern of mutation. *Human Mutation*, **16**, 31–36.

271 Lozier, J.N. *et al.* (1990) Factor IX New London: substitution of proline for glutamine at position 50 causes severe hemophilia B. *Blood*, **75**, 1097–1104.

272 Tartary, M. *et al.* (1993) Detection of a molecular defect in 40 of 44 patients with haemophilia B by PCR and denaturing gradient gel electrophoresis. *British Journal of Haematology*, **84**, 662–669.

273 Winship, P.R. and Dragon, A.C. (1991) Identification of haemophilia B patients with mutations in the two calcium binding domains of factor IX: importance of a β-OH Asp 64 → Asn change. *British Journal of Haematology*, **77**, 102–109.

274 Derian, C.K. *et al.* (1989) Inhibitors of 2-ketoglutarate-dependent dioxygenases block aspartyl β-hydroxylation of recombinant human factor IX in several mammalian expression systems. *The Journal of Biological Chemistry*, **264**, 6615–6618.

275 Nelson, R.M. *et al.* (1991) β-Hydroxyaspartic acid and

β-hydroxyasparagine residues in recombinant human protein S are not required for anticoagulant cofactor activity or for binding to C4b-binding protein. *The Journal of Biological Chemistry*, **266**, 20586–20589.

276 Harris, R.J., Ling, V.T., Spellman, M.W. (1992) O-linked fucose is present in the first epidermal growth factor domain of factor XII but not protein C. *The Journal of Biological Chemistry*, **267**, 5102–5107.

277 Dinchuk, J.E. *et al.* (2002) Absence of post-translational aspartyl β-Texthydroxylation of epidermal growth factor domains in mice leads to developmental defects and an increased incidence of intestinal neoplasia. *The Journal of Biological Chemistry*, **277**, 12970–12977.

278 Lavaissiere, L. *et al.* (1996) Overexpression of human aspartyl(asparaginyl)β-hydroxylase in hepatocellular carcinoma and cholangiocarcinoma. *The Journal of Clinical Investigation*, **98**, 1313–1323.

279 Handford, P.A. *et al.* (1991) Key residues involved in calcium-binding motifs in EGF-like domains. *Nature*, **351**, 164–167.

280 Gronke, R.S. *et al.* (1989) Aspartyl β-hydroxylase: *in vitro* hydroxylation of a synthetic peptide based on the structure of the first growth factor-like domain of human factor IX. *Proceedings of the National Academy of Sciences of the United States of America*, **86**, 3609–3613.

281 Stenflo, J. *et al.* (1989) Hydroxylation of aspartic acid in domains homologous to the epidermal growth factor precursor is catalyzed by a 2-oxoglutarate-dependent dioxygenase. *Proceedings of the National Academy of Sciences of the United States of America*, **86**, 444–447.

282 Jia, S. *et al.* (1992) cDNA cloning and expression of bovine aspartyl (asparaginyl) β-hydroxylase. *The Journal of Biological Chemistry*, **267**, 14322–14327.

283 Korioth, F., Gieffers, C., Frey, J. (1994) Cloning and characterization of the human gene encoding aspartyl β-hydroxylase. *Gene*, **150**, 395–399.

284 McNeill, L.A. *et al.* (2002) Hypoxia-inducible factor asparaginyl hydroxylase (FIH-1) catalyses hydroxylation at the β-carbon of asparagine-803. *The Biochemical Journal*, **367**, 571–575.

285 Kivirikko, K.I. and Pihlajaniemi, T. (1998) Collagen hydroxylases and the protein disulfide isomerase subunit of prolyl 4-hydroxylases. *Advances in Enzymology and Related Areas of Molecular Biology*, **72**, 325–398.

286 Astermark, J. *et al.* (1991) Structural requirements for Ca^{2+} binding to the γ-carboxyglutamic acid and epidermal growth factor-like regions of factor IX: studies using intact domains isolated from controlled proteolytic digests of bovine factor IX. *The Journal of Biological Chemistry*, **266**, 2430–2437.

287 Persson, K.E.M. *et al.* (1998) Calcium binding to the first EGF-like module of human factor IX in a recombinant fragment containing residues 1-85: mutations V46E and Q50E each manifest a negligible increase in calcium affinity. *FEBS Letters*, **421**, 100–104.

288 Lenting, P.J. *et al.* (1996) Ca^{2+} binding to the first epidermal growth factor-like domain of human blood coagulation factor IX promotes enzyme activity and factor VIII light chain binding. *The Journal of Biological Chemistry*, **271**, 25332–25337.

289 Rees, D.J. *et al.* (1988) The role of β-hydroxyaspartate and adjacent carboxylate residues in the first EGF domain of human factor IX. *The EMBO Journal*, **7**, 2053–2061.

290 Öhlin, A.-K., Linse, S., Stenflo, J. (1988) Calcium binding to the epidermal growth factor homology region of bovine protein C. *The Journal of Biological Chemistry*, **263**, 7411–7417.

291 Mather, T. *et al.* (1996) The 2.8 Å crystal structure of Gla-domainless activated protein C. *The EMBO Journal*, **15**, 6822–6831.

292 Öhlin, A.-K. *et al.* (1988) β-Hydroxyaspartic acid in the first epidermal growth factor-like domain of protein C: its role in Ca^{2+} binding and biological activity. *The Journal of Biological Chemistry*, **263**, 19240–19248.

293 Gerlitz, B. *et al.* (1990) Effect of mutation of Asp71 on human protein C activation and function. *Journal of Cellular Biochemistry*, **44**, 201.

294 Geng, J.P., Cheng, C.H. and Castellino, F.J. (1996) Functional consequences of mutations in amino acid residues that stabilize calcium binding to the first epidermal growth factor homology domain of human protein C. *Thrombosis and Haemostasis*, **76**, 720–728.

295 Cheng, C.H., Geng, J.P. and Castellino, F.J. (1997) The functions of the first epidermal growth factor homology region of human protein C as revealed by a charge-to-alanine scanning mutagenesis investigation. *Biological Chemistry*, **378**, 1491–500.

296 Martinez, A. (2007) Marine-derived drugs in neurology. *Current Opinion in Investigational Drugs*, **8**, 525–530.

297 Nielsen, C.K. *et al.* (2005) Anti-allodynic efficacy of the chi-conopeptide, Xen2174, in rats with neuropathic pain. *Pain*, **118**, 112–124.

298 Obata, H., Conklin, D. and Eisenach, J.C. (2005) Spinal noradrenaline transporter inhibition by reboxetine and Xen2174 reduces tactile hypersensitivity after surgery in rats. *Pain*, **113**, 271–276.

299 Sharpe, I.A. *et al.* (2003) Inhibition of the norepinephrine transporter by the venom peptide chi-MrIA: site of action, Na^+ dependence, and structure-activity relationship. *The Journal of Biological Chemistry*, **278**, 40317–40323.

300 Nilsson, K.P. *et al.* (2005) Solution structure of χ-conopeptide MrIA, a modulator of the human norepinephrine transporter. *Biopolymers*, **80**, 815–823.

301 Zhang, J.-H. *et al.* (1999) Development of a carbon dioxide-capture assay in microtiter plate for aspartyl-beta-hydroxylase. *Analytical Biochemistry*, **271**, 137–142.

302 Spitzer, S.G. *et al.* (1990) Factor $IX_{Hollywood}$: substitution of Pro^{55} by Ala in the first epidermal growth factor-like domain. *Blood*, **76**, 1530–1537.

303 Denton, P.H. *et al.* (1988) Hemophilia B Durham: a mutation in the first EGF-like domain of factor IX that is characterized by polymerase chain reaction. *Blood*, **72**, 1407–1411.

10
C-Terminal α-Amidation

Nozer M. Mehta, Sarah E. Carpenter, and Angelo P. Consalvo

10.1
Introduction

Peptidylglycine α-amidating monooxygenase (PAM; E.C. 1.14.17.3) is the sole known enzyme responsible for the bioconversion of glycine terminal prohormones into the *des*-glycine α-amidated products and glyoxylate. PAM is a bifunctional enzyme, and the conversion of a glycine-extended peptide into the corresponding peptide amide is carried out by the sequential action of two independent catalytic domains, peptidylglycine α-hydroxylating monooxygenase (PHM) and peptidyl-α-hydroxyglycine α-amidating lyase (PAL). α-Amidation is a downstream post-translational modification necessary for the full biological activity of many known α-amidated peptide hormones. A catalytic cascade of sequence specific proteolytic events takes place to produce the glycine-extended hormone precursor from a larger primary translation product (Figure 10.1). The first cleavage of the primary translation products is carried out by proprotein convertases. These enzymes are a class of subtilisin-like serine proteases discovered in the early 1990s [1]. The prohormone consensus sequences required for processing contain either one or more basic residues, or basic residues separated by an internal sequence of up to six amino acids [2]. Peptides whose fate is to become a C-terminal glycine precursor will contain a glycine residue that is N-terminal to the proprotein convertases' recognition site. Excision of the basic residues is carried out by an exopeptidase, primarily carboxypeptidase E, although carboxypeptidase D has also been shown to be involved [3, 4]. The last step in the maturation cascade is the conversion of the glycine-extended peptide precursor into the corresponding peptide amide by the action of PAM [5].

10.2
Substrate Specificity of PAM

PAM exhibits broad substrate specificity; the enzyme catalyzes the *in vivo* α-amidation of various glycine-extended peptides. *In vitro* α-amidation of glycine-

Post-translational Modification of Protein Biopharmaceuticals. Edited by Gary Walsh

ISBN: 978-3-527-32074-5

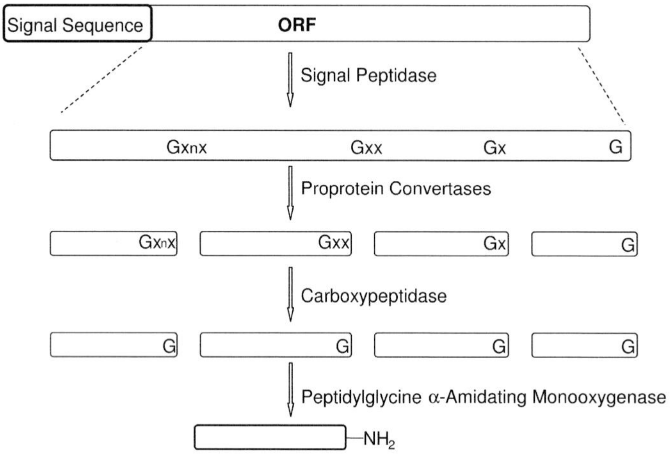

Figure 10.1 Post-translational modifications for the conversion of prohormones into the mature α-amidated peptide hormones. A proteolytic cascade results in the production of the mature C-terminally α-amidated peptides. The proprotein convertases cleave at single basic, dibasic or basic residues separated by an internal sequence (n) of up to six amino acids. Basic residues are denoted by an "x"; the liberated basic residues are removed exoproteolytically by carboxypeptidase E or carboxypeptidase D. Hormones with a C-terminal glycine residue that are thus exposed then form the substrates for PAM. Conversion of the glycine-extended peptides into the corresponding peptide amides is the final and rate-limiting step in post-translational processing.

extended fatty acids, glycine-conjugated bile acids [6], the N-substituted glycines, namely *N*-acyl and *N*-thioacylglycines [7, 8], and the glycine-extended aspirin metabolites salicyluric and gentisuric acids [9] has also been demonstrated. In addition to α-amidation, alternate PAM activities such as sulfoxidation of benzylthioacetic acid derivatives to the thionylacetic acids, and amine *N*-dealkylation and *O*-dealkylation [10], have also been reported *in vitro*. Ultimately, it is the C-terminal glycine residue that is necessary for conversion of the peptide PAM substrate into the corresponding α-amidated product and glyoxylate, while the attached moieties determine substrate kinetic parameters [11].

10.3 Activity of PAM

10.3.1 Assays for Measurement of PAM Activity

Various assays exist for measuring the activity of PAM *in vitro*. Independent assays exist for both monooxygenase and lyase activities as well as complete amidation. Monooxygenase activity can be detected by either amperometric O_2 consumption or detection of the α-hydroxylated product. A Clark-type oxygen electrode monitors the amount of dissolved oxygen (DO) in solution, and has been a widely employed method for the kinetics of monooxygenase activity [12, 13]. Peptide substrates derivatized with chromogenic labels allow for the sensitive detection of substrate and product by reversed-phase HPLC (RP-HPLC). Initial robust assays utilized a fluorescently labeled tripeptide substrate with RP-HPLC separation of glycine-extended substrate and α-amidated product [14]. Modifications of these RP-HPLC methods allow for independent measurement of monooxygenase and lyase activity as resolution of both the α-hydroxylated and α-amidated species was achieved [15–17]. As an alternative to the RP-HPLC assays, PAL/PAM activity can be independently assayed via detection of glyoxylate production by methods including spectrophotometric quantification of the phenylhydrazone adduct [18–20], and enzymatically by coupling to LDH activity and NADH consumption [21], or coupling to the production of a foramzan dye [22].

10.3.2 Mechanism of Action

PAM is a bifunctional, copper type-II metallo-monooxygenase that consists of two independent catalytic domains, PHM and PAL. The PAM reaction is a two-step process. Initially, PHM removes the *pro*-S hydrogen to allow for hydroxylation at the α-glycyl carbon, resulting in a α-hydroxylated intermediate [23–27]. PHM is a molecular oxygen, Cu(II) and ascorbate (reductant) dependent enzyme. Empirically defined as a monooxygenase, substrate activation catalyzes dioxygen cleavage, resulting in the insertion of one oxygen into the product and reduction of the other oxygen to water. The use of catalase increases the catalytic efficacy of PHM as it protects the enzyme from damage caused by reactive oxygen species (ROS) generated *in situ* by ascorbate and Cu(II) in the presence of O_2. Other enzymes such as horseradish peroxidase and superoxide dismutase also demonstrate a protective effect on PAM at various concentrations. The second catalytic domain, PAL, dealkylates the α-hydroxylated intermediate, yielding the α-amidated product and glyoxylate. Figure 10.2 shows schematically the two-step mechanism of the PAM reaction. Bifunctional PAM is the only known enzyme that catalyzes this unique form

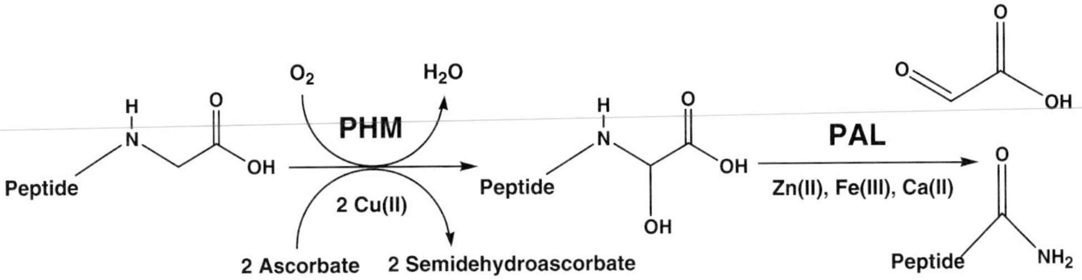

Figure 10.2 Peptide α-amidation reaction catalyzed by bifunctional PAM. PHM catalyzes the first step in the reaction mechanism and requires copper, ascorbate and molecular oxygen. The final step in the formation of the α-amidated product is catalyzed by the PAL moiety.

of post-translational modification; many of the glycine-extended peptide hormones remain relatively inactive prior to α-amidation [28–31].

To date, the glycine-extended peptide substrates have been the most widely studied PAM substrates, as approximately 50% of all mammalian peptide hormones contain an α-amide moiety at their C-terminus [29]. PAM is both the rate-limiting and final step within the cascade of proteolytic events that take place to produce the active α-amidated peptides (Figure 10.1). Other post-translational modifications of peptides, such as acylation, sulfation and phosphorylation, also occur. Published data have demonstrated that inhibition of PAM activity in both cultured mammalian cells, and in rats, leads to the accumulation of the glycine-extended prohormones, and coincidently a decreased concentration of the active α-amidated peptides. Moreover, in some cases the α-amidation is absolutely necessary for peptide bioactivation [28].

PHM (E.C.1.14.17.3) has been the most extensively studied domain of the bifunctional enzyme both structurally and mechanistically, in part on account of its reactivity and structural homology to dopamine-beta-monooxygenase (DβM) [32, 33]. DβM and rat PHM exhibit a 32% sequence identity over the conserved copper-binding domain composed of 291 amino acids, suggesting a divergent evolutionary link between these two enzymes. The catalytic core of PHM (PHMcc) is refined to residues 42–356, and is contained within the N-terminal fragment of bifunctional PAM [34]. PHM is the ascorbate, copper and O_2 dependent domain responsible for hydroxylation of the substrate α-glycyl carbon. PHM contains ten cysteine residues, each occupied in a disulfide linkage. The three-dimensional structure of PHMcc is defined as a prolate ellipsoid that consists of two α-sandwich domains (domains I and II) closely associated through hydrophobic forces. Domains I and II are tethered by a solvent accessible linker region, with each domain containing one of the two copper atoms within its center [35]. The solvent filled inter-domain cleft separates these two essential yet non-coupled copper atoms by 11 Å [36]. CuH of domain I is bound to enzyme via three histidines, and CuM of domain II is bound by two histidines and one water [37].

PHM catalysis is dependent upon the redox cycling of oxidized cupric 2-Cu(II) to the activated cuprous 2-Cu(I), via an ascorbate-dependent, two single-electron reduction process. Reduction occurs via a ping-pong mechanism for ascorbate with an overall one to one stoichiometry with copper. Dioxygen activation begins by a sequential mechanism upon substrate binding. Following substrate C–H bond cleavage, C–O bond formation occurs by either radical recombination [38, 39] or a concerted rebinding step [40–42] of the copper-oxygen intermediate, resulting in hydroxylated product release [43]. The tightly coupled nature of copper reoxidation by O_2 and substrate hydroxylation puts the DβM and PHM mechanism in a class apart from that of the well-studied cytochrome P450 iron heme-monooxygenases [44].

Upon formation of the hydroxylated intermediate, the second domain of the PAM holoenzyme is responsible for carbinolamide dealkylation. PAL (E.C. 4.3.2.5), also known as peptidylamidoglycolate lyase (PGL), contains zinc, calcium and iron. PAL activity resembles that of ureidoglycolate lyase (UGL, E.C. 4.3.2.3) as both enzymes catalyze a stereospecific dealkylation of (*S*)-hydroxyglycine substrates [45, 46]. UGL catalyzes the conversion of ureidoglycolate into urea and glyoxylate, yet displays little sequence homology to PAL. Various mutagenesis, thermal denaturation and proteolysis studies on PAL purified from a CHO cell overexpression system identified the catalytic core and metal content of PAL [47]. Residues Asp^{498}-Val^{820} containing two disulfide bonds comprise the catalytic core of PAL (PALcc); it is a 33 kDa monomer bound to the C-terminus of PHM. Inductively coupled plasma optical emission spectrometry identified the presence of stoichiometric quantities of both calcium and zinc in 1 : 1 ratio to PAL; both metals appear to play a structural role, and perhaps for Zn(II) a catalytic role. The most recent advance was the identification of an Fe(III) phenolate charge transfer complex within the active site of PAL. Site-directed mutagenesis studies exposed a catalytic and structural metal binding site located within an essential tyrosine residue. Data obtained from UV/Vis spectra revealed weak absorption at 560 nm, which is a signature of the Fe(III)-Zn(II) binuclear center of the purple acid phosphatases. Striking similarities in UV/Vis absorption spectra provide convincing evidence that PAL may utilize a similar tyrosine bridged Fe(III)-Zn(II) complex as the purple acid phosphatases; namely, an Fe(III)-Fe(II) or Fe(III)-Zn(II) dinuclear center [47]. The true mechanism of this enzyme remains to be elucidated as data also illustrate that PAL may contain alternate binding sites. Recent data have positioned PAL as a unique variant within the mixed-valent dinuclear complex dependent metalloenzymes.

10.3.3
Species Distribution of α-Amidated Peptides and PAM

In mammals, several alternately spliced isoforms of PAM are formed from a common primary transcript. However, in more primitive organisms PAM is found as two monofunctional precursors that form the catalytically independent domains PHM and PAL. Mammalian PAM is found in high levels within the neurosecretory vesicles of many neuronal and endocrine cells, with high abundance in the pituitary gland and in cardiac atria [48]. C-terminal α-amidated peptides are widely distributed and are present in mammals [29, 49], insects [50, 51], fish [52], cnidarians [53], mollusks [54] and plants [55]. Although a TRH-like tripeptide (pyroGlu-Tyr-Pro amide) in alfalfa has been reported, a plant PAM isozyme has not been identified. The absence of a PAM-like enzyme suggests that plant α-amidated peptides are produced via a PAM independent pathway.

10.4
Genomic Structure and Processing of PAM

10.4.1
Organization of the PAM Gene

In rat PAM, both PHM and PAL are encoded within a complex single copy gene that spans over 160 kb of genomic DNA [56]. Alternate splicing of the PAM primary RNA transcript results in several forms of the bifunctional enzyme that are either integral membrane-bound or soluble proteins. A total of 27 different exons are distributed over the 160 kb of genomic DNA. Twelve PHM exons spanning 76 kb of genomic DNA can combine by post-transcriptional processing to encode soluble monofunctional PHM. An additional eight exons cover the coding region of PAL and these span at least 19 kb of genomic DNA. Ultimately, the expression of PAM, either as a bifunctional enzyme or as its constituent PHM and PAL domains, is tissue specific and developmentally regulated.

10.4.2
Tissue-Specific Forms of PAM

One very important feature of the PAM holoenzyme is that no substrate channeling has been observed between the PHM and PAL domains [57]. A unique feature of PAM is that each domain of enzyme can function independently [56]. Early investigations revealed several different molecular forms of PAM in eukaryotes, and research continues to demonstrate further alternately spliced forms of PAM within specific tissues [58–61]. The enzyme is found as both membrane-bound and soluble, and in many cases is *N*-glycolsylated. Overall, the conserved regions consist of the initial N-terminal signal sequence, PHM and the PAL region. Variations arise mainly in the presence of the linker region between PHM and PAL, and in the region C-terminal to PAL. Interestingly, dibasic cleavage sites present within the linker region of some forms of PAM result in an endoproteolytic dependent pathway for liberating PAM as its two separate domains. Correlation between the extensive processing of both PAM and its substrates reveals the interesting prospect that both enzyme and substrate are processed by a similar endoproteolytic pathway. This affords PAM with yet another level of regulation within the formation of bioactive peptides via these multiple forms of PAM with variable substrate specificities [62].

10.5
Structure–Activity Relationships (SAR) for Rat PAM Activity

PAM can produce peptide-amides that terminate with all 20 amino acids [28, 48]. While the C-terminal glycine residue is essential for tight binding to PAM, other structural features modulate the interaction between enzyme and substrate. The influence of the amino acid at the penultimate position (-X-Gly-OH) was examined in early structure–activity relationships (SAR) studies. A series of tripeptides of the form Gly-X-Gly-OH were ranked for their ability to inhibit (K_i) α-amidation of Dansyl-Tyr-Val-Gly-OH [63]. Substitution at the penultimate position with the 20 commonly occurring amino acids resulted in a 1300-fold range in K_i for those peptides that were shown to cause detectable inhibition. The resultant inhibition constants were ranked with respect to the characteristics of the amino acid side chain. A decrease in competitor potency was observed through the following grouping of amino acid side chains: sulfur containing > aromatic ≥ histidine > non-polar > polar > glycine > charged (basic and acidic).

In a related SAR study, kinetic parameters were obtained for PAM substrates of the form *N*-dansyl-(Gly)$_4$-X-Gly-OH, where X is each of the 20 commonly occurring amino acids [64]. The α-amidated products *N*-dansyl-(Gly)$_4$-X-NH$_2$ were separated and quantified by RP-HPLC utilizing fluorescent detection. The proximity of the *N*-dansyl group was initially investigated, since this organic moiety was previously shown to dramatically increase binding of peptides to rat PAM [14]. The use of hexapeptide substrates was demonstrated to be the minimal length required to effectively eliminate the influence of the dansyl group on peptide binding. The *N*-dansyl-(Gly)$_6$-OH was found to interact with enzyme with an apparent affinity similar to that observed for the non-dansylated peptide.

The dansylated hexapeptides were ranked in order of decreasing affinity into four distinct groups. Table 10.1 presents these data. An approximately 130-fold range in K_m was observed for those peptides that were α-amidated at sufficient rates to enable determination of kinetic parameters. In contrast, the V_{max} values showed only an eleven-fold variation. The ability of the amino acid side chain at position X to stabilize the enzyme–substrate complex decreased through the series X = planar aromatic or sulfur containing > neutral aliphatic > polar and basic > cyclic aliphatic or acidic.

Table 10.1 Kinetic parameters for enzymatic α-amidation of model peptides of the form *N*-dansyl-(Gly)$_4$-X-Gly-OH.

Group	Penultimate residue (X)	K_m (μM)	V_{max} (nmol min^{-1} mg^{-1})
1	Phe	4	50
	Tyr	5	40
	Met	7	23
	Cys	11	10
2	Ile	20	55
	His	41	10
	Ala	46	6
	Val	49	48
	Leu	54	22
	Trp	58	58
3	Asn	83	7
	Ser	196	9
	Arg	200	15
	Lys	206	4
	Gln	308	17
	Thr	334	50
4	Glu	449	5
	Pro	618	22
	Gly	—	<1
	Asp	—	No activity detected

Further analyses reveal some interesting comparisons. For example, the dramatic difference in the K_m for which X = Gly or Ala; these data clearly demonstrate the importance of the side-chain interaction with the enzyme. A proline residue at the penultimate position causes an apparent disruption of the peptide backbone, resulting in an approximately ten-fold decrease in affinity compared to other peptides containing neutral aliphatic side-chains. Introduction of a negative charge at the penultimate position results in destabilization of the enzyme–substrate complex, which can be seen by comparing Asp and Glu to the corresponding uncharged residues Asn and Gln. Aromatic or sulfur-containing residues at the penultimate position confer the highest recognition for rat PAM.

A survey of 47 different α-amidated peptides identified in mammalian tissue was conducted and compared to the substrate affinity profile in Table 10.1. Such a comparison is reasonable considering the high nucleotide sequence homology of PAM derived from divergent species. Interestingly, 36% of these peptides resided in group 1, which are those peptides that contain planar aromatic or methionine residues at the C-terminus. The binding kinetics observed for these model peptides only partially explains the interaction of PAM with peptide substrates. These SAR data are broadly applicable and can be used to predict the efficiency of *in vitro* α-amidation of peptide substrates. However, little is known about the structural relationship of amino acids adjacent to the penultimate position or further upstream in the peptide chain on enzyme recognition.

10.6
α-Amidation of Glycine-Extended Peptides

10.6.1
In Vitro α-Amidation

PAM exhibits broad substrate specificity; the enzyme can catalyze various substrates, including glycine-extended peptides, glycine-extended fatty acids, N-substituted acyl-glycines and glycine-conjugated bile acids. Table 10.2 shows standard reaction components for *in vitro* α-amidation of glycine-extended peptides.

Typically, small-scale reactions are conducted with a final volume of 0.25–1.0 mL. Extensive optimization using a small-scale model is normally required; nearly all reaction components must be titrated to achieve optimal conversion of substrate into product. α-Hydroxylation of a glycine-extended peptide and subsequent dealkylation to the corresponding peptide-amide by PAM results in a change in the net charge of the peptide (loss of one negative charge, carboxyl group). As a result, ion exchange HPLC is ideally suited to monitor α-amidation. Such assays can be used to either estimate catalysis by determining the percent conversion or accurately quantify the amount of product formed for calculation of kinetic parameters.

Oxygen is a co-substrate for PAM; a stoichiometric amount of oxygen compared to glycine-extended peptide is required to achieve α-amidation; however, small-scale reactions do not require oxygen supplementation. Reactive oxygen species (ROS) are

Table 10.2 Standard reaction components for *in vitro* α-amidation.

Reagent	Purpose
Peptide-Gly	Substrate
MES or HEPES	Buffer
Cupric sulfate	Co-factor
PAM	Enzyme
Sodium ascorbate	Reductant (co-substrate)
Catalase (*Aspergillus niger*)	Protects PAM from peroxide-mediated inactivation
Ethanol	Protects catalase from inactivation
Oxygen	Co-substrate

formed during α-amidation and rapidly inactivate PAM. Catalase is used in the reactions to protect PAM from ROS mediated inactivation. Ethanol protects catalase from inactivation and it is routinely used at concentrations as high as 5% without deleterious effects on PAM. Furthermore, PAM has been shown to be fully active in other organic solvents such as 5% acetonitrile. It is recommended to use high quality catalase from a non-animal source (i.e., *Aspergillus niger*) since preparations with low specific activity often contain proteases that can degrade the peptide substrate and product.

10.6.2 Optimization of the PAM Reaction *In Vitro*

Conversion of glycine-extended peptides into their corresponding peptide amides by PAM is largely dependent on kinetic parameters such as the K_m and V_{max}. Notwithstanding the importance of these parameters, the efficiency of *in vitro* α-amidation is greatly influenced by substrate purity. The source of the glycine-extended peptides can be either chemically synthesized or recombinant. Poor α-amidation can result even from low level impurities present in the substrate preparations. Process related impurities can inhibit PAM directly or indirectly through copper chelation and/or ascorbate oxidation. The likelihood of these process related impurities affecting the α-amidation efficiency is greatly diminished when high purity substrates are utilized. Nearly all preparations of PAM contain bound Cu(II) as trace amounts of the metal can be found in most water sources. However, the addition of exogenous Cu(II) is usually required for *in vitro* α-amidation. Typically, the addition of 0.5 μM copper sulfate is sufficient to achieve efficient α-amidation, with higher levels of Cu(II) required to overcome chelation by process related impurities. This is illustrated in Figure 10.3, where 2.0 μM Cu(II) was required to achieve nearly quantitative α-amidation of recombinant Ser8-GLP1(7–36)Gly37-OH.

Other reaction variables such as pH and temperature must be optimized for each glycine-extended substrate. PAM retains some level of activity over a relatively wide pH range and the enzyme is functional in various buffers. However, small variations in the final reaction pH can have a marked effect on the amount of

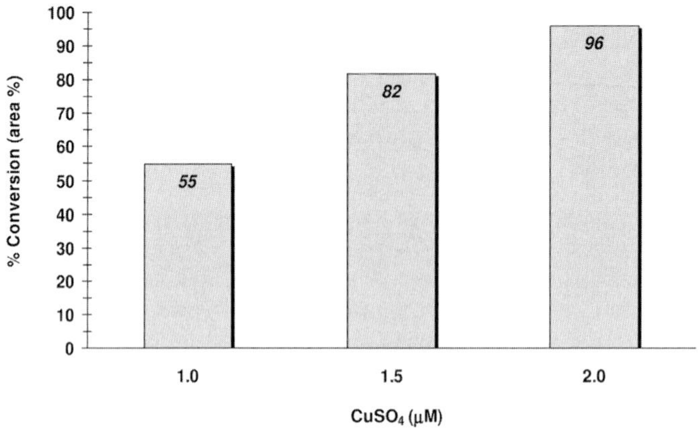

Figure 10.3 Copper sulfate titration. α-Amidation of Ser8-GLP1 (7–36)Gly37-OH to Ser8-GLP1(7–36)-NH_2 by recombinant PAM (rPAM). All reactions were incubated at 37 °C for 3 h.

peptide-amide produced. This can be attributed to the pH optimum of PAM in the context of the differences in the physicochemical properties of the glycine-extended substrates. Important physicochemical properties such as solubility, isoelectric point, propensity to aggregate and stability must be considered before selecting a reaction pH. Generally, *in vitro* α-amidation reactions are carried out between pH 5.0–7.5. Figure 10.4 shows the influence of reaction pH on α-amidation efficiency. α-Amidation of rhPTH(1–34)Gly35-OH proceeds to only 48% at pH 4.8, whereas the percent conversion increases to >87% between pH 5.8–6.5, with a maximum percent conversion at pH 5.8.

Enzyme reactions are usually performed at 37 °C for not more than 4–6 h, with most reactions complete within 2–4 h. α-Amidation below 37 °C is particularly useful for substrates that are unstable or aggregate at high temperatures. Substrates with

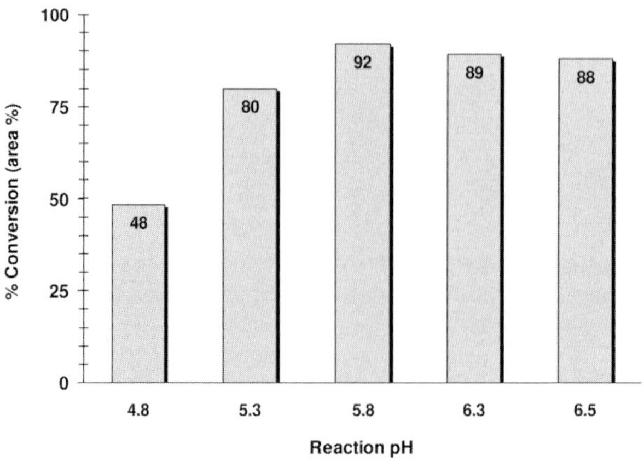

Figure 10.4 Influence of pH on α-amidation of rhPTH(1–34) Gly35-OH to rhPTH(1–34)-NH_2 by rPAM. All reactions were incubated at 37 °C for 4 h.

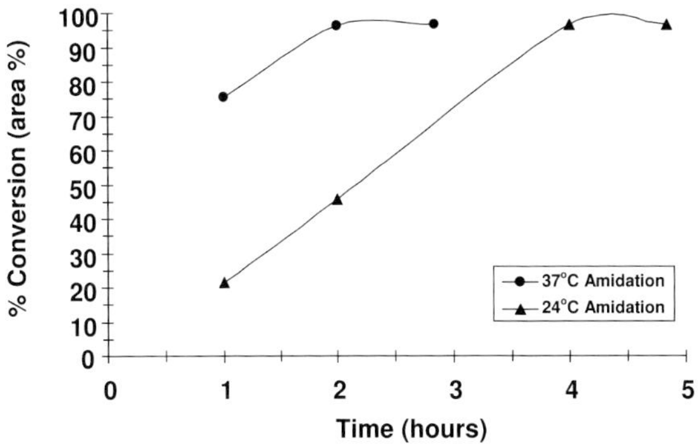

Figure 10.5 Influence of temperature on the enzymatic α-amidation of rhPTH(1–34)Gly35-OH to rhPTH(1–34)-NH_2 at 24 and 37 °C.

poor catalytic efficiency usually cannot be α-amidated effectively below 37 °C, since PAM is inactivated before a sufficient level of peptide-amide is produced. Nearly quantitative conversion of rhPTH(1–34)Gly35-OH into rhPTH(1–34)-NH_2 by rPAM can be achieved at either 24 or 37 °C. Figure 10.5 depicts these data.

The α-amidation of recombinant glycine-extended salmon calcitonin, rsCT(1–31) Gly32-OH, has been studied extensively and optimized using a small-scale model. The optimized conditions developed at the 1 mL scale have been used to successfully scale the α-amidation reaction to 17 L (17 000-fold) and further to 410 L (410 000-fold) without a significant loss in enzyme efficiency. Figure 10.6 depicts these results. The reaction conditions at the three different scales (1 mL, 17 L and 410 L) were essentially identical with the exception of DO; both the 17 L and 410 L reactions were supplemented with oxygen. Oxygen supplementation is generally required for large reaction volumes. The level of DO must be maintained at or near saturation during the reaction to ensure efficient α-amidation. At large-scale, the rate of oxygen consumption by PAM is faster than the mass transfer rate for dissolution of oxygen back into the reaction mixture. A high level of DO is routinely maintained by sparging with either pure oxygen or oxygen-enriched air. Oxygen supplementation at these scales results in faster turnover; this is clearly illustrated in Figure 10.6, where the 1 mL reaction proceeds at a significantly slower rate than either the 17 L or 410 L reactions. The line curves for the 17 L and 410 L reactions are essentially superimposable, clearly demonstrating the scalability of the enzyme reaction.

10.7 Cloning and Expression of Various Forms of PAM

Several expression systems have been described for PHM, PHMcc, PAL and bifunctional PAM derived from various species. The levels of expression (where reported) have generally been fairly low, and vary depending on the host cell, the expression system and the fragment of PAM being expressed.

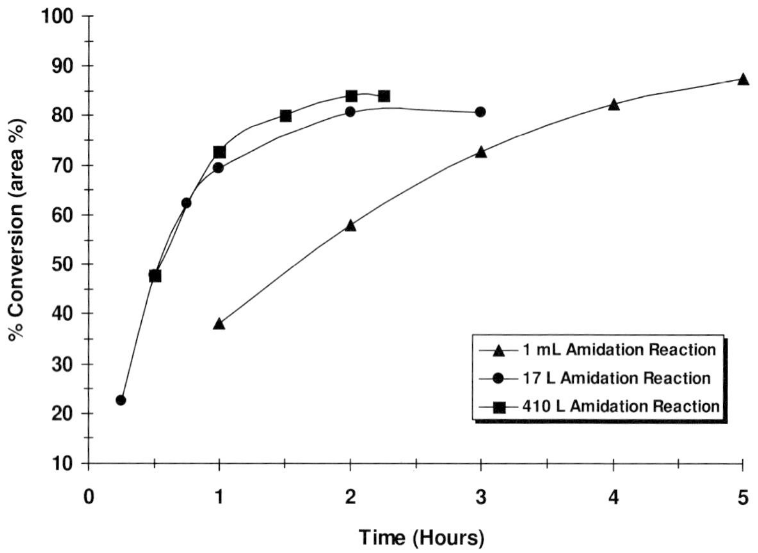

Figure 10.6 Time course of enzymatic conversion of rsCT(1–31) Gly32-OH into rsCT(1–31)-NH_2 by rPAM at three different reaction scales. All reactions were incubated at 37 °C. The 17 L and 410 L α-amidation reactions were supplemented with oxygen.

10.7.1 PHM, PHMcc and PAL

PHM derived from frog skin (*Xenopus laevis*) has been successfully cloned and expressed in insect cells [27] and *Escherichia coli* [65]. However, improper folding of PHM following expression in *E. coli*, most likely due to mismatching of the disulfide linkages, resulted in an enzyme with extremely low specific activity. Rat PHMcc was cloned and expressed in CHO cells, where large-scale production of enzyme was achieved in two different types of bioreactors [66]. Rat PHMcc has also been cloned and expressed in yeast (*Saccharomyces cerevisiae*) in an effort to elucidate the process of copper loading in the secretory pathway [67]. A 50 kDa form of PAL containing the C-terminal transmembrane and hydrophilic domains that was identified in bovine neuro-intermediate pituitary granules was cloned and expressed in hEK293 cells [68].

10.7.2 Bifunctional PAM

Bifunctional frog PAM, lacking the membrane-spanning domain and C-terminal cytosolic tail, has been produced in insect cells (BoMo-15AIIc) and silkworm larvae using the baculovirus expression system [69]. Bifunctional PAM derived from frog has also been overexpressed in CHO cells, where the effect of culture temperature on productivity was investigated [70]. Human bifunctional PAM has been cloned and expressed in CHO [71] and COS cell lines [72]. Rat bifunctional PAM was overexpressed in mouse C127 cells and the recombinant enzyme was purified to homogeneity and characterized [73]. Additionally, bifunctional rat PAM has been cloned and expressed in CHO cells [74]; this enzyme was subsequently used in the development of a large-scale production process for recombinant salmon calcitonin [75].

10.7.3
Co-expression of PAM with Glycine-Extended Peptides

Bioactive α-amidated peptides have been produced *in vivo* by co-expression of bifunctional PAM with glycine-extended peptides. Production of biologically active BmK ITa1, a Chinese scorpion neurotoxin, was attained by co-expression with rat PAM in insect cells [76]. Similarly, biologically active salmon calcitonin was produced in *Streptomyces lividans* by co-expression with rat PAM [77]. In this case, *S. lividans* was engineered to produce and secrete α-amidated salmon calcitonin into the culture medium.

10.8
A Process for Recombinant Production of α-Amidated Peptides

Although great strides have been made in cost and scalability for production of peptides by chemical synthesis, for larger peptides (25 amino acids or greater) recombinant technology offers the potential for large-scale production that is more cost-effective, readily scalable and environmentally acceptable. Production of peptide hormones in bacteria or yeast is the most cost-effective method. However, the relatively small size and lack of tertiary structure of most peptides makes them susceptible to rapid degradation in the cytoplasm. This problem can be circumvented by expression of these peptides with a larger protein as a fusion partner. This allows the resulting fusion protein to accumulate relatively non-degraded in a soluble or insoluble form in the cytoplasm [78]. Although large amounts of the fusion protein can be made in this manner to give yields on the order of $g\,L^{-1}$, fusion protein expression also has several disadvantages. Liberation of the fusion partner from the peptide hormone by chemical or enzymatic cleavage can be difficult and/or expensive. Moreover, the resulting yield of the peptide is greatly diminished, since the peptide represents only a fraction of the entire fusion protein. Also, purification from the large number of cytoplasmic proteins before and after cleavage of the fusion may require several steps, which can increase the cost of production and reduce the overall yield of the peptide hormone.

A further complication associated with producing peptide hormones in bacteria or yeast is that PAM is not present in prokaryotes and, therefore, peptide hormones that are produced in *E. coli* are not C-terminally α-amidated. An efficient and cost-effective solution to the production of amidated peptides in bacteria is afforded by a dual recombinant direct expression technology [75, 79]. This direct expression process is shown in the schematic in Figure 10.7.

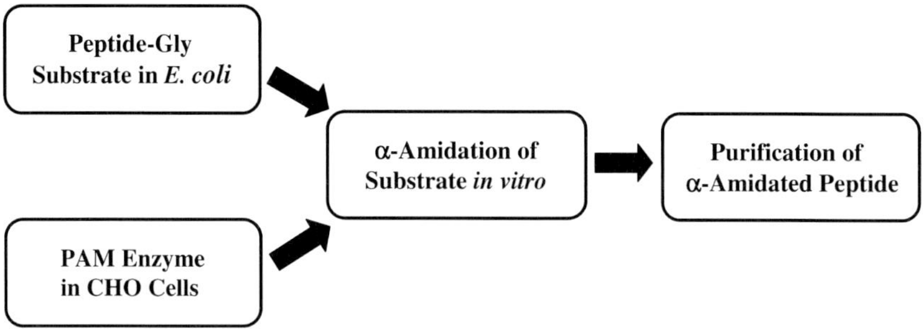

Figure 10.7 A dual recombinant expression process for the production of α-amidated peptides. Glycine-extended peptide precursors are expressed in *E. coli* without a fusion partner. The peptide is secreted and accumulates in the conditioned growth medium, from which it is harvested and purified. Separately, bifunctional PAM is expressed and secreted from CHO cells and purified to a high specific activity. α-Amidation of the purified glycine-extended peptide to the corresponding peptide amide by PAM is carried out in a highly optimized and scalable *in vitro* reaction. The bioactive α-amidated peptide is then further purified to homogeneity.

10.8.1 Expression of Glycine-Extended Peptides in *E. coli* by a Direct Expression Process

A glycine-extended precursor of the desired peptide is produced in recombinant *E. coli* cells. The glycine-extended peptide is produced with an upstream signal sequence that translocates the peptide from the cytoplasm to the periplasm. The signal sequence is cleaved during transport to the periplasm. Owing to a combination of specific growth conditions and components of the growth medium, the peptide is secreted through the outer membrane of the *E. coli* cell into the growth medium. High levels of peptide are obtained due to several desirable features of the system. These include the use of a unique plasmid vector with multiple expression cassettes, a protease-deficient host cell and a fermentation protocol that allows for high density cell mass. Since *E. coli* secretes very few endogenous proteins, the glycine-extended peptide hormone can be recovered from the conditioned medium in a relatively enriched form, and this offers advantages in purification. Cell lysis is not required to recover the peptide product. Therefore, the peptide does not need to be purified away from the large number of cytoplasmic and periplasmic proteins, and high molecular weight DNA, that would be liberated from lysed cells. Since no fusion partner is used, no chemical or enzymatic cleavage is required to liberate the peptide from the fusion partner.

10.8.2 Purification of the Glycine-Extended Peptides

At the end of the fermentation process, the intact *E. coli* cells are removed by centrifugation or tangential flow ultrafiltration. The glycine-extended peptide is initially captured from the conditioned medium using an appropriate ion-exchange column. Further purification is typically achieved by a two-column procedure. At this stage, the glycine-extended peptide is >80% pure and exchanged into a buffer that is suitable for the *in vitro* α-amidation reaction.

10.8.3 Expression of PAM in CHO Cells

As stated previously, multiple forms of PAM have been detected in different tissues and species. In rat tissues and cell lines, the major soluble PAM species detected has a molecular weight of approximately 75 kDa [80]. Sequence analyses of multiple rat cDNA clones revealed two major DNA types, designated as type A and type B [81, 82]. Both types contained the PHM and PAL domains, followed by a membrane-spanning domain and a cytoplasmic tail, and differed primarily by the presence or absence of a 315 bp exon. The type A sequence was chosen for the generation of a recombinant mammalian cell line, and the sequence was further modified to introduce a stop codon upstream of the membrane-spanning domain. This results in the expression of a 75 kDa bifunctional PAM enzyme in the recombinant Chinese hamster ovary (CHO) cells. A readily scalable fed-batch bioreactor production protocol has been developed where the recombinant CHO cells are grown in a medium that is free of protein and animal-sourced components, and soluble PAM is secreted into the growth medium [79].

10.8.4 Purification of Recombinant PAM (rPAM)

The steps involved in the purification of rPAM are shown in the schematic in Figure 10.8. Tangential flow ultrafiltration (TFF #1) is utilized to concentrate and diafilter rPAM prior to anion exchange (AEX) chromatography. The clarified harvest is concentrated five-fold and the conductivity is decreased to facilitate binding. The AEX column provides a gross purification of rPAM from conditioned CHO medium components and protein impurities. DNA is effectively removed at this stage. Hydrophobic interaction chromatography (HIC) provides further purification of PAM; HIC removes most protein impurities remaining following the AEX step. Since prolonged storage in high salt HIC elution buffer can lead to enzyme inactivation, the HIC output is immediately processed to minimize inactivation. A second TFF step is utilized to concentrate and diafilter rPAM into a suitable buffer prior to the final step, virus filtration. This step provides clearance of potential virus contaminants intrinsic to the cell line and/or adventitious viruses introduced during bioprocessing. After virus filtration, rPAM is aliquoted and maintained at −20 °C for long-term storage.

Table 10.3 shows an example of production and purification of rPAM at an intermediate scale (30 L bioreactor). The downstream process affords approximately an eight-fold purification of rPAM. Typically, a final specific activity of 8.01×10^6 units per mg protein is achieved with an overall yield of 49%. The rPAM purified by this process is free of contaminating proteolytic activities that could degrade the enzyme or substrate, and displays a 75 kDa band on SDS-PAGE.

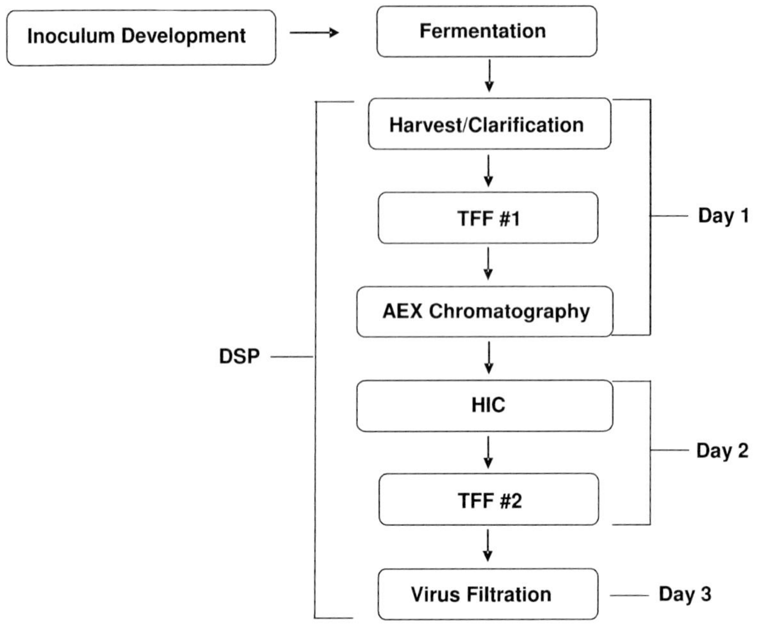

Figure 10.8 Schematic of a production process for recombinant bifunctional rPAM in CHO cells. rPAM is expressed and secreted into the growth medium, from which it is purified by a combination of anion exchange chromatography and HIC (hydrophobic interaction chromatography).

10.8.5 Post-Amidation Purification

Following α-amidation, the bioactive peptide is separated from any remaining glycine-extended precursor and other minor contaminants by one or more chromatography steps. At this stage there are little or no process related contaminants such as *E. coli* protein, DNA and endotoxin in the final product. The yield of peptide following

Table 10.3 Purification of rPAM from a 30 L bioreactor. A unit of PAM activity is defined as the conversion of 1 pmol min^{-1} of Dansyl-Tyr-Val-Gly-OH into Dansyl-Tyr-Val-NH_2 at 37 °C.

Process step	10^{-6}×Specific activity (U mg^{-1})	Step yield (%)	Overall yield (%)
Bioreactor	1.20	—	—
Clarified harvest	1.32	95	95
TFF #1	1.43	94	89
AEX chromatography	4.23	84	75
HIC	7.64	90	67
TFF #2	6.71	92	62
Virus filtration	8.01	79	49

purification and α-amidation is generally at least 50%, and the percent purity of the α-amidated peptide is typically >98%. The purified peptide can be lyophilized and stored at −80 °C until formulation into the final product.

10.8.6 Expression Levels of Peptides by Direct Expression Technology

The fermentation protocol used in the direct expression technology has been shown to be readily scalable for several recombinant cell lines with no significant loss of productivity. The yield of peptide from fermentation varies, depending on the peptide. α-Amidated peptides such as sCT, analogs of parathyroid hormone (PTH), glucose regulatory peptide analogs, secretin and growth hormone releasing factor are examples of peptides that have been expressed and purified with high yields using this direct expression technology. For some peptides, yields greater than $1\,g\,L^{-1}$ of intact peptide have been achieved. As mentioned above, the direct expression fermentation technology, as well as the purification steps, which utilize low-pressure chromatography, is readily scalable. Peptides have been expressed by direct expression in *E. coli* at 1000 L and 10 000 L with no loss in productivity. Expression of rPAM in bioreactors has been carried out at scales of up to 2500 L.

10.9 Marketed Peptides

Most peptide compounds do not follow Lipinski's rules for the "drugability" of active compounds [83]. Despite this, peptide based drugs represent a growing number of pharmaceutical molecules. Peptide drugs offer several advantages over small molecule based drugs, such as high specificity, high potency or specific activity, and low toxicity. In 2005, there were 67 peptide drugs on the market, with another 20 in Phase III trials and 87 in Phase II trials [84]. Among the approved agents are natural peptides such as insulin, oxytocin, exendin-4, parathyroid hormone (PTH) and calcitonin. Several synthetic or derivatized peptides or peptide analogs have also been developed into successful products, such as Fuseon, Integrilin, DDAVP, Sandostatin, Lupron and Symlin. One form of insulin alone, Lilly's Humalog, enjoyed sales of $1.1 billion in 2004.

10.9.1 Marketed α-Amidated Peptides

Approximately 50% of all known peptide hormones are α-amidated at the C-terminus, and, in many cases, biological activity is greatly diminished or even eliminated in the absence of α-amidation [31, 48, 85, 86]. Even for peptides that are not naturally α-amidated and are therefore active in the absence of α-amidation, the activity or receptor binding or half-life of the peptide can be improved by the α-amidation of the C-terminus. Table 10.4 shows examples of currently marketed

Table 10.4 Examples of currently marketed α-amidated peptide hormones.

Drug name (Company)	Compound	Delivery	Dose	Indications
Zoladex (AstraZeneca)	Decapeptide analog of LHRH	Subcutaneous, 28-day slow release	3.6 mg	Prostate cancer, endometriosis, advanced breast cancer, endometrial thinning
Lupron (TAP)	Nonapeptide analog of LHRH	Subcutaneous, daily	1.0 mg daily	Advanced prostate central, precocious puberty
DDAVP (Sanofi-Aventis)	Analog of 8-Arg vasopressin	Oral tablets	0.1–0.8 mg daily	Central diabetes insipidus, primary nocturnal enuresis
Cortrosyn (Organon)	α-(1–24)-corticotropin	I.m. or i.v. injections or i.v. infusion	250 μg	Diagnostic agent for adrenocortical deficiency
Sandostatin (Novartis)	Cyclic octapeptide analog of somatostatin	S.c. or i.v. injections	50–1500 μg daily	Acromegaly, carcinoid tumors, VIPomas
Thyrel TRH (Ferring)	Synthetic tripeptide	I.v. injection	500 μg	Diagnostic assessment of thyroid function
Miacalcin (Novartis)	Salmon calcitonin	S.c. injection	100 IU daily	Postmenopausal osteoporosis, Paget's disease, hypercalcemia
Fortical (Upsher-Smith)	Salmon calcitonin	Nasal spray	200 IU daily	Postmenopausal osteoporosis
Geref (Serono)	GHRF(1–29)-NH_2	S.c. injection	30 μg kg^{-1} daily	Pediatric, idiopathic growth hormone deficiency
Acthrel (Ferring)	Ovine CRF(1–41)-NH_2	I.v. injection	1.0 μg kg^{-1} single injection	Differentiates pituitary and ectopic production of ACTH in Cushing's syndrome
Secretin-Ferring (Ferring)	Porcine secretin(1–27)-NH_2	Slow i.v. injection	1–2 CU kg^{-1}	Testing for pancreatic function and gastrinoma
Byetta (Amylin)	Exenatide(1–39)-NH_2 (Exendin-4)	S.c. injection	5–10 μg b.i.d.	Adjunctive therapy in Type 2 diabetes
Fuseon (Roche)	T-20 peptide	S.c. injection	90 mg daily	Treatment of HIV-1 infection in combination with antiretroviral agents
Symlin (Amylin)	Analog of Amylin(1–37)-NH_2	S.c. injection	15–60 μg for Type 1 and 60–120 μg preprandial for Type 2 diabetes	Adjunct treatment for Type 1 and Type 2 diabetes
Forteo (Lilly)	PTH(1–34)-OH	S.c. injection	20 μg daily	Postmenopausal osteoporosis with high risk of fracture

α-amidated peptides. These and other peptides are efficacious for various disease indications, such as osteoporosis, allergy/asthma, arthritis, cancer, diabetes, growth impairment, cardiovascular diseases, inflammation and analgesia.

10.10 Conclusions

PAM is a ubiquitous and highly conserved enzyme whose presence is essential for the survival of an organism. Although it is a complex, bifunctional enzyme with many catalytic requirements, it has been successfully cloned and expressed in mammalian and other cell lines. The availability of recombinant PAM enables recombinant production of peptides that require a C-terminal amide group for biological activity. α-Amidation of the glycine-extended precursors of such peptides can be carried out in a scalable *in vitro* reaction where the mass ratio of substrate to enzyme is very high. In turn, recombinant production of α-amidated peptides by a scalable and cost-effective production scheme enables the development of alternate delivery routes for peptide based drugs (oral, nasal, inhaled, etc.), where the bioavailability is relatively low compared to parenteral formulations.

References

1 Steiner, D.F. (1998) The proprotein convertases. *Current Opinion in Chemical Biology*, **2**, 31–39.

2 Fricker, L.D. (2005) Neuropeptide-processing enzymes: applications for drug discovery. *The AAPS Journal*, **07** (2), E449–E455.

3 Lindberg, I.M. and Hutton, J.C. (1991) Peptide processing proteinases with selectivity for paired basic residues, in *Peptide Biosynthesis and Processing*, (ed L.D. Fricker), CRC Press, Boca Raton, FL, pp. 141–174.

4 Wei, S. *et al.* (2003) Neuropeptide-processing carboxypeptidases. *Life Sciences*, **73**, 655–662.

5 Kreil, G. (1985) Late reaction in the processing of peptide precursors: stepwise cleavage of dipeptides and formation of terminal amides, in *Enzymology of Post-Translational Modification of Proteins*, Vol. **2** (eds R.B. Freedman and H.C. Hawkins), Academic Press, New York, pp. 41–51.

6 King, L. *et al.* (2000) The enzymatic formation of novel bile acid primary amides. *Archives of Biochemistry and Biophysics*, **374** (2), 107–117.

7 Wilcox, B.J. *et al.* (1999) *N*-Acylglycine amidation: implications for the biosynthesis of fatty acid primary amides. *Biochemistry*, **38** (11), 3235–3245.

8 McIntyre, N.R. *et al.* (2006) Thiorphan, tiopronin, and related analogs as substrates and inhibitors of peptidylglycine alpha-amidating monooxygenase (PAM). *FEBS Letters*, **580** (2), 521–532.

9 DeBlassio, J.L. *et al.* (2000) Amidation of salicyluric acid and gentisuric acid: a possible role for peptidylglycine alpha-amidating monooxygenase in the metabolism of aspirin. *Archives of Biochemistry and Biophysics*, **383**, 46–55.

10 Katopodis, A.G. *et al.* (1990) Novel substrates and inhibitors of peptidylglycine alpha-amidating

monooxygenase. *Biochemistry*, **29** (19), 4541–4548.
11 Prigge, S.T. *et al.* (2000) New insights into copper monooxygenases and peptide amidation: structure, mechanism, and function. *Cellular and Molecular Life Sciences*, **57**, 1236–1259.
12 Wilmasena, D.S. *et al.* (2002) Plausible molecular mechanism for activation by fumarate and electron transfer of the dopamine β-mono-oxygenase reaction. *The Biochemical Journal*, **367**, 77–85.
13 Merkler, D.J. *et al.* (1995) The irreversible inactivation of two copper-dependent monooxygenases by sulfite: peptidylglcine alpha-amidating enzyme and dopamine β-monooxygenase. *FEBS Letters*, **366**, 165–169.
14 Jones, B.N. *et al.* (1988) A fluorometric assay for peptidyl alpha-amidation activity using high- performance liquid chromatography. *Analytical Biochemistry*, **168**, 272–279.
15 Consalvo, A.P. *et al.* (1992) Rapid fluorimetric assay for the detection of the peptidyl α-amidating enzyme intermediate using high performance liquid chromatography. *Journal of Chromatography*, **607**, 25–29.
16 Moore, A.B. and May, S.W. (1999) Kinetic and inhibition studies on substrate channeling in the bifunctional enzyme catalyzing C-terminal amidation. *The Biochemical Journal*, **341**, 33–40.
17 Chikuma, T. *et al.* (1991) A colorimetric assay for measuring peptidylglycine α-amidating monooxygenase using high performance liquid chromatography. *Analytical Biochemistry*, **198**, 263–267.
18 Schryver, S.B. (1910) The photochemical formation of formaldehyde in green plants, in *Proceedings of the Royal Society of London. Series B*, **82**, 226.
19 Pentz, E.I. (1969) Adaptation of the Rimini-Schryver reaction for the measurement of allantoin in urine to the auto-analyzer: allantoin and taurine excretion following neutron irradiation. *Analytical Biochemistry*, **27** (2), 333–342.
20 Katsuki, H. *et al.* (1961) The determination of alpha-ketoglutaric acid by 2,4-dinitrophenylhydrazine: salting out extraction method. *Analytical Biochemistry*, **2**, 421–432.
21 McFadden, B.A. and Howes, W.V. (1960) The determination of glyoxylic acid in biological systems. *Analytical Biochemistry*, **1**, 240–248.
22 Carpenter, S.E. and Merkler, D.J. (2003) An enzyme-coupled assay for glyoxylic acid. *Analytical Biochemistry*, **323** (2), 242–246.
23 Tajima, M. *et al.* (1990) The reaction product of peptidylglycine alpha-amidating enzyme is a hydroxyl derivative at the alpha-carbon of the carboxy-terminal glycine. *The Journal of Biological Chemistry*, **265** (17), 9620–9605.
24 Takahashi, K. *et al.* (1990) Peptidylglycine alpha-amidating reaction: evidence for a two-step mechanism involving a stable intermediate at neutral pH. *Biochemical and Biophysical Research Communications*, **169** (2), 524–530.
25 Katopodis, A.G., Ping, D. and May, S.W. (1990) A novel enzyme from bovine neurointermediate pituitary catalyzes dealkylation of alpha-hydroxyglycine derivatives, thereby functioning sequentially with peptidylglycine alpha-amidating monooxygenase in peptide amidation. *Biochemistry*, **29** (26), 6115–6120.
26 Young, S.D. and Tamburini, P.P. (1989) Enzymatic peptidyl alpha-amidation proceeds through formation of an alpha-hydroxyglycine intermediate. *Journal of the American Chemical Society*, **111** (5), 1933–1934.
27 Suzuki, K. *et al.* (1990) Elucidation of amidating reaction mechanism by frog amidating enzyme, peptidylglycine alpha-hydroxylating monooxygenase, expressed in insect cell culture. *The EMBO Journal*, **9** (13), 4259–4265.
28 Merkler, D.J. (1994) C-terminal amidated peptides: production by the *in vitro*

enzymatic amidation of glycine-extended peptides and the importance of the amide to bioactivity. *Enzyme and Microbial Technology*, **16**, 450–456.

29 Eipper, B.A. and Mains, R.E. (1988) Peptide α-amidation. *Annual Review of Physiology*, **50**, 333–344.

30 Bradbury, A.F. *et al.* (1982) Mechanism of C-terminal amide formation by pituitary enzymes. *Nature*, **298**, 686–688.

31 Eipper, B.A., Mains, R.E. and Glembotski, C.C. (1983) Identification in pituitary tissue of a peptide alpha-amidation activity that acts on glycine-extended peptides and requires molecular oxygen, copper, and ascorbic acid. *Proceedings of the National Academy of Sciences*, **80**, 5144–5148.

32 Klinman, J.P. (2006) The copper-enzyme family of dopamine β-monooxygenase and peptidylglycine alpha-hydroxylating monooxygenase: resolving the chemical pathway for substrate hydroxylation. *The Journal of Biological Chemistry*, **281** (6), 3013–3016.

33 Southan, C. and Kruse, L.I. (1989) Sequence similarity between dopamine beta-hydroxylase and peptide alpha-amidating enzyme: evidence for a conserved catalytic domain. *FEBS Letters*, **255** (1), 116–120.

34 Prigge, S.T. (1997) *et al.* Amidation of bioactive peptides: the structure of peptidylglycine alpha-hydroxylating monooxygenase. *Science*, **278** (5341), 1300–1305.

35 Kolhekar, A.S. *et al.* (1997) Peptidylglycine alpha-hydroxylating monooxygenase: active site residues, disulfide linkages, and a two-domain model of the catalytic core. *Biochemistry*, **36** (36), 10901–10909.

36 Bollinger, J.M. and Krebs, C. (2007) Enzymatic C-H activation by metal superoxo intermediates. *Current Opinion in Chemical Biology*, **11** (2), 151–158.

37 Siebert, X. *et al.* (2005) The catalytic copper of peptidylglycine alpha-hydroxylating monooxygenase also plays a critical structural role. *Biophysical Journal*, **89** (5), 3312–3319.

38 Klinman, J.B. *et al.* (1984) Evidence for two copper atoms/subunit in dopamine beta-monooxygenase catalysis. *The Journal of Biological Chemistry*, **259** (6), 3399–3402.

39 Ash, D.E. *et al.* (1984) Kinetic and spectroscopic studies of the interaction of copper with dopamine beta-hydroxylase. *The Journal of Biological Chemistry*, **259** (6), 3395–3398.

40 Kulathila, R. *et al.* (1994) Bifunctional peptidylglycine alpha-amidating enzyme requires two copper atoms per maximum activity. *Archives of Biochemistry and Biophysics*, **311** (1), 191–5.

41 Prigge, S.T. *et al.* (2004) Dioxygen binds end-on to mononuclear copper in a precatalytic enzyme complex. *Science*, **304** (5672), 864–867.

42 Prigge, S.T. *et al.* (1999) Substrate-mediated electron transfer in peptidylglycine alpha-hydroxylating monooxygenase. *Nature Structural Biology*, **6** (10), 976–983.

43 Bauman, A. *et al.* (2006) The hydrogen peroxide reactivity of peptidylglycine monooxygenase supports a Cu (II)-superoxo catalytic intermediate. *The Journal of Biological Chemistry*, **281** (7), 4190–4098.

44 Evans, J.P., Ahn, K. and Klinman, J.P. (2003) Evidence that dioxygen and substrate activation are tightly coupled in dopamine alpha-monooxygenase. *The Journal of Biological Chemistry*, **278** (50), 49691–49698.

45 Ping, D.E., Mounier, C.E. and May, S.W. (1995) Reaction versus subsite stereospecificity of peptidylglycine alpha-monooxygenase and peptidylamidoglycolate lyase, the two enzymes involved in peptide amidation. *The Journal of Biological Chemistry*, **270**, 29250–29255.

46 McIninch, J.K., McIninch, J.D. and May, S.W. (2003) Catalysis, stereochemistry, and inhibition of ureidoglycolate lyase. *The Journal of Biological Chemistry*, **278**, 50091–50100.

47 Mithu, D. *et al.* (2006) Role for an essential tyrosine in peptide amidation. *The Journal of Biological Chemistry*, **281**, 20873–20882.

48 Braas, K.M. *et al.* (1992) Expression of peptidyl-glycine alpha-amidating monooxygenase; an *in situ* hybridization and immunocytochemical study. *Endocrinologica*, **1305** (5), 2778–2788.

49 Eipper, B.A., Stoffers, D.A. and Mains, R.E. (1992) The biosynthesis of neuropeptides: peptide alpha-amidation. *Annual Review of Neuroscience*, **15**, 57–85.

50 Bendena, W.G., Donly, B.C. and Tobe, S.S. (1999) Allatostatins: a growing family of neuropeptides with structural and functional diversity. *Advances in Optical Biopsy and Optical Mammography*, **897**, 311–29.

51 Shoofa, L. *et al.* (1997) Peptides in locusts, *Locusta migratoria* and *Schistorcera grearia*. *Peptides*, **18**, 145–156.

52 McDonald, J.K., Klein, K. and Noe, B.D. (1995) Distribution of peptidyl-glycine alpha-amidating monooxygenase immunoreactivity in the brain, pituitary and islet organ of the anglerfish (*Lophius americanus*). *Cell and Tissue Research*, **280** (1), 159–170.

53 Grimmelikhuijzen, C.J.P., Leviv, I. and Carstensen, K. (1996) Peptides in the nervous system of Cnidarians: structure function and biosynthesis. *International Review of Cytology*, **167**, 37–89.

54 Spijker, S. *et al.* (1999) A molluscan peptide alpha-amidating enzyme precursor that generates five distinct enzymes. *The FASEB Journal*, **6**, 735–48.

55 Lackey, D.B. (1992) Isolation and structure determination of novel TRH-Like tripeptide, pyroGlu-Tyr- Pro amide from Alfalfa. *The Journal of Biological Chemistry*, **267**, 17508–17511.

56 Ouafik, L.H. *et al.* (1992) The multifunctional peptidylglycine alpha-amidating monooxygenase gene: exon/intron organization of catalytic, processing, and routing domains. *Molecular Endocrinology*, **6** (10), 1571–1584.

57 Moore, A.B. and May, S.W. (1999) Kinetic and inhibition studies on substrate channeling in the bifunctional enzyme catalyzing C-terminal amidation. *The Biochemical Journal*, **341** (pt 1), 33–40.

58 Bradbury, A.F. and Smyth, D.G. (1988) Peptide amidation: evidence for multiple molecular forms of the amidating enzyme. *Biochemical and Biophysical Research Communications*, **154** (3), 1293–1300.

59 Stoffers, D.A., Green, C.B. and Eipper, B.A. (1989) Alternative mRNA splicing generates multiple forms of peptidyl-glycine α-amidating monooxygenase in rat atrium. *Proceedings of the National Academy of Sciences*, **86**, 735–739.

60 Williamson, M., Hauser, F. and Grimmelikhuijzen, C.J. (2000) Genomic organization and splicing variants of a peptidylglycine alpha-hydroxylating monooxygenase from sea anemones. *Biochemical and Biophysical Research Communications*, **277** (1), 7–12.

61 Jiménez, N. *et al.* (2003) Androgen-independent expression of adrenomedullin and peptidylglycine alpha-amidating monooxygenase in human prostatic carcinoma. *Molecular Carcinogenesis*, **38** (1), 14–24.

62 Eipper, B.A. *et al.* (1992) Alternative splicing and endoproteolytic processing generate tissue-specific forms of pituitary peptidylglycine α-amidating monooxygenase (PAM). *The Journal of Biological Chemistry*, **267** (6), 4008–4015.

63 Tamburini, P.P. *et al.* (1988) Structure-activity relationships for glycine-extended peptides and the α-amidating enzyme derived from medullary thyroid CA-77 cells. *Archives of Biochemistry and Biophysics*, **267**, 623–631.

64 Tamburini, P.P. *et al.* (1990) Peptide substrate specificity of α-amidating enzyme isolated from rat medullary thyroid CA-77 cells. *International Journal of Peptide and Protein Research*, **35**, 153–156.

65 Mizuno, K. *et al.* (1987) Cloning and sequence of cDNA encoding a peptide C-terminal α-amidating enzyme from *Xenopus laevis*. *Biochemical and Biophysical Research Communications*, **148** (2), 546–552.

66 Bauman, A.T., Ralle, M. and Blackburn, N.J. (2007) Large-scale production of the copper enzyme peptidylglycine monooxygenase using an automated bioreactor. *Protein Expression and Purification*, **51**, 34–38.

67 Meshini, R.E. *et al.* (2003) Supplying copper to the cuproenzyme peptidylglycine α-amidating monooxygenase. *The Journal of Biological Chemistry*, **278** (14), 12278–12284.

68 Eipper, B.A. *et al.* (1991) Peptidyl-alpha-hydroxyglycine alpha-amidating lyase. Purification, characterization and expression. *The Journal of Biological Chemistry*, **266** (12), 7827–7833.

69 Kobayashi, J. *et al.* (1992) High level expression of a frog α-amidating enzyme, AEII, in cultured cells and silkworm larvae using *Bombyx mori* nuclear polyhedrosis virus expression vector. *Cytotechnology*, **8**, 103–108.

70 Furukawa, K. and Ohsuye, K. (1999) Enhancement of productivity of recombinant α-amidating enzyme by low temperature culture. *Cytotechnology* **31**, 85–94.

71 Santini, M. *et al.* (2003) Expression and characterization of human bifunctional peptidylglycine α-amidating monooxygenase. *Protein Expression and Purification*, **28**, 293–302.

72 Glauder, J. *et al.* (1990) Human peptidylglycine alpha-amidating monooxygenase: cDNA, cloning and functional expression of a truncated form in COS cells. *Biochemical and Biophysical Research Communications*, **169** (2), 551–558.

73 Beaudry, G.A. *et al.* (1990) Purification and characterization of functional recombinant α-amidating enzyme secreted from mammalian cells. *The Journal of Biological Chemistry*, **265** (29), 17694–17699.

74 Miller, D.A. *et al.* (1992) Characterization of a bifunctional peptidylglycine α-amidating enzyme expressed in Chinese hamster ovary cells. *Archives of Biochemistry and Biophysics*, **298** (2), 380–388.

75 Ray, M.V.L. *et al.* (2002) Production of salmon calcitonin by direct expression of a glycine-extended precursor in *Escherichia coli*. *Protein Expression and Purification*, **26**, 249–259.

76 Lui, Z. *et al.* (2003) Cloning, co-expression with an amidating enzyme and activity of the scorpion toxin BmK ITa1 cDNA in insect cells. *Molecular Biotechnology*, **24** (1), 21–26.

77 Hong, B., Wu, B. and Li, Y. (2003) Production of C-terminal amidated recombinant salmon calcitonin in *Streptomyces lividans*. *Applied Biochemistry and Biotechnology*, **110** (2), 113–123.

78 Ray, M.V.L. *et al.* (1993) Production of recombinant salmon calcitonin by *in vitro* amidation of an *Escherichia coli* produced precursor peptide. *Biotechnology*, **11**, 64–70.

79 Mehta, N.M. (2004) Oral delivery and recombinant production of peptide hormones. Part II: recombinant production of therapeutic peptides. *Biochemical Pharmacology International*, **17**, 44–46.

80 Mehta, N.M. *et al.* (1988) Purification of a peptidylglycine alpha-amidating enzyme from transplantable rat medullary thyroid carcinomas. *Archives of Biochemistry and Biophysics*, **261** (1), 44–54.

81 Gilligan, J.P. *et al.* (1989) Multiple forms of peptidyl alpha-amidating enzyme: purification from rat medullary thyroid carcinoma CA-77 cell-conditioned medium. *Endocrinologica*, **124** (6), 2729–2736.

82 Bertelsen, A.H. *et al.* (1990) Cloning and characterization of two alternatively spliced rat alpha-amidating enzyme cDNAs from rat medullary thyroid carcinoma. *Archives of Biochemistry and Biophysics*, **279** (1), 87–96.

83 Lipinski, C.A. *et al.* (2001) Experimental and computational approaches to estimate solubility and permeability in drug discovery and development settings. *Advanced Drug Delivery Reviews*, **46** (1–3), 3–26.

84 Sehgal, A. (2006) New applications in discovery, manufacturing and therapeutics, *Peptides*, Report 9214, D&MD Publications.

85 Eipper, B.A. *et al.* (1993) Peptidylglycine α-amidating monooxygenase: a multifunctional protein with catalytic processing and routing domains. *Protein Science*, **2**, 489–497.

86 Merkler, D.J. (1994) C-terminal amidated peptides: the importance of the amide to bioactivity and production by *in vitro* enzymatic amidation of glycine extended peptides. *Enzyme and Microbial Technology*, **16**, 450–456.

11
Disulfide Bond Formation

Hayat El Hajjaji and Jean-François Collet

11.1
Introduction

To be active and stable, newly synthesized proteins have to fold correctly. However, the cytoplasm is such a crowded environment (the average macromolecule concentration of the *Escherichia coli* cytoplasm can reach 300–400 mg mL^{-1} [1]) that reaching a proper three-dimensional structure is a real challenge. As they emerge from the ribosome tunnel, nascent polypeptides expose hydrophobic surfaces and are therefore prone to aggregation. This is especially true for multidomain and overexpressed recombinant proteins. To prevent this fatal process, cells possess molecular chaperones that restrain nascent proteins from folding into off-pathway intermediates and help them to reach their native conformation. For instance, the *E. coli* cytoplasm contains a complex network of molecular chaperones that include trigger factor, which acts as a cradle at the exit of the ribosome, the DnaK(Hsp70)/DnaJ/GrpE system and the GroEL(Hsp60)/GroES system (reviewed in Reference [2]).

In addition to molecular chaperones, both eukaryotic and prokaryotic cells possess protein folding catalysts that accelerate rate-limiting steps in the folding process of a protein. There are two classes of protein folding catalysts: the peptidyl-proline cis-trans isomerases (PPIases) and proteins that are involved in the formation of a disulfide bond between two cysteine residues.

PPIases are ubiquitous proteins expressed in prokaryotic and eukaryotic cells that facilitate the *cis–trans* isomerization of peptide bonds N-terminal to proline residues within polypeptide chains [3]. This isomerization is a rather slow process and needs to be catalyzed to ensure fast and efficient protein folding in the cell.

Disulfide bonds are rarely found in cytoplasmic proteins. In contrast, disulfide bond formation is important for numerous secreted proteins, including many proteins of pharmaceutical importance such as insulin, scFv antibodies, human tissue plasminogen activator (tPA) and human nerve growth factor. Failure to form a disulfide bond or formation of an incorrect disulfide will prevent these proteins from folding correctly, which often leads to protein misfolding and aggregation or degradation by proteases. In eukaryotes, disulfide bond formation takes place in the

Post-translational Modification of Protein Biopharmaceuticals. Edited by Gary Walsh

ISBN: 978-3-527-32074-5

endoplasmic reticulum, where it is catalyzed by a machinery involving, among others, the protein disulfide isomerase (PDI), the first discovered disulfide bond formation catalyst [4]. In prokaryotes, disulfide bonds are introduced into secreted proteins by the Dsb proteins family [5]. In Gram-negative bacteria, this process takes place in the periplasm.

In this chapter we review the pathways involved in the formation of the correct disulfide bonds in prokaryotes and eukaryotes. As *E. coli* is the host organism that is the most widely used for protein expression, the pathways of protein disulfide bond formation in this bacterium are reviewed in more detail.

11.2
Disulfide Bonds have a Stabilizing Effect

Disulfide bonds are formed from the thiol group of two cysteine residues, with the concomitant release of two protons and two electrons. They are almost exclusively found in membrane and secreted proteins in which they play a vital role by stabilizing the protein three-dimensional structure. The main effect of the formation of a disulfide bond is to decrease the conformational entropy of the denatured state of the protein. The stabilizing effect has been estimated to be up to $\approx 4\,\text{kcal mol}^{-1}$ per disulfide formed [6, 7].

11.3
Disulfide Bond Formation is a Catalyzed Process

Disulfide bonds can form spontaneously in the presence of molecular oxygen. For instance, exposure of a cysteine solution to air leads to the formation of a thin-layer of cystine at the air–liquid interface. However, spontaneous oxidation by air oxygen is a slow process. For instance, it has been shown that formation of the native disulfide bonds required for the folding of a protein such as RNAse A, which presents four disulfides in its native conformation, takes several hours *in vitro*. This slow air-oxidation process cannot account for the rapid rates of protein folding required by the cell, where disulfide bond formation usually occurs within minutes after synthesis of a protein. For instance, it takes less than 2 min to oxidatively fold RNAse A in the endoplasmic reticulum. It is the discrepancy between these *in vivo* and *in vitro* rates that led Anfinsen to the discovery of the first catalyst of disulfide bond formation, the protein disulfide isomerase [8].

11.4
Disulfide Bond Formation in the Bacterial Periplasm

In Gram-negative bacteria, such as *Escherichia coli*, disulfide bonds are introduced in the periplasm by proteins from the Dsb (disulfide bond) family: DsbA and DsbB catalyze disulfide bond formation, whereas DsbC and DsbD are involved in disulfide bond isomerization.

11.4.1
The Oxidation Pathway: DsbA and DsbB

11.4.1.1 DsbA, a very Oxidizing Protein

DsbA is a soluble monomeric protein of 21 kDa that is the primary catalyst of disulfide bond formation in the *E. coli* periplasm [9]. Like many other oxidoreductases, DsbA possesses a CXXC catalytic site motif, which is present within a thioredoxin-like fold [10].

In agreement with the function of DsbA as an oxidase, the catalytic cysteine residues are found oxidized *in vivo*. The disulfide bond present in the catalytic site is very unstable and is rapidly transferred from DsbA to target proteins (Figure 11.1). DsbA has a redox potential of −120 mV [11], which makes it the strongest thiol oxidant among known oxidoreductases. Structural and biophysical studies have shown that the oxidizing redox potential of DsbA arises from the stabilization of

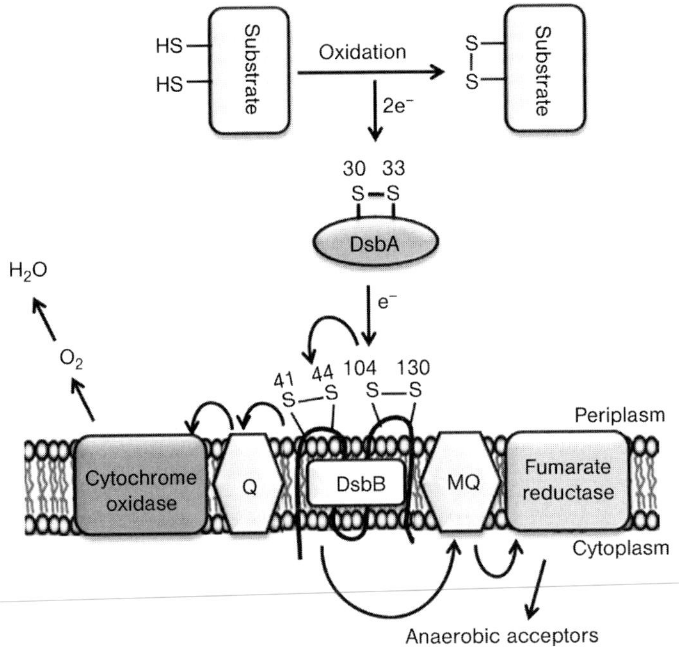

Figure 11.1 Oxidative pathway for protein disulfide bond formation. The direction of electron flow is shown by black arrows. DsbA reacts with a newly translocated protein and is then reoxidized by the inner membrane protein DsbB. In aerobic conditions, DsbB passes electrons to ubiquinone (Q), then electrons flow to terminal oxidases and finally to molecular oxygen. Under anaerobic conditions, DsbB passes electrons from DsbA to menaquinone (MQ), which is up-regulated upon oxygen depletion. Anaerobic oxidoreductases such as fumarate reductase serve to reoxidize menaquinone.

the reduced form versus the oxidized form of the protein. The first cysteine residue of the CXXC motif (Cys30) has been shown to play a key role in this stabilization. This residue has an unusually low pK_a of about 3.5 [12, 13], which is significantly lower than that of free cysteines (pK_a = 8.8). Cys30 is therefore entirely in a thiolate anion state under physiological conditions. The structure of DsbA shows that the thiolate form of Cys30 is stabilized by structural elements of the thioredoxin fold [14]. The stabilization of the thiolate anion drives therefore the reaction towards the reduction of DsbA and the transfer of the disulfide bond to the folding proteins.

DsbA plays a central role in the periplasm and catalyzes the oxidative folding of numerous periplasmic proteins. Over 300 proteins secreted to the *E. coli* periplasm have at least two cysteine residues and are therefore potential DsbA substrates. It is therefore not surprising that *dsbA*$^-$ strains have a pleiotropic phenotype (reviewed in Reference [15]): *dsbA*$^-$ mutants are hypersensitive to benzylpenicillin, dithiothreitol and metals such as Cd^{2+}. They also lack a functional flagellar motor, which results in a loss of motility [16]. About 25 DsbA substrates have been identified so far experimentally, either by trapping of mixed disulfides with DsbA mutants [17] or by proteomics experiments using 2D-gels analysis [18, 19] or 2D-LC-MS/MS [20].

11.4.1.2 **DsbB**

After transfer of the catalytic disulfide to target proteins, DsbA needs to be reoxidized to stay active in the periplasm. The protein that reoxidizes DsbA is the inner-membrane protein DsbB (Figure 11.1) [21]. The interaction between DsbA and DsbB was first suggested by the fact that *dsbB*$^-$ strains accumulate DsbA in the reduced state and exhibit the same phenotypes as *dsbA* mutants. This was later confirmed *in vitro* by showing that a purified preparation of DsbB can catalyze the reoxidation of DsbA in the presence of molecular oxygen [22].

DsbB is a small protein of 21 kDa, which has four transmembrane segments and two small hydrophilic regions that are exposed to the periplasm. Each of these periplasmic segments contains a conserved pair of cysteine residues (Cys41 and Cys44 in domain 1 and Cys104 and Cys130 in domain 2). These cysteines can undergo oxidation–reduction cycles and are essential for the activity of DsbB [23].

The mechanism by which DsbB is able to keep DsbA oxidized has been characterized [24]. According to the current model, the first step is the attack of the 104–130 disulfide from domain 2 of DsbB by Cys30 of DsbA, to give a mixed disulfide complex between DsbA and DsbB. The formation of this disulfide has been observed *in vivo* [25] and the structure of the DsbA-DsbB mixed-disulfide complex has been solved recently using DsbACys33Ala and DsbBCys130Ser mutants [26]. The mixed-disulfide is then resolved by attack of Cys33 of DsbA, which results in the oxidation of DsbA and the reduction of Cys104 and Cys130. Then, electrons flow from these latter two cysteine residues to the second pair of cysteine residues present in DsbB, Cys41 and Cys44 (Figure 11.1).

11.4.1.3 **DsbB is Reoxidized by the Electron Transport Chain**

Cys41 and Cys44 of DsbB need to pass the electrons to another electron acceptor to recycle DsbB and restore its activity. Early experiments performed on *E. coli* strains defective in quinone and heme biosynthesis showed that these mutants accumulate DsbB in a reduced state, suggesting that electrons flow from DsbB to quinones [27]. This hypothesis was later confirmed by Bardwell and coworkers, who reconstituted the complete disulfide bond formation system *in vitro* [28]. Using purified components, they showed that electrons flow from DsbB to ubiquinone then onto cytochromes oxidases *bd* and *bo* and finally to molecular oxygen. Under anaerobic conditions, electrons are transferred from DsbB to menaquinone and then to fumarate or nitrate reductase (Figure 11.1) [28, 29]. DsbB is therefore a unique enzyme that catalyzes the efficient formation of a disulfide bond de novo from the reduction of a membrane-bound quinone.

How does DsbB manage to catalyze the reoxidation of such an oxidizing oxidoreductase as DsbA? It was tempting to speculate that it is the presence of quinone molecules with very oxidizing redox potentials (+110 mV for ubiquinone and −74 mV for menaquinone) that drives the reaction towards the oxidation of DsbA. However, elegant work by Inaba and coworkers showed that quinone-free DsbB prepared from cells unable to synthesize ubiquinone and menaquinone is still able to oxidize reduced DsbA up to a stoichiometry of 0.4 M [30]. This indicates that DsbB has some intrinsic properties that allow it to reoxidize DsbA even in the absence of quinones. Intriguingly, these properties do not seem to be related to the redox potential of DsbB's cysteine pairs as determination of the redox potential of the Cys104-Cys130 and Cys41-Cys44 catalytic disulfides using quinone free-DsbB showed that these cysteine pairs are less oxidizing than DsbA (−227 and −204 mV, respectively). The oxidation of the CXXC motif of DsbA ($E'_0 = -119$mV) by the 104–130 disulfide of DsbB is therefore an energetic uphill reaction. The solution to this fascinating problem came from the recent determination of the crystal structure of a DsbA-DsbB mixed-disulfide complex [26]. This structure revealed that DsbB undergoes a structural transition upon forming the mixed disulfide with DsbA, which leads to the relocation of Cys130 in the proximity of the Cys41-Cys44 disulfide. This relocation prevents Cys130 attacking the DsbBCys104-DsbACys30 mixed disulfide to reverse the reaction. The mixed-disulfide is then predominantly resolved by attack of Cys33 of DsbA, releasing oxidized DsbA and Cys104 and Cys130 of DsbB reduced. This reduction is followed by a rapid disulfide exchange reaction between the Cys104-Cys130 and Cys41-Cys44 cysteine pairs of DsbB. The latter cysteine residues are then rapidly reoxidized by quinones.

11.4.1.4 Engineering of a New Oxidation Pathway

In collaboration with others, we have designed a new pathway for the formation of disulfide bonds in the periplasm [31]. By imposing an evolutionary pressure, we managed to select *E. coli* thioredoxin mutants that can catalyze disulfide bond formation in the periplasm independently of the action of DsbA and DsbB.

Thioredoxin is a small 12 kDa monomeric protein that also has a CXXC active site motif. However, in contrast to DsbA, thioredoxin is present in the cytoplasm, where it reduces disulfide bonds. In agreement with the function of thioredoxin as a reductase, the CXXC motif is found reduced in the cytoplasm.

We fused thioredoxin to a leader peptide to export it to the periplasm via the Tat-secretion pathway and the CXXC active site motif was mutagenized by random PCR. The resulting library was expressed in a *dsbB⁻dsbA⁻* strain and we screened for clones that were able to restore disulfide bond formation in the periplasm.

We isolated three mutants and noticed that they all had a third cysteine present in the catalytic site, replacing the CXXC motif with CXCC. We showed that the presence of this additional cysteine has a dramatic effect on thioredoxin, leading the protein to dimerize and to acquire a [2Fe-2S] iron sulfur cluster. Export of the iron-sulfur cluster bridged dimer to the periplasm was required to restore disulfide bond formation.

We solved the structure of one of these mutants and observed that the first and the second cysteine of the CXCC motif are involved in cluster coordination [31, 32].

11.4.2
Disulfide Isomerization Pathway

DsbA is a powerful oxidant but apparently lacks proofreading activity. DsbA oxidizes cysteine residues present on secreted proteins as they enter the periplasm. If the native disulfide bond pattern involves cysteine residues that are consecutive in the protein sequence, DsbA forms disulfides correctly. However, when secreted proteins have disulfides that need to be formed between non-consecutive cysteine residues, DsbA has the possibility of introducing non-native disulfides into the folding protein, which leads to protein misfolding and degradation by proteases [33]. Notably, our recent work on RNase 1, a periplasmic protein with one non-consecutive disulfide, showed that DsbA is able to correctly fold this protein, indicating that DsbA may be more specific than generally assumed [34].

To correct non-native disulfide bonds, *E. coli* possesses a "disulfide isomerization system," which involves a periplasmic protein disulfide isomerase, DsbC, and a membrane protein, DsbD (Figure 11.2).

11.4.2.1 DsbC, a Periplasmic Protein Disulfide Isomerase

DsbC is a soluble periplasmic protein of 23.3 kDa that shares several characteristics with DsbA. For instance, DsbC possesses a CXXC catalytic site motif present in a thioredoxin fold [35] and the first cysteine residue of the CXXC motif, Cys98, has a low pK_a (4.1). This cysteine is therefore in the thiolate form at neutral pH and, like for DsbA, stabilization of the thiolate state of this residue renders DsbC a rather oxidizing protein ($E_0' = -119$mV) [36]. However, despite these similarities between the structural and redox properties of DsbA and DsbC, the catalytic cysteine residues of DsbC are kept reduced in the periplasm [37], in contrast with the cysteine residues of DsbA that are found oxidized.

The fact that the catalytic cysteine residues are kept reduced allows DsbC to perform a nucleophilic attack on non-native disulfides, which is the first step of the isomerization reaction (Figure 11.2). This results in the formation of a mixed disulfide complex between DsbC and the substrate protein. This mixed-disulfide will be resolved either by attack of another cysteine present in the sequence of the

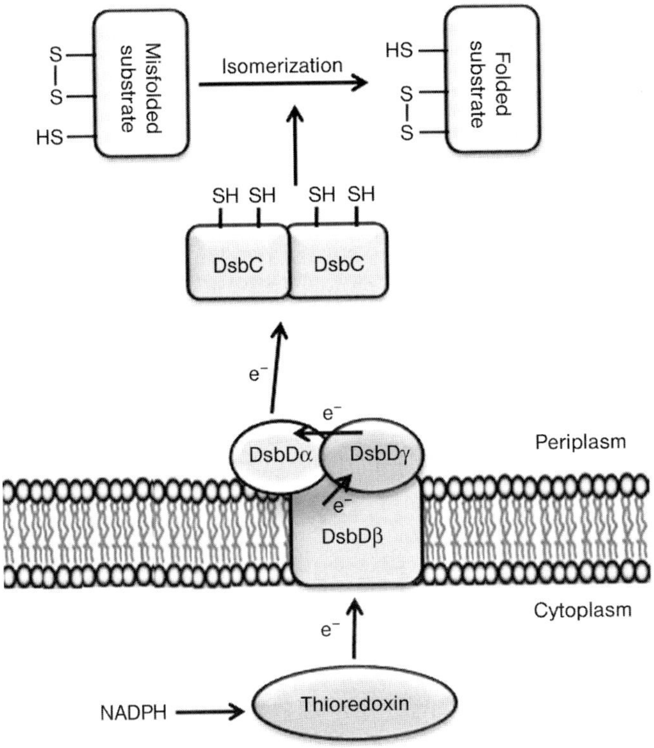

Figure 11.2 Isomerization pathway in the *E. coli* periplasm. The direction of electron flow is indicated by the arrows. Disulfide-bond rearrangement is catalyzed by the thiol-disulfide oxidoreductase DsbC, which is maintained in a reduced state by the membrane protein DsbD. DsbD receives electrons from the cytoplasmic thioredoxin system. Within DsbD, electrons flow from the β domain to the γ and α domains, successively.

substrate, resulting in the formation of a new disulfide in the substrate and the release of DsbC in the reduced state, or by attack of the mixed disulfide by Cys101, the second cysteine of the CXXC motif. In this case, DsbC is released in an oxidized state and will need to be reduced to stay active in the periplasm (see below).

In contrast to the central role played by DsbA in the periplasm where DsbA oxidizes a large number of secreted proteins, DsbC seems to be required for the correct folding of a rather limited number of *E. coli* proteins that contain nonconsecutive disulfides, such as the endopeptidase MepA or the phytase AppA [18, 20, 33]. Unsurprisingly, therefore, *dsbC*$^{-}$ strains have a milder phenotype than *dsbA*$^{-}$ mutants (reviewed in Reference [15]). Interestingly, DsbC can assist the folding of eukaryotic recombinant proteins with multiple disulfide bonds expressed in *E. coli*. These proteins include RNase A, bovine pancreatic trypsin inhibitor (BPTI) and urokinase [38, 39]. Moreover, DsbC also exhibits some chaperone activity *in vitro* and can assist the refolding of denaturated proteins like lysozyme or glyceraldehyde-3-phosphate dehydrogenase [40].

11.4.2.2 **DsbC is a Dimeric Protein**

The structure of DsbC has been solved [35]. DsbC appears as a V-shaped homodimeric molecule. Each DsbC monomer forms one arm of the V and consists of two domains: a C-terminal domain with a thioredoxin fold and a N-terminal dimerization domain [35]. The CXXC catalytic site is found within the thioredoxin domain, where the first cysteine residue is partially exposed to the solvent. The thioredoxin domain and the dimerization domain are connected by a α-helix linker.

The fact that DsbC is a dimeric protein seems to directly affect the properties of the protein. First, the dimerization of DsbC has been shown to be important for both the isomerase and chaperone activity of DsbC as truncated versions of DsbC lacking the dimerization domain are inactive as isomerase and chaperone [39]. Moreover, fusion of the N-terminal domain of DsbC to other proteins with thioredoxin-fold such as DsbA leads to protein dimerization and confers some isomerase and chaperone activity [41]. Second, the dimerization domain seems to prevent the oxidation of DsbC by DsbB as truncated versions of DsbC lacking the dimerization domain can be oxidized by DsbB both *in vivo* and *in vitro* and can substitute for DsbA [42]. Such an oxidation would prevent DsbC acting as a disulfide isomerase in the periplasm. However, recent work by Segatori and coworkers showed that the dimerization of DsbC is not enough to prevent oxidation by DsbB. It appears that more subtle conformational features that determine the geometry and orientation of the active sites in the overall structure of DsbC also prevent the oxidation of DsbC by DsbB [41, 43]. The α-helix that connects the N-terminal dimerization domain to the C-terminal thioredoxin seems to play a particularly important role.

11.4.2.3 **DsbC is Kept Reduced by DsbD**

With a redox potential around −165 mV, the periplasm is a rather oxidizing environment [34]. To stay active as an isomerase in this cellular compartment, DsbC has to be kept reduced. The protein that keeps DsbC reduced is the inner-membrane protein DsbD [44]. In agreement with DsbD's function in recycling DsbC, mutants lacking DsbD accumulate DsbC in an oxidized state.

DsbD is a 60 kDa protein that has three different domains: an amino-terminal periplasmic domain (α), a membranous domain with eight transmembrane segments (β) and a carboxy-terminal periplasmic domain (γ) (Figure 11.2). Each domain contains one pair of strictly conserved cysteine residues that are essential for activity [45]. *In vivo* and *in vitro* studies have shown that DsbD transfers electrons from the cytoplasmic pool of NADPH and the thioredoxin system to DsbC in the periplasm [38]. Notably, the electrons coming from the cytoplasm are also transferred by DsbD to DsbG (see below) and to CcmG, a membrane-tethered thioredoxin-like protein involved in cytochrome *c* maturation [46, 47].

Several residues that play an important role in DsbD's mechanism have been identified recently [48, 49], but the precise mechanism used by DsbD to transport electrons from one side of the membrane to the other is still unclear. *In vivo* and *in vitro* experiments suggest, however, that electrons are transferred via a succession of disulfide bond exchange reactions involving all six conserved cysteine residues of DsbD [50, 51]. According to the current model, electrons are successively transferred from thioredoxin to DsbD-β, then to DsbD-γ, DsbD-α and finally to DsbC (Figure 11.2). The two redox-active cysteine residues of the membranous β domain appear to be located at the center of the membrane [52], enabling them to be accessible from both the cytoplasmic and periplasmic side of the membrane. The transmembrane electron flow from thioredoxin to DsbC is kinetically and thermodynamically driven [51, 53].

So far, only the structures of the α and γ periplasmic domains have been solved. The γ domain of DsbD [54] has a typical thioredoxin-fold and contains a CXXC catalytic motif, whereas the α domain has an immunoglobulin-like fold [55]. The structure of a trapped reaction intermediate between the α and γ domains of DsbD [53] and the structures of mixed-disulfide complexes between the α domain and DsbC [55, 56] or CcmG [46] have also been solved.

11.4.2.4 DsbC can Function Independently of DsbD

Recently, we showed that the simultaneous absence of DsbA and DsbC leads to a decreased integrity of the cell envelope and affects the global protein content of the periplasm [20]. Surprisingly, we observed that strains lacking both DsbA and DsbD do not share these characteristics. These results indicate therefore that DsbC can cooperate with DsbA in a DsbD-independent manner to ensure the correct folding of *E. coli* envelope proteins. This suggests that DsbC is able to function in both the oxidation and isomerization pathways. We proposed that when DsbC gets oxidized upon reduction of a non-native disulfide it is reduced either by DsbD or by transferring its disulfide to a reduced protein (Figure 11.3). DsbC may therefore possibly be acting as a stand-alone protein folding catalyst that is able to cycle from the reduced to the oxidized state upon substrate oxidation and substrate reduction, respectively.

11.4.2.5 DsbG, a Controversial Protein Disulfide Isomerase

The *E. coli* periplasm contains another Dsb protein called DsbG. DsbG is a 25.7 kDa dimeric protein that shares 24% sequence identity with DsbC. Like DsbA and DsbC, DsbG has a catalytic CXXC motif that is present in a thioredoxin-like fold [57]. DsbG has a redox potential similar to this of DsbC (about −126 mV) [57]. The catalytic cysteine residues of DsbG are also kept reduced in the periplasm. Moreover, like DsbC, DsbG exhibits some chaperone activity and can assist the *in vitro* folding of luciferase and citrate synthase [58]. In many aspects, DsbG resembles DsbC, which suggests that DsbG may be another periplasmic protein disulfide isomerase, possibly with a more restricted substrate specificity, as suggested by some structural features. This assumption is further supported by the fact that overexpression of DsbG is able to restore the ability of *dsbC*$^-$ mutants to express some heterologous proteins containing multiple disulfide bonds [57]. However, the role of this protein in the *E. coli* periplasm is still unclear. For instance, in contrast to *dsbC*$^-$ mutants, *dsbG* null mutants have no defect in the folding of heterologous proteins containing multiple disulfide bonds and no *in vivo* DsbG substrates have been identified so far.

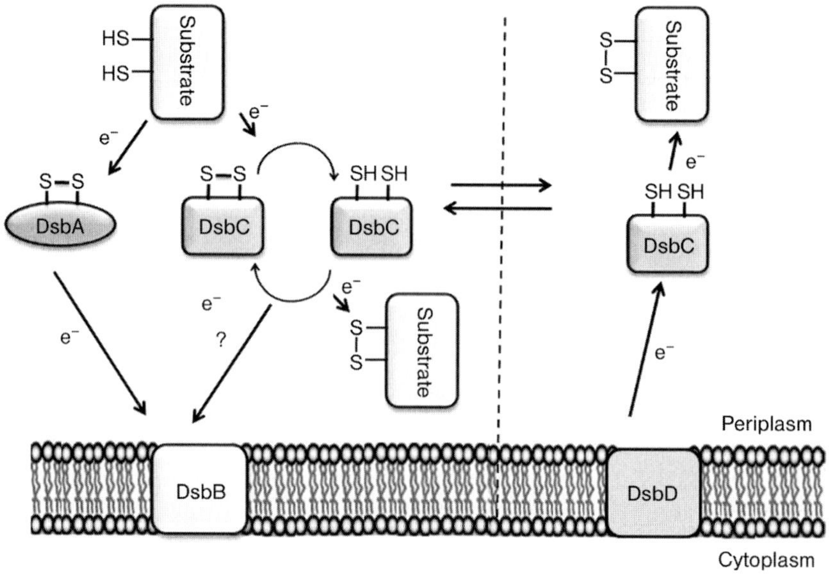

Figure 11.3 Revised model for the formation of disulfide bonds in the *E. coli* periplasm. Disulfide bonds are introduced by the DsbA/DsbB pathway. Non-native disulfides are corrected by DsbC, which is recycled by DsbD. Both pathways are kinetically isolated. DsbC is able to function on the other side of the barrier, where it assists DsbA in a DsbD-independent manner. DsbC may be acting as a stand-alone protein folding catalyst that cycles from the reduced to the oxidized state upon substrate oxidation and reduction, respectively. Kinetic data showed that DsbC is not a good substrate for DsbB. However, we cannot exclude that a slow oxidation of DsbC by DsbB may play a more significant role in the absence of DsbA.

Moreover, DsbG is unable to catalyze disulfide bond rearrangements *in vitro*. For instance, in contrast with DsbC, DsbG is unable to catalyze the oxidative refolding of proteins with multiple disulfide bonds such as hirudin or the bovine pancreatic trypsin inhibitor [57, 59].

11.5
Disulfide Bond Formation in the Cytoplasm

Structural disulfide bonds do not usually form in cytoplasmic proteins, except under oxidative stress conditions or as part of the catalytic cycle of certain enzymes, such as nucleotide reductase. In the *E. coli* cytoplasm, two partially overlapping pathways function to maintain cysteine residues in the reduced state [60]. The thioredoxin system involves two thioredoxins, Trx1 and Trx2, which are kept reduced by thioredoxin reductase (*trxB*) at the expense of NADPH. The glutaredoxin system includes four glutaredoxins, glutathione and glutathione reductase (*gor*). This latter enzyme also uses the reducing power of NADPH to recycle glutathione. Notably, thioredoxins and glutaredoxins all assume a thioredoxin fold and possess a catalytic CXXC motif, like most of the proteins of the Dsb family (Figure 11.4).

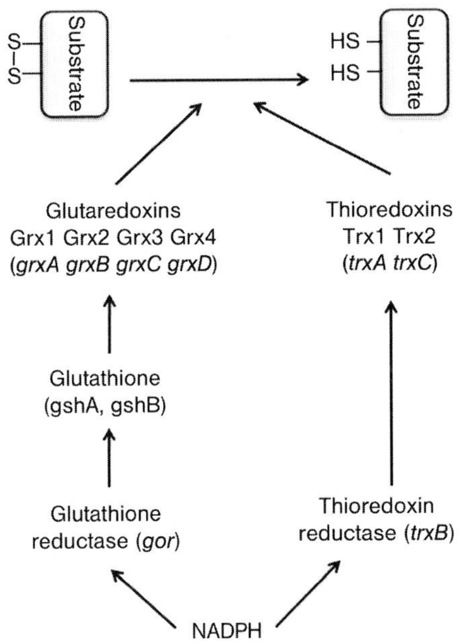

Figure 11.4 The thioredoxin and glutaredoxin systems keep cytoplasmic proteins reduced. The direction of electron flow is indicated by the arrows. Both systems use the reducing power of NADPH.

Interestingly, several studies have shown that structural disulfide bonds can be formed in the cytoplasm of *E. coli* strains lacking certain components of these reducing pathways [61, 62]. For instance, inactivation of the gene coding for thioredoxin reductase (*trxB*) allows expression of high levels of active alkaline phosphatase, an enzyme that needs two disulfide bonds to fold in its active conformation [61]. In $trxB^-$ strains, Trx1 and Trx2 accumulate in an oxidized state and directly promote the formation of disulfide bonds in alkaline phosphatase [63].

Strains lacking components of both pathways grow very poorly under aerobic conditions, unless a reductant such as dithiothreitol is present. However, Bessette and coworkers isolated a strain lacking both the thioredoxin reductase and glutathione reductase genes that is able to grow at normal rates due to the presence of a suppressor mutation [62]. Interestingly, this strain allows efficient formation of structural disulfide bonds in the cytoplasm and gives high yields of properly oxidized proteins. Even proteins that have a very complex pattern of disulfide bonds, such as tissue plasminogen activator (vtPA), have been successfully overexpressed and folded in this strain [62].

The suppressor mutation has been identified. It is a single amino acid insertion in the bacterial peroxiredoxin AhpC, a cysteine-dependent peroxidase [64]. The mutation converts this enzyme into a disulfide reductase, AhpC*, that has lost its peroxidase activity but is now able to channel electrons from the cytoplasmic pool of NADH into the glutaredoxin pathway [65].

11.6
Formation of Protein Disulfide Bond in Heterologous Proteins Expressed in *E. coli*

Escherichia coli is one of the most widely used host organisms for the expression of eukaryotic proteins, the folding of which often involves the formation of one or more disulfide bonds. Because expression in the cytoplasm of wild-type *E. coli* cells does not allow formation of these disulfides, there have been numerous attempts to express these proteins in the oxidizing environment of the periplasm. However, the bacterial periplasm is a very different environment compared to the endoplasmic reticulum, the compartment where disulfides are formed in eukaryotic proteins (see below). For instance, there is no real glutathione redox buffer in the periplasm (although there is glutathione) and no ATP-regulated chaperones. Unsurprisingly, therefore, eukaryotic proteins with multiple disulfides do not generally fold correctly in the periplasm and accumulate in a misfolded state. The bacterial production of therapeutic proteins with a complex disulfide bond pattern is therefore a real technological challenge.

However, the discovery and identification of the Dsb proteins opened the door to new techniques that dramatically improve the yield of production of therapeutic proteins with complex disulfide bond patterns, such as human tissue plasminogen activator (tPA). tPA, a serine protease that converts plasminogen into plasmin, is a 527 amino acid protein with 35 cysteine residues forming 17 disulfides. Expression of tPA in the periplasm of *E. coli* results in the accumulation of a misfolded protein that is completely inactive. However, co-expression of tPA with DsbC allows the production of a correctly folded protein with a specific activity nearly identical to that of the purified protein [66]. Other examples of proteins that have been successfully folded in the *E. coli* periplasm thanks to the co-expression of DsbA and/or DsbC have been reported [67]. For certain proteins, such as human pro-insulin, fusion of the protein of interest with DsbA also has a positive effect on the yields of production [68].

11.7
Disulfide Bond Formation in the Endoplasmic Reticulum

In eukaryotes, disulfide bond formation is essential for the structure and function of many proteins that travel through the secretory pathway to the cell surface or to intracellular membrane-bound compartments. The oxidative folding of these proteins takes place in the endoplasmic reticulum (ER). The most abundant oxido-reductase present in the ER is the protein disulfide isomerase (PDI).

11.7.1
PDI Functions both as an Oxidase and an Isomerase

In contrast with prokaryotes, where the pathways of disulfide bond formation and isomerization are separated, eukaryotic PDI appears to play a central role in both the formation and the isomerization of disulfide bonds. Depending on the redox state of its active site, PDI apparently acts *in vivo* as either an oxidase or an isomerase/reductase (reviewed in References [4, 69, 70]).

PDI consists of four thioredoxin domains, two inactive (b and b′) and two active (a and a′), that are arranged in a twisted U shape with the two catalytically active thioredoxin domains facing each other [71]. The structure of PDI resembles the V-shaped structure of DsbC and DsbG. The a′ domain is thought to mainly function as a disulfide oxidase and the a domain as a disulfide isomerase.

Other oxidoreductases with thioredoxin folds are present in the ER: for instance, there are three other oxidoreductases that are structurally homologous to PDI in the yeast ER and more than ten in the human ER [72]. These oxidoreductases differ with regard to substrate specificity, tissue expression and the number and structural arrangement of their thioredoxin-like domains.

11.7.2
PDI is Reoxidized by Ero1

Biochemical and genetics studies using the yeast *Saccharomyces cerevisiae* showed that PDI is reoxidized by the membrane associated oxidoreductase Ero1p (Figure 11.5), in a manner analogous to the reoxidation of DsbA by DsbB [73].

Ero1 (for ER oxidoreductin) is a FAD-bound protein that is closely associated with the lumenal face of the ER membrane. Ero1 possesses four catalytic cysteine residues. Two cysteines are in close proximity to the FAD cofactor and constitute the catalytic site of the protein [74]. The other two cysteine residues are involved in the direct reoxidation of PDI. These cysteines are located on a flexible loop that shuttles between PDI and the cysteine residues of the catalytic site. After oxidizing PDI, the shuttle cysteines are reoxidized through an internal disulfide exchange reaction with the catalytic cysteine residues. These latter residues are then reoxidized by transfer of the electrons to the FAD cofactor. The electrons finally flow to molecular oxygen [75], which results in the production of hydrogen peroxide. Thus, like DsbB, Ero1 generates disulfide bonds de novo by catalyzing transfer of two electrons from the dithiol motif in its active site to other electron acceptors.

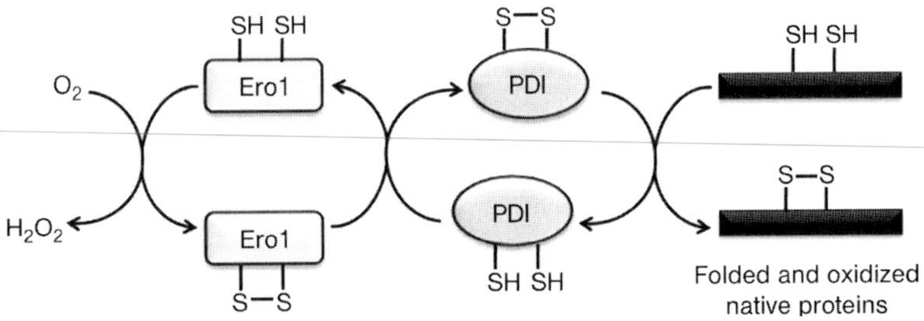

Figure 11.5 Disulfide bond formation in the endoplasmic reticulum (ER). Protein disulfide isomerase (PDI) oxidizes newly synthesized proteins folding in the ER. PDI is then recycled by the membrane-associated protein Ero1. When PDI is reduced, it acts as a disulfide isomerase.

11.7.3 Regulation of Ero1

An intriguing question that arises for eukaryotes is how PDI can be directed to either the disulfide formation or isomerization pathway. In an elegant study, Sevier and coworkers recently showed that four additional cysteine residues that are present in Ero1 play a regulatory role by adapting the level of Ero1 activity to the redox state of the ER and the oxidative folding needs of the cell [76]. These additional cysteine residues regulate Ero1's activity by controlling the range of motion of the flexible loop that contains the shuttle cysteines. Under oxidizing ER conditions, the regulatory cysteines become oxidized, which restricts the movement of the shuttle cysteines by tethering the flexible loop to the enzyme core. Shifting the ER environment to more reducing conditions leads to the reduction of the regulatory cysteines, which releases the flexible loop and increases Ero1 activity.

11.8 Conclusions

The study of disulfide bond formation began with *in vitro* experiments that lead to the discovery of PDI in the early 1960s. Since then, significant progress has been made in the characterization of the mechanisms that govern the formation of correct disulfide bonds into nascent proteins in both eukaryotes and prokaryotes. In particular, the structures of the yeast PDI and of the *E. coli* DsbB–DsbA complex have considerably improved our understanding of how these oxidoreductases function and interact. Interestingly, all these findings have also paved the way to new technological developments that have significantly improved the yields of production of proteins with complex disulfide bond patterns. This further highlights how the study of fundamental biochemical and cellular mechanisms often impacts on pharmacological and industrial processes.

References

1 Ellis, R.J. and Minton, A.P. (2003) Cell biology: join the crowd. *Nature*, **425** (6953), 27–28.

2 Hartl, F.U. and Hayer-Hartl, M. (2002) Molecular chaperones in the cytosol: from nascent chain to folded protein. *Science*, **295** (5561), 1852–1858.

3 Shaw, P.E. (2002) Peptidyl-prolyl isomerases: a new twist to transcription. *EMBO Reports*, **3** (6), 521–526.

4 Sevier, C.S. and Kaiser, C.A. (2002) Formation and transfer of disulphide bonds in living cells. *Nature Reviews. Molecular Cell Biology*, **3** (11), 836–847.

5 Messens, J. and Collet, J.F. (2006) Pathways of disulfide bond formation in *Escherichia coli*. *The International Journal of Biochemistry & Cell Biology*, **38** (7), 1050–1062.

6 Clarke, J. and Fersht, A.R. (1993) Engineered disulfide bonds as probes of the folding pathway of barnase: increasing the stability of proteins against the rate of

denaturation. *Biochemistry*, **32** (16), 4322–4329.

7 Pantoliano, M.W. *et al.* (1987) Protein engineering of subtilisin BPN': enhanced stabilization through the introduction of two cysteines to form a disulfide bond. *Biochemistry*, **26** (8), 2077–2082.

8 Goldberger, R.F., Epstein, C.J. and Anfinsen, C.B. (1963) Acceleration of reactivation of reduced bovine pancreatic ribonuclease by a microsomal system from rat liver. *The Journal of Biological Chemistry*, **238**, 628–635.

9 Bardwell, J.C., McGovern, K. and Beckwith, J. (1991) Identification of a protein required for disulfide bond formation *in vivo*. *Cell*, **67** (3), 581–589.

10 Martin, J.L., Bardwell, J.C. and Kuriyan, J. (1993) Crystal structure of the DsbA protein required for disulphide bond formation *in vivo*. *Nature*, **365** (6445), 464–468.

11 Zapun, A., Bardwell, J.C. and Creighton, T.E. (1993) The reactive and destabilizing disulfide bond of DsbA, a protein required for protein disulfide bond formation *in vivo*. *Biochemistry*, **32** (19), 5083–5092.

12 Nelson, J.W. and Creighton, T.E. (1994) Reactivity and ionization of the active site cysteine residues of DsbA, a protein required for disulfide bond formation *in vivo*. *Biochemistry*, **33** (19), 5974–5983.

13 Grauschopf, U. *et al.* (1995) Why is DsbA such an oxidizing disulfide catalyst? *Cell*, **83** (6), 947–955.

14 Martin, J.L. *et al.* (1993) Crystallization of DsbA, an *Escherichia coli* protein required for disulphide bond formation *in vivo*. *Journal of Molecular Biology*, **230** (3), 1097–1100.

15 Collet, J.F. and Bardwell, J.C. (2002) Oxidative protein folding in bacteria. *Molecular Microbiology*, **44** (1), 1–8.

16 Dailey, F.E. and Berg, H.C. (1993) Mutants in disulfide bond formation that disrupt flagellar assembly in *Escherichia coli*. *Proceedings of the National Academy of Sciences of the United States of America*, **90** (3), 1043–1047.

17 Kadokura, H. *et al.* (2004) Snapshots of DsbA in action: detection of proteins in the process of oxidative folding. *Science*, **303** (5657), 534–537.

18 Hiniker, A. and Bardwell, J.C. (2004) *In vivo* substrate specificity of periplasmic disulfide oxidoreductases. *The Journal of Biological Chemistry*, **279** (13), 12967–12973.

19 Leichert, L.I. and Jakob, U. (2004) Protein thiol modifications visualized *in vivo*. *PLoS Biology*, **2** (11), e333.

20 Vertommen, D. *et al.* (2008) The disulphide isomerase DsbC cooperates with the oxidase DsbA in a DsbD-independent manner. *Molecular Microbiology*, **67** (2), 336–349.

21 Bardwell, J.C. *et al.* (1993) A pathway for disulfide bond formation *in vivo*. *Proceedings of the National Academy of Sciences of the United States of America*, **90** (3), 1038–1042.

22 Bader, M. *et al.* (1998) Reconstitution of a protein disulfide catalytic system. *The Journal of Biological Chemistry*, **273** (17), 10302–10307.

23 Jander, G., Martin, N.L. and Beckwith, J. (1994) Two cysteines in each periplasmic domain of the membrane protein DsbB are required for its function in protein disulfide bond formation. *The EMBO Journal*, **13** (21), 5121–5127.

24 Inaba, K. and Ito, K. (2008) Structure and mechanisms of the DsbB-DsbA disulfide bond generation machine. *Biochimica et Biophysica Acta*, **1783**, 520–529.

25 Guilhot, C. *et al.* (1995) Evidence that the pathway of disulfide bond formation in *Escherichia coli* involves interactions between the cysteines of DsbB and DsbA. *Proceedings of the National Academy of Sciences of the United States of America*, **92** (21), 9895–9899.

26 Inaba, K. *et al.* (2006) Crystal structure of the DsbB-DsbA complex reveals a mechanism of disulfide bond generation. *Cell*, **127** (4), 789–801.

27 Kobayashi, T. *et al.* (1997) Respiratory chain is required to maintain oxidized

states of the DsbA-DsbB disulfide bond formation system in aerobically growing *Escherichia coli* cells. *Proceedings of the National Academy of Sciences of the United States of America*, **94** (22), 11857–11862.

28 Bader, M. *et al.* (1999) Oxidative protein folding is driven by the electron transport system. *Cell*, **98** (2), 217–227.

29 Takahashi, Y.H., Inaba, K. and Ito, K. (2004) Characterization of the menaquinone-dependent disulfide bond formation pathway of *Escherichia coli*. *The Journal of Biological Chemistry*, **279** (45), 47057–47065.

30 Inaba, K., Takahashi, Y.H. and Ito, K. (2005) Reactivities of quinone-free DsbB from *Escherichia coli*. *The Journal of Biological Chemistry*, **280** (38), 33035–33044.

31 Masip, L. *et al.* (2004) An engineered pathway for the formation of protein disulfide bonds. *Science*, **303** (5661), 1185–1189.

32 Collet, J.F. *et al.* (2005) The crystal structure of TrxA(CACA): Insights into the formation of a [2Fe-2S] iron-sulfur cluster in an *Escherichia coli* thioredoxin mutant. *Protein Science: A Publication of the Protein Society*, **14** (7), 1863–1869.

33 Berkmen, M., Boyd, D. and Beckwith, J. (2005) The nonconsecutive disulfide bond of *Escherichia coli* phytase (AppA) renders it dependent on the protein-disulfide isomerase, DsbC. *The Journal of Biological Chemistry*, **280** (12), 11387–11394.

34 Messens, J. *et al.* (2007) The oxidase DsbA folds a protein with a nonconsecutive disulfide. *The Journal of Biological Chemistry*, **282**, 31302–31307.

35 McCarthy, A.A. *et al.* (2000) Crystal structure of the protein disulfide bond isomerase, DsbC, from *Escherichia coli*. *Nature Structural Biology*, **7** (3), 196–199.

36 Zapun, A. *et al.* (1995) Structural and functional characterization of DsbC, a protein involved in disulfide bond formation in *Escherichia coli*. *Biochemistry*, **34** (15), 5075–5089.

37 Joly, J.C. and Swartz, J.R. (1997) *In vitro* and *in vivo* redox states of the *Escherichia coli* periplasmic oxidoreductases DsbA and DsbC. *Biochemistry*, **36** (33), 10067–10072.

38 Rietsch, A. *et al.* (1997) Reduction of the periplasmic disulfide bond isomerase, DsbC, occurs by passage of electrons from cytoplasmic thioredoxin. *Journal of Bacteriology*, **179** (21), 6602–6608.

39 Sun, X.X. and Wang, C.C. (2000) The N-terminal sequence (residues 1–65) is essential for dimerization, activities, and peptide binding of *Escherichia coli* DsbC. *The Journal of Biological Chemistry*, **275** (30), 22743–22749.

40 Chen, J. *et al.* (1999) Chaperone activity of DsbC. *The Journal of Biological Chemistry*, **274** (28), 19601–19605.

41 Segatori, L. *et al.* (2004) Engineered DsbC chimeras catalyze both protein oxidation and disulfide-bond isomerization in *Escherichia coli*: Reconciling two competing pathways. *Proceedings of the National Academy of Sciences of the United States of America*, **101** (27), 10018–10023.

42 Bader, M.W. *et al.* (2001) Turning a disulfide isomerase into an oxidase: DsbC mutants that, imitate DsbA. *The EMBO Journal*, **20** (7), 1555–1562.

43 Segatori, L. *et al.* (2006) Conserved role of the linker alpha-helix of the bacterial disulfide isomerase DsbC in the avoidance of misoxidation by DsbB. *The Journal of Biological Chemistry*, **281** (8), 4911–4919.

44 Rietsch, A. *et al.* (1996) An *in vivo* pathway for disulfide bond isomerization in *Escherichia coli*. *Proceedings of the National Academy of Sciences of the United States of America*, **93** (23), 13048–13053.

45 Stewart, E.J., Katzen, F. and Beckwith, J. (1999) Six conserved cysteines of the membrane protein DsbD are required for the transfer of electrons from the cytoplasm to the periplasm of *Escherichia coli*. *The EMBO Journal*, **18** (21), 5963–5971.

46 Stirnimann, C.U. *et al.* (2005) Structural basis and kinetics of DsbD-dependent cytochrome c maturation. *Structure (London, England: 1993)*, **13** (7), 985–993.

47 Thony-Meyer, L. (1997) Biogenesis of respiratory cytochromes in bacteria. *Microbiology and Molecular Biology Reviews*, **61** (3), 337–376.

48 Hiniker, A. *et al.* (2006) Evidence for conformational changes within DsbD: possible role for membrane-embedded proline residues. *Journal of Bacteriology*, **188** (20), 7317–7320.

49 Cho, S.H. and Beckwith, J. (2006) Mutations of the membrane-bound disulfide reductase DsbD that block electron transfer steps from cytoplasm to periplasm in *Escherichia coli*. *Journal of Bacteriology*, **188** (14), 5066–5076.

50 Katzen, F. and Beckwith, J. (2000) Transmembrane electron transfer by the membrane protein DsbD occurs via a disulfide bond cascade. *Cell*, **103** (5), 769–779.

51 Collet, J.F. *et al.* (2002) Reconstitution of a disulfide isomerization system. *The Journal of Biological Chemistry*, **277** (30), 26886–26892.

52 Cho, S.H. *et al.* (2007) Redox-active cysteines of a membrane electron transporter DsbD show dual compartment accessibility. *The EMBO Journal*, **26** (15), 3509–3520.

53 Rozhkova, A. *et al.* (2004) Structural basis and kinetics of inter- and intramolecular disulfide exchange in the redox catalyst DsbD. *The EMBO Journal*, **23** (8), 1709–1719.

54 Kim, J.H. *et al.* (2003) Crystal structure of DsbDgamma reveals the mechanism of redox potential shift and substrate specificity(1). *FEBS Letters*, **543** (1–3), 164–169.

55 Goulding, C.W. *et al.* (2002) Thiol-disulfide exchange in an immunoglobulin-like fold: structure of the N-terminal domain of DsbD. *Biochemistry*, **41** (22), 6920–6927.

56 Haebel, P.W. *et al.* (2002) The disulfide bond isomerase DsbC is activated by an immunoglobulin-fold thiol oxidoreductase: crystal structure of the DsbC-DsbDalpha complex. *The EMBO Journal*, **21** (18), 4774–4784.

57 Bessette, P.H. *et al.* (1999) *In vivo* and *in vitro* function of the *Escherichia coli* periplasmic cysteine oxidoreductase DsbG. *The Journal of Biological Chemistry*, **274** (12), 7784–7792.

58 Shao, F. *et al.* (2000) DsbG, a protein disulfide isomerase with chaperone activity. *The Journal of Biological Chemistry*, **275** (18), 13349–13352.

59 Hiniker, A. *et al.* (2007) Laboratory evolution of one disulfide isomerase to resemble another. *Proceedings of the National Academy of Sciences of the United States of America*, **104** (28), 11670–11675.

60 Ritz, D. and Beckwith, J. (2001) Roles of thiol-redox pathways in bacteria. *Annual Review of Microbiology*, **55**, 21–48.

61 Derman, A.I. *et al.* (1993) Mutations that allow disulfide bond formation in the cytoplasm of *Escherichia coli*. *Science*, **262** (5140), 1744–1747.

62 Bessette, P.H. *et al.* (1999) Efficient folding of proteins with multiple disulfide bonds in the *Escherichia coli* cytoplasm. *Proceedings of the National Academy of Sciences of the United States of America*, **96** (24), 13703–13708.

63 Stewart, E.J., Aslund, F. and Beckwith, J. (1998) Disulfide bond formation in the *Escherichia coli* cytoplasm: an *in vivo* role reversal for the thioredoxins. *The EMBO Journal*, **17** (19), 5543–5550.

64 Ritz, D. *et al.* (2001) Conversion of a peroxiredoxin into a disulfide reductase by a triplet repeat expansion. *Science*, **294** (5540), 158–160.

65 Yamamoto, Y. *et al.* (2008) Mutant AhpC peroxiredoxins suppress thiol-disulfide redox deficiencies and acquire deglutathionylating activity. *Molecular Cell*, **29** (1), 36–45.

66 Qiu, J., Swartz, J.R. and Georgiou, G. (1998) Expression of active human tissue-type plasminogen activator in *Escherichia coli*. *Applied and Environmental Microbiology*, **64** (12), 4891–4896.

67 Georgiou, G. and Segatori, L. (2005) Preparative expression of secreted proteins in bacteria: status report and future prospects. *Current Opinion in Biotechnology*, **16** (5), 538–545.

68 Winter, J. *et al.* (2001) Increased production of human proinsulin in the periplasmic space of *Escherichia coli* by fusion to DsbA. *Journal of Biotechnology*, **84** (2), 175–185.

69 Sevier, C.S. and Kaiser, C.A. (2006) Conservation and diversity of the cellular disulfide bond formation pathways. *Antioxidants & Redox Signaling*, **8** (5–6), 797–811.

70 Heras, B. *et al.* (2007) The name's bond. . .disulfide bond. *Current Opinion in Structural Biology*, **17** (6), 691–698.

71 Tian, G. *et al.* (2006) The crystal structure of yeast protein disulfide isomerase suggests cooperativity between its active sites. *Cell*, **124** (1), 61–73.

72 Kleizen, B. and Braakman, I. (2004) Protein folding and quality control in the endoplasmic reticulum. *Current Opinion in Cell Biology*, **16** (4), 343–349.

73 Frand, A.R. and Kaiser, C.A. (1999) Ero1p oxidizes protein disulfide isomerase in a pathway for disulfide bond formation in the endoplasmic reticulum. *Molecular Cell*, **4** (4), 469–477.

74 Gross, E. *et al.* (2004) Structure of Ero1p, source of disulfide bonds for oxidative protein folding in the cell. *Cell*, **117** (5), 601–610.

75 Tu, B.P. and Weissman, J.S. (2002) The FAD- and O(2)-dependent reaction cycle of Ero1-mediated oxidative protein folding in the endoplasmic reticulum. *Molecular Cell*, **10** (5), 983–994.

76 Sevier, C.S. *et al.* (2007) Modulation of cellular disulfide-bond formation and the ER redox environment by feedback regulation of Ero1. *Cell*, **129** (2), 333–344.

Part Three
Engineering of PTMS

Post-translational Modification of Protein Biopharmaceuticals. Edited by Gary Walsh

ISBN: 978-3-527-32074-5

12
Glycoengineering of Erythropoietin

Steve Elliott

12.1
Introduction

Red blood cell (RBC) number is primarily regulated by erythropoietin (Epo), a glycoprotein hormone produced by the kidneys in response to low oxygen (O_2) tension. Epo stimulates proliferation and blocks apoptosis of erythroid precursor cells. It also promotes differentiation of precursor cells into mature erythrocytes. These activities of Epo are mediated by binding to an Epo receptor (EpoR) present on the surface of cells. Epo binding induces a conformational change in EpoR that activates Jak2 whose phosphorylation triggers subsequent events in the signal transduction pathway.

Recombinant human erythropoietin (rHuEpo; epoetin alfa) is typically produced by expression in mammalian cells and it is commonly used for the treatment of anemia associated with chronic kidney disease [1–3] and for the treatment of anemia associated with myelosuppressive cancer chemotherapy, HIV infection, and for use in surgical situations to reduce allogeneic blood transfusion requirements. However, treatment often requires frequent injections, placing a burden on both patients and caregivers. Thus, there was a desire for new erythropoiesis stimulating molecules, allowing for new treatment options including flexible or less frequent dosing.

Secreted proteins, including erythropoietin, are often glycosylated during transit through the secretory apparatus in eukaryotic cells. These carbohydrates can be attached to the hydroxyl group on a serine or threonine (O-linked glycosylation) or to the amide of an asparagine via an *N*-glycosidic bond (N-linked glycosylation). While the biological function (binding and activation of an Epo receptor) is determined by the protein component, carbohydrate can play a role in molecular stability, solubility, *in vivo* activity, serum half-life and immunogenicity.

Manipulation of carbohydrate structures or content can affect the properties of proteins such as erythropoietin in clinically relevant ways [4]. This chapter examines the structure and function of the carbohydrate attached to Epo and methods that have been used to manipulate carbohydrate number and structures to make new molecules that may more closely match desired clinical uses.

Post-translational Modification of Protein Biopharmaceuticals. Edited by Gary Walsh

ISBN: 978-3-527-32074-5

12.2
Endogenous Epo, rHuEpo and their Attached Carbohydrates

Epo, whether from endogenous or recombinant sources used commercially, is synthesized as a 193-amino-acid precursor peptide. During secretion of the protein, a 27-amino-acid signal peptide and C-terminal arginine are removed, resulting in secretion of a protein containing 165 amino acids (Figure 12.1) [5–7]. During transit through the secretory apparatus, attachment of sugar chains to three N-linked glycosylation sites and one O-linked glycosylation site results in approximately 40% of its mass being composed of carbohydrate [5, 8].

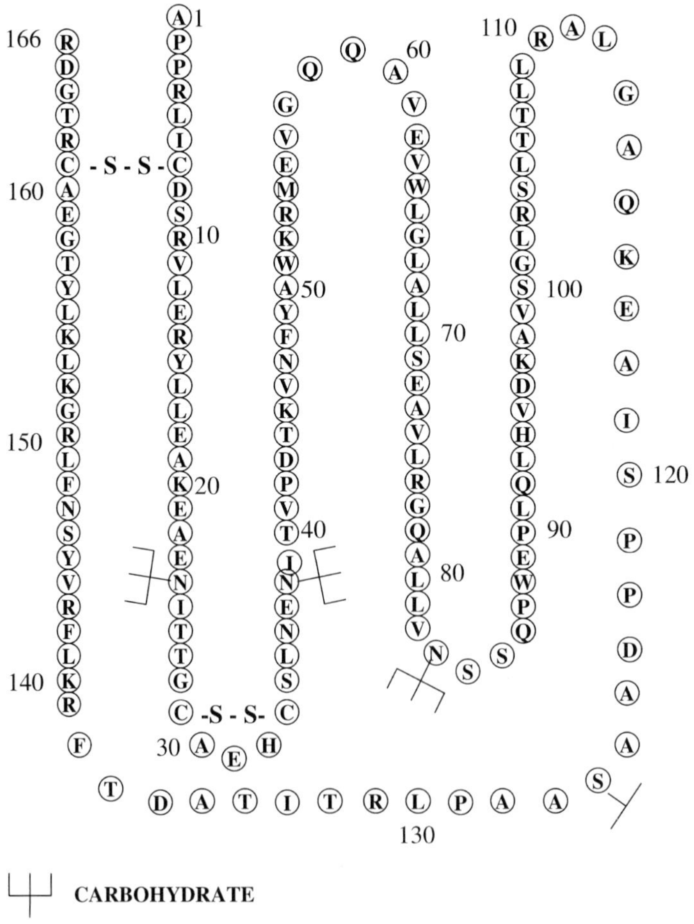

Figure 12.1 Amino acid sequence of rHuEpo. The peptide backbone contains two disulfide bonds linking Cys residues 7 to 161 and Cys 29 to 33. The three N-linked and one O-linked glycosylation sites are also indicated.

Glycopeptide mapping of carbohydrate structures on rHuEpo made in Chinese hamster ovary (CHO) cells revealed typical complex carbohydrates with >50 different forms identified [9–11]. Structures could be separated according to differences in number of sialic acids (NeuAc) as well as number of lactosamine repeats [10]. There were also differences in degree of branching, with bi-, tri- and tetra-antennary structures identified.

The relative content of the various glycoforms attached at the different sites on rHuEpo also varied. The O-linked carbohydrate at Ser126 contained two different oligosaccharide forms, both consisting of a SA-Gal-GalNAc core with either one or two attached sialic acids [10]. Carbohydrates attached to Asn83 were more homogeneous than those at other positions. These carbohydrates consisted primarily of tetra-antennary chains lacking *N*-acetyllactosamine. Carbohydrates attached to Asn 24, in contrast, were more heterogeneous and on average smaller due to an increased proportion of bi- and triantennary forms. There were also differences in content of O acetylated NeuAc forms in the various carbohydrate chains, with primarily mono and di-*O*-acetyl forms attached to Asn83 while Asn-24 and -38 contained mono-*O*-acetyl forms [9].

Endogenous Epo (eEpo) and rHuEpo also have a natural variation in charge due to the presence of a variable number of sialic acids (up to four) on each of the three N-linked carbohydrate chains and up to two sialic acids on the single O-linked carbohydrate chain. Thus, both eEpo and rHuEpo may have up to 14 sialic acids in total [12]. In complex carbohydrates from mammalian cells, sialic acid is typically attached to galactose in NeuAcα2 $\rightarrow$ 6 or 2 $\rightarrow$ 3 linkages. Recombinant human Epos (such as Epoetin alfa and beta) are made from CHO cells. As these cells lack a sialotransferase with NeuAcα2 $\rightarrow$ 6 activity only NeuAcα2 $\rightarrow$ 3 linkages are present in rHuEpo made from these cells [13].

Recombinant human Epo has the same amino acid sequence and is functionally similar to endogenous Epo (eEpo) derived from urine (uEpo) [14]. However, eEpo and rHuEpo are not identical, with differences in molecular weight and electrostatic charge described, primarily because of differences in the microheterogeneity of attached carbohydrates [15, 16]. The sugars that make up the carbohydrates on the two forms are not identical. For example, rHuEpo made in CHO cells can contain traces of *N*-glycolylneuraminic acid in addition to the typical Neu5Ac found in urinary Epo [17, 18]. rHuEpo produced in baby hamster kidney (BHK) cells contains a type 1 lactosamine chain that is not expressed on the rHuEpo (Galβ1 $\rightarrow$ β $\rightarrow$ Galβ1 $\rightarrow$ 3) group [10, 11].

Urinary Epo (uEpo) and serum Epo (sEpo) from humans also differ from rHuEpo in that both forms have a substantially higher content of sulfate. About 3% of the carbohydrate chains on Epoetin alfa contain sulfate, which is typically limited to one per chain [19]. As sulfate attachment can be extensive on eEpo, with as many as three per carbohydrate chain, this post-translational modification may have a profound affect on the physicochemical character of the protein, at least in part by conferring a negative charge at physiological pH. Thus, eEpo can be considerably more negative (acidic) than rHuEpo [19, 20]. The potential effects of this difference between natural and rHuEpo are unknown.

Recombinant HuEpo appears to be largely unmodified during circulation and filtration into urine. However, there was some differential clearance of the most basic isoforms of rHuEpo [12], resulting in a modest enrichment of the more acidic species over time [16, 21–23]; the pattern of administered rHuEpo compared with rHuEpo excreted in urine was not substantially altered. As a result, eEpo as secreted in the urine can be differentiated from administered rHuEpo by IEF gels.

12.2.1
Rules for Attachment of Carbohydrates

The nature of the signal for carbohydrate addition is partially understood. N-linked carbohydrate addition is mediated by oligosaccharide transferase and occurs at asparagine residues that are part of the consensus sequence Asn-Xxx-Ser/Thr where Xxx can be any amino acid, except proline [24–27]. The attached oligosaccharide then undergoes enzymatic trimming and maturation via the action of glycosidases and glycosyltransferases that trim and append additional sugar units to the terminal mannose.

The observation that not all consensus sequences in secreted proteins are glycosylated, suggests that there are additional sequence or conformational requirements essential for efficient carbohydrate attachment [28–30]. While at least 12–14 amino acids must be synthesized and have entered the luminal surface of the endoplasmic reticulum for carbohydrate addition, the synthesis of the protein need not be completed for glycosylation to take place [31, 32]. This suggests that the structures for carbohydrate addition are recognized in partially folded molecules [33].

The sequence context of the glycosylation site has also been shown to influence the efficiency of glycosylation. For example, it has been reported that Pro at the +3 position relative to Asn or Asp at +1 can inhibit glycosylation [25, 29, 34, 35]. Distance from the C-terminus may also affect glycosylation [36].

To understand the rules for N-linked carbohydrate addition on rHuEpo and to aid in construction of new molecules with additional carbohydrates, an *in vitro* mutagenesis approach was used to introduce N or O-linked glycosylation sites into various positions on the amino acid backbone of rHuEpo [28, 37]. The modified genes were introduced into mammalian cells and the secreted rHuEpo analogs were examined for addition of N- and O-linked carbohydrate. The effects of adjacent amino acids on glycosylation were also examined.

Sixty two different rHuEpo glycosylation analogs were constructed with N-linked consensus sequences (Asn-Xxx-Ser/Thr) introduced into 47 different positions and the effects on carbohydrate addition were examined. Additional N-linked carbohydrates were detected with analogs containing consensus sequences introduced at 15 of the 47 positions examined (32%). This proportion was similar to the proportion of glycosylated consensus sequences compared to total N-linked consensus sequences in secreted proteins [29, 30]. The fact that additional carbohydrate addition was not detected at most consensus sequence sites (68%) indicated that the presence of a consensus sequence is necessary but not sufficient for N-linked carbohydrate addition. There must be a sequence or structural context present in functional glycosylation sites that is compatible with carbohydrate addition.

Analysis of the specific sites at which carbohydrate addition occurred in rHuEpo hyperglycosylated analogs (those containing additional numbers of carbohydrate chains) indicated that glycosylation was able to take place within some regions in alpha helices, loops and bends within the tertiary structure of the molecule. It is likely that this is a reflection of the particular orientation and spacing of the asparagine relative to serine or threonine residues within the glycosylation consensus sequence. For example, β or Asn-X turns are secondary structures adopted by functional N-linked glycosylation sites [38] and these structures may be contained within some positions in an alfa helix. A likely structure is a β- or Asx turn that positions the hydroxyl group a particular distance from the amide.

A threonine at the third position of introduced consensus sequences resulted in increased efficiency of carbohydrate addition for N-linked carbohydrate addition compared to a serine residue in agreement with other reports [34, 39–42]. These results suggest that the hydroxyl group may play a direct role in oligosaccharide transfer to the protein. In this scenario, the hydroxyl in a threonine residue is much more effective at catalyzing the transfer. Alternatively, the threonine may affect the local stereochemistry of the glycosylation site, thus affecting recognition and transfer of oligosaccharides to the protein by the oligosaccharyl transferase [38].

It has been observed previously that a proline residue in the Xxx position of the Asn-Xxx-Ser/Thr consensus sequence reduces the likelihood of N-linked carbohydrate addition [34]. In addition, a Pro at the −1 position can inhibit carbohydrate addition. Notably however, there exist glycosylation sites with a Pro at −1 relative to Asn in consensus sequences in proteins that are glycosylated [25]. With introduced consensus sequences, Pro at −1 was shown to inhibit carbohydrate addition and removal by *in vitro* mutagenesis increased oligosaccharyl transfer [28].

Consensus sequences added at or near the O-linked glycosylation site in rHuEpo (Ser126) did not attract additional N-linked glycosylation. This region is exposed on the surface of the molecule so the lack of additional carbohydrate addition could not be explained by lack of exposure of this region to the glycosylation machinery. Instead the lack may be due to the extended conformation and flexibility of this region [43].

With both effective and ineffective consensus sequences there was no correlation between predicted surface exposure of the asparagine in the consensus sequence and the presence of additional carbohydrate in the mature glycoprotein. This suggests that carbohydrate addition is co-translational (i.e., precedes folding). This is further supported by the observation that molecules with carbohydrate attached to sites normally buried in the secreted molecule were destabilized because the attached carbohydrate could not be accommodated without a change in structure [28].

O-linked carbohydrates are added in the Golgi to Ser or Thr residues within a less defined sequence context than that determined for N-linked glycosylation sites. To define the structural requirements for addition of O-linked glycosylation *in vivo*, 33 different O-linked rHuEpo glycosylation analogs were constructed by introducing Ser/Thr residues into the rHuEpo molecule by *in vitro* mutagenesis. Most positions that had an introduced Ser or Thr had no evidence of O-linked carbohydrate addition.

However, variants with Thr mutations at amino acid positions 123 and 125 contained additional O-linked carbohydrate, which suggests that several positions around the existing O-linked glycosylation site (Ser 126) contain the necessary information for O-linked carbohydrate addition. Two forms of the Thr125 variant were identified. One form was glycosylated only at residue 125, and a second form was glycosylated at both the introduced Thr125 and existing Ser126, indicating that multiple *O*-glycosylated forms could be constructed. Prolines at −1 and +1 relative to the *O*-glycosylation site enhanced *O*-glycosylation.

12.2.2
Carbohydrate Glycoforms and their Effect on Structure and Activity

The carbohydrate on rHuEpo is essential for *in vivo* activity. Removal of one or more N-linked chains by either enzymatic deglycosylation or construction of glycosylation analogs that removed glycosylation sites by *in vitro* mutagenesis, thereby generating molecules lacking specific chains, resulted in a decrease in *in vivo* activity that was proportional to the number of carbohydrates removed [11, 44–47]. *In vivo* activity of *Escherichia coli* produced rHuEpo entirely lacking carbohydrate was reduced by about 1000-fold. The effect was not site specific as removal of any of the chains resulted in decreased *in vivo* activity.

In concert with carbohydrate removal, molecules with increased content or increased numbers of carbohydrate chains had increased *in vivo* activity and the increase was proportional to the number of chains added [48]. "Carbohydrate shuffling" experiments were performed whereby a new chain was added to a molecule that initially lacked one of the three N-linked chains by introduction of a functional N-linked glycosylation site into molecules that had deletions of an existing glycosylation site [28]. These all had similar *in vivo* potencies, indicating that the benefit of added carbohydrate was largely independent of the position to which a new carbohydrate chain was added but, rather, was dependent on the number of total glycans.

The mechanism by which carbohydrate affected *in vivo* activity has been the subject of considerable study. One hypothesis was that carbohydrate enhanced binding to the EpoR, explaining the increased *in vivo* activity. However, glycosylation analogs that had decreased *in vivo* activity due to removal of carbohydrates or partial enzymatic deglycosylation had increased receptor binding and *in vitro* activity [44, 49, 50]. Expression of rHuEpo in *E. coli* where no carbohydrate is added resulted in a >10-fold increase in both receptor binding and *in vitro* biological activity and this rHuEpo derivative had the lowest *in vivo* activity [49, 51]. In concert with these results, analogs with increased numbers of N-linked chains due to introduction of functional N-linked glycosylation sites (see below) had increased *in vivo* activity despite decreased receptor binding and *in vitro* activity. Overall there was a direct relationship between receptor binding activity and *in vitro* activity but an inverse relationship between receptor binding activity and *in vivo* activity. Thus, a positive effect of carbohydrate on the interaction of Epo to EpoR could not explain the effect of carbohydrate on *in vivo* activity.

The loss of *in vivo* activity with rHuEpo analogs engineered to lack carbohydrate was partially explained by some loss of protein stability [51–55]. However, the molecules completely lacking carbohydrate retained a normalized conformation, as detected in receptor binding and *in vitro* bioassays [51]. In addition, increased stability did not explain the increased *in vivo* activity described for molecules with increased carbohydrate content [49, 56, 57]. Thus the inverse relationship between carbohydrate and *in vivo* activity of ESAs cannot be attributed to structural stability.

The sialic acid component of carbohydrate of rHuEpo was found to play an inhibitory role on clearance, as evidenced by an almost complete loss of *in vivo* biological activity following neuraminidase treatment [58, 59]. Demonstration that the exposed sugars following desialylation (i.e., galactose) were rapidly bound by a liver expressed asialoglycoprotein receptor led to the hypothesis that a role of sialic acid may be to prevent clearance of rHuEpo by this receptor [60]. While sialic acid may inhibit binding to ASGR this does not explain the increased activity of glycosylation analogs containing additional carbohydrate.

Studies on glycoforms of rHuEpo containing different sialic acid contents demonstrated a direct relationship between increased sialic acid content and increased *in vivo* activity [48, 57]. The increased *in vivo* activity was due to an increased serum half-life of the molecule and not increased receptor affinity [48, 57]. Thus it became apparent that increased *in vivo* activity was a consequence of increased serum half-life effected by increased carbohydrate content and resulting in prolonged exposure to the agent [49]. The prolonged stimulation of erythropoiesis resulted in increased hemoglobin response and thus increased apparent activity.

To determine if it was carbohydrate *per se* or the amount of sialic acid on carbohydrate that was responsible for the increased activity, rHuEpo was separated into populations of molecules according to charge and thus sialic acid content. These "isoform" preparations were examined for their *in vivo* potency in mice. Increasing sialic acid content was found to have a direct and positive effect on *in vivo* activity [48]. There was an inverse relationship between carbohydrate content in general and sialic acid content specifically and receptor binding activity [46, 48, 49]. The reduced receptor binding activity was a consequence of charge repulsion between negatively charged sialic acid residues and negatively charged residues on EpoR at the Epo–EpoR interface [49, 61]. The serum half-lives of rHuEpo preparations that had differences in sialic acid content were also examined and there was a direct and positive relationship between sialic acid content and serum half-life [48].

12.2.3
Glycoengineering of New Molecules – Darbepoetin Alfa

Darbepoetin alfa was created by increasing content of sialic acid-containing carbohydrate on rHuEpo, resulting in increased *in vivo* activity. The theoretical maximum number of sialic acids on rHuEpo is 14 (up to four sialic acids for each of the three N-linked carbohydrates and up to two sialic acids for the O-linked carbohydrate) [9]. Thus it was hypothesized that introduction of consensus sequences into the rHuEpo peptide backbone might result in new molecules with additional carbohydrate and thus increased *in vivo* potency [56]. Each new N-linked chain could add up to four additional sialic acids. It became apparent that simply adding an N-linked consensus sequence to rHuEpo would not be sufficient. The changes needed to be introduced in such a way that the resultant molecule was efficiently glycosylated and retained activity, conformation and stability.

To increase the likelihood of success, the amino-acid changes were introduced into regions of the molecule distal to the receptor-binding site to ensure that the molecule could efficiently engage and activate the EpoR. This effort was aided by structure–function studies that defined the active sites of rHuEpo and identified amino acids important for maintenance of structure [62, 63].

Sixty two different analogs with N-linked consensus sequences introduced into 48 different positions were constructed and examined to determine if they met the criteria of efficient glycosylation and retention of *in vitro* activity [28, 56]. While most introduced consensus sequences were not glycosylated efficiently, some of those that were glycosylated had poor biological activity due to reduced ability to bind to the Epo receptor. One explanation was that the introduced sequences were at or near to the receptor binding site, thereby compromising the ability of the analog to bind to the Epo receptor. Alternatively, the amino acid substitutions altered conformation or protein stability, thereby reducing the ability to attach to an Epo receptor. However, some introduced sequences at some positions were glycosylated efficiently, and retained *in vitro* biological activity (e.g., A30N/H32T). These results indicated that it is possible to construct glycosylation analogs that have efficient glycosylation and retention of *in vitro* activity.

Another important criterion for analog selection was retention of an Epo-like conformation. The monoclonal antibody 9G8a binds an Epo sequence that is partially buried in the interior of the properly folded molecule. Unfolding or aggregation of rHuEpo increased antibody binding to the partially denatured protein [43, 64, 65]. 9G8a binding was also increased with certain amino acid substitutions even when they had no effect on *in vitro* bioactivity, indicating that this antibody could detect subtle changes in conformation. 9G8a binding was therefore used as a sensitive surrogate assay for conformation and stability of the Epo glycosylation analogs. Some of the glycosylation analogs had low or near normal binding to 9G8a, including the A30N/H32T analog. Thus this analog met all the preferred criteria of efficient glycosylation, normal *in vitro* activity and near normal conformation/stability [56].

One analog with a consensus sequence at position 88, Asn88 Thr90, was not glycosylated at this position. However, an additional P87/S substitution (Ser87 Asn88 Thr90) was efficiently glycosylated and retained *in vitro* bioactivity (Table 12.1), indicating that it was possible to convert an inactive consensus sequence into a functional N-linked glycosylation site [28, 56]. However, the protein with this sequence had a substantial increase in 9G8A immunoreactivity relative to the rHuEpo standard (Table 12.1). Other analogs with additional substitutions in this region were examined to determine if any would meet both the retention of bioactivity and low 9G8a immunoreactivity selection criteria. Several analogs, including V87N88T90 and A87N88T90, N30T32, N55T57 and N114T116, had low

Table 12.1 Naturally occurring amino acids in rHuEpo are shown in the top row. Substitutions for each analog are shown. A twofold increase in 9G8A immunoreactivity compared to rHuEpo is suggestive of an altered conformation. Val87 substitutions allow carbohydrate addition at position 88 and normalization of the conformation.

Analog/mutations									N-linked chains	RIA-9G8A (% of control)
Ala 30	His 32	Glu 55	Gly 57	Pro 87	Trp 88	Pro 90	Ala114	Lys116*	3	100
					Asn 88	Thr 90			3	390
				Ser 87	**Asn 88**	Thr 90			4	780
				Val 87	**Asn 88**	Ser 90			3.5	280
				Ala 87	**Asn 88**	Thr 90			4	180
				Val 87	**Asn 88**	Thr 90			4	120
				Aranesp					5	80
				AMG205					7	36

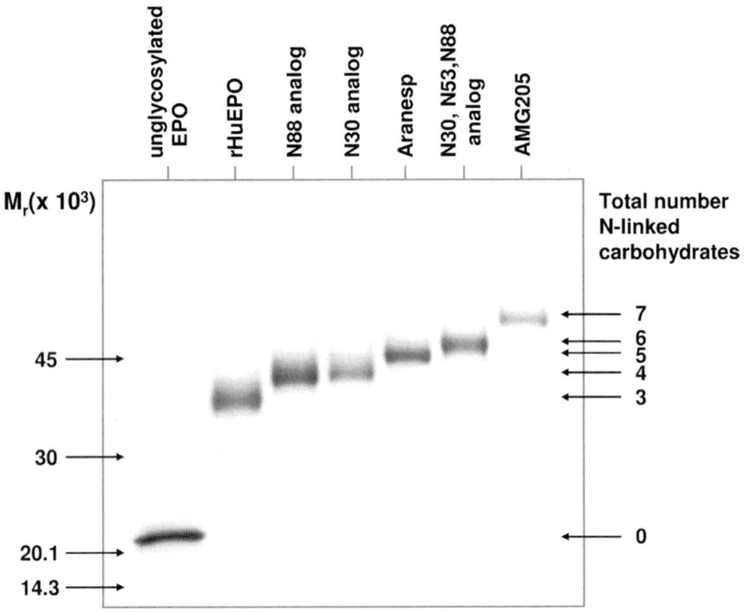

Figure 12.2 Glycoengineered rHuEpo glycosylation analogs were expressed in Chinese hamster ovary cells and the secreted ESA was purified and subjected to gel electrophoresis. Unglycosylated Epo was prepared by expression in *E. coli* followed by purification and refolding.

9G8a immunoreactivity, retained near normal *in vitro* and receptor binding activity and were glycosylated efficiently (Table 12.1) and were evaluated further.

Combinations of these and other consensus sequences into one molecule were examined. The size of the hyperglycosylated erythropoiesis stimulating agent (ESA) increased with increasing numbers of glycosylation sites, and analogs with up to four additional N-linked carbohydrates were identified (Figure 12.2). Two particular consensus sequences (Asn30 Thr32 and Val87 Asn88 Thr90) were combined to generate a new molecule with two additional N-linked carbohydrates. This molecule (Aranesp, darbepoetin alfa) had near-normal *in vitro* activity, was glycosylated efficiently and had a similar conformation and stability to rHuEpo. The carbohydrate content was increased from 40% to 51%, the size from approximately 30 400 Da to approximately 37 100 Da and the maximum number of sialic acids was increased from 14 to 22 [56]. The carbohydrate structures found on the new N-linked chains had the same composition as was found on the naturally occurring glycosylation sites. Thus the engineered sites were recognized and glycosylated similarly.

Clinical testing was performed and darbepoetin alfa had an approximately three-fold longer mean terminal half-life than rHuEpo (25.3 vs. 8.5 h, respectively) in rHuEpo-naive patients receiving peritoneal dialysis, when administered IV [66].

There was also a more than twofold greater AUC (291 ± 8 vs. 132 ± 8 ng h mL^{-1}), and a 2.5-fold lower clearance (1.6 ± 0.3 vs. 4.0 ± 0.3 mL h^{-1} kg^{-1}), which was biphasic. The volume of distribution was similar for epoetin alfa and darbepoetin alfa (48.7 ± 2.1 and 52.4 ± 2.0 mL kg^{-1}, respectively). The mean terminal half-life for darbepoetin alfa SC was longer than for IV administration (~49 h); C_{max} averaged about 10% of the IV value, T_{max} averaged 54 ± 5 h, and the mean relative bioavailability was 37%.

With repeated dosing up to once monthly, administration and DA was able to stably maintain Hb levels [67, 69]. The safety and adverse event profiles of darbepoetin alfa and Epoetin alfa were similar, suggesting that the glycoengineering did not adversely affect the molecule or its mechanism of action. The increased dosing intervals resulted in increased convenience for patients and caregivers.

12.2.4
Glycoengineering of New Molecules – AMG114

It was found that carbohydrate content could be increased even further than that contained on darbepoetin alfa. AMG205 an rHuEpo glycosylation analog with four additional carbohydrates compared to epoetin alfa had low 9G8a immunoreactivity, indicating intact conformation (Table 12.1). Its *in vivo* potency and serum half-life were increased compared to DA. In rats, AMG205 had nearly twice the half-life of DA (IV, 31 vs. 17.2 h; SC, 28.2 vs. 13.2 h, respectively) [70]. However, AMG205 had an O-linked glycosylation deficit, which could contribute to low production yield of fully glycosylated protein. Another candidate, AMG114, with a Ser126 to Thr substitution, exhibited increased efficiency of O-linked carbohydrate addition. It also had low 9G8a immunoreactivity, which is consistent with a similar structure/conformational stability to rHuEpo, as well as retention of *in vitro* activity. The serum half-life in rats was similar to that of AMG205 (16.7 h). In dogs the serum half life of AMG114 administered IV was 68.9 h compared to 20.4 h for darbepoetin alfa.

The carbohydrates on AMG114 (and AMG205) are opposite to the two chains found on darbepoetin alfa (Figure 12.3). The areas of the molecule changed are distal to the region of rHuEpo that is involved in the interaction with the Epo receptor, explaining the retention of *in vitro* activity of both darbepoetin alfa and AMG114.

Clinical testing in CIA patients with AMG114 revealed an approximately twofold increase in serum half-life compared to darbepoetin alfa (131 vs. 73.7 h) and a dose-dependent increase in Hb levels when administered at Q3W dosing intervals [71]. Thus serum half-life could be increased in humans even further than that observed with darbepoetin alfa with more carbohydrates, and such molecules could maintain Hb levels in humans.

AMG114 has not been developed for use clinically primarily because existing molecules, including Epoetin alfa & darbepoetin alfa, can meet clinical needs. In addition there was concern that molecules with very long half-lives might be more difficult to use clinically because of a potential increase in swings in Hb levels.

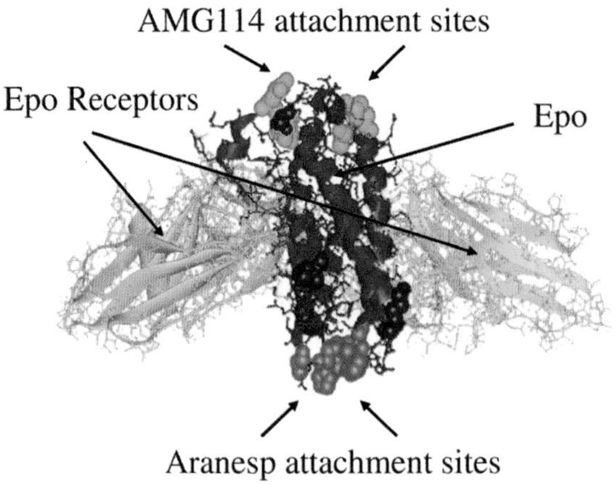

Figure 12.3 A three-dimensional model of rHuEpo (red) based on the crystal structure of Epo bound to the extracellular domains of two EpoR (yellow). Naturally occurring carbohydrate addition sites on Epo are purple. Amino acids changed to make AMG114 and darbepoetin alfa are green and dark green, respectively. The AMG114 glycosylation sites are at the end opposite the added glycosylation sites in darbepoetin alfa. Both regions are largely devoid of exiting carbohydrates, allowing "room" for additional carbohydrates. All the new N-linked carbohydrates in AMG114, including those in darbepoetin alfa, are distal from the Epo receptor binding sites, resulting in minimal interference with intrinsic activity.

12.2.5
Glycoengineered ESAs and Biological Activity

Darbepoetin alfa had an approximate 2.2-fold increase in *in vivo* activity in the mouse polycythemic ^{59}Fe uptake assay [72]. This result indicated that additional carbohydrates, beyond that found naturally, could increase the *in vivo* activity of glycoengineered rHuEpo analogs. The relative potency of darbepoetin alfa compared with rHuEpo increased when the dosing interval was extended. At a TIW dosing schedule, three times more rHuEpo was needed than darbepoetin alfa to maintain or elicit a similar erythropoietic response in normal mice [57, 72]. When the dosing interval was extended to QW administration, the difference increased to 13-fold. With a single injection, the dose of rHuEpo needed to be 30–40 times higher to match the effect of the lower dose of darbepoetin alfa. In humans receiving rHuEpo or darbepoetin alfa QW, the dose of rHuEpo needed to be 45% higher than darbepoetin alfa to maintain Hb levels in CKD patients [73]. The differences in relative potencies appeared to be explained by the observed threefold longer serum half-life of darbepoetin alfa, allowing for a prolonged period of time for erythroid cell exposure to the ESA [49].

AMG114 also had increased *in vivo* potency compared to rHuEpo and darbepoetin alfa. When the activity of AMG114 was compared with darbepoetin alfa in mice, a single injection of 10 ug kg^{-1} of AMG114 was comparable to 100 ug kg^{-1} of darbepoetin alfa. Thus there was a further increase in *in vivo* potency in this assay when the number of carbohydrate chains was increased compared to darbepoetin alfa.

Compared with rHuEpo, darbepoetin alfa has an approximately four- to fivefold lower Epo receptor binding activity and a similar decrease in *in vitro* biological activity [4, 49, 57]. The observation that the difference in receptor binding activity between rHuEpo and darbepoetin alfa could be eliminated by desialylation indicated that the different receptor binding activity was due to different sialic acid content and not amino acid, conformation or carbohydrate differences [49]. Sialic acid may reduce EpoR binding by altering electrostatic interactions between Epo and the EpoR [61]. The interface between Epo and EpoR includes contacts between positively charged amino acids on EPO and negatively charged amino acids on EpoR [64]. The fact that sialic acid is negatively charged at physiological pH suggests that sialic acid may decrease binding via charge repulsion.

The apparent paradox, that glycosylation analogs with lower receptor binding activity had increased *in vivo* activity, is explained by the counteracting effects of sialic acid-containing carbohydrate on clearance [49, 57, 72]. Slower clearance *in vivo* results in an increase in relative serum concentration compared to that of a faster clearing molecule, and the relative concentration differences between slow and fast clearing molecules will increase with time. The prolonged exposure of the target cell to the drug thus results in increased *in vivo* activity. Because reduced receptor binding affinity is overcome by increased concentration of the ligand, slower clearance will more than compensate for its reduced receptor binding affinity.

12.3 Effect of Carbohydrate on Clearance, Mechanism of Clearance

Clearance of ESAs was first thought to be mediated primarily through the liver and kidney or via the EpoR on receptor-expressing cells. Desialylated rHuEpo is rapidly cleared by hepatocytes via an asialoglycoprotein receptor (ASGR) that is expressed in the liver [60, 74, 75], suggesting that a role of sialic acid may be to prevent clearance by this receptor. Indeed perfusion experiments demonstrated that asialoEpo could be efficiently bound following a single pass through the liver, demonstrating that a high capacity clearance pathway was in effect. Clearance of rHuEpo via ASGR would require a two-step process, desialylation and subsequent clearance by ASGR in liver, but there is no evidence that desialylation of rHuEpo occurs *in vivo*. Furthermore, hepatectomy experiments in sheep did not reduce clearance of rHuEpo [76], liver perfusion experiments with intact rHuEpo showed minimal clearance [75, 76] and patients with liver disease showed no difference in rHuEpo clearance rates compared with healthy volunteers [77]. Consequently, the role of the liver in rHuEpo clearance is thought to be minimal and the correlation between carbohydrate content and clearance was not explained by its impact on clearance via ASGR.

rHuEpo and other ESAs are found in the urine, suggesting that the kidney is a clearance pathway [16, 21, 22]. However, initial data suggesting a major elimination pathway for rHuEpo through the kidneys [87] were not supported by subsequent studies showing that rHuEpo clearance is independent of kidney function [76, 88, 89]. In healthy men, only a small amount of intact radiolabelled epoetin beta (<5% of the dose) was found to be excreted in the urine, suggesting that rHuEpo is degraded elsewhere in the body [90]. One such pathway may be degradation or metabolism in the interstitium, such as via cells involved in the reticuloendothelial scavenging pathway or lymphatic system [91]. Consistent with this hypothesis, the lymphatic system plays an important role in the reduced bioavailability after SC administration of proteins. In addition, only small peptides or free ^{125}I and not intact material are detected in tissues following IV administration of ^{125}I-darbepoetin alfa, suggesting that degradation may occur in tissue [86]. Hyperglycosylation increased hydrodynamic size due to attached hydrated carbohydrate and the reduced clearance of hyperglycosylated molecules might be explained by steric factors. Thus a possible pathway impacted by hyperglycosylation may be the transport of ESAs from the blood to the interstitial fluid where degradation takes place and this transport may be reduced as the size of the molecule increases.

Binding of Epo to the EpoR can lead to cellular internalization or to dissociation of the molecule from the receptor [78–81]. Once the ligand–receptor complex is internalized, the ligand may be degraded, or it may be recycled intact to the cell surface and released. Consistent with Epo receptor-mediated endocytosis and degradation in the bone marrow is the observation that darbepoetin alfa, which has a lower affinity for the receptor than rHuEpo, has reduced clearance [4, 57]. Reports of a more substantial role for this pathway *in vivo* is supported by indirect evidence, mostly derived from chemotherapy experiments [69, 82–84].

However, chemotherapy may result in nonspecific inhibition of elimination pathways, such as destruction via macrophages or neutrophils that may mediate some clearance of proteins. Furthermore, the hypothesis that ESAs are cleared by an Epo receptor-mediated pathway is contradicted by data from Piroso and colleagues [85], who showed no difference in the life span of rHuEpo in rats with hypo- and hyperplastic bone marrows, concluding that it was unlikely that erythroid mass, and therefore Epo receptors, played a dominant role in rHuEpo catabolism.

Darbepoetin alfa has other biophysical characteristics, such as increased size and carbohydrate content, suggesting that the reduced clearance could be explained by other mechanisms. To evaluate the role of Epo receptor-mediated metabolism on clearance, an engineered rHuEpo analog (NM385) that was devoid of detectable receptor binding activity but retained similar structure and carbohydrate content to rHuEpo was examined [86]. Following a single IV dose of $30\,\mu g\,kg^{-1}$ in rats, NM385 had a slightly longer terminal half-life but similar clearance compared to rHuEpo. However, the impact on clearance of complete elimination of receptor binding activity on NM385 was substantially less than that observed with hyperglycosylated ESAs. Indeed, the addition of one additional N-linked carbohydrate chain on rHuEpo impacted clearance to a greater degree than complete loss of receptor binding activity. Although the available data, taken together, suggest that Epo receptor-mediated pathways may play some role in clearance of ESAs, non-Epo receptor-mediated pathways play a substantial role in the clearance of these proteins and account for most of the elimination from the body.

In analyzing the effects of hyperglycosylation of ESAs it was apparent that it was not reduced receptor binding alone that reduced clearance but rather addition of size. Reduced binding affinity of darbepoetin alfa and PEG-rHuEpo was a casualty of adding size due to the modifications. If receptor-mediated endocytosis is not the major mechanism of ESA clearance, then it is apparent that other pathway(s) contribute to the elimination of ESAs *in vivo*.

Summary and Conclusion

Secreted proteins are often glycosylated during transit through the secretory apparatus in eukaryotic cells. These carbohydrates can be attached to the hydroxyl group on a serine or threonine (O-linked glycosylation), or the amine of an asparagine via an *N*-glycosidic bond (N-linked glycosylation). Human erythropoietin is a 165 amino acid glycoprotein with three N-linked carbohydrates attached to asparagines at amino acid positions 24, 38 and 83, and one O-linked carbohydrate attached to Ser126 [14, 49]. The addition of carbohydrate chains to the polypeptide backbone of rHuEpo may have an impact on the structure, solubility, antigenicity, folding, secretion and stability of the protein. The structural impact of carbohydrate addition can be deduced by analyzing glycosylation analogs. In this manner structural changes in O-linked and N-linked glycosylation of rHuEpo have been determined through experimentation. The importance of the carbohydrate on rHuEpo has been demonstrated by the ability to increase the *in vivo* activity due to an increase in serum half-life of the protein by adding more N-linked carbohydrates to the molecule. Novel hyperglycosylated ESAs based on this approach (darbepoetin alfa and AMG114) have been tested and/or marketed.

References

1 Winearls, C.G., Forman, E., Wiffen, P. and Oliver, D.O. (1989) Recombinant human erythropoietin treatment in patients on maintenance home haemodialysis. *Lancet*, **2**, 569.

2 Eschbach, J.W., Egrie, J.C., Downing, M.R., Browne, J.K. and Adamson, J.W. (1987) Correction of the anemia of end-stage renal disease with recombinant human erythropoietin. Results of a combined phase I and II clinical trial. *New England Journal of Medicine*, **316**, 73–78.

3 Eschbach, J.W., Kelly, M.R., Haley, N.R., Abels, R.I. and Adamson, J.W. (1989) Treatment of the anemia of progressive renal failure with recombinant human erythropoietin. *New England Journal of Medicine*, **321**, 158–163.

4 Sinclair, A.M. and Elliott, S. (2005) Glycoengineering: the effect of glycosylation on the properties of therapeutic proteins. *Journal of Pharmaceutical Sciences*, **94**, 1626–1635.

5 Lai, P.H., Everett, R., Wang, F.F., Arakawa, T. and Goldwasser, E. (1986) Structural characterization of human erythropoietin. *Journal of Biological Chemistry*, **261**, 3116–3121.

6 Lin, F.K., Suggs, S., Lin, C.H., Browne, J.K., Smalling, R., Egrie, J.C., Chen, K.K., Fox, G.M., Martin, F. and Stabinsky, Z. (1985) Cloning and expression of the human erythropoietin gene. *Proceedings of the National Academy of*

Sciences of the United States of America, **82**, 7580–7584.

7 Lin, F.K., Lin, C.H., Lai, P.H., Browne, J.K., Egrie, J.C., Smalling, R., Fox, G.M., Chen, K.K., Castro, M. and Suggs, S. (1986) Monkey erythropoietin gene: cloning, expression and comparison with the human erythropoietin gene. *Gene*, **44**, 201–209.

8 Browne, J.K., Cohen, A.M., Egrie, J.C., Lai, P.H., Lin, F.-K., Strickland, T., Watson, E. and Stebbing, N. (1986) Erythropoietin: gene cloning, protein structure, and biological properties. *Cold Spring Harbor Symposia on, Quantitative Biology* **51**, 693–702.

9 Rush, R.S., Derby, P.L., Smith, D.M., Merry, C., Rogers, G., Rohde, M.F. and Katta, V. (1995) Microheterogeneity of erythropoietin carbohydrate structure. *Analytical Chemistry*, **67**, 1442–1452.

10 Sasaki, H., Bothner, B., Dell, A. and Fukuda, M. (1987) Carbohydrate structure of erythropoietin expressed in Chinese hamster ovary cells by a human erythropoietin cDNA. *Journal of Biological Chemistry*, **262**, 12059–12076.

11 Takeuchi, M. and Kobata, A. (1991) Structures and functional roles of the sugar chains of human erythropoietins. *Glycobiology*, **1**, 337–346.

12 Egrie, J. and Browne, J. (2002) Darbepoetin alfa is more potent in vivo and can be administered less frequently than rHuEPO. *British Journal of Cancer*, **87**, 476–477.

13 Takeuchi, M., Takasaki, S., Miyazaki, H., Kato, T., Hoshi, S., Kochibe, N. and Kobata, A. (1988) Comparative study of the asparagine-linked sugar chains of human erythropoietins purified from urine and the culture medium of recombinant Chinese hamster ovary cells. *Journal of Biological Chemistry*, **263**, 3657–3663.

14 Egrie, J.C., Strickland, T.W,, Lane, J., Aoki, K., Cohen, A.M., Smalling, R., Trail, G., Lin, F.K., Browne, J.K. and Hines, D.K. (1986) Characterization and biological effects of recombinant human erythropoietin. *Immunobiology*, **172**, 213–224.

15 Wide, L. and Bengtsson, C. (1990) Molecular charge heterogeneity of human serum erythropoietin. *British Journal of Haematology*, **76**, 121–127.

16 Lasne, F., Martin, L., Crepin, N. and De Ceaurriz, J. (2002) Detection of isoelectric profiles of erythropoietin in urine: Differentiation of natural and administered recombinant hormones. *Analytical Biochemistry*, **311**, 119–126.

17 Noguchi, A., Mukuria, C.J., Suzuki, E. and Naiki, M. (1996) Failure of human immunoresponse to *N*-glycolylneuraminic acid epitope contained in recombinant human erythropoietin. *Nephron*, **72**, 599–603.

18 Hokke, C.H., Bergwerff, A.A., Van Dedem, G.W., van, O.J., Kamerling, J.P. and Vliegenthart, J.F. (1990) Sialylated carbohydrate chains of recombinant human glycoproteins expressed in Chinese hamster ovary cells contain traces of *N*-glycolylneuraminic acid. *FEBS Letters*, **275**, 9–14.

19 Strickland, T., Adler, B., Aoki, K., Asher, S., Derby, P., Goldwasser, E. and Rogers, G. (1992) Occurrence of sulfate on the N-linked oligosaccharides of human erythropoietin. *Journal of Cellular Biochemistry*, (Supplement 16D), 167.

20 Kawasaki, N., Haishima, Y., Ohta, M., Itoh, S., Hyuga, M., Hyuga, S. and Hayakawa, T. (2001) Structural analysis of sulfated N-linked oligosaccharides in erythropoietin. *Glycobiology*, **11**, 1043–1049.

21 Breidbach, A., Catlin, D.H., Green, G.A., Tregub, I., Truong, H. and Gorzek, J. (2003) Detection of recombinant human erythropoietin in urine by isoelectric focusing. *Clinical Chemistry*, **49**, 901–907.

22 Catlin, D.H., Breidbach, A., Elliott, S. and Glaspy, J. (2002) Comparison of the isoelectric focusing patterns of darbepoetin alfa, recombinant human erythropoietin, and endogenous erythropoietin from human urine. *Clinical Chemistry*, **48**, 2057–2059.

23 Lasne, F., Martin, L., Martin, J.A. and De Ceaurriz, J. (2007) Isoelectric profiles of human erythropoietin are different in serum and urine. *International Journal of Biological, Macromolecules*, **41**, 354–357.

24 Bause, E. (1983) Structural requirements of *N*-glycosylation of proteins. Studies with proline peptides as conformational probes. *Biochemical Journal*, **209**, 331–336.

25 Roitsch, T. and Lehle, L. (1989) Structural requirements for protein *N*-glycosylation. Influence of acceptor peptides on cotranslational glycosylation of yeast invertase and site-directed mutagenesis around a sequon sequence. *European Journal of Biochemistry*, **181**, 525–529.

26 Imperiali, B. and Shannon, K.L. (1991) Differences between Asn-Xaa-Thr-containing peptides: a comparison of solution conformation and substrate behavior with oligosaccharyltransferase. *Biochemistry*, **30**, 4374–4380.

27 Berg, D.T. and Grinnell, B.W. (1993) Pro to Gly (P219G) in a silent glycosylation site results in complete glycosylation in tissue plasminogen activator. *Protein Science*, **2**, 126–127.

28 Elliott, S., Chang, D., Delorme, E., Eris, T. and Lorenzini, T. (2004) Structural requirements for additional N-linked carbohydrate on recombinant human erythropoietin. *Journal of Biological Chemistry*, **279**, 16854–16862.

29 Gavel, Y. and von Heijne, G. (1990) Sequence differences between glycosylated and non-glycosylated Asn-X-Thr/Ser acceptor sites: implications for protein engineering. *Protein Engineering*, **3**, 433–442.

30 Apweiler, R., Hermjakob, H. and Sharon, N. (1999) On the frequency of protein glycosylation, as deduced from analysis of the SWISS-PROT database. *Biochimica et Biophysica Acta*, **1473**, 4–8.

31 Nilsson, I.M. and von Heijne, G. (1993) Determination of the distance between the oligosaccharyltransferase active site and the endoplasmic reticulum membrane. *Journal of Biological Chemistry*, **268**, 5798–5801.

32 Chen, W., Helenius, J., Braakman, I. and Helenius, A. (1995) Cotranslational folding and calnexin binding during glycoprotein synthesis. *Proceedings of the National Academy of Sciences of the United States of America*, **92**, 6229–6233.

33 Kornfeld, R. and Kornfeld, S. (1985) Assembly of asparagine-linked oligosaccharides. *Annual Review of Biochemistry*, **54**, 631–664. A review with 283 references.

34 Bause, E. and Hettkamp, H. (1979) Primary structural requirements for *N*-glycosylation of peptides in rat liver. *FEBS Letters*, **108**, 341–344.

35 Shakin-Eshleman, S.H., Spitalnik, S.L. and Kasturi, L. (1996) The amino acid at the X position of an Asn-X-Ser sequon is an important determinant of N-linked core-glycosylation efficiency. *Journal of Biological Chemistry*, **271**, 6363–6366.

36 Shakin-Eshleman, S.H., Wunner, W.H. and Spitalnik, S.L. (1993) Efficiency of N-linked core glycosylation at asparagine-319 of rabies virus glycoprotein is altered by deletions C-terminal to the glycosylation sequon. *Biochemistry*, **32**, 9465–9472.

37 Elliott, S., Bartley, T., Delorme, E., Derby, P., Hunt, R., Lorenzini, T., Parker, V., Rohde, M.F. and Stoney, K. (1994) Structural requirements for addition of O-linked carbohydrate to recombinant erythropoietin. *Biochemistry*, **33**, 11237–11245.

38 Imperiali, B. and Hendrickson, T.L. (1995) Asparagine-linked glycosylation: specificity and function of oligosaccharyl transferase. *Bioorganic & Medicinal Chemistry*, **3**, 1565–1578.

39 Breuer, W., Klein, R.A., Hardt, B., Bartoschek, A. and Bause, E. (2001) Oligosaccharyltransferase is highly specific for the hydroxy amino acid in Asn-Xaa-Thr/Ser. *FEBS Letters*, **501**, 106–110.

40 Kasturi, L., Eshleman, J.R., Wunner, W.H. and Shakin-Eshleman, S.H. (1995) The hydroxy amino acid in an Asn-X-Ser/Thr sequon can influence N-linked core glycosylation efficiency and the level of

expression of a cell surface glycoprotein. *Journal of Biological Chemistry*, **270**, 14756–14761.

41 Nishikawa, A. and Mizuno, S. (2001) The efficiency of N-linked glycosylation of bovine DNase I depends on the Asn-Xaa-Ser/Thr sequence and the tissue of origin. *Biochemical Journal*, **355**, 1–8.

42 Nicolaes, G.A., Villoutreix, B.O. and Dahlback, B. (1999) Partial glycosylation of Asn2181 in human factor V as a cause of molecular and functional heterogeneity. Modulation of glycosylation efficiency by mutagenesis of the consensus sequence for N-linked glycosylation. *Biochemistry*, **38**, 13584–13591.

43 Cheetham, J.C., Smith, D.M., Aoki, K.H., Stevenson, J.L., Hoeffel, T.J., Syed, R.S., Egrie, J. and Harvey, T.S. (1998) NMR structure of human erythropoietin and a comparison with its receptor bound conformation. *Nature Structural Biology*, **5**, 861–866.

44 Delorme, E., Lorenzini, T., Giffin, J., Martin, F., Jacobsen, F., Boone, T. and Elliott, S. (1992) Role of glycosylation on the secretion and biological activity of erythropoietin. *Biochemistry*, **31**, 9871–9876.

45 Tsuda, E., Kawanishi, G., Ueda, M., Masuda, S. and Sasaki, R. (1990) The role of carbohydrate in recombinant human erythropoietin. *European Journal of Biochemistry*, **188**, 405–411.

46 Yamaguchi, K., Akai, K., Kawanishi, G., Ueda, M., Masuda, S. and Sasaki, R. (1991) Effects of site-directed removal of N-glycosylation sites in human erythropoietin on its production and biological properties. *Journal of Biological Chemistry*, **266**, 20434–20439.

47 Dordal, M.S., Wang, F.F. and Goldwasser, E. (1985) The role of carbohydrate in erythropoietin action. *Endocrinology*, **116**, 2293–2299.

48 Egrie, J.C. and Browne, J.K. (2001) Development and characterization of novel erythropoiesis stimulating protein (NESP). *Nephrology Dialysis Transplantation*, **16** (Suppl-13), 3–13.

49 Elliott, S., Egrie, J., Browne, J., Lorenzini, T., Busse, L., Rogers, N. and Ponting, I. (2004) Control of rHuEPO biological activity: the role of carbohydrate. *Experimental Hematology*, **32**, 1146–1155.

50 Higuchi, M., Oh-eda, M., Kuboniwa, H., Tomonoh, K., Shimonaka, Y. and Ochi, N. (1992) Role of sugar chains in the expression of the biological activity of human erythropoietin. *Journal of Biological Chemistry*, **267**, 7703–7709.

51 Narhi, L.O., Arakawa, T., Aoki, K., Wen, J., Elliott, S., Boone, T. and Cheetham, J. (2001) Asn to Lys mutations at three sites which are *N*-glycosylated in the mammalian protein decrease the aggregation of Escherichia coli-derived erythropoietin. *Protein Engineering*, **14**, 135–140.

52 Toyoda, T., Itai, T., Arakawa, T., Aoki, K.H. and Yamaguchi, H. (2000) Stabilization of human recombinant erythropoietin through interactions with the highly branched *N*-glycans. *Journal of Biochemistry*, **128**, 731–737.

53 Toyoda, T., Arakawa, T. and Yamaguchi, H. (2002) *N*-Glycans stabilize human erythropoietin through hydrophobic interactions with the hydrophobic protein surface: studies by surface plasmon resonance analysis. *Journal of Biochemistry*, **131**, 511–515.

54 Narhi, L.O., Arakawa, T., Aoki, K.H., Elmore, R., Rohde, M.F., Boone, T. and Strickland, T.W. (1991) The effect of carbohydrate on the structure and stability of erythropoietin. *Journal of Biological Chemistry*, **266**, 23022–23026.

55 Endo, Y., Nagai, H., Watanabe, Y., Ochi, K. and Takagi, T. (1992) Heat-induced aggregation of recombinant erythropoietin in the intact and deglycosylated states as monitored by gel permeation chromatography combined with a low-angle laser light scattering technique. *Journal of Biochemistry*, **112**, 700–706.

56 Elliott, S., Lorenzini, T., Asher, S., Aoki, K., Brankow, D., Buck, L., Busse, L., Chang, D., Fuller, J., Grant, J., Hernday, N.,

Hokum, M., Hu, S., Knudten, A., Levin, N., Komorowski, R., Martin, F., Navarro, R., Osslund, T., Rogers, G., Rogers, N., Trail, G. and Egrie, J. (2003) Enhancement of therapeutic protein in vivo activities through glycoengineering. *Nature Biotechnology*, **21**, 414–421.

57 Egrie, J.C., Dwyer, E., Browne, J.K., Hitz, A. and Lykos, M.A. (2003) Darbepoetin alfa has a longer circulating half-life and greater in vivo potency than recombinant human erythropoietin. *Experimental Hematology*, **31**, 290–299.

58 Takeuchi, M., Takasaki, S., Shimada, M. and Kobata, A. (1990) Role of sugar chains in the in vitro biological activity of human erythropoietin produced in recombinant Chinese hamster ovary cells. *Journal of Biological Chemistry*, **265**, 12127–12130.

59 Goldwasser, E., Kung, C.K. and Eliason, J. (1974) On the mechanism of erythropoietin-induced differentiation. 13. The role of sialic acid in erythropoietin action. *Journal of Biological Chemistry*, **249**, 4202–4206.

60 Spivak, J.L. and Hogans, B.B. (1989) The in vivo metabolism of recombinant human erythropoietin in the rat. *Blood*, **73**, 90–99.

61 Darling, R.J., Kuchibhotla, U., Glaesner, W., Micanovic, R., Witcher, D.R. and Beals, J.M. (2002) Glycosylation of erythropoietin affects receptor binding kinetics: role of electrostatic interactions. *Biochemistry*, **41**, 14524–14531.

62 Elliott, S., Lorenzini, T., Chang, D., Barzilay, J. and Delorme, E. (1997) Mapping of the active site of recombinant human erythropoietin. *Blood*, **89**, 493–502.

63 Elliott, S., Lorenzini, T., Chang, D., Barzilay, J., Delorme, E., Giffin, J. and Hesterberg, L. (1996) Fine-structure epitope mapping of antierythropoietin monoclonal antibodies reveals a model of recombinant human erythropoietin structure. *Blood*, **87**, 2702–2713.

64 Syed, R.S., Reid, S.W., Li, C., Cheetham, J.C., Aoki, K.H., Liu, B., Zhan, H., Osslund, T.D., Chirino, A.J., Zhang, J., Finer-Moore, J., Elliott, S., Sitney, K., Katz, B.A., Matthews, D.J., Wendoloski, J.J., Egrie, J. and Stroud, R.M. (1998) Efficiency of signalling through cytokine receptors depends critically on receptor orientation. *Nature*, **395**, 511–516.

65 Elliott, S., Chang, D., Delorme, E., Dunn, C., Egrie, J., Giffin, J., Lorenzini, T., Talbot, C. and Hesterberg, L. (1996) Isolation and characterization of conformation sensitive antierythropoietin monoclonal antibodies: effect of disulfide bonds and carbohydrate on recombinant human erythropoietin structure. *Blood*, **87**, 2714–2722.

66 Macdougall, I.C., Gray, S.J., Elston, O., Breen, C., Jenkins, B., Browne, J. and Egrie, J. (1999) Pharmacokinetics of novel erythropoiesis stimulating protein compared with epoetin alfa in dialysis patients. *Journal of the American Society of, Nephrology*, **10**, 2392–2395.

67 Nissenson, A.R. (2001) Novel erythropoiesis stimulating protein for managing the anemia of chronic kidney disease. *American Journal of Kidney Diseases*, **38**, 1390–1397.

68 Nissenson, A.R., Swan, S.K., Lindberg, J.S., Soroka, S.D., Beatey, R., Wang, C., Picarello, N., McDermott-Vitak, A. and Maroni, B.J. (2002) Randomized, controlled trial of darbepoetin alfa for the treatment of anemia in hemodialysis patients. *American Journal of Kidney Diseases*, **40**, 110–118.

69 Glaspy, J., Henry, D., Patel, R., Tchekmedyian, S., Applebaum, S., Berdeaux, D., Lloyd, R., Berg, R., Austin, M., Rossi, G. and Darbepoetin, A. (2005) Effects of chemotherapy on endogenous erythropoietin levels and the pharmacokinetics and erythropoietic response of darbepoetin alfa: a randomised clinical trial of synchronous versus asynchronous dosing of darbepoetin alfa. *European Journal of Cancer*, **41**, 1140–1149.

70 Elliott, S., Aoki, K., Agoram, B., Brankow, D., Doshi, S. and Molineux, G. (2006) Evaluation of hyperglycosylated erythropoiesis stimulating proteins developed using glycoengineering. *Proceedings of the American Association for Cancer Research*, **47**, abstract 2176.

71 Osterborg, A., De Boer, R., Clemens, M., Renczes, G., Kotasek, D., Prausova, J., Marschner, N., Hedenus, M., Hendricks, L. and Amado, R. (2006) A novel erythropoiesis-stimulating agent (AMG114) with 131-hour half-life effectively treats chemotherapy-induced anemia when administered as 200 mcg every 3 weeks. *Journal of Clinical Oncology*, **24**, 8626.

72 Sasu, B.A., Hartley, C., McElroy, T., Khaja, R., Elliott, S., Egrie, J.C., Browne, J.K., Begley, C.G. and Molineux, G. (2004) Has the Unit of erythropoietic activity outlived its usefulness? *Blood*, **102**, 4386.

73 Tolman, C., Richardson, D., Bartlett, C. and Will, E. (2005) Structured conversion from thrice weekly to weekly erythropoietic regimens using a computerized decision-support system: a randomized clinical study. *Journal of the American Society of Nephrology* **16**, 1463–1470.

74 Fukuda, M.N., Sasaki, H., Lopez, L. and Fukuda, M. (1989) Survival of recombinant erythropoietin in the circulation: the role of carbohydrates. *Blood*, **73**, 84–89.

75 Dinkelaar, R.B., Engels, E.Y., Hart, A.A., Schoemaker, L.P., Bosch, E. and Chamuleau, R.A. (1981) Metabolic studies on erythropoietin (EP): II. The role of liver and kidney in the metabolism of Ep. *Experimental Hematology*, **9**, 796–803.

76 Widness, J.A., Veng-Pedersen, P., Schmidt, R.L., Lowe, L.S., Kisthard, J.A. and Peters, C. (1996) In vivo 125I-erythropoietin pharmacokinetics are unchanged after anesthesia, nephrectomy and hepatectomy in sheep. *Journal of Pharmacology & Experimental Therapeutics*, **279**, 1205–1210.

77 Jensen, J.D., Jensen, L.W., Madsen, J.K. and Poulsen, L. (1995) The metabolism of erythropoietin in liver cirrhosis patients compared with healthy volunteers. *European Journal of Haematology*, **54**, 111–116.

78 Sawyer, S.T., Krantz, S.B. and Goldwasser, E. (1987) Binding and receptor-mediated endocytosis of erythropoietin in Friend virus-infected erythroid cells. *Journal of Biological Chemistry*, **262**, 5554–5562.

79 Sawada, K., Krantz, S.B., Sawyer, S.T. and Civin, C.I. (1988) Quantitation of specific binding of erythropoietin to human erythroid colony-forming cells. *Journal of Cellular Physiology*, **137**, 337–345.

80 Fraser, J.K., Lin, F.K. and Berridge, M.V. (1988) Expression of high affinity receptors for erythropoietin on human bone marrow cells and on the human erythroleukemic cell line, HEL. *Experimental Hematology*, **16**, 836–842.

81 Gross, A.W. and Lodish, H.F. (2006) Cellular trafficking and degradation of erythropoietin and novel erythropoiesis stimulating protein (NESP). *Journal of Biological Chemistry*, **281**, 2024–2032.

82 Chapel, S., Veng-Pedersen, P., Hohl, R.J., Schmidt, R.L., McGuire, E.M. and Widness, J.A. (2001) Changes in erythropoietin pharmacokinetics following busulfan-induced bone marrow ablation in sheep: evidence for bone marrow as a major erythropoietin elimination pathway. *Journal of Pharmacology & Experimental Therapeutics*, **298**, 820–824.

83 Veng-Pedersen, P., Chapel, S., Al-Huniti, N.H., Schmidt, R.L., Sedars, E.M., Hohl, R.J. and Widness, J.A. (2004) Pharmacokinetic tracer kinetics analysis of changes in erythropoietin receptor population in phlebotomy-induced anemia and bone marrow ablation. *Biopharmaceutics & Drug Disposition*, **25**, 149–156.

84 Hartley, C., Elliott, S., Begley, C.G., McElroy, P., Sutherland, W., Khaja, R., Heatherington, A.C., Graves, T., Schultz, H., Del Castillo, J. and Molineux, G. (2003) Kinetics of haematopoietic recovery after dose-intensive chemo/radiotherapy in mice: optimized erythroid support with darbepoetin alpha.

British Journal of Haematology, **122**, 623–636.

85 Piroso, E., Erslev, A.J., Flaharty, K.K. and Caro, J. (1991) Erythropoietin life span in rats with hypoplastic and hyperplastic bone marrows. *American Journal of Hematology*, **36**, 105–110.

86 Agoram, B., Molineux, G., Jang, G., Aoki, K., Gegg, L., Narhi, L. and Elliott, S. (2007) Effects of altered receptor binding activity on the clearance of erythropoiesis-stimulating proteins: a minor role of erythropoietin receptor-mediated pathways. *Nephrology Dialysis Transplantation*, **21**, iv303–iv304.

87 Fu, J.-S., Lertora, J.J.L., Brookins, J., Rice, J.C. and Fisher, J.W. (1988) Pharmacokinetics of erythropoietin in intact and anephric dogs. *Journal of Laboratory and Clinical, Medicine* **111**, 669–676.

88 Kindler, J., Eckardt, K.U., Ehmer, B., Jandeleit, K., Kurtz, A., Schreiber, A., Scigalla, P. and Sieberth, H.G. (1989) Single-dose pharmacokinetics of recombinant human erythropoietin in patients with various degrees of renal failure. *Nephrology Dialysis Transplantation*, **4**, 345–349.

89 Macdougall, I.C., Roberts, D.E., Coles, G.A. and Williams, J.D. (1991) Clinical pharmacokinetics of epoetin (recombinant human erythropoietin). *Clinical Pharmacokinetics*, **20**, 99–113.

90 Yoon, W.H., Park, S.J., Kim, I.C. and Lee, M.G. (1997) Pharmacokinetics of recombinant human erythropoietin in rabbits and 3/4 nephrectomized rats. *Research Communications in Molecular Pathology & Pharmacology*, **96**, 227–240.

91 Flaharty, K.K. (1990) Clinical pharmacology of recombinant human erythropoietin (r-HuEPO). *Pharmacotherapy*, **10**, 9S–14S.

13
Glycoengineering: Cerezyme as a Case Study

Scott M. Van Patten and Tim Edmunds

13.1
Introduction

A major hurdle in developing effective biotherapeutics is efficient targeting of the therapeutic molecule to the specific tissue and cell type where it is needed to reverse the pathology being treated. Protein therapies are generally administered intravenously and face several obstacles, such as the endothelial barrier itself, in reaching their desired physiological target. In addition, the body has multiple mechanisms for the clearance of proteins (such as the asialoglycoprotein receptor in liver) that can make it difficult to maintain sufficient concentrations in serum for long enough periods of time to be therapeutically effective.

Ceredase and Cerezyme are enzyme therapies that have been developed to replace the enzyme that is deficient in the rare genetic disorder known as Gaucher disease. They were the first of what are now several enzyme replacement therapies for the class of diseases referred to as lysosomal storage diseases. Of particular relevance in the current context is the fact that Ceredase was also the first biotherapeutic on which the carbohydrate portion was glycoengineered to target the enzyme to a specific cell type to improve its efficacy. This chapter describes the development of these two enzyme therapies and discusses possible alternative glycoengineering strategies that have been considered in developing future treatments for this disease.

13.2
Basis for Glycan-Directed Enzyme Replacement Therapy for LSDs

13.2.1
Lysosomal Storage Diseases

Lysosomal storage diseases (LSDs) are a group of over 40 progressive, inherited diseases that result from a deficiency in the activity of one or more of the proteins involved in the catabolic pathway within the lysosome, typically a lysosomal

Post-translational Modification of Protein Biopharmaceuticals. Edited by Gary Walsh

ISBN: 978-3-527-32074-5

enzyme [1–3]. While each of these diseases is considered rare, as a group their incidence is about 1 in 7700 live births [4]. Each can result from various mutations in the gene for the deficient enzyme, leading to a build-up of the substrate for that enzyme in the lysosomes of multiple tissues. The resulting pathology varies, depending on the organs and cell types in which this accumulation occurs. For example, Fabry disease is caused by a deficiency in α-galactosidase A, leading to accumulation of globotriaosylceramide (GL-3) primarily in capillary endothelial cells within the kidney, heart and skin. In Pompe disease, in contrast, deficiency in the lysosomal enzyme acid α-glucosidase results in a build-up of glycogen in multiple tissues, the pathology being associated primarily with accumulation in myocytes of skeletal and cardiac muscle. Thus, for each of these diseases the organ(s) and cell-type (s) that must be targeted for therapy differs and must be defined.

13.2.2 Gaucher Disease

Gaucher disease was first described in 1882 by Philippe Gaucher in a 32-year old female patient having an enlarged spleen [5]. It is an autosomal-recessive genetic disorder and is the most common of the lysosomal storage diseases. It is nevertheless a very rare disease (<10 000 cases worldwide) but is particularly prevalent in the Ashkenazi Jewish population, where it occurs with a frequency as high as one in every 400 live births [6]. Early work identified the cause of this disease to be lysosomal accumulation of glucosylceramide, a membrane glycolipid that is deposited in the lysosome primarily as a result of the breakdown of senescent circulating red and white blood cells phagocytosed by macrophages [7]. However, it was not until the early 1960s that the pioneering efforts of Roscoe Brady and colleagues at the NIH, as well as Desmond Patrick in the United Kingdom, identified the cause of this accumulation to be a deficiency in the activity of the lysosomal enzyme acid β-glucosidase (also known as glucocerebrosidase or GCase) [8, 9]. This also gave the first hints at a possible therapy for the disease.

Clinically, Gaucher disease can present with varying degrees of severity, depending on residual levels of enzyme activity, and is classified as one of three types based on symptoms and age of onset (Table 13.1). The most severe forms of the disease are referred to as type 2 and type 3 (acute and sub-acute neuronopathic, respectively) and involve pathological accumulation of substrate both viscerally and in the central nervous system. Type 1 is a less severe but more common form of the disease that has no primary CNS involvement (non-neuronopathic). The major sites of substrate accumulation in type 1 Gaucher disease are liver, spleen and bone (and less frequently lung), leading to hepatosplenomegaly, anemia, cytopenias, growth retardation and bone pain and fractures. This occurs primarily in cells of the reticuloendothelial system (macrophages) within these tissues: Kupffer cells, splenic macrophages and osteoclasts, respectively. Cells can become so engorged that the substrate deposits take on a fibrillar and almost crystalline appearance. These cells are referred to as Gaucher cells [10] and are the primary target in developing an effective therapy for this disease.

Table 13.1 Gaucher disease variants. (Reproduced with permission from Reference [26].)

Clinical features	Non-neuronopathic	Neuronopathic	
	Type 1	Type 2	Type 3
Age at onset	Childhood/adulthood	Infancy	Childhood
Lifespan	Infancy to 80+ years	Infancy	Childhood-middle age
Hepatosplenomegaly	Yes	Yes	Yes
Skeletal disease	Yes	No	Yes
Primary CNS disease	No	Yes	Yes
Predominant ethnic group	Ashkenazi Jewish	Pan-ethnic	Pan-ethnic

13.2.3 Glucocerebrosidase

Glucocerebrosidase (GCase) is the final exoglycosidase in the removal of carbohydrate from glycosphingolipids in the lysosome, catalyzing the breakdown of glucosylceramide to glucose and ceramide. GCase is a 497 amino acid membrane-associated monomeric glycoprotein of ~67 kDa and has a pH optimum for enzymatic activity of ~5.5, as would be expected for a lysosomal enzyme. The active site residues directly responsible for catalysis have been shown to be glutamic acids at positions 235 and 340 [11, 12]. Purified GCase is activated by negatively charged phospholipids *in vitro* and corresponding membrane lipids are thought to play a role in the enzyme's activity *in vivo* [6]. Normal physiological activity of GCase *in vivo* also requires the presence of a small molecular weight (9 kDa) heat-stable protein cofactor, saposin C. Mutations in saposin C can lead to glucosylceramide accumulation and a very rare form of Gaucher disease [13].

The X-ray crystal structure of glucocerebrosidase has now been solved in various contexts by several groups [14–19]. It reveals a classic $(\beta/\alpha)_8$ TIM barrel catalytic core (Domain III) typical of the GH-A group of lysosomal glycosidases as well as two additional domains composed entirely of β-sheets. The smaller of these domains contains the N-terminus of the protein (Domain I) while the larger of the two is an IgG-like domain containing the C-terminus (Domain II). Using a structure-based docking model it was recently suggested that the saposin C activator interacts with GCase through two of its TIM barrel helices and the IgG-like domain [20]. The enzyme contains seven cysteines, four of which are involved in two disulfide bonds within Domain I. The three remaining cysteines are all near the active site but two (C126 and C248) have been shown not to be involved in catalytic activity based on mutagenesis studies [17].

More than 200 mutations in the gene for GCase (GBA) have been identified in Gaucher patients [5]. These mutations lead to amino acid changes throughout the structure of the protein [14], indicating the critical nature of the entire protein sequence in the normal function of this enzyme. The effect of 52 of these mutations on the enzymatic behavior of GCase *in vitro* (activity and response to inhibitors/activators) as well as its sensitivity to protease was recently evaluated using baculovirus-expressed proteins [17]. The most common mutation found in Gaucher patients results in the substitution of a serine for an asparagine at amino acid residue 370 in the hinge α-helix between the catalytic and IgG-like domains (N370S). This mutation accounts for ~70% of the mutant alleles in the Ashkenazi Jewish population (~45% in non-Jewish patients) and is associated with the less severe type 1 form of the disease. A second common mutation results in the substitution of a proline for a leucine at position 444 within the hydrophobic core of the IgG-like domain (L444P) and is associated with the more severe neuronopathic forms. It has repeatedly been shown that there is substantial variability in phenotype for any given genotype however [21, 22].

Glucocerebrosidase has five potential glycosylation sites within its amino acid sequence, only four of which are utilized (Asn19, Asn59, Asn146 and Asn270). Glycosylation of N19 in Domain I is required for correct folding of the protein into an active form of the enzyme; however, once it has folded it is possible to remove all carbohydrate and maintain enzymatic activity [23, 24]. Enzyme isolated from human placenta was shown to contain a mixture of bi-and triantennary complex oligosaccharides with partially sialylated termini [25]. In addition, the placental enzyme contained ~20% oligomannose structures and it was later demonstrated that this oligomannose was present exclusively on the N19 glycosylation site, primarily as Man_6 [26].

Most lysosomal proteins traffic to the lysosome as a result of the addition of mannose 6-phosphate (M6P) to one or more of their glycosylation sites in the Golgi [27]; however, GCase is an exception to this rule. It is able to reach the lysosome in I-cell disease fibroblasts despite their inability to transfer M6P to lysosomal proteins [28] and GCase purified from human tissue does not contain M6P [25, 29]. In addition, its trafficking to the lysosome takes place independent of N-linked glycosylation [30, 31]. Recently it was demonstrated that the transporter responsible for delivery of GCase to the lysosome is the integral membrane protein LIMP-2 [32]. These studies showed that in LIMP-2 deficient cells GCase is secreted rather than sorted to the lysosome, and lysosomal transport could be rescued by expression of LIMP-2.

13.2.4
Enzyme Replacement as a Therapy for LSDs

The first reference to the possibility of treating lysosomal storage diseases by replacing the defective enzyme has been attributed to Christian de Duve [5, 33, 34], who reasoned that any substance endocytosed by a cell was likely to end up in the lysosome. It was shortly after this that the connection between accumulation of glucosylceramide in Gaucher disease and a deficiency of glucocerebrosidase activity led to the suggestion that enzyme replacement therapy might serve as an effective treatment for Gaucher disease [35]. As indicated above, the key to such a therapy is to deliver active enzyme to the lysosomes of affected cells in the appropriate tissues for that particular LSD.

One strategy that has been successful in this regard is to direct enzyme therapies to the lysosome via binding to mannose 6-phosphate receptor (MPR) on the surface of target cells. As discussed above, most lysosomal enzymes acquire mannose 6-phosphate in the Golgi and are transported to the lysosome as a result of their binding to this receptor. Two different MPRs have been described: the cation-dependent MPR (CD-MPR; 46 kDa) and the cation-independent MPR (CI-MPR; $\sim$300 kDa) [27, 36]. Both are involved in trafficking of enzymes to the lysosome, with the CI-MPR handling most of this traffic. Both MPRs also cycle through the plasma membrane [37, 38] but only the CI-MPR is capable of binding and internalizing lysosomal enzymes [39]. This receptor is expressed on various cell types and has been effectively employed in targeting enzyme replacement therapy for Fabry disease [40, 41] as well as Pompe disease [42, 43]. As previously discussed, glucocerebrosidase does not sort to the lysosome via the MPR pathway and does not contain M6P. This, together with the fact that the target cell for therapy is the macrophage, suggests that this strategy is not appropriate for targeting enzyme replacement therapy in Gaucher disease.

13.3
Use of Placental Glucocerebrosidase for ERT

13.3.1
Initial Clinical Studies with Unmodified GCase Isolated from Placenta

Although enzyme replacement therapy (ERT) was suggested shortly after the identification of the deficient enzyme in 1965 [35], purification of glucocerebrosidase from tissue proved to be very challenging, primarily due to its hydrophobicity. It was not until eight years later that Brady and coworkers were able to isolate enzyme of sufficient purity from human placenta [44] to allow dosing of the first two Gaucher patients. Despite the relatively small amount of enzyme given to these patients there was significant reduction in liver glucosylceramide after only one dose [45].

To generate enough material for larger trials, a modified purification procedure was developed [46]; however, subsequent clinical studies with this material were disappointing (reviewed in Reference [47]). While studies on patient biopsy samples indicated the purified enzyme could be taken up by the affected tissues and cause breakdown of glucosylceramide [48], animal studies showed that the vast majority was being taken up *in vivo* by hepatocytes rather than the desired target in the liver, Kupffer cells [49]. This was likely due to incomplete sialylation of the complex carbohydrate chains on the enzyme and the resulting uptake into hepatocytes through binding of the exposed galactose to asialoglycoprotein receptor on these cells.

Other groups attempted different strategies for delivering purified glucocerebrosidase to the affected macrophages of Gaucher patients. Using their own purification procedure [50] Beutler *et al.* packaged the enzyme into erythrocyte ghosts coated with gamma globulin in an effort to direct it to macrophages, with limited clinical success [51]. Gregoriadis *et al.* used liposome encapsulation of the enzyme to promote macrophage uptake, again with limited success [52, 53].

13.3.2
Identification of Mannose Receptor and its Role in Macrophage Uptake

The breakthrough in terms of more efficiently targeting glucocerebrosidase to the appropriate cells for therapy came with the identification of a cell-surface receptor for terminal mannose residues of glycoproteins. This receptor was identified on macrophages, both in the lung [54] and in the liver [55], where it was suggested this might serve as a useful tool in delivering enzyme to affected cells of the reticuloendothelial system in Gaucher patients.

Based on this finding, multiple strategies were developed in an attempt to redirect glucocerebrosidase to macrophages via its mannose receptor. One such strategy focused on the covalent attachment of structures containing terminal mannose to the purified enzyme. In the first instance, linear pentamannosyl chains were attached using cyanoborohydride chemistry [56]; however, this modification provided no improvement in macrophage uptake in animals. A second approach involved attachment of the small, synthetic glycopeptide trimannosyldilysine (Man_3Lys_2) to multiple primary amino groups in the enzyme (lysine) via carbodiimide chemistry. This did result in some increase in Kupffer cell uptake of the modified placental enzyme in rat liver [57, 58] but this modest increase was deemed insufficient to provide consistent clinical benefit [47].

13.3.3
Glycoengineering via Sequential Removal of Glycans

The targeting strategy that did ultimately provide a path to a clinically effective therapy was to modify the carbohydrate structures already present on the purified placental enzyme in a way that redirected it to the macrophage mannose receptor. Scott Furbish and John Barranger [59] used exoglycosidases to sequentially remove monosaccharides from the complex carbohydrate structures on purified GCase, administered the modified enzyme to animals and looked at the effect these modifications had on uptake of the enzyme into different cell-types within the liver.

As mentioned above, unmodified glucocerebrosidase is taken up primarily by hepatocytes, likely due to incomplete sialylation and thus binding of the exposed galactose to asialoglycoprotein receptor on these cells. Not surprisingly, removal of sialic acid on the placental enzyme by treatment with neuraminidase increased targeting to hepatocytes and decreased the amount taken up by non-parenchymal cells, primarily endothelial and Kupffer cells (Table 13.2). However, subsequent removal of galactose using β-galactosidase increased non-parenchymal cell uptake dramatically and this was further enhanced by removal of *N*-acetylglucosamine with hexosaminidase. The use of these three carbohydrate-remodeling enzymes therefore provided the conceptual basis for redirecting glucocerebrosidase and creating an effective therapy for Gaucher disease (Figure 13.1).

Table 13.2 Effect of carbohydrate remodeling on liver cell uptake. (Reproduced with permission from Reference [26].)

	Cellular distribution (units GCase activity per 10^6 cells)	
Treatment	**Non-parenchymal**	**Hepatocytes**
Control (no enzyme)	6.6	104
None	48	256
Neuraminidase	16	337
Neuraminidase and galactosidase	170	380
Neuraminidase, galactosidase and hexoseaminidase	245	290

13.3.4 Development of First ERT for Gaucher Disease – Ceredase

Once it had been established which form of glucocerebrosidase would be used for development of a therapeutic, several challenges remained. Since at that point the most practical source of the enzyme was human tissue, a source of human placenta sufficient to support clinical trials and commercial production was needed. Ultimately ~20 000 placentas were required to generate enough enzyme to treat one patient for one year [26], so this was not a trivial concern.

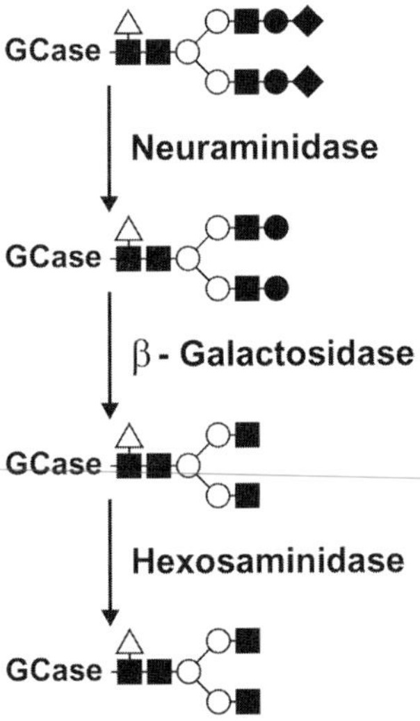

Figure 13.1 Scheme for sequential remodeling of complex carbohydrate chains on glucocerebrosidase. *N*-acetylglucosamine (■), mannose (○), galactose (●), sialic acid (♦) and fucose (△).

A second major challenge was scaling up the extraction and purification of glucocerebrosidase from placenta to the level needed for clinical trials and production. The initial purification process developed by Brady *et al.* [46] was modified to improve both the purification efficiency and the purity of the final product [26]. Purity was important not only from the standpoint of removal of other human proteins, but also removal or inactivation of any virus that might be present in the large pool of placental tissue. Contaminating human proteins were somewhat less of an issue than with current recombinant protein therapies made in non-human organisms because they would be less likely to be immunogenic.

The carbohydrate remodeling process that had been shown to be so critical for efficient targeting of the enzyme [59] also had to be scaled up. This required large-scale production of the three enzymes involved in the remodeling (neuraminidase, β-galactosidase and hexosaminidase) as well as modification of the purification process to ensure their removal. This removal was particularly important for the remodeling enzymes since they are derived from non-human organisms and would be more likely to be immunogenic. Sensitive assays were developed to ensure that this was the case, and the effectiveness of the final purification design at removing these and any other possible immunogenic proteins was later demonstrated by the low level of immune response in the clinic, even in patients who had received multiple doses of greater than 100 mg [60].

A final challenge was formulating the carbohydrate-remodeled glucocerebrosidase in such a way as to keep it stable for long periods of time. Since it is normally a membrane-associated lysosomal protein, GCase can lose activity and aggregate under conditions where other enzymes might be stable, such as neutral pH. It was therefore formulated at a reduced pH of 5.9 with 1% human serum albumin to improve stability and was packaged as Ceredase (aglucerase).

Initial clinical studies on type 1 Gaucher patients dosed with carbohydrate-remodeled enzyme were very encouraging, particularly in the youngest of the patients [47, 61]. A dose–response study measuring hepatic glucosylceramide reduction following a single dose of enzyme suggested that the appropriate dose for a pivotal efficacy trial was 60 IU kg^{-1} of body weight every other week [47]. The pivotal trial with Ceredase was conducted on 12 Gaucher patients with type 1 disease using this dosing regimen over a 9–12-month period [62]. The trial included both male and female patients, ranging in age from 7 to 38 years old. During the trial the dosing frequency was increased to once a week for two of the most severely affected children. The results of the trial were remarkably positive for such a small patient group, with all patients having significantly decreased splenic volume and increased hemoglobin within six months. Other clinical benefits included decreased serum acid phosphatase activity (an LSD biomarker) in ten patients, decreased liver volume in five, and early signs of skeletal improvement in three. Side effects from the treatment were minimal and no antibodies to the enzyme were detected during the study [63]. Based on the results of this trial, Ceredase was approved by the US Food and Drug Administration (FDA) for treatment of type 1 Gaucher disease in 1991.

13.4
Development of a Second-Generation ERT using Recombinant Technology

13.4.1
Production of a CHO Cell-Expressed Recombinant Human GCase – Cerezyme

It was clear even during the development of Ceredase that it would ultimately not be possible to provide an adequate supply of this drug through purification of the enzyme from human placenta due its limited availability. There was also growing concern of the potential for viral contamination in therapeutics derived from human-source material, although this risk is limited in the case of Ceredase due to various factors, including the use of detergents in the tissue extraction process. Thus development of a recombinant form was initiated prior to the approval of Ceredase.

To assure correct folding and activity of the enzyme, a cDNA for human glucocerebrosidase was cloned into a mammalian cell line, the commonly used Chinese hamster ovary (CHO) cell. Highly-expressing clones of these cells were generated using a methtrexate/DHFR selection system and an anchorage-dependent serum-free microcarrier spinner cell culture system was developed to improve productivity. The cell culture process was scaled up to progressively larger bioreactors, ultimately leading to the 2000L production bioreactors that are currently used. These bioreactors were designed to be operated in continuous perfusion mode to maximize efficiency of expression and minimize damage to the enzyme as a result of cell lysis.

The purification process developed for the recombinantly-produced enzyme was similar to that used for Ceredase; however, since it was being expressed in CHO cells, modifications were required to remove CHO cell proteins, including hamster GCase. As with other therapeutics produced in non-human cells it was also necessary to develop particularly sensitive assays to detect these host cell proteins, due to their potential for immunogenicity. The carbohydrate remodeling procedure that had been incorporated into this process (involving sequential removal of sialic acid, galactose and *N*-acetylglucosamine) was also similar to that used for Ceredase. In developing a formulation for the enzyme, human serum albumin was eliminated due to the potential for viral contamination from human-source material. Instead, lyophilization and Tween were used to stabilize the enzyme and prolong its shelf-life. This material was approved for use in the clinic in 1994 as Cerezyme (imiglucerase).

13.4.2
Biochemical Comparison of Cerezyme to Ceredase

An extensive comparison of the recombinantly-expressed carbohydrate-remodeled enzyme (Cerezyme) to its counterpart derived from placenta (Ceredase) was performed [26]. One particularly helpful tool in this comparison was peptide mapping, where the proteins are enzymatically digested (in this case with trypsin) and the resulting peptides are separated using reversed-phase HPLC. This gives a "fingerprint" of the protein that can reveal sequence differences and post-translational modifications to the protein that might affect its activity or immunogenicity. Mass spectrometry (in its infancy in terms of protein analysis when this comparison was carried out) has become a very powerful tool in identifying and characterizing these differences. The peptide maps for these two forms of GCase were nearly identical – however, differences were noted [26]. Mass analysis of the maps revealed that these differences could all be attributed to a single amino acid sequence difference between the two at position 495, histidine in the case of the recombinant enzyme and arginine in the placental form. It was later determined that this difference occurred due to a cloning error in the cDNA sequence from which the cell line expressing GCase was developed [64, 65]. This single amino acid difference between the two forms of GCase does not have any effect on their activity, as the enzymatic properties of the two are indistinguishable (Table 13.3). Similar results have been obtained by others looking at the effect of this amino acid difference [17].

Of particular importance in this comparison was the characterization of glycosylation, since this directs the targeting of the enzyme in its remodeled form of each. As discussed above, the purified placental enzyme contains primarily complex carbohydrate (both bi- and triantennary) as well as ~20% oligomannose structures [25]. The recombinant enzyme was found to contain only complex structures, with no oligomannose [26]. Analysis of glycans released from isolated glycopeptides showed that the oligomannose on the placental form (primarily Man_6) was located at Asn19. A higher level of proximal fucosylation at Asn146 in the recombinant form was also noted in this analysis.

Analysis of the total glycans released from the recombinant and placental forms before carbohydrate remodeling revealed significant differences between the two (Figure 13.2; lanes 1 and 3). After remodeling, however, the glycan pattern for the two is very similar (Figure 13.2; lanes 2 and 4), with the main differences (upper bands in lane 4) being due to the oligomannose present in the placental enzyme. In the carbohydrate-remodeled version of each the predominant oligosaccharides are the fucosylated and non-fucosylated Man_3 core structures. This is what would be expected if the complex carbohydrate chains are completely trimmed back by the three remodeling enzymes. Monosaccharide analysis of each form before and after remodeling confirmed the loss of galactose and GlcNAc, with mannose and fucose being unchanged (Table 13.4). Thus, with the exception of the oligomannose difference on Asn19, Ceredase and Cerezyme were found to be very similar enzymatically and in their glycosylation.

Table 13.3 Kinetic parameters[a] for Ceredase and Cerezyme for the hydrolysis of *p*-nitrophenyl β-D-glucopyranoside. (Reproduced with permission from Reference [26].)

Sample	K_m (mM)	V_{max} (μmol min^{-1} mg^{-1})
Ceredase	1.00 ± 0.04	48.1 ± 2.0
Cerezyme	1.01 ± 0.09	48.4 ± 3.9

[a]Mean ± standard deviation.

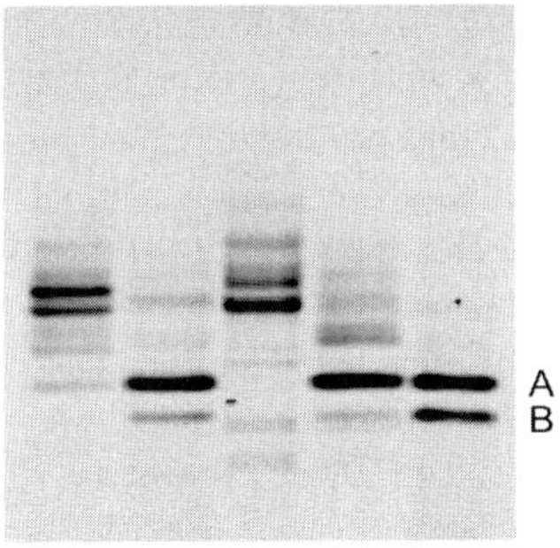

Figure 13.2 Fluorophore-assisted carbohydrate electrophoresis (FACE) of released glycans from glucocerebrosidase before and after carbohydrate modification. Lane 1, recombinant human glucocerebrosidase; lane 2, Cerezyme; lane 3, placental glucocerebrosidase; lane 4, Ceredase; lane 5, labeled oligosaccharide standards: A = $Man_3GlcNAc_2Fuc$, B = $Man_3GlcNAc_2$. (Reproduced with permission from Reference [26].)

13.4.3
Comparison of Ceredase to Cerezyme *In Vivo*

The behavior of Cerezyme *in vivo* was compared to that of Ceredase in terms of pharmacokinetics and biodistribution using normal Balb/c mice [66]. Serum half-life was indistinguishable between the two at 2.6 ± 0.6 min for Ceredase and 3.0 ± 0.4 min for Cerezyme. Biodistribution was also compared 20 min after a bolus IV does of 40 IU kg^{-1}. Again, the behavior of the two was comparable with respect to uptake into liver, spleen, lung and kidney; ~95% of the dose went to the liver in each case. Distribution of enzyme among cell types within the liver was also compared and ~75% of the enzyme was found in parenchymal cells (hepatocytes) for both forms, likely because these represent the vast majority of cells in liver. Interestingly, approximately twice as much Cerezyme as Ceredase (22% vs. 11%) was taken up by Kupffer cells. The remainder was found in endothelial cells.

Table 13.4 Monosaccharide compositions of placental and recombinant glucocerebrosidase before and after carbohydrate remodeling. (Reproduced with permission from Reference [26].)

	Mole per mole of protein			
Sample	**Fuc**	**GlcNAc**	**Gal**	**Man**
Placental glucocerebrosidase	1.6	14.8	8.9	15.2
Carbohydrate-remodeled (Ceredase)	1.4	7.4	0.4	15.2
Recombinant human glucocerebrosidase	2.1	15.0	8.6	12.0
Carbohydrate-remodeled (Cerezyme)	2.1	7.9	0.9	12.3

Cerezyme was also compared to Ceredase in the clinic, both directly [67] and after both drugs had been in the clinic for several years [68]. No differences in safety or efficacy were noted; however, in early studies the incidence of IgG antibody formation was somewhat higher for Ceredase than for Cerezyme (40% vs. 20%) [67]. More recent studies indicate that seroconversion following Cerezyme administration is actually even lower, at ~15% in 1134 patients evaluated between 1994 and 2005 [69].

13.5
Alternative Strategies for Glycoengineered GCase

13.5.1
Overview of Possible Strategies for Targeting Mannose Receptor

It has been clearly demonstrated that targeting GCase to macrophages via the mannose receptor is an effective means of delivery of the enzyme in replacement therapy for Gaucher disease. The sequential remodeling of oligosaccharides used in making Cerezyme and Ceredase is just one strategy to accomplish this; other approaches could be used to target the enzyme to this receptor. As mentioned above, initial attempts at targeting this receptor by addition of mannose chains to the protein in various ways yielded only very limited success. An alternative strategy would be to block normal intracellular remodeling of the oligomannose that is added to glycosylation sites upstream of the first addition of *N*-acetylglucosamine (GlcNAc) in the formation of complex carbohydrate structures. This could be done either by inhibiting the remodeling enzymes (e.g., α-mannosidase I) or by expression in mutant cell lines deficient in one or more of the remodeling enzymes. Another strategy would be to use a different expression system (e.g., a non-mammalian expression system) where the normal intracellular machinery does not remodel oligosaccharides to complex structures but leaves them as some form of oligomannose. Any of these strategies would be expected to result in exposed mannose that may allow targeting of the mannose receptor on macrophages; however, they would also result in other differences relative to the current therapy that would need to be considered.

13.5.2
Use of Mannosidase Inhibitors

Oligosaccharides are added to the N-linked glycosylation sites of proteins as Man_9Glc_3 structures in the ER. Following removal of the terminal glucose residues, the exoglycosidase α-mannosidase I is responsible for trimming back the mannose chain to Man_5 in the ER and Golgi prior to the addition of the first GlcNAc. Inhibition of this enzyme in cells should cause the oligosaccharides of all cellular glycoproteins to remain as Man_9 chains rather than being remodeled to complex structures. This strategy is being evaluated as a possible means for targeting recombinant glucocerebrosidase to macrophages via the mannose receptor in ERT for Gaucher disease [70].

In studies where human glucocerebrosidase was expressed in CHO cells in the presence of the α-mannosidase I inhibitor kifunensine, GCase containing primarily Man_9 structures was obtained along with somewhat smaller oligomannose structures and a small amount of complex structures [71]. This material (referred to as kGCase) was compared to carbohydrate-remodeled GCase (Cerezyme) in terms of receptor binding, macrophage uptake/half-life and pharmacokinetics/biodistribution in normal mice. The D409V Gaucher mouse model was also used to compare these with respect to uptake into macrophages in the liver and spleen *in vivo*. While the kGCase did bind to mannose receptor, and was taken up by macrophages both *in vitro* and *in vivo*, its binding and uptake were similar to the carbohydrate-remodeled GCase that contains Man_3 structures (Figure 13.3). In other words, the longer mannose chains did not improve binding to mannose receptor or targeting of macrophages. This was also true for uptake into splenic macrophages *in vivo* (Figure 13.4).

One difference that was noted in comparing kGCase to Cerezyme was significantly greater binding of kGCase to serum mannose binding protein (mannose-binding lectin; MBL) (Figure 13.5) [71]. This soluble protein is a collectin involved in the innate immune response and is present in serum at concentrations of 1–5 $\mu g\,mL^{-1}$. Its role is to eliminate infectious microorganisms by binding to mannose-terminal carbohydrate structures on their surface and activating the host complement system to kill the invading cell. MBL has a structure quite different from that of the

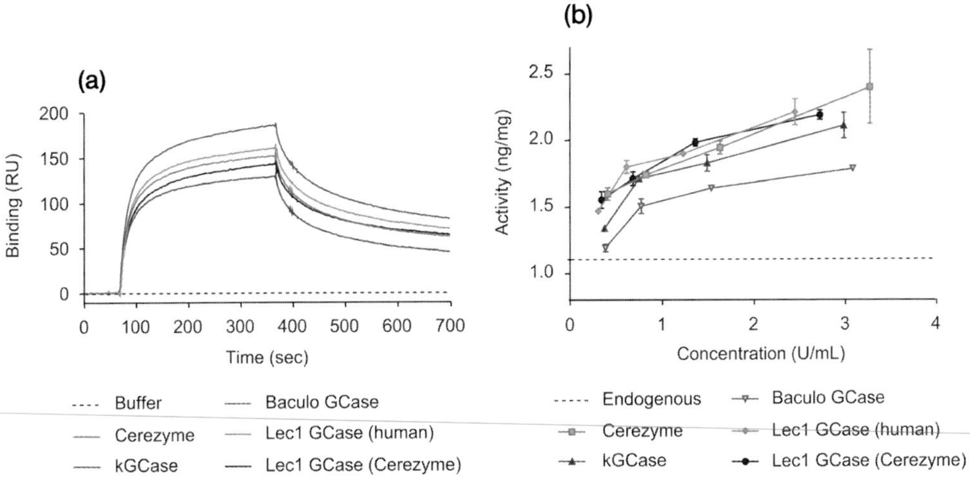

Figure 13.3 (a) Mannose receptor binding of various forms of glucocerebrosidase (100 nM) measured using surface plasmon resonance (SPR) with immobilized receptor [71]. kGCase = GCase expressed in CHO cells in the presence of kifunensine; Baculo GCase = GCase expressed in a baculovirus-based insect cell system; Lec1 GCase = GCase expressed in the Lec1 CHO cell mutant (lacking GlcNAc-TI; human versus Cerezyme refers to the arginine versus histidine at position 495 as described in the text). Binding was performed in the presence of 10 mM $CaCl_2$. (b) Uptake of these forms of glucocerebrosidase into cultured NR8383 rat lung alveolar macrophages (Reproduced with permission from Reference [71].)

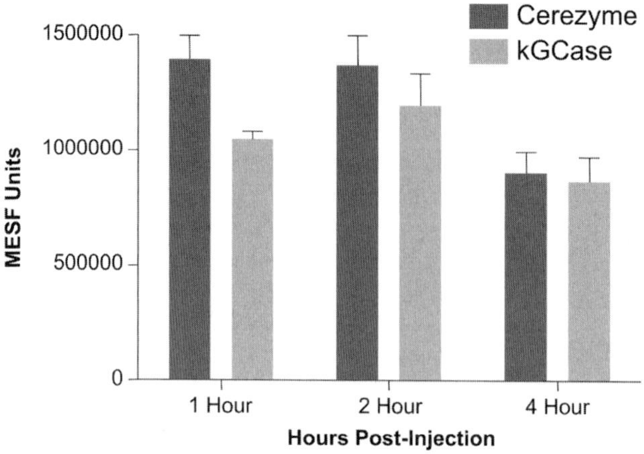

Figure 13.4 Uptake into macrophages *in vivo* at various times following intravenous administration of Cerezyme or kGCase to D409V/null Gaucher mice at 5 mg kg^{-1}. GCase activity was measured in macrophages isolated from spleen based on the turnover of the cell-permeable fluorescent substrate PFB-FDGlu in CD11b-F4/80 positive cells by fluorescence-assisted cell sorting (FACS). (Reproduced with permission from Reference [71].)

macrophage mannose receptor, and it has been reported that proteins with higher levels of mannose exposure bind more tightly to MBL, in contrast with mannose receptor [72]. This group also found that MBL could block uptake of mannosylated liposomes by macrophages [73]. This could be a concern in administering GCase containing Man_9 structures as a therapeutic.

Man_9 oligosaccharides could also be a substrate for the GlcNAc-phosphotransferase that adds mannose 6-phosphate (M6P) to oligosaccharides for sorting to the lysosome, as discussed above. As suggested by Grabowski [6] this could redirect targeting of GCase away from the mannose receptor to the mannose 6-phosphate receptor (MPR), expressed on various non-macrophage cells. Although the level of

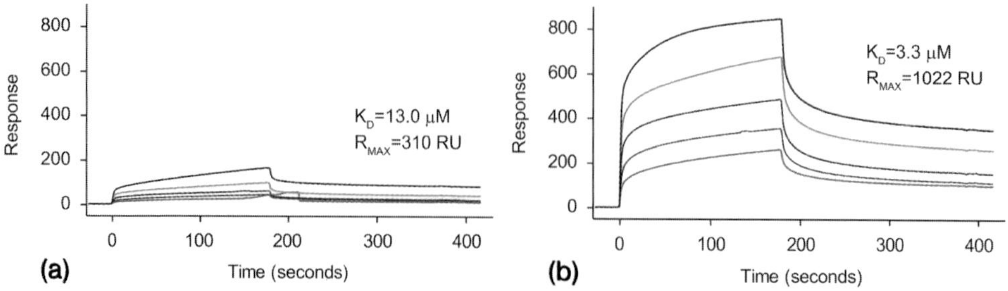

Figure 13.5 Binding of Cerezyme (a) and kGCase (b) to serum mannose-binding lectin (MBL) using SPR with immobilized MBL. Concentration of enzyme was varied from 50 to 800 mg mL^{-1} in the presence of 1 mM $CaCl_2$. (Reproduced with permission from Reference [71].)

M6P was relatively low in the kGCase described above, this may vary with expression system and could also be a concern in using this strategy.

13.5.3
Use of Mutant Cell Lines

A second alternative approach to creating mannose-terminated oligosaccharides on GCase would be to express the enzyme in a mutant cell line where carbohydrate remodeling in the ER and Golgi is blocked due to the absence of one of the remodeling glycosidases upstream of the conversion into complex oligosaccharides. The Lec1 CHO cell mutant is deficient in the enzyme *N*-acetylglucosaminyltransferse I (GlcNAc-TI) [74], which adds the first GlcNAc in the formation of complex carbohydrate chains. This results in glycoproteins that contain Man_5 oligosaccharide structures.

When GCase was expressed in these cells, either in the Arg495 or His495 form, [71] the results were similar to the Man_9 form of GCase (kGCase) discussed above. The slightly longer mannose chains provided no significant benefit in terms of mannose receptor binding or macrophage uptake relative to Cerezyme (Figure 13.3). Binding to MBL was increased somewhat relative to Cerezyme but was less than with kGCase (not shown). It appears that extending the mannose chains on GCase does not offer any advantage in targeting macrophages. In addition, GCase expression in these cells was relatively poor, possibly due to the continuing absence of complex oligosaccharides on all cellular proteins.

13.5.4
Use of Alternative Expression Systems

While mammalian proteins typically have primarily complex carbohydrate structures on their secreted glycoproteins, several non-mammalian organisms lack the enzymes needed to convert oligomannose into complex structures. Expression of GCase in one of these non-mammalian systems would be expected to produce mannose-terminated glycans that could allow targeting of the enzyme to macrophages. It has been suggested that expression in yeast, plant or insect systems might take advantage of this approach to targeting the mannose receptor.

The comparison of GCase with varying mannose chain lengths described above included a form of GCase expressed in baculovirus [71]. Insect cells tend to make glycoproteins containing short Man_3 core structures similar to carbohydrate-remodeled mammalian glucocerebrosidase. However, they can also make smaller pauci-mannose structures (Man_2) [75] not found in mammals and these were found to be present in the baculovirus-expressed GCase. The enzyme expressed in these insect cells did appear to have slightly higher affinity for mannose receptor than the other forms tested (Figure 13.3A) but it was taken up into macrophages less well (Figure 13.3B). Insect glycoproteins are also known to contain glycan variants not found in mammals such as core $\alpha(1,3)$-linked fucosylation, which can be immunogenic in mammals [76]. It has been suggested that these modifications may need to be engineered out for insect cells to be a viable expression system for human therapeutics [77].

Plant or yeast expression systems could also be used to produce GCase with mannose-terminated glycans in the absence of remodeling. This approach is currently being explored using a carrot cell expression system [78]; however, this system requires additional amino acids at the C-terminus of the GCase (DLLVDTM) to target it to the storage vacuole and also results in additional amino acids at the N-terminus due to modification of the signal sequence. Similar to the insect proteins, plant cell glycoproteins often contain core α-(1,3)-fucose and pauci-mannose structures not found in humans [79]. In addition, they are likely to contain the non-mammalian monosaccharide α-(1,2)-xylose. All of these structures were found in the carrot cell-expressed GCase, raising questions of possible immunogenicity in humans [80]. Glycoproteins expressed in yeast have similar issues in that they contain particularly large N-linked oligomannose structures as well as non-mammalian O-linked structures that tend to make them immunogenic [81], thereby limiting their therapeutic usefulness.

13.6
Summary

Roscoe Brady's groundbreaking efforts with enzyme replacement as a therapeutic strategy coupled with the carbohydrate remodeling approach of Furbish and Barranger for targeting glucocerebrosidase provided for the development of an effective therapy for Gaucher patients. Ceredase and Cerezyme now have a proven track record of safety and efficacy in the treatment of type 1 Gaucher disease that spans nearly 20 years. It has been stated that "The successes of ERT in Gaucher disease established a viable medical and industrial model for the development of ERTs for other LSDs" [82].

While other strategies could be used to express glucocerebrosidase with terminal mannose residues in the absence of carbohydrate remodeling, the data suggest that none of these would improve targeting of the enzyme to macrophages relative to the carbohydrate-remodeled enzyme. Furthermore. each of these approaches carries with it additional risks that must be taken into account in evaluating it as potential therapy. Expression systems that result in longer oligomannose chains may alter targeting of the enzyme due to binding to non-mannose receptor proteins such as MBL, whereas expression in non-mammalian systems raises issues of immunogenicity. Given the lack of any obvious biological advantage, it is difficult to justify this level of risk for a therapeutic that has such a long history of safety and effectiveness in patients [69], particularly if the only potential benefit is a slight reduction in manufacturing cost.

References

1 Gieselmann, V. (1995) Lysosomal storage diseases. *Biochimica et Biophysica Acta*, **1270**, 103–136.

2 Neufeld, E.F. (1991) Lysosomal storage diseases. *Annual Review of Biochemistry*, **60**, 257–280.

3 Winchester, B., Vellodi, A. and Young, E. (2000) The molecular basis of lysosomal storage diseases and their treatment. *Biochemical Society Transactions*, **28**, 150–154.

4 Meikle, P.J. *et al.* (1999) Prevalence of lysosomal storage disorders. *The Journal of the American Medical Association*, **281**, 249–254.

5 Beutler, E. and Grabowski, G.A. (2001) Gaucher Disease, in *The Metabolic and Molecular Bases of Inherited Disease* (eds C.R. Scriver, A.L., Beaudet, W.S. Sly and D. Valle), McGraw-Hill Inc, New York, pp. 3635–3668.

6 Grabowski, G.A. (2006) Delivery of lysosomal enzymes for therapeutic use: glucocerebrosidase as an example. *Expert Opinion on Drug Delivery*, **3**, 771–782.

7 Kattlove, H.E. *et al.* (1969) Gaucher cells in chronic myelocytic leukemia: an acquired abnormality. *Blood*, **33**, 379–390.

8 Brady, R.O., Kanfer, J.N. and Shapiro, D. (1965) The metabolism of glucocerebrosides: II. Evidence of an enzymatic deficiency in Gaucher's disease. *Biochemical and Biophysical Research Communications*, **18**, 221–225.

9 Patrick, A.D. (1965) Short communications: A deficiency of glucocerebrosidase in Gaucher's disease. *The Biochemical Journal*, **97**, C17–C18.

10 Boven, L.A. *et al.* (2004) Gaucher cells demonstrate a distinct macrophage phenotype and resemble alternatively activated macrophages. *American Journal of Clinical Pathology*, **122**, 359–369.

11 Fabrega, S. *et al.* (2000) Human glucocerebrosidase: heterologous expression of active site mutants in murine null cells. *Glycobiology*, **10**, 1217–1224.

12 Miao, S. *et al.* (1994) Identification of Glu340 as the active-site nucleophile in human glucocerebrosidase by use of electrospray tandem mass spectrometry. *The Journal of Biological Chemistry*, **269**, 10975–10978.

13 Christomanou, H., Kleinschmidt, T. and Braunitzer, G. (1987) N-terminal amino-acid sequence of a sphingolipid activator protein missing in a new human Gaucher disease variant. *Biological Chemistry Hoppe-Seyler*, **368**, 1193–1196.

14 Dvir, H. *et al.* (2003) X-ray structure of human acid-beta-glucosidase, the defective enzyme in Gaucher disease. *EMBO Reports*, **4**, 1–6.

15 Premkumar, L. *et al.* (2005) X-ray structure of human acid-beta-glucosidase covalently-bound to conduritol-B-epoxide: Implications for Gaucher disease. *The Journal of Biological Chemistry*, **280**, 23815–23819.

16 Brumshtein, B. *et al.* (2006) Structural comparison of differently glycosylated forms of acid-beta-glucosidase, the defective enzyme in Gaucher disease. *Acta Crystallographica. Section D, Biological Crystallography*, **62**, 1458–1465.

17 Liou, B. *et al.* (2006) Analyses of variant acid beta-glucosidases: effects of Gaucher disease mutations. *The Journal of Biological Chemistry*, **281**, 4242–4253.

18 Brumshtein, B. *et al.* (2007) Crystal structures of complexes of *N*-butyl- and *N*-nonyl-deoxynojirimycin bound to acid beta-glucosidase: insights into the mechanism of chemical chaperone action in Gaucher disease. *The Journal of Biological Chemistry*, **282**, 29052–29058.

19 Lieberman, R.L. *et al.* (2007) Structure of acid beta-glucosidase with pharmacological chaperone provides insight into Gaucher disease. *Nature Chemical Biology*, **3**, 101–107.

20 Atrian, S. *et al.* (2008) An evolutionary and structure-based docking model for glucocerebrosidase-saposin C and glucocerebrosidase-substrate interactions – relevance for Gaucher disease. *Proteins*, **70**, 882–891.

21 Zhao, H. and Grabowski, G.A. (2002) Gaucher disease: perspectives on a prototype lysosomal disease. *Cellular and Molecular Life Sciences*, **59**, 694–707.

22 Amato, D. *et al.* (2004) Gaucher disease: variability in phenotype among siblings. *Journal of Inherited Metabolic Disease*, **27**, 659–669.

23 Berg-Fussman, A. *et al.* (1993) Human acid beta-glucosidase. *N*-glycosylation site occupancy and the effect of glycosylation on enzymatic activity. *The Journal of Biological Chemistry*, **268**, 14861–14866.

24 Grace, M.E. and Grabowski, G.A. (1990) Human acid beta-glucosidase: glycosylation is required for catalytic activity. *Biochemical and Biophysical Research Communications*, **168**, 771–777.

25 Takasaki, S. *et al.* (1984) Structure of the *N*-asparagine-linked oligosaccharide units of human placental beta-glucocerebrosidase. *The Journal of Biological Chemistry*, **259**, 10112–10117.

26 Edmunds, T. (2006) β-Glucocerebrosidase: Ceredase and Cerezyme, in *Directory of Therapeutic Enzymes* (eds B.M. McGrath and G. Walsh), CRC Press, Taylor & Francis Group, Boca Raton, FL, pp. 117–133.

27 Dahms, N.M., Lobel, P. and Kornfeld, S. (1989) Mannose 6-phosphate receptors and lysosomal enzyme targeting. *The Journal of Biological Chemistry*, **264**, 12115–12118.

28 van Dongen, J.M. *et al.* (1985) The subcellular localization of soluble and membrane-bound lysosomal enzymes in I-cell fibroblasts: a comparative immunocytochemical study. *European Journal of Cell Biology*, **39**, 179–189.

29 Aerts, J.M. *et al.* (1988) Glucocerebrosidase, a lysosomal enzyme that does not undergo oligosaccharide phosphorylation. *Biochimica et Biophysica Acta*, **964**, 303–308.

30 Rijnboutt, S. *et al.* (1991) Mannose 6-phosphate-independent membrane association of cathepsin D, glucocerebrosidase, and sphingolipid-activating protein in HepG2 cells. *The Journal of Biological Chemistry*, **266**, 4862–4868.

31 Leonova, T. and Grabowski, G.A. (2000) Fate and sorting of acid beta-glucosidase in transgenic mammalian cells. *Molecular Genetics and Metabolism*, **70**, 281–294.

32 Reczek, D. *et al.* (2007) LIMP-2 is a receptor for lysosomal mannose-6-phosphate-independent targeting of beta-glucocerebrosidase. *Cell*, **131**, 770–783.

33 De Duve, C. (1964) From cytases to lysosomes. *Federation Proceedings*, **23**, 1045–1049.

34 Brady, R.O. (2006) Enzyme replacement for lysosomal diseases. *Annual Review of Medicine*, **57**, 283–296.

35 Brady, R.O. (1966) The sphingolipidoses. *The New England Journal of Medicine*, **275**, 312–318.

36 Kornfeld, S. (1992) Structure and function of the mannose 6-phosphate/insulinlike growth factor II receptors. *Annual Review of Biochemistry*, **61**, 307–330.

37 Duncan, J.R. and Kornfeld, S. (1988) Intracellular movement of two mannose 6-phosphate receptors: return to the Golgi apparatus. *The Journal of Cell Biology*, **106**, 617–628.

38 Puertollano, R. *et al.* (2001) Sorting of mannose 6-phosphate receptors mediated by the GGAs. *Science*, **292**, 1712–1716.

39 Stein, M. *et al.* (1987) Mr 46,000 mannose 6-phosphate specific receptor: its role in targeting of lysosomal enzymes. *The EMBO Journal*, **6**, 2677–2681.

40 Schiffmann, R. *et al.* (2000) Infusion of alpha-galactosidase A reduces tissue globotriaosylceramide storage in patients with Fabry disease. *Proceedings of the National Academy of Sciences of the United States of America*, **97**, 365–370.

41 Lee, K. *et al.* (2003) A biochemical and pharmacological comparison of enzyme replacement therapies for the glycolipid storage disorder Fabry disease. *Glycobiology*, **13**, 305–313.

42 Raben, N. *et al.* (2003) Enzyme replacement therapy in the mouse model of Pompe disease. *Molecular Genetics and Metabolism*, **80**, 159–169.

43 Zhu, Y. *et al.* (2005) Carbohydrate-remodelled acid alpha-glucosidase with higher affinity for the cation-independent mannose 6-phosphate receptor demonstrates improved delivery to

muscles of Pompe mice. *The Biochemical Journal*, **389**, 619–628.
44 Pentchev, P.G. *et al.* (1973) Isolation and characterization of glucocerebrosidase from human placental tissue. *The Journal of Biological Chemistry*, **248**, 5256–5261.
45 Brady, R.O. *et al.* (1974) Replacement therapy for inherited enzyme deficiency. Use of purified glucocerebrosidase in Gaucher's disease. *The New England Journal of Medicine*, **291**, 989–993.
46 Furbish, F.S. *et al.* (1977) Enzyme replacement therapy in Gaucher's disease: large-scale purification of glucocerebrosidase suitable for human administration. *Proceedings of the National Academy of Sciences of the United States of America*, **74**, 3560–3563.
47 Brady, R.O., Murray, G.J., Barton, N.W. (1994) Modifying exogenous glucocerebrosidase for effective replacement therapy in Gaucher disease. *Journal of Inherited Metabolic Disease*, **17**, 510–519.
48 Pentchev, P.G. *et al.* (1978) Incorporation of exogenous enzymes into lysosomes. A theoretical and practical means for correcting lysosomal blockage, in *Glycoproteins and Glycolipids in Disease Processes* (ed. E.F.J. Walborg), The American Chemical Society, Washington, DC, pp. 150–159.
49 Furbish, F.S. *et al.* (1978) The uptake of native and desialylated glucocerebrosidase by rat hepatocytes and Kupffer cells. *Biochemical and Biophysical Research Communications*, **81**, 1047–1053.
50 Dale, G.L. and Beutler, E. (1976) Enzyme replacement therapy in Gaucher's disease: a rapid, high-yield method for purification of glucocerebrosidase. *Proceedings of the National Academy of Sciences of the United States of America*, **73**, 4672–4674.
51 Beutler, E. *et al.* (1977) Enzyme replacement therapy in Gaucher's disease: preliminary clinical trial of a new enzyme preparation. *Proceedings of the National Academy of Sciences of the United States of America*, **74**, 4620–4623.
52 Belchetz, P.E. *et al.* (1977) Treatment of Gaucher's disease with liposome-entrapped glucocerebroside: beta-glucosidase. *Lancet*, **2**, 116–117.
53 Gregoriadis, G. *et al.* (1980) Experiences after long-term treatment of a type I Gaucher disease patient with liposome-entrapped glucocerebroside: beta-glucosidase. *Birth Defects Original Article Series*, **16**, 383–392.
54 Stahl, P.D. *et al.* (1978) Evidence for receptor-mediated binding of glycoproteins, glycoconjugates, and lysosomal glycosidases by alveolar macrophages. *Proceedings of the National Academy of Sciences of the United States of America*, **75**, 1399–1403.
55 Achord, D.T. *et al.* (1978) Human beta-glucuronidase: in vivo clearance and in vitro uptake by a glycoprotein recognition system on reticuloendothelial cells. *Cell*, **15**, 269–278.
56 Brady, R.O. and Furbish, F.S. (1982) Enzyme replacement therapy: specific targeting of exogenous enzymes to storage cells, in *Membranes and Transport* (ed. A.N. Martonosi), Plenum Press, New York, pp. 587–592.
57 Doebber, T.W. *et al.* (1982) Enhanced macrophage uptake of synthetically glycosylated human placental beta-glucocerebrosidase. *The Journal of Biological Chemistry*, **257**, 2193–2199.
58 Murray, G.J. *et al.* (1985) Targeting of synthetically glycosylated human placental glucocerebrosidase. *Biochemical Medicine*, **34**, 241–246.
59 Furbish, F.S. *et al.* (1981) Uptake and distribution of placental glucocerebrosidase in rat hepatic cells and effects of sequential deglycosylation. *Biochimica et Biophysica Acta*, **673**, 425–434.
60 Richards, S.M., Olson, T.A. and McPherson, J.M. (1993) Antibody response in patients with Gaucher disease after repeated infusion with macrophage-targeted glucocerebrosidase. *Blood*, **82**, 1402–1409.

61 Barton, N.W. *et al.* (1990) Therapeutic response to intravenous infusions of glucocerebrosidase in a patient with Gaucher disease. *Proceedings of the National Academy of Sciences of the United States of America*, **87**, 1913–1916.

62 Barton, N.W. *et al.* (1991) Replacement therapy for inherited enzyme deficiency--macrophage-targeted glucocerebrosidase for Gaucher's disease. *The New England Journal of Medicine*, **324**, 1464–1470.

63 Murray, G.J. *et al.* (1991) Gaucher's disease: lack of antibody response in 12 patients following repeated intravenous infusions of mannose terminal glucocerebrosidase. *Journal of Immunological Methods*, **137**, 113–120.

64 Sorge, J. *et al.* (1985) Molecular cloning and nucleotide sequence of human glucocerebrosidase cDNA. *Proceedings of the National Academy of Sciences of the United States of America*, **82**, 7289–7293.

65 Sorge, J. *et al.* (1986) Correction to: molecular cloning and nucleotide sequence of human glucocerebrosidase cDNA. *Proceedings of the National Academy of Sciences of the United States of America*, **83**, 3567.

66 Friedman, B. *et al.* (1999) A comparison of the pharmacological properties of carbohydrate remodeled recombinant and placental-derived beta-glucocerebrosidase: implications for clinical efficacy in treatment of Gaucher disease. *Blood*, **93**, 2807–2816.

67 Grabowski, G.A. *et al.* (1995) Enzyme therapy in type 1 Gaucher disease: comparative efficacy of mannose-terminated glucocerebrosidase from natural and recombinant sources. *Annals of Internal Medicine*, **122**, 33–39.

68 Weinreb, N.J. *et al.* (2002) Effectiveness of enzyme replacement therapy in 1028 patients with type 1 Gaucher disease after 2 to 5 years of treatment: a report from the Gaucher Registry. *The American Journal of Medicine*, **113**, 112–119.

69 Starzyk, K. *et al.* (2007) The long-term international safety experience of imiglucerase therapy for Gaucher disease. *Molecular Genetics and Metabolism*, **90**, 157–163.

70 Zimran, A. *et al.* (2007) A pharmacokinetic analysis of a novel enzyme replacement therapy with Gene-Activated human glucocerebrosidase (GA-GCB) in patients with type 1 Gaucher disease. *Blood Cells, Molecules & Diseases*, **39**, 115–118.

71 Van Patten, S.M. *et al.* (2007) Effect of mannose chain length on targeting of glucocerebrosidase for enzyme replacement therapy of Gaucher disease. *Glycobiology*, **17**, 467–478.

72 Opanasopit, P. *et al.* (2001) In vivo recognition of mannosylated proteins by hepatic mannose receptors and mannan-binding protein. *American Journal of Physiology. Gastrointestinal and Liver Physiology*, **280**, G879–G889.

73 Opanasopit, P. *et al.* (2002) Serum mannan binding protein inhibits mannosylated liposome-mediated transfection to macrophages. *Biochimica et Biophysica Acta*, **1570**, 203–209.

74 Chaney, W. and Stanley, P. (1986) Lec1A Chinese hamster ovary cell mutants appear to arise from a structural alteration in *N*-acetylglucosaminyltransferase I. *The Journal of Biological Chemistry*, **261**, 10551–10557.

75 Kulakosky, P.C., Hughes, P.R. and Wood, H.A. (1998) N-linked glycosylation of a baculovirus-expressed recombinant glycoprotein in insect larvae and tissue culture cells. *Glycobiology*, **8**, 741–745.

76 Tretter, V. *et al.* (1993) Fucose alpha 1,3-linked to the core region of glycoprotein *N*-glycans creates an important epitope for IgE from honeybee venom allergic individuals. *International Archives of Allergy and Immunology*, **102**, 259–266.

77 Harrison, R.L. and Jarvis, D.L. (2006) Protein *N*-glycosylation in the baculovirus-insect cell expression system and engineering of insect cells to produce

"mammalianized" recombinant glycoproteins. *Advances in Virus Research*, **68**, 159–191.

78 Shaaltiel, Y. *et al.* (2007) Production of glucocerebrosidase with terminal mannose glycans for enzyme replacement therapy of Gaucher's disease using a plant cell system. *Plant Biotechnology Journal*, **5**, 579–590.

79 Twyman, R.M. *et al.* (2003) Molecular farming in plants: host systems and expression technology. *Trends in Biotechnology*, **21**, 570–578.

80 Bardor, M. *et al.* (2003) Immunoreactivity in mammals of two typical plant glyco-epitopes, core alpha(1,3)-fucose and core xylose. *Glycobiology*, **13**, 427–434.

81 Cereghino, G.P. *et al.* (2002) Production of recombinant proteins in fermenter cultures of the yeast Pichia pastoris. *Current Opinion in Biotechnology*, **13**, 329–332.

82 Burrow, T.A. *et al.* (2007) Enzyme reconstitution/replacement therapy for lysosomal storage diseases. *Current Opinion in Pediatrics*, **19**, 628–635.

14
Engineering in a PTM: PEGylation

Gian Maria Bonora and Francesco Maria Veronese

14.1
Protein PEGylation

The importance of proteins with structural, signaling or enzymatic functions has always been recognized but the significance of their post-translational modifications and the role of such post-translational modifications in the pathological process was only more recently demonstrated. Through the use of post-translational modifications (e.g., phosphorylation, methylation, glycosylation), biological systems have managed to naturally evolve effective chemical strategies for the modulation of protein biophysical properties [1, 2]. What nature discovered and selected by evolution can now effectively be mimicked in the laboratory by using chemical tools to link synthetic molecules, generally polymers such as PEG [poly(ethylene glycol)], to native protein, yielding conjugates with novel functional properties (Figure 14.1).

The practice of covalent coupling of PEG to pharmaceutical proteins, commonly named PEGylation, can be an extremely useful procedure to overcome many problems faced in the development and use of protein drugs. The continuing interest in PEGylation is well documented by the number of papers and patents that have appeared since the discovery of this methodology by Davis and Abuchowski in the late 1970s [3]. Indeed, PEGylation has become the dominant protein drug delivery system for the biotech industry, with sales of PEGylated protein drugs having surpassed $4 billion [4]. It is probable that this strategy is not exclusive for PEG, as recent studies using polymers of natural origin such as polysaccharides [5, 6] and synthetic polymers [7, 8] are demonstrating. Certainly the future will see interesting discoveries in this field of delivery and therapy, which now are probably limited solely by the availability of alternative polymers, a situation similar to that of several years ago at the beginning of PEG development when only a few laboratories could produce this polymer in a reproducible and pure form.

The first pioneering studies in the field of protein–polymer conjugation were carried out with dextran [9]. Since then, several alternative polymers have been

Post-translational Modification of Protein Biopharmaceuticals. Edited by Gary Walsh

ISBN: 978-3-527-32074-5

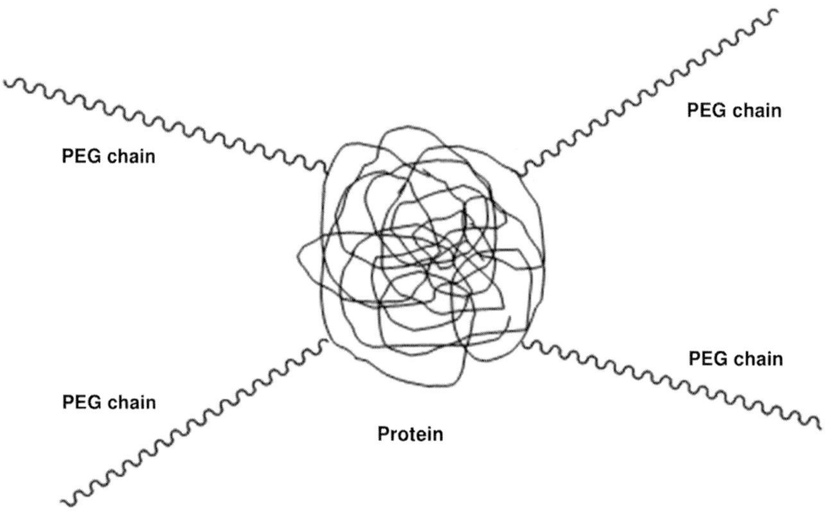

Figure 14.1 Schematic drawing of a PEGylated protein.

explored for polymer conjugation, with poly(ethylene glycol) (PEG) emerging as the most prominent candidate for protein modification [10, 11]. Indeed, its leading position is reflected by the fact that most conjugates on the market are PEGylated products and that many other PEGylated products are in advanced clinical investigation. Polymer conjugates are applicable for the treatment of several severe diseases, demonstrating that the potential of this technique is not restricted to few therapeutic areas. PEG–protein conjugates have all the benefits described above for low-molecular-weight drug conjugates and, also, some additional advantages can be achieved – namely, the protection of sites of proteolysis and the reduction or prevention of immunogenicity and antigenicity [12, 13]. These are the result of the shielding effect of polymer chains on the protein surface, which prevents the approach of other proteins or enzymes by steric hindrance (Figure 14.1). In general this also influences the activity of conjugated protein, which is usually lower than that of the native protein in *in vitro* tests. However, this is usually more than compensated for *in vivo* due to the significant improvement in pharmacokinetic parameters [14, 15]. One of the critical challenges for the clinical development of PEGylated derivatives relates to the capability of scaling-up conjugation reactions and purification of products, especially in the case of proteins containing several reactive groups [16, 17]. Reproducible preparation and separation procedures of conjugates, as well as methods for full characterization of the derivatives, are key factors in guaranteeing the product consistency required for moving from laboratory discoveries to the development necessary to bring the final product to clinical investigation and use.

Thorough and up to date information regarding PEG and PEGylated products is reported in two books [18, 19], while a complete overview of this subject can be found in a series of dedicated issues published in *Advanced Drug Delivery Reviews* [20–22].

14.2
General Properties of PEG

PEG is synthesized by ring-opening polymerization of ethylene oxide using methanol or water as initiator. The reaction gives polymers with one or two end chain hydroxyl groups, termed monomethoxy-PEG (mPEG-OH) or diol-PEG (HO-PEG-OH), respectively. The polymerization process produces a family of PEG molecules with a wide range of molecular weights. In this form PEG is not suitable for use in drug conjugation, although it is largely employed in pharmaceutical technology as an excipient. As with any polymer, the biological properties of PEG, such as kidney excretion rate or protein surface coverage, are based on its size. It is therefore necessary to use PEG preparations that are as homogeneous as possible. Chromatographic fractionation techniques may be used to achieve narrow polydispersivity and to exclude the presence of the diol form (formed from traces of water in the polymerization reaction) as an impurity in monomethoxy PEG batches.

Although PEG has been long been used as an excipient in many pharmaceutical or cosmetic formulations [23], studies in the field of drug conjugation and delivery started only 30 years ago. This polymer presents unique properties such as (i) lack of immunogenicity, antigenicity and toxicity; (ii) high solubility in water and in many organic solvents; (iii) high hydration and flexibility of the chain; and (iv) approval by FDA for human use. Another property, important for pharmaceutical applications, is its availability in a low polydispersity form, with an Mw/Mn spanning from 1.01 for PEG of 5000 Da up to 1.1 for PEG as high as 50 kDa. The lone hydroxyl group in the case of the methoxy form, or the two in the case of PEG diol, can be modified to be reactive towards different chemical groups by several activation strategies. Therefore, several activated PEGs are commercially available. These, as used for protein modification, are usually monofunctional PEGs, either in their linear or branched structure. Alternatively, PEG diols, PEG dendrons [24–26] or dendrimers [27] are generally used for small drug delivery due to the possibility of reaching higher drug to polymer ratios. In fact, an increased loading value is needed in therapeutic agents with low biological activity, which would otherwise require the administration of a large amount of conjugate with consequent high viscosity of the solution.

The contribution of PEG to the final hydrodynamic volume of conjugates is also of particular relevance. This behavior is due to PEGs ability to coordinate a large number of water molecules per ethylene unit and to its highly chain flexibility [28, 29], thus giving PEG an apparent molecular weight 5–10× higher than that of a globular protein of a comparable mass, as verified by gel permeation chromatography [30, 31]. Additionally, PEG is considered a non-biodegradable polymer. Only slow degradation by alcohol dehydrogenase [32], aldehyde dehydrogenase [33] and cytochrome P-450 [34] has been documented, especially in the case of PEG oligomers. Therefore, its body clearance depends upon its molecular weight: below 20 kDa it is easily excreted into the urine, while higher molecular weight PEGs are eliminated more slowly and clearance via the liver becomes predominant. The threshold for kidney filtration is about 40–60 kDa (a hydrodynamic radius of approximately 45 Å [35]), which represents the albumin excretion limit. Over this limit the polymer remains in circulation even longer and accumulates in the liver. The plasma kinetics of PEGs depend on both the molecular weight and site of injection [36]. It was demonstrated that using PEGs of increasing molecular weight increased both the AUC and the half-life. In particular, the half-life displayed a sigmoidal relationship with polymer molecular weight, which agrees with theoretical models of renal excretion of macromolecules, based on the pore sizes of the glomerular capillary wall [37]. However, for safety reasons, PEGs with a molecular weight below 40 kDa are the most appropriate choice when designing a new drug–polymer conjugate, to avoid accumulation in the body. This is the PEG molecular weight employed in two successful products, "Pegasys" and "Macugen," belonging to the PEG–protein and PEG–small drug conjugate classes, respectively. Notably, in this regard, this molecular size may be obtained by a unique, linear polymeric chain or by a "branched" form; the latter can give a better biological performance through an umbrella-like effect [38]. In any case, several years of PEG use as an excipient in foods, cosmetics and pharmaceuticals, without toxic effects, demonstrates safety.

14.3 Chemically Activated PEGs and the Process of PEGylation

PEGylation at the level of protein amino groups may be carried out using PEGs having various different reactive groups at the end of the chain. Although the coupling reaction is based on the same chemistry (for instance acylation), the products obtained are often different. The difference may relate to the number of PEG chains linked per protein molecule, to the specific amino acids involved or to the chemical bond between PEG and drug. Products available on the market so far mainly display random PEGylation; the FDA approve these conjugate mixtures upon demonstration of reproducibility. The activated PEG for amino linking can be chosen from a range of commercially available polymers, the most common being PEG succinimidyl succinate (SS-PEG), PEG succinimidyl carbonate (SC-PEG), PEG *p*-nitrophenyl carbonate (pNPC-PEG), PEG benzotriazolyl carbonate (BTC-PEG), PEG trichlorophenyl carbonate (TCP-PEG), PEG imidazolyl carbonate (CDI-PEG), PEG tresylate (Tres-PEG), PEG dichlorotriazine (DCT-PEG) and PEG aldehyde (AL-PEG) (Figure 14.2).

The difference between these PEGs lies in the kinetic rate of amino coupling and in the resulting link between polymer and drug. The derivatives with slower reactivity, such as the carbonate PEGs (pNPC-PEG, CDI-PEG and TCP-PEG) or the aldehyde PEGs, allow a certain degree of selective conjugation within the amino groups present in a protein, according to their nucleophilicity or accessibility [39]. An important difference in reactivity is usually observed between the ε-amino and the α-amino group in proteins due to their pK_a: 9.3–9.5 for the ε-amino residue of lysine and 7.6–8 for the α-amino group. This has been exploited for the α-amino modification reached by a conjugation at pH 5.5–6.0, as is the case for G-CSF with PEG-aldehyde [40]. The ε-amino groups of lysine, which possess high nucleophilicity at high pH, are instead the preferred site of conjugation at pH 8.5–9.

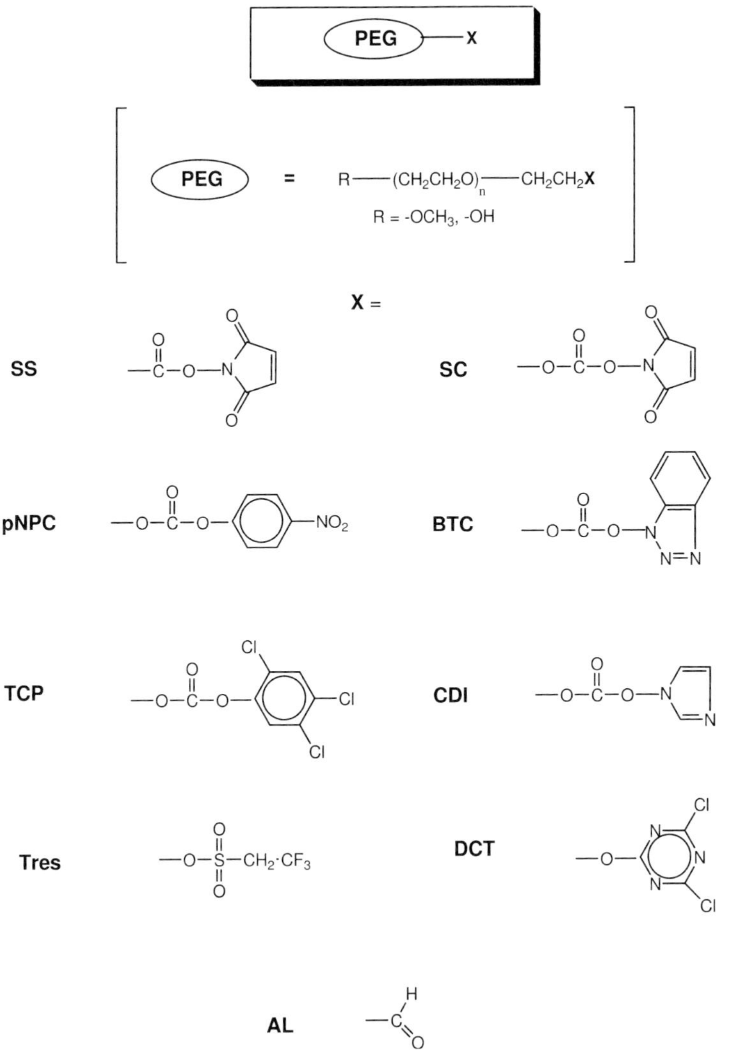

Figure 14.2 PEG activated towards amino residues.

Notably, conjugation performed using PEG dichlorotriazine, PEG tresylate and PEG aldehyde (the latter after sodium cyanoborohydride reduction) maintains the same total charge on the native protein surface, since these derivatives react through an alkylation reaction, yielding a secondary amine. In contrast, PEGylation conducted with acylating PEGs (i.e., SS-PEG, SC-PEG, pNPC-PEG, CDI-PEG and TCP-PEG) gives weakly acidic amide or carbamate linkages with loss of the positive charge. The PEG derivatives described above may sometimes undergo side reactions involving the hydroxyl groups of serine, threonine and tyrosine and the secondary amino group of histidine. These linkages, however, are generally hydrolytically unstable. The reaction conditions or particular conformational disposition may enhance the

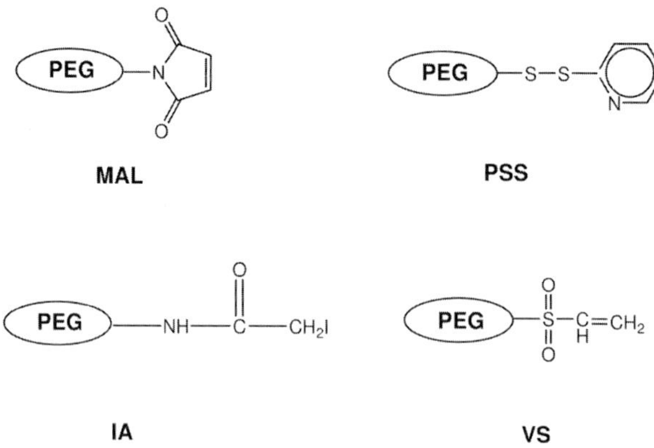

Figure 14.3 PEG activated towards thiol residues.

percentage of these unusual PEGylations. For example, α-interferon was found to be conjugated by SC-PEG or BTC-PEG also at His34 under slightly acidic conditions [41] (the pK_a of histidine is between those of the α- and ε-amino groups). Moreover, PEG was found to be linked to the hydroxyl groups of serine in a LHRH analogue [42] or those of tyrosine in epidermal growth factor (EGF) [43].

The presence of a free cysteine residue represents an optimal opportunity to achieve site-directed modification, because they more rarely occur in proteins. PEG derivatives having specific reactivity towards the thiol group, that is, PEG-maleimide (MAL-PEG), PEG-pyridyl disulfide (PSS-PEG), PEG-iodoacetamide (IA-PEG) and PEG-vinylsulfone (VS-PEG), are commercially available and allow thiol coupling with a good yield. However, there are differences among the manufacturers in terms of protein–polymer linkage achieved and reaction conditions employed (Figure 14.3).

Even if the thiol reaction rate of IA-, MAL- or VS-PEGs is very rapid, some degree of amino coupling may also take place, especially if the reaction is carried out at pH > 8. In contrast, the reaction with OPSS-PEG is very specific for thiol groups, but the conjugates obtained may be cleaved in the presence of reducing agents such as simple thiols or glutathione (present *in vivo*). PEGylation at the level of cysteine allows easier purification of the reaction mixture since the presence of only one or a few derivatizable sites (free cysteines) avoids the formation of a large number of positional isomers or products with different degrees of substitution. The potential of thiol PEGylation may be further exploited by genetic engineering, which allows the introduction of an additional cysteine residue in a protein sequence or the switching of a nonessential amino acid to cysteine. An example of this strategy was the PEGylation of the human growth hormone (hGH). To overcome the common problems of random amino PEGylation, cysteine muteins were synthesized by recombinant DNA technology. Among all the possible mutations described in the literature and patents, cysteine addition at the C-terminus of hGH leads to a fully active product that allows a site-specific PEGylation using the thiol-reactive PEG-maleimide (PEG-MAL, 8 kDa). It was necessary to treat the rhGH mutant with 1,4-dithio-DL-threitol (DTT) before the coupling step, to maintain the C-terminal cysteine in the reduced form and to prevent the formation of scrambled disulfide bridges [44].

PEGylation at the level of protein carboxylic groups needs their activation for the reaction with an amino PEG. This procedure, however, is not without limitations since undesired intra- or intermolecular crosslinks may occur by reaction with the amino groups of the protein itself. To circumvent this problem it is possible to use PEG-hydrazide (PEG-CO-NH-NH_2) instead of the usual PEGamino and to carry out the coupling at pH 4–5, thanks to the low pK_a of PEG-hydrazide. In this case, the protein's COOH groups, activated by water-soluble carbodiimide, do not react with the protein amino groups (which at low pHs are protonated) but with the amino group of PEG-hydrazide only [45]. An alternative method for C-terminal-specific PEGylation is based on the Staudinger reaction [46]. The protocol, as applied with a truncated thrombomodulin mutant [47], begins with the expression by *Escherichia coli* of a mutated protein containing a C terminal azido-methionine. This reacts specifically with an engineered PEG derivative, methyl-PEG-triarylphosphine, leading to a C-terminal mono-PEGylated protein. This method, however, involves the preparation of a gene encoding a protein with a C-terminal linker ending with methionine. Unfortunately, this method is applicable only to the rare case of proteins lacking methionine in the sequence; otherwise they will stop the protein transduction because the azido-analog does not permit the linking of the following amino acid.

As described above, the most common methods of protein PEGylation rely on chemical conjugation through reactive side-chain groups. More recently, enzymatic methods have been explored for devising novel, mild strategies of site-directed PEGylation of pharmaceutical proteins [48, 49]. These methods effectively mimic the post-translational modification of proteins commonly performed by nature to modify and improve the bioactivity of those molecules. One such method involves the enzymatic GalNAc glycosylation at specific serine and threonine residues in proteins. This enzymatic method, termed GlycoPEGylation, was applied to three clinically important proteins and it was demonstrated that these proteins were modified at the *O*-glycosylation sites of the native proteins. Therefore, GlycoPEGylation appears to be a very useful methodology, enabling production of long-acting PEGylated protein drugs with enhanced structural homogeneity as compared with similar conjugated obtained by other chemical methods. Additionally, a very promising enzymatic method for the covalent binding of PEG moieties to pharmaceutical proteins involving the use of transglutaminase, or TGase, has been successfully used to produce several PEGylated proteins of clinical interest [50–52]. In this regard, it has been shown recently that TGase acts on the polypeptide substrate in analogy to a protease [53].

14.4
Potential Effects of PEGylation that are of Therapeutic Relevance

As previously observed, PEGylation is currently considered one of the most successful approaches to prolonging the residence time of protein drugs in the bloodstream [54–56]. In a few cases polymer conjugation was also demonstrated to confer targeting properties to the disease site such as tumor masses by passive diffusion. Notably, the biopharmaceutical properties of protein–poly(ethylene glycol) conjugates depend strictly on the physicochemical and biological properties of the constructs (polymer and protein) as well as on the properties of the whole conjugate. Therefore, knowledge of the biological fate of both polymer and protein is of relevance in the design of PEGylation strategies useful to obtain derivatives with predictable pharmacokinetic and distribution behavior. For example, while glycosylation often does not alter binding affinities, PEGylation often leads to a decrease in activity as it can hamper binding to ligand. These differences can be understood, in part at least, if one considers that the structural composition of PEG in solution is generally a compact globular form, in contrast to the semi-linear nature of glycans [57–59]. Moreover, PEGylation, contrary (in some instances at least) to natural post-translational modification, does not facilitate any recognition effect towards specific biological receptors. An advance in this context may be offered by new heterobifunctional PEGs [60, 61] that (similarly to the application of functionalized liposomes) will permit the introduction at the other end of the polymeric chain of a further molecule that will be specifically recognized by a suitable receptor.

Generally, PEGs and PEGylated proteins, following intravenous administration, display bicompartmental pharmacokinetic behavior, indicating that they undergo peripheral distribution. The distribution properties of macromolecules depend mainly on their molecular weight, structure and size and on morphological characteristics of the vasculature and tissue disposition. Furthermore, specific or non-specific interactions with tissues and circulating cells may occur. Usually, it is difficult to accurately predict the pharmacokinetic behavior of native proteins when their structural properties are well known. Prediction becomes practically impossible in the case of PEG conjugates, where many new variables are introduced, such as the effect of molecular weight and shape of the polymer, extent of modification and site of PEG attachment. All these parameters affect the biological processes involved in the *in vivo* fate of conjugate, for example glomerular filtration, stability, immunogenicity, tissue localization, cell uptake, liver excretion, and so on. Charge masking may be an additional consequence of the chemical modification of proteins. In particular, the most common procedures for producing protein–PEG derivatives involve the conversion of ionizable amino groups into amides with the consequent loss of positive charges [62]. Since anionic macromolecules have been found to be cleared by renal ultrafiltration more slowly than neutral or positive ones, it could be expected that PEG conjugation to amino groups prolongs the permanence in the bloodstream [63]. In their pioneering studies on PEGylation, Abuchowski *et al.* demonstrated that PEG conjugation also suppresses protein immunogenicity [64]. This beneficial effect of PEGylation was confirmed by several studies [65]. In a few cases PEG conjugation was also found to switch immunogenic proteins into tolerogenic ones [66]. However, the tolerogenic properties of conjugates are often unpredictable because they depend upon a set of parameters, including protein, polymer and whole construct physicochemical properties [67].

14.5 Some Specific PEGylated Biopharmaceuticals

The practical importance of PEGylation is clearly underlined by the number of protein–PEG conjugates now on the market or in advanced clinical trials. Examples of commercialized PEGylated products include Adagen, Oncaspar, PEGIntron, PEGASYS, Neulasta and Somavert (Table 14.1); a dozen other PEG–proteins are now in advanced clinical trials [68].

Type 1 interferons are structurally and functionally related proteins belonging to the cytokine family. The first study of their PEGylation was the modification of IFN-α-2a with linear succinimidyl carbonate PEG (SC–PEG; 5 kDa) via a urea linkage. The coupling, performed at an equimolar ratio of protein to polymer, mainly led to mono-PEGylated isomers and, in small amounts, di-PEGylated conjugates, as well as free IFN [69]. Characterization of the conjugates indicated that lysine residues were the site of PEGylation [70]. Although a once a week schedule was possible, while the unconjugated protein was usually given three times a week, the desired pharmacokinetic/activity profile was still not fully achieved.

Granulocyte colony-stimulating factor (G-CSF) is a cytokine that controls proliferation, differentiation and functional activation of neutrophilic granulocytes [71]. Conventional PEGylation of G-CSF yielded a mixture of conjugates with prolonged blood residence time and resistance towards enzyme degradation. A reductive alkylation strategy with PEG–aldehyde/sodium cyanoborohydride in acidic buffer solution was proposed to selectively label the low-pK_a amino groups [72]. Under these conditions, modification limited to the N-terminal methyonine of r-metHuG-CSF was obtained [73, 74]. The resulting PEG 20 kDa conjugate presented an improved pharmacokinetic profile, mainly due to reduced kidney excretion. The normal relatively rapid clearance of G-CSF is also due to an internalization process of the receptor–ligand complex in neutrophil cells and is in some way related to the number of circulating neutrophils. PEG–G-CSF, which remains in the bloodstream for a prolonged time, stimulates the proliferation of neutrophils and consequently causes its own clearance when the therapeutic target is achieved. The PEG–G-CSF conjugate pegfilgastrim has been on the market since 2002.

Thrombopoietin or megakaryocyte growth and development factor (MGDF) is a key regulator of thrombopoiesis [75]. It stimulates the expansion and maturation of megakaryocyte progenitor cells, promoting an increase in platelet counts [76]. As the native form is eliminated very quickly via the urine, a PEG–rHuMGDF was synthesized [77] with the same chemical strategy already proposed in G-CSF modification. The PEG derivative was as active as the glycosylated full-length native thrombopoietin and much more active than both the free rHuMGDF and the non-glycosylated full protein. Interestingly, in this research the truncated active form of the protein and not the native form was elected for PEGylation, obtaining a conjugate with comparable activity and improved pharmacokinetics with respect to the full-length hormone. This method could open up new strategies and potential for protein PEGylation.

Human growth hormone (hGH) is a protein that has several effects, including enhancing linear body growth, tissue growth, activation of macrophages, lactation,

Table 14.1 PEGylated proteins on the market.

Trade name	Conjugated protein	Pathological target	Year on the market	Proprietary company
Adagen	Adenosine deaminase	Severe combined immunodeficiency disease	1990	Enzon
Oncaspar	Aspariginase	Acute lymphoblastic leukemia	1994	Enzon
PEGintron	Interferon α2b	Hepatitis C, multiple sclerosis, AIDS	2000	Schering Plough
PEGASYS	Interferon α2a	Hepatitis C	2002	Roche
Neulasta	G-CSF	Neutropenia (from chemotherapy)	2002	Amgen
Somavert	Growth hormone receptor agonist	Acromegaly	2002	Pfizer
Cymzia	Anti-TNF Fab	Rheumatoid arthritis, Crohn's disease	2008	UCB

insulin-like and diabetogenic effects [78]. One of the first studies in this respect involved a random PEGylation of hGH with low molecular weight PEG-*N*-hydroxysuccinimide (PEG-NHS; 5 kDa), leading to a mixture of derivatives. After purification it was possible to isolate hGH conjugates with up to seven PEG chains [79]. Binding studies demonstrated an inverse correlation between the number of PEG chains and receptor affinity, as PEG reduced the association rate for the receptor/adduct formation. Conversely, the higher the number of PEG chains linked the higher the *in vivo* potency, due to reduced kidney clearance. It was found that hGH conjugated to five PEG chains was a good compromise for therapeutic purposes, presenting tenfold higher potency than the unmodified hormone.

An alternative treatment for GH deficiency involves the use of growth hormone-releasing hormone (GHRH) [80] or its analogues (commonly termed GRF). However, clinical use of this truncated form of GRF is widely limited by the short biological half-life (10–20 min in humans), due to both rapid kidney excretion and to N-terminal enzymatic degradation by endogenous aminopeptidases. PEG-GRF conjugates showed an improved pharmacokinetic profile that finally resulted in a more favorable pharmacodynamic response on the growth hormone-insulin grow factor 1 (GH-IGF-1) axis, as reported by the increased GH levels and the number of peaks in pig and rat plasma [81].

Interleukin-2 is a powerful immunoregulatory lymphokine [82] produced by lectin- or antigen-activated T cells that, like IFN, enhances their natural killer (NK) cell activity and therefore may find a role in the treatment of cancer. An interesting modification strategy for a recombinant IL-2 was proposed where the native glycosylation site was substituted by a PEG chain [83]. To reach this goal a cysteine residue was introduced at the glycosylation site, thus obtaining Cys3-rIL-2 that was later conjugated to PEG–MAL. The derivative retained the full activity of the parent protein and presented a fourfold increase in blood residence time. In this case PEG–Cys3-rIL-2 mimics the native IL-2, in fact the PEG chain replaces the sugar moiety. This method may better preserve the activity since the glycosylation site area, due to the steric bulk of the sugar moiety, is often not involved in receptor binding.

Increasing interest is centered on antibodies for the treatment of several diseases and some such products have already reached the market [84], with many more currently undergoing clinical evaluation [85]. PEGylation was initially applied to these proteins to reduce the high immunogenicity of murine monoclonal antibodies. Later, humanized and human antibodies seemed to be the solution to the problem, but the high production costs and the low expression levels somewhat limited their use. The reduced binding affinity of PEGylated antibodies was either due to modification near the antigen binding domain, generating steric hindrance, or to reduction of the number of free lysine residues involved in the ionic interactions that initiate the antigen binding. A different approach used was based on engineered antibody fragments, such as Fab′, with at least one free cysteine residue located far from the antigen binding site. This cysteine residue was reacted specifically with PEG–maleimide, leading to mono-PEGylated antibody forms. Studies based on this strategy, employing a PEG of molecular weight up to 40 kDa, demonstrated that the binding affinity was completely retained for all PEG-engineered Fab′ conjugates, whereas the random PEGylation of Fab′ NH2 groups led to up to 50% loss of binding activity [86–88].

Microbial arginine deiminase has been studied as a potential anticancer enzyme as it degrades arginine, an essential nutrient for some tumors. However, its use is limited because of its short half-life and its high immunogenicity. The enzyme was modified with PEGs of various molecular weights and structures (branched or linear) and with different coupling chemistries [89]. It was demonstrated that 50% of the enzyme activity was maintained with up to 40% amino group modification. With a PEG molecular weight up to 20 kDa, linearity was found between PEG weight, pharmacokinetic and pharmacodynamic properties. Conjugates with a prolonged blood residence time, obtained using higher molecular weight PEGs, showed an improved pharmacodynamic response.

Hemoglobin (Hb) has been extensively investigated as an oxygen-carrying therapeutic agent, but limitations for its clinical use resulted from its high vasoactivity, due to extravasation into interstitial spaces and scavenging of nitric oxide. PEGylation has been performed to prevent such Hb extravasation. After several unsuccessful random PEGylation attempts [90], a site-specific modification was made at Cys93 with maleimidophenyl PEG (5, 10 and 20 kDa), leading to PEGylated Hb carrying two chains of the polymer per Hb tetramer [91]. The colligative properties of the derivatives suggested that PEG helped to eliminate the Hb vasoactivity [92].

Erythropoietin (EPO) is a glycoprotein that increases the production of reticulocytes and red blood cells by stimulation of bone marrow cells. Recombinant EPOs have been prepared to increase the number of glycosylation sites, and mono-PEGylated conjugates of the new rEPO showed higher *in vivo* potency than unmodified EPO [93, 94]. The PEG conjugates are administered only once a week, instead of the three times a week in the case of the native molecule.

Leptin (Ob proteins) is primarily secreted from adipose tissue and plays an important role in body weight homeostasis by regulating food intake and energy expense. As a result, it could potentially be used to treat obesity. A modification of Ob protein with branched PEG2-NHS has also been carried out and the purified mono-PEGylated form was tested in mice for its ability to reduce food intake and body weight [95].

14.6 Conclusions

In this chapter we have provided an updated account of the influence of PEGylation on proteins and on their pharmacological properties. At present research on PEGylation is mainly focused on its application to different proteins of the conjugation methods already developed. This continuing interest in protein PEGylation is well documented and several hundred relevant patents have been filed so far.

References

1 Mann, M. and Jensen, O.N. (2003) Proteomic analysis of post-translational modifications. *Nature Biotechnology*, **21**, 255–261.

2 Walsh, C.T., Garneau-Tsodikova, S. and Gatto, G.J. (2005) Protein post-translational modifications: the chemistry of proteome diversifications. *Angewandte Chemie. International Edition*, **44**, 7342–7372.

3 Abuchowski, A., McCoy, J.R., Palczuk, N.C., van Es, T. and Davis, F.F. (1977) Effect of covalent attachment of polyethylene glycol on immunogenicity and circulating life of bovine liver catalase. *The Journal of Biological Chemistry*, **252**, 3582–3586.

4 Maggon, K. (2007) R&D paradigm shift and billion-dollar biologics, in *Handbook of Pharmaceutical and Biotechnology*, John Wiley & Sons, Inc, pp. 161–198.

5 Gregoriadis, G., Jain, S., Papaioannou, I. and Laing, P. (2005) Improving the therapeutic efficacy of peptides and proteins: a role for polysialic acids. *International Journal of Pharmaceutics*, **300**, 125–130.

6 Fernandes, A.I. and Gregoriadis, G. (2001) The effect of polysialylation on the immunogenicity and antigenicity of asparaginase: implication in its pharmacokinetics. *International Journal of Pharmaceutics*, **217**, 215–224.

7 Veronese, F.M., Caliceti, P. and Schiavon, O. (1977) New synthetic polymers for enzymes and lyposomes modification, in Poly(ethylene glycol) Chemistry and Biological Applications, ACS Symposium Series, vol. **680** (eds M.J. Harris and S. Zalipski), pp. 182–192.

8 Miyamoto, M., Naka, K., Shiozaki, M., Chujo, Y. and Saegusa, T. (1990) Preparation and enzymatic activity of poly [(*N*-acylimino) ethylene] modified catalase. *Macromolecules*, **23**, 3201–3205.

9 Torchilin, V.P., Voronkov, J.I. and Mazoev, A.V. (1982) The use of immobilized streptokinase. (Streptodekaza) for the therapy of thromboses. *Therap. Arkhiv (Russ)*, **54**, 21–25.

10 Abuchowski, A., Van Es, T., Palczuk, N.C. and Davis, F.F. (1977) Alteration of immunological properties of bovine serum albumin by covalent attachment of polyethylene glycol. *The Journal of Biological Chemistry*, **252**, 3578–3581.

11 Abuchowski, A., McCoy, J.R., Palczuk, N.C., Van Es, T. and Davis, F.F. (1977) Effect of covalent attachment of polyethylene glycol on immunogenicity and circulating life of bovine liver glutaminase. *The Journal of Biological Chemistry*, **252**, 3582–3586.

12 Kopecek, J., Kopeckova, P., Minko, T. and Lu, Z. (2000) HPMA copolymer-anticancer drug conjugates: Design, activity, and mechanism of action. *European Journal of Pharmaceutics and Biopharmaceutics*, **50**, 61–81.

13 Hoste, K., De Winne, K. and Schacht, E. (2004) Polymeric prodrugs. *International Journal of Pharmaceutics*, **277**, 119–131.

14 Bailon, P., Palleroni, A., Schaffer, C.A., Spence, C.L., Fung, W.J., Porter, J.E., Ehrlich, G.K., Pan, W., Xu, Z.X., Modi, M.W., Farid, A. and Berthold, W. (2001) Rational design of a potent, long-lasting form of interferon: A 40 kDa branched polyethylene glycol-conjugated interferon α-2a for the treatment of hepatitis C. *Bioconjugate Chemistry*, **12**, 195–202.

15 Delgado, C., Francis, G.E. and Fisher, D. (1992) The uses and properties of PEG-linked proteins. *Critical Reviews in Therapeutic Drug Carrier Systems*, **9**, 249–304.

16 Hooftman, G., Herman, S. and Schacht, E. (1996) Poly(ethylene glycol)s with reactive endgroups. II. Practical consideration for the preparation of protein–PEG conjugates. *Journal of Bioactive and Compatible Polymers*, **11**, 135–159.

17 Delgado, C., Francis, G.E., Malik, F., Fisher, D. and Parkes, V. (1997) Polymer-derivatized proteins: analytical and preparative problems. *Pharmaceutical Sciences*, **3**, 59–66.

18 Harris, J.M. (ed.) (1992) *Poly(ethylene glycol) Chemistry-Biotechnical and Biomedical Applications*, Topics in Applied Chemistry, Plenum Press, New York.

19 Harris, J.M. and Zalipsky, S. (eds) (1997) *Poly(ethylene glycol): Chemistry and Biological Applications*, ACS Symposium Series, Vol **680**, American Chemical Society, Washington DC.

20 Veronese, F.M. and Harris, J.H. (Theme Eds.) (2002) Issue on peptide and protein PEGylation. *Advanced Drug Delivery Reviews*, **54** (4).

21 Harris, J.M. and Veronese, F.M. (Theme Eds.) (2003) Issue on PEGylation of peptides and proteins. II: clinical evaluation. *Advanced Drug Delivery Reviews*, **55** (10).

22 Veronese, F.M. and Harris, J.H. (Theme Eds) (2008) Issue on peptide and protein PEGylation. III: Advances in chemistry and clinical applications. *Advanced Drug Delivery Reviews*, **60** (1).

23 Working, P.K., Newman, S.S., Johnson, J. and Cornacoff, J.B. (1997) Safety of poly (ethylene glycol) derivatives, in *Poly (ethylene glycol): Chemistry and Biological Applications* (eds J.M. Harris and S. Zalipsky), ACS Symposium Series 680, American Chemical Society, Washington DC, pp. 45–54.

24 Choe, Y.H., Conover, C.D., Wu, D., Royzen, M., Gervacio, Y. and Borowski, V. (2002) Anticancer drug delivery systems: multiloaded N4-acyl poly(ethylene glycol) prodrugs of ara-C: II. Efficacy in ascites and solid tumors. *Journal of Controlled Release*, **79**, 55–70.

25 Schiavon, O., Pasut, G., Moro, S., Orsolini, P., Guiotto, A. and Veronese, F.M. (2004) PEG–Ara-C conjugates for controlled release. *European Journal of Medicinal Chemistry*, **39**, 123–133.

26 Pasut, G., Scaramuzza, S., Schiavon, O., Mendichi, R. and Veronese, F.M. (2005) PEG–epirubicin conjugates with high loading. *Journal of Bioactive and Compatible Polymers*, **20**, 13–30.

27 Berna, M., Dalzoppo, D., Pasut, G., Manunta, M., Izzo, L. and Jones, A.T. (2006) Novel monodisperse PEG–dendrons as new tools for targeted drug delivery: synthesis, characterization and cellular uptake. *Biomacromolecules*, **7**, 146–153.

28 Israelachvili, J. (1997) The different faces of poly(ethylene glycol). *Proceedings of the National Academy of Sciences of the United States of America*, **94**, 8378–8379.

29 Harris, J.M. and Chess, R.B. (2003) Effect of PEGylation on pharmaceuticals. *Nature Reviews. Drug Discovery*, **2**, 214–221.

30 Manjula, B.N., Tsai, A., Updhya, R., Perumalsamy, K., Smith, P.K. and Malavalli, A. (2003) Site-specific PEGylation of hemoglobin at Cys-93(b): correlation between the colligative properties of the PEGylated protein and the length of the conjugated PEG chain. *Bioconjugate Chemistry*, **14**, 464–472.

31 Basu, A., Yang, K., Wang, M., Liu, S., Chintala, M. and Palm, T. (2006) Structure–function engineering of interferon-beta-1b for improving stability, solubility, potency, immunogenicity, and pharmacokinetic properties by site-selective mono-PEGylation. *Bioconjugate Chemistry*, **17**, 618–630.

32 Kawai, F. (2002) Microbial degradations of polyethers. *Applied Microbiology and Biotechnology*, **58**, 30–38.

33 Mehvar, R. (2000) Modulation of the pharmacokinetics and pharmacodynamic SOD proteins by polyethylene glycol conjugation. *Journal of Pharmaceutical Sciences*, **3**, 123–125.

34 Beranova, M., Wasserbauer, R., Vancurova, D., Stifter, M., Ocenaskova, J. and Mara, M. (1990) Effect of cytochrome P-450 inhibition and stimulation on intensity of polyethylene degradation in

microsomial fraction of mouse and rat livers. *Biomaterials*, **11**, 521–524.

35 Petrak, K. and Goddard, P. (1989) Transport of macromolecules across the capillary walls. *Advanced Drug Delivery Reviews*, **3**, 191–214.

36 Yamaoka, T., Tabata, Y. and Ikada, Y. (1994) Distribution and tissue uptake of poly (ethylene glycol) with different molecular weights after intravenous administration to mice. *Journal of Pharmaceutical Sciences*, **83**, 601–606.

37 Yamaoka, T., Tabata, Y. and Ikada, Y. (1995) Fate of water-soluble administered via different routes. *Journal of Pharmaceutical Sciences*, **84**, 349–354.

38 Monfardini, C., Schiavon, O., Caliceti, P., Morpurgo, M., Harris, J.M. and Veronese, F.M. (1995) A branched monomethoxypoly(ethylene glycol) for protein modification. *Bioconjugate Chemistry*, **6**, 62–69.

39 Veronese, F.M. and Morpurgo, M. (1999) Bioconjugation in pharmaceutical chemistry. *Il Farmaco*, **54**, 497–516.

40 Kinstler, O., Molineux, G., Treheit, M., Ladd, D. and Gegg, C. (2002) Mono-N-terminal poly(ethylene glycol)-protein conjugates. *Advanced Drug Delivery Reviews*, **54**, 477–L 485.

41 Lee, S. and McNemar, C.(November 16 1999) Substantially pure histidine-linked protein polymer conjugates. (ENZON Inc.), US Patent 5,985,263.

42 El Tayar, N., Zhao, X. and Bentley, M.D.(April 11 1999) PEG-LHRH analog conjugates. (Applied Research System), World Patent 9,955,376.

43 Orsatti, L. and Veronese, F.M. (1999) An unusual coupling of poly(ethylene glycol) to tyrosine residues in epidermal growth factor. *Journal of Bioactive and Compatible Polymers*, **14**, 429–436.

44 Cox, G.N.(January 28 1999) Derivatives of growth hormone and related proteins, (Bolder Biotechnology Inc.), World Patent 9,903,887.

45 Zalipsky, S. and Menon-Rudolph, S. (1998) Hydrazine derivatives of poly(ethylene glycol) and their conjugates, in *Poly(ethylene glycol) Chemistry: Biotechnical and Biomedical Applications* (ed. J.M. Harris), Plenum Press, New York, pp. 319.

46 Saxon, E. and Bertozzi, C.R. (2000) Cell surface engineering by a modified Staudinger reaction. *Science*, **287**, 2007–2010.

47 Cazalis, C.S., Haller, C.A., Sease-Cargo, L. and Chaikof, E.L. (2004) C-terminal site-specific PEGylation of a truncated thrombomodulin mutant with retention of full bioactivity. *Bioconjugate Chemistry*, **15**, 1005–1012.

48 Sato, H. (2002) Enzymatic procedure for site-specific PEGylation of proteins. *Advanced Drug Delivery Reviews*, **54**, 487–504.

49 De Frees, S., Wang, Z.G. and Xing, R. (2006) GlycoPEGylation of recombinant therapeutic proteins produced in *Escherichia coli*. *Glycobiology*, **16**, 833–843.

50 Sato, H., Ikeda, M., Suzuki, K. and Hirayama, K. (1996) Site-specific modification of interleukin-2 by the combined use of genetic engineering techniques and transglutaminase. *Biochemistry*, **35**, 13072–13080.

51 Sato, H., Yamamoto, Y., Hayashi, E. and Takahara, Y. (2000) Transglutaminase-mediated dual and site-specific incorporation of poly(ethylene glycol) derivatives into a chimeric interleukin-2. *Bioconjugate Chemistry*, **11**, 502–509.

52 Sato, H., Hayashi, E., Yamada, N., Yatagai, M. and Takahara, Y. (2001) Further studies on the site-specific protein modification by microbial transglutaminase. *Bioconjugate Chemistry*, **12**, 701–710.

53 Fontana, A., Spolaore, B., Mero, A. and Veronese, F.M. (2008) Site-specific modification and PEGylation of pharmaceutical proteins mediated by transglutaminase. *Advanced Drug Delivery Reviews*, **60**, 13–28.

54 Harris, J.M., Martin, N.E. and Modi, M. (2000) Pegylation: a novel process for modifying pharmacokinetics. *Clinical Pharmacokinetics*, **40**, 539–551.

55 Delgado, C., Francis, G.E. and Fisher, D. (1992) The uses and properties of PEG-linked proteins. *Critical Reviews in Therapeutic Drug Carrier Systems*, **9** 249–304.

56 Kartre, K. (1993) The conjugation of proteins with poly(ethylene glycol) and other polymers. *Advanced Drug Delivery Reviews*, **10**, 91–114.

57 Manjula, B.N., Tsai, S., Upadhya, R., Perumalsamy, K., Smith, P.K., Malavalli, A., Vandegriff, K., Winslow, R.M., Intaglietta, M. and Prabhakaran, M. (2003) Site-specific PEGylation of hemoglobin at cys-93(beta): Correlation between the colligative properties of the PEGylated protein and the length of the conjugated PEG chain. *Bioconjugate Chemistry*, **14**, 464–472.

58 Hu, T., Prabhakaran, M., Acharya, S.A. and Manjula, B.N. (2005) Influence of the chemistry of conjugation of poly(ethylene glycol) to Hb on the oxygen-binding and solution properties of the PEG-Hb conjugated. *Biochemical Journal*, **392** (3), 555–564.

59 Solà, R.J., Rodriguez-Martinez, J.A. and Griebenow, K. (2007) Modulation of protein biophysical properties by chemical glycosylation: biochemical insights and biomedical implications. *Cellular and Molecular Life Sciences*, **64**, 2133–2215.

60 Campaner, P., Drioli, S. and Bonora, G.M. (2006) Synthesis of selectively end-modified high-molecular weight polyethyleneglycol. *Letters in Organic Chemistry*, **10**, 773–779.

61 Bonora, G.M., Campaner, P. and Drioli, S. (2005) Bifunctional derivatives of polyethylene glycol, their preparation and use, WO 2005/123139.

62 Veronese, F.M. (2001) Peptide and protein PEGylation: a review of problems and solutions. *Biomaterials*, **22**, 405–417.

63 Takakura, Y., Fujita, T., Hashida, M. and Sezaki, H. (1990) Disposition of macromolecules in tumor bearing mice. *Pharmaceutical Research*, **7**, 339–346.

64 Abuchowski, A.A., van Es, T., Palczuk, N.C. and Davis, F.F. (1977) Alteration of immunological properties of bovine serum albumin by covalent attachment of polyethylene glycol. *The Journal of Biological Chemistry*, **252**, 3578–3581.

65 Davis, F.F., Kazo, G.M., Nucci, M.L. and Abuchowski, A. (1991) Reduction of immunogenicity and extension of circulating half-life of peptides and proteins, in *Peptide and Protein Drug Delivery* (ed. V.H.L. Lee), Marcel Dekker, New York, pp. 831.

66 Sehon, A.H. (1999) Suppression of antibody responses by conjugates of antigens and monomethoxypoly(ethylene glycol), polymers. *Bioconjugate Chemistry*, **10**, 638–646.

67 Caliceti, P., Veronese, F.M. and Jonak, Z. (1999) Immunogenic and tolerogenic properties of monomethoxypoly(ethylene glycol) dismutase modified by monomethoxy(polyethylene glycol) conjugated proteins. *Il Farmaco*, **54**, 430–437.

68 Pasut, G. and Veronese, F.M. (2007) Polymer drug conjugation, recent achievements and general strategies. *Progress in Polymer Science*, **32**, 933–961.

69 Hoffmann-La Roche, (1994) EP-0593868.

70 Monkarsh, S., Ma, Y. and Aglione A. (1997) Positional isomers of mono-pegylated interferon α-2a: Isolation, characterization, and biological activity. *Analytical Biochemistry*, **247**, 434–440.

71 Pepinsky, R.B., Le Page, D.J., Gill, A., Chakraborty, A., Vaidyanathan, S., Green, M., Baker, D.P. and and Martin, P. (2001) Improved phamacokinetic properties of polyethylene glycol-modified form of interferon-(-1a with preserved in vitro bioactivity. *The Journal of Pharmacology and Experimental Therapeutics*, **297** (3), 1059–1066.

72 Kinstler, O., Moulinex, G. and Treheit, M. (2002) Mono-N-terminal poly(ethylene glycol)-protein conjugates. *Advanced Drug Delivery Reviews*, **54**, 477–485.

73 Welte, K., Gabrilove, J., Bronchud, M., Platzer, E. and Morstyn, G. (1996) Filgrastrim (R-Methugcsf): the first 10 years. *Blood*, **88**, 1907–1929.

74 Amgen, Inc . (1999) US5985265.

75 Eliason, J., Greway, A. and Tare, N. (2000) Extended activity in cynomolgus monkeys of a granulocyte colony-stimulating factor mutein conjugated with high molecular weight polyethylene glycol. *Stem Cells (Dayton, Ohio)*, **18** (1), 40–45.

76 Lok, S., Kaushansky, K. and Holly, R. (1994) Cloning and expression of murine thrombopoietin cdna and stimulation of platelet production, in vivo. *Nature*, **369**, 565–568.

77 Amgen, Inc. , (1998) US5795569.

78 Guerra, P., Acklin, C., Kosky, A., Davis, J., Treuheit, M. and Brems, D. (1998) Pegylation prevents the N-terminal degradation of megakaryocyte growth and development factor. *Pharmaceutical Research*, **15**, 1822–1827.

79 Chawla, R., Parks, J. and Rudman, D. (1983) Structural variants of human growth hormone: biochemical, genetic, and clinical aspects. *Annual Review of Medicine*, **34**, 519–547.

80 Clark, R., Olson, K., Fuh, G. and Marian, M. (1996) Long-acting growth hormones produced by conjugation with polyethylene glycol. *The Journal of Biological Chemistry*, **271** (36), 21969–21977.

81 Piquet, G., Gatti, M., Barbero, L., Traversa, S., Caccia, P. and Esposito, P. (2002) Set-up of a large laboratory scale chromatographic separation of poly (ethylene glycol) derivatives of the growth hormone-releasing factor 1–29 analogue. *Journal of Chromatography A*, **944**, 141–148.

82 Parkinson, C., Scarlett, J. and Trainer, P. (2003) Pegvisomant in the treatment of acromegaly. *Advanced Drug Delivery Reviews*, **55**, 1303–1314.

83 Smith, K. (1988) Interleukin-2: inception, impact, and implications. *Science*, **240**, 1169–1176.

84 King, D. (1998) Preparation, structure and function of monoclonal antibodies, in *Application and Engineering of Monoclonal Antibodies*, Taylor and Francis, London, pp. 1–26.

85 King, D. and Adair, J. (1999) Recombinant antibodies for the diagnosis and therapy of human disease. *Current Opinion in Drug Discovery & Development*, **2**, 110–117.

86 Anderson, W. and Tomasi, T. (1988) Polymer modification of antibody to eliminate immune complex and Fc binding. *Journal of Immunological Methods*, **109**, 37–42.

87 Celltech Therapeutics Ltd., (2001) EP-1160255.

88 Celltech Therapeutics Ltd., (1999) WO9964460.

89 Hinds, K. and Kim, S. (2002) Effects of PEG conjugation on insulin properties. *Advanced Drug Delivery Reviews*, **54**, 505–530.

90 Enzon, Inc., (1995) US5386014.

91 Manjula, B., Tsai, A. and Upadhya, R. (2003) Site-specific PEGylation of hemoglobin at Cys-93(β): correlation between the colligative properties of the PEGylated protein and the length of the conjugated PEG chain. *Bioconjugate Chemistry*, **14**, 464–472.

92 Eistein Coll. Med., (1996) US5585484.

93 Schurig, H. (2002) US2002115833.

94 Amgen, Inc., (2003) US6586398.

95 Hoffmann-La Roche, (2000) US6025324.

Index

a

activated protein C 7, 11
– schematic diagram 7
adaptor molecules 80
alien tissue 80
– CHO 80
– NS0 80
– Sp2/0 cells 80
AMG114 molecule 307
– carbohydrates on 307
– clinical testing in CIA patients 307
– glycoengineering 307
α-amidated peptide 253, 254, 256, 257, 260, 265, 266, 269–271
– dual recombinant expression process 266
– recombinant production 265
– species distribution of 257
α-amidated peptide hormones 254, 270
– conversion of prohormones into mature 254
– disease indications 271
α-amidating enzyme (α-AE) 12
α-amidation 260, 262
– C-terminus 269
– *in vitro* 261
– influence of pH 262
– reactive oxygen species (ROS) 260
– standard reaction components 261
amino acid side chains 87, 258
8-aminonaphthalene-1,3,6-trisulfonic acid (ANTS) 35
amplifiable gene 52
– dihydrofolate reductase (DHFR) 52
– glutamine synthetase (GS) 52
animal cell cultures 51
antibody(ies) 9, 34–36, 65, 68, 79, 80, 86, 91, 92, 94–100, 123, 156, 157, 159, 172, 227, 277, 351
– anti-CD20 monoclonal antibody 159
– anti-CD28 antibody 99
– anti-CD33 antibody 98
– anti-D antibodies 93
– based products 9
– basic structure/function 80–83
– fragments 99
– glycoforms 100
– immunoglobulin isotypes 84
– TGN 1412 99
– therapeutics 79
antibody-dependent cellular cytotoxicity (ADCC) 57, 68, 89, 157
– activity 94, 95
anti-parallel β-sheets 81
ascorbic acid 233
Ashkenazi Jewish population 320
asialoglycoprotein receptor (ASGR) 309
asialylglycoprotein receptor (ASGPR) 91, 92
Aspergillus niger 261

b

baby hamster kidney (BHK) cells 60, 61, 65, 299
bacterial glycoproteins 205
– bacterial polysaccharides 205
bacterial glycosylation technologies 193–195, 197
– advantage 197
– characteristics 203
– crystal structures 197
– 3D design 197
– eukaryotes 197
– protein glycosylation 194
bacterial polysaccharide metabolizing enzymes 193
bacterial systems 149
baculovirus 174, 178, 322

Post-translational Modification of Protein Biopharmaceuticals. Edited by Gary Walsh

ISBN: 978-3-527-32074-5

- expressed proteins 322
- *ie* promoter(s) 174, 178

baculovirus encoded glycosyltransferases 175
- *N*-acetylglucosaminyltransferase I 175
- *N*-acetylglucosaminyltransferase II 176
- galactosyltransferase 175
- sialyltransferase 175
- *trans*-sialidase (TS) 176

baculoviruses encoded sugar processing genes 176
- *N*-acetylmannosamine kinase 176
- CMP-sialic acid synthetase 177
- sialic acid synthetase 177
- UDP-GlcNAc 2-epimerase 176

baculovirus expression vector system (BEVS) 165, 166
- core α1,3 fucosyltransferase 172
- derived *N*-glycoprotein product 166, 170
- encoded firefly luciferase gene 184
- lack of mannose-6-phosphate 172
- *N*-glycan processing enzymes 171
- processing β-*N*-acetylglucosaminidase 171
- produced recombinant *N*-glycoproteins 170

baculovirus expression vector system *N*-glycosylation 174
- baculovirus encoded glycosyltransferases 175
- baculoviruses encoded sugar processing genes 176
- baculoviruses use 174
- promoter choice 174

baculovirus expression vector system product pipeline 166
- Chimigen vaccines 167
- FluBlok 167
- influenza virus-like particles 167
- Provenge 167
- recombinant *N*-glycoproteins 166
- Specifid™ 168

β-barrel structure 81

batch culture 61, 62
- γ-interferon production 62

beta-elimination process 37

binary expression system 165

bioconjugate production 204
- *Escherichia coli* 204

biologically active proteins 149

biological molecules 34
- nucleic acids 34
- proteins 34

biopharmaceuticals 2, 7, 9, 57, 165
- glycosylated 9
- IgG proteins 57
- insect cell glycosylation patterns 165
- products profiles 2
- products sub-categories 2
- PTMs 7
- PEGylated 13

blood factor(s) 2, 9, 11, 12
- Asp64 233
- blood clotting 221
- factor IX (FIX) 2, 221
- factor VII (FVII) 2, 221
- factor VIII 2
- β-hydroxylation of Asp64 221
- in liver 221
- treatment of hemophilia B 221

blood clotting mechanisms 18

Bombyx mori larvae 169

bovine pancreatic trypsin inhibitor (BPTI) 283

bovine prothrombin 215
- Ca^{2+}-loaded Gla domain 215

building enzymes 22
- glycosyltransferases 22

Burkitt's lymphoma 98

c

Campylobacter jejuni 194, 195

Candida albicans 152

capillary affinity electrophoresis (CAE) 39

capillary electrophoresis (CE) 39
- oligosaccharides mapping 39
- oligosaccharides separation 39

carbohydrate addition 300
- AMG114 307–308
- darbepoetin alfa 303–307
- effect 309
- glycoengineered ESAs 308–309
- glycoforms 302–303
- mechanism of clearance 309–311
- *O*-linked 301
- rules 300–302
- threonine 301

carbohydrate remodeling process 326
- effect 325

γ-carboxy glutamate (Gla) 211
- biosynthesis of 217
- blood clotting protein 212
- catalytic (serine protease) domain 213
- Cys residues 214
- epidermal growth factor-like domain (EGF1) 213
- γ-carboxylated vertebrate proteins 214
- NMR structures 216
- *N*-terminus of mature polypeptides 212

– protein-protein interactions 215
– vitamin K-dependent polypeptides 217
– vitamin K redox cycle 218
– X-ray crystallography 216
γ-carboxylated biopharmaceuticals 219
– factor IX (FIX) 219
– factor VII (FVII) 221
– protein C/activated protein C 222
– prothrombin 223
γ-carboxylated proteins 212, 226
– analytical characterization of 227–228
– cone snails 213
– enzyme assays 228
– metal binding-induced structural changes 228
– metal content 227
– methods for detecting Gla 227
– phospholipid membrane binding assays 228
– purification 226
– roles in 212
γ-carboxylation process 211
– biological function of 211–214
– biopharmaceuticals 219–224
– biosynthesis of Gla 217–219
– cellular carboxylation capacity 224–226
– conotoxins 224
– Gla domain 214–217
– Glu residues 215, 219
– in Ca^{2+} binding 212
– of glutamyl (Glu) 211
– recognition site 217
carboxyl-terminal segment 110
catabolic pathway 319
catalytic cysteine residues 282
– DsbC 282
cDNA, sequence 111, 327, 328
cell culture process 327
cell signaling system 56
cell surface membrane 109
cell surface receptor(s) 172, 324
cellular carboxylation capacity 224
– γ-carboxylation machinery 225
– enhancement of 224
– inhibition of calumenin expression 226
– propeptide/propeptidase engineering 226
cellular glutamine metabolism 65
cellular mannose receptor (MR) 94
cellular splicing mechanisms 183
Chaperone molecules 25
– calnexin 25
– calreticulin 25
charge distribution profile 129
chimeric proteins 167
Chimigen technology 167
Chinese hamster ovary (CHO) cells 51–53, 56, 59–61, 63, 66, 68, 80, 94, 119, 130, 221, 327, 331
– cell culture 59
– cell-derived gonadotropins 130
– cell line(s) 80, 95
– cell proteins 327
– complex type *N*-linked glycans 130
– core-fucosylation 52
– glycosylation metabolism 56
– glycosylation pathways 52
– mutagenesis 52
– mutants 52
– *O*-glycan synthesis 53
– *O*-linked glycans 53
– recombinant bifunctional rPAM 268
choriogonadotropin (CG) hormones 109
– clearance 122
– deglycosylation 123
– induced fluorescence resonance energy 123
– β-subunit genes 111, 121
chromatographic fractionation techniques 343
clonal cells lines 178
clone selection process 99
CMP-sialic acid 175, 179, 182
– synthetase enzyme 177
complement dependent cytotoxicity (CDC) 89
complex cell-type *in vitro* assay systems 124
complex glycan structures 19
– formation 19
– α/β glycosidic bonds 19
computer-assisted algorithms 36
congenital disorders of glycosylation (CDG) 79
conotoxins 234
– biopharmaceutical potential 234
Conus marmoreus 234
copper sulfate titration 262
core-fucosylation pathway 57
correlated spectroscopy (COSY) 40
cost of goods (CoG) 79
cost of treatment (CoT) 79
Crohn's disease 18
cystine knot proteins 112, 113
– crystal structures 113
cytoplasmic proteins 265, 287
– gene coding 287
– glutaredoxin systems 287
– thioredoxin systems 287

d

d-arabinofuranose (d-Araf) 201
– alpha-1,5-linked 201
darbepoetin, *see* human erythropoietin (EPO)
Desulfovibrio desulfuricans 194
dextran ladder, *see* glucose polymers
2D gel electrophoresis 66
2,6-diaminopyridine (DAP) 35
differential scanning micro-calorimetry (DSC) 87
dimeric protein 284
direct expression technology 269
– expression levels of peptides 269
– fermentation protocol 269
dissolved oxygen (DO) 66
disulfide bond (Dsb) 278, 286
– oxidation-reduction cycles 279, 280, 284
– pathway 286
– dimeric protein 284
disulfide bond formation 10, 278, 286
– catalyzed process 278
– cytoplasm 286
– disulfide isomerization pathway 282
– DsbA/DsbB oxidizing protein 279, 280
– *E. coli* thioredoxin mutants 281
– ER oxidoreductin 289
– Ero1 regulatory role 290
– in bacterial periplasm 278
– in endoplasmic reticulum 288
– new oxidation pathway 281
– PDI functions 288
– revised model for formation 286
– role 10
– thiol group 278
disulfide isomerization pathway 282
– periplasmic protein disulfide isomerase 282
dolichol oligosaccharide precursor 24
– building 24
– positioning 24
dopamine-beta-monooxygenase (DβM) 256
double-stranded RNA (dsRNA) 183
– encoding construct 184
Drosophila melanogaster 173
– FDL protein 171
DsbC *see* dimeric protein
D409V Gaucher mouse model 331

e

Edman degradation 235
effector activities 91–96
effector function profile 99
electron transport chain 280
– Cys41/Cys44 280
– DsbB 280
electrospray ionization (ESI) 41
endocrine-governed mechanism 127
endoplasmic reticulum (ER) 113, 150, 288, 289
– disulfide bond formation 289
enzymatic α-amidation 259, 263
– kinetic parameters 259
– temperature influence 263
enzyme replacement therapy (ERT) 319, 322, 323, 330
– Ceredase 319
– Cerezyme 319
– clinical studies 323
– glycoengineering 324
– mannose receptor identification 324, 330
– placental glucocerebrosidase use 323–327
– role in macrophage uptake 324
epidermal growth factor (EGF) 229, 346
– Ca^{2+}-binding 229
– contain Hya/Hyn 231
– FIX/FX 230
– functional importance 231
– β-hydroxylation of 231
– schematic of the Ca^{2+}-binding site 230
erythro-β-hydroxyasparagine (Hyn) 211, 232
– biosynthesis 232
erythro-β-Hydroxyaspartic acid (Hya) 211, 232
– Asp/Asn residue 232
– biosynthesis 232
erythropoiesis stimulating agent (ESA) 306, 309
– clearance 309
erythropoietin (EPO) 67, 159, 299, 352
– endogenous 299
– heterogeneity analysis 66
– receptor-mediated metabolism 310
– role 310
Escherichia coli 7, 12, 99, 149, 160, 178, 193, 264, 283, 286, 288
– disulfide bonds 286
– β-galactosidase 178
– glycosylated protein 7
– isomerization pathway 283
– periplasm 283, 286
eukaryotic protein processing 166
eukaryotic recombinant proteins 283
eukaryotic systems 149, 165, 196
European Union (EU) 3
– approved biopharmaceutical products 3–5
eye-specific marker 181

f

Fabry's disease 320
– galactosidase A deficiency 320
fast atom bombardment (FAB) 41
Fcg receptors (FcgR) 88
– IIIa receptor 159
– human receptor 86
fed-batch culture 63, 64
– nutrient consumption 64
– product accumulation 64
– strategies for culture studies 63
fibroblast growth factor-1 (FGF-1) 156
fluorescence-assisted cell sorting (FACS) 332
fluorophore-assisted carbohydrate electrophoresis (FACE) 39, 329
– oligosaccharides mapping 39
– oligosaccharides separation 39
follicle-stimulating hormone (FSH) 67, 109, 110, 121, 129, 131
– α-subunit 118
– analogs 132
– binding site 113
– chromatofocusing separation 125
– glycan structure 126
– glycoforms 124, 128
– hormone glycosylation 121
– *in vitro* B/I ratio 129
– isoforms 122
– mediated signal transduction 123
– oligosaccharides 129
– plasma disappearance curves 125
– production 128
– receptor 123
– recombinant human 125
– single-chain 124
– structure 124
– wild type 120
fragment antigen binding (IgG-Fab) regions 82, 92, 96–98
– glycoforms 92
– glycosylation 96–98
– region 96
fragment crystallizable activities 94
– bisecting *N*-acetylglucosamine Influence 94
– fucose influence 94
fragment crystallizable effector functions 88, 93
– inflammatory cascades 88–90
fragment crystallizable (IgG-Fc) regions 82, 84, 87, 89–99
– binding proteins 89, 95
– definition 82
– diantennary oligosaccharide structures 83
– drug-fusion proteins 91
– fragment 86
– galactosylation 84, 92
– glycoform profile 84, 87, 91–96, 98
– normal human 83, 85, 93
– oligosaccharides HPLC profile 84, 85, 97, 98
– oligosaccharides sialylation 92
– receptor(s) 82, 89, 90, 99
– regions 82
– structure 92, 93
α1,6 fucosyltransferase (α1,6FT) 57
– nascent glycoprotein 57
functional glycomics 67–68
fungal protein expression systems 159

g

galactose-fed cultures 63
β-galactosidase 324
galactosylated anti-D antibody 93
galactosylation pathway 65
– galactose feeding 65
gas-liquid chromatography (GLC) 37, 41
– monosaccharides Linkage Position Determination 41
Gaucher's disease 13, 172, 319–322, 325, 334
– ERT development 325
GDP-mannose 4–6 dehydratase (GDM) 57
gene encoding 225
– cloning of 225
genetic engineering 2
genetic methods 155
genome sequencing 132
globotriaosylceramide (GL-3) 320
glucocerebrosidase (GCase) 320–322, 324–326, 328–331, 333, 334
– carbohydrate chains remodeling 325
– carbohydrate-remodeled 331
– carrot cell-expressed 334
– mannose receptor binding 331
– monosaccharide compositions 329
– purification process 326
– X-ray crystal structure 321
glucose polymers 38
glucose units (GU) 38
glutamine fructose 6-phosphate amidotransferase (GFAT) 64
γ-glutamyl carboxylase 217
– polypeptide substrate 217
– propeptide mediates binding 217
glutamyl residue (Glu) 211
– post-translational conversion 211
glutathione reductase (*gor*) 286
glycan 18, 20, 39, 41, 121

– chains termination 19, 33
– fluorescently-labeled 39
– glycozymes synthesis 22
– mass-to-charge ratio (m/z) 41
– monosaccharides units 18–20
– role 121
glycan detection 34
– lectin binding 34
– periodic acid-schiff (PAS) reaction 34
– pulsed amperometric detection (PAD) 35
glycan-directed enzyme replacement therapy 319–323
– Gaucher's disease 320–321
– glucocerebrosidase 321–322
– lysosomal storage diseases 319
glycan heterogeneity 55
– *N*-acetyl glucosaminyltransferases 55–57
– fucosylation 57
– sialylation 58–60
glycan labeling 35
– fluorescent labels 35
– radiolabels 35
glycan motifs 32, 33
– ABO blood group antigens 33
– Cad/Sd blood group antigens 33
– Lewis blood group antigens 33
glycan profiling 35
– microarrays use 35
glycan structures 17, 23, 51, 68, 124
glycine-extended fatty acids 254
– *in vitro* α-amidation 253
glycine-extended peptides 260, 266
– α-amidation 260–263
– direct expression process 266
– α-hydroxylation 260
– *in vitro* α-amidation 260
– precursors 266
– purification 266
– upstream signal sequence 266
glyco-catch approach 35
glycoconjugate vaccines 204
glycoengineered erythropoiesis stimulating agent 308
– AMG114 308
– darbepoetin alfa 308
– rHuEpo 309
glycoengineered glucocerebrosidase 330
– alternative expression systems use 333
– mannosidase inhibitors use 330
– mutant cell lines use 333
– targeting mannose receptor strategies 330
glycoengineered rHuEpo glycosylation analogs 306
glycoengineering 319, 324
– cerezyme case study 319
– strategies 319
– targeting strategy 324
glycoform profile 97
glycome exploring 36
– one/two-dimensional gel electrophoresis 36
glycoprotein 17, 177, 178, 204
– chemical release 36
– containing humanized glycans 204
– glycans structural complexity 19
– glycoprotein of interest (GOI) 174
– polypeptide sequences 17
– production 51
– synthesis 204
glycoprotein hormone 111–115, 117, 121, 125, 129, 132
– glycosylation 115, 132
– *in vitro/in vivo* bioactivity 125
– physicochemical properties 121
– processing 132
– producing tissues 113
– receptors 109
– role 114
– structurally-related 129
– sulfation 115
glycoproteins separation method 36
– HPLC 36
– mass spectroscopy 36
– SDS-PAGE 36
glycosidic linkage formation 21
glycosylation process 2, 8, 51, 54, 56, 58, 84, 86, 87
– case study 51
– central reaction network 56
– growth effect 54
– impact on stability 87
– impact on structure 86
– importance 109
– normal human IgG 84–86
– pathway 59
– profiles 80
– protein production rate effect 54
glycosylation culture parameters 61–67
– ammonia 65
– fed-batch cultures 63
– galactose supplement 65
– glucosamine supplement 64
– nutrient depletion 61
– oxygen 66
– pH 66
glycosyltranferase 22, 23, 52, 173
– *N*-acetylglucosaminyltransferases II-IV 173

– families 22
– galactosyl/sialyltransferase 173
– glycosyltransferase donor substrates 174
– lack of 173
– reactions 52
– transglycosylation reactions 23
Golgi apparatus (GA) 22–25, 86, 150, 154, 159
– factors 23
– *N*-glycosylation Step 24
– *N*-linked oligosaccharides processing 25
– role 23
– secretory pathway 23, 24
gonadotropin 109, 118, 121–122, 124, 126, 130, 131
– binding 123
– degree of charge heterogeneity 118
– function 121, 124, 126
– glycoforms 126
– glycosylation 114–121
– *in vivo* biological potency 122
– metabolic clearance rate 122
– microheterogeneity 132
– role in folding 121
– role in secretion 121
– role in subunit assembly 121
– signal transduction 123
– structure 111–114
– therapeutic applications 130–132
gonadotropin glycosylation regulation 127
– androgens effects 128
– estrogens effects 127
– gonadotropin-releasing hormone effects 129
gonadotropin-releasing hormone (GnRH) 128, 129
– role 129
– triggered alterations 130
G protein-coupled receptors 109
– heterotrimeric 111
granulocyte colony-stimulating factor (G-CSF) 349
growth hormone-insulin grow factor 1 (GH-IGF-1) 351

h

Haemophilus influenza 205
Halobacterium salinarium 194
healthcare biotechnology 1
α-helical transmembrane domains 110
hepatocyte receptors 122
heterologous proteins 288
– in *E. coli* 288
high-performance anion exchange chromatography (HPAEC) 39
high-performance liquid chromatography (HPLC) 37, 38
– normal phase HPLC (NP-HPLC) 38
– oligosaccharide mapping 38
– oligosaccharide separation 38
– weak anion exchange high-performance liquid chromatography (WAX-HPLC) 38
high pH anion exchange chromatography, *see* High-performance anion exchange chromatography (HPAEC)
homology modeling 110
– human FSHR 110
horseradish peroxidase (HRP) 172
– epitope 172, 184
human B cell lymphoproliferative disease 98
human cells 18
– glycosylation building blocks 18–19
– line(s) 57, 99, 150
human embryonic kidney (HEK293) cells 223
human erythropoietin (EPO) 13, 60, 68
– Aranesp 13
human FSH (hFSH) glycoform glycosylation 116
– receptor 124
human glycoprotein hormone 114
– glycosylation sites 114
human granulocyte-macrophage colony-stimulating factor 167
human growth hormone (hGH) 349
human immunoglobulin 90
– catabolism 90
– glycoform profile 100
– myeloma IgG proteins 92, 97
– pharmacokinetics 90
– placental transport 90
human placental free-α lacks *O*-glycans 114
human tissue plasminogen activator (t-PA) 54, 178
hydrophobic interaction chromatography (HIC) 267
β-hydroxylase enzyme assays 235
hydroxylated amino acids 231
– biosynthesis 231, 232
hydroxylated biopharmaceuticals 233
– conotoxins 234
– factor IX (FIX) 233
– protein C/activated protein C 233
β-hydroxylated proteins 235
– analytical characterization 235
– β-hydroxylase enzyme assays 235
hypogonadotropic hypogonadism 130
hypothalamic-releasing factors 132

i

immune response 82
immunogenic proteins 326
immunoglobulin (Ig) 67, 81
– alpha carbon backbone structure 83
– classes 67, 81, 84
– dissolved oxygen Effect 67
– Fc region 88
– four-chain structure 81
– heavy chains 82
– IgG antibody 68, 81, 84, 95, 99
– IgG glycoform profile 67
– IgG heavy chain 88
– IgG molecule(s) 81–83, 86, 88, 96
– IgG protein(s) 84, 168
– IgG-like domain(s) 321, 322
– light chains 82
in vitro fertilization protocols 112
in vitro models 126
influenza virus-like particles 167
insect cell(s) 166, 175, 182
– *N*-glycosylation patterns 166
– derived glycoproteins 168
– lines 170
– mammalian epimerase/kinase enzyme 182
– lepidopteran cell(s) 165, 182, 183
– lines 172
– *N*-glycosylation pathway 182
insect glycoprotein *N*-glycan structure 168
– hybrid/complex *N*-glycans 170
– sialylated *N*-glycans 170
intact oligosaccharides 37
– chemical release 37
– enzymatic release 37
γ-interferon (IFNγ) 60, 61, 169
– recombinant 169
interleukin-2 351
– immunoregulatory lymphokine 351

j

JAR choriocarcinoma cells 126

k

2-keto-3-deoxy-D-glycero-D-galacto-nononic acid (KDN) 177
Kitasatosporia kifunense 68
Kluyveromyces lactis 152
Kupffer cells 122, 320, 323, 324

l

lipid-linked oligosaccharides (LLOs) 196
– PglB 197
Lipinski's rules 269
lipoarabinomannan (LAM) polymers 201
luteinizing hormone (LH) 109, 111
– CGβ-subunit genes 111
– glycan structure 126
– glycoforms 124
– glycosylation 130
– isoforms 128
luteinizing hormone and CG bind to same receptor (LHCGR), 111
– green fluorescent protein 123
– yellow fluorescent protein chimeras 123
Lymantria dispar cells 169
lysosomal membrane glycoprotein-2 (LAMP-2) 54, 322
– deficient cells 322
– polylactosamine formation 54
lysosomal storage diseases (LSDs) 319, 322
– glycan-directed enzyme replacement therapy 319–323
– serum acid phosphatase activity 326

m

mammalian cell 159
– case study 51
– Chinese hamster ovary (CHO) 154
– glucocerebrosidase 333
– glycosyltransferases 177
– human embryonic kidney (HEK) 293 154
– lines 51, 154
– *O*-linked glycosylation 153
mannan binding lectin (MBL) 89, 331
mannose 6-phosphate (M6P) 322, 332
– modification 172
mannose 6-phosphate receptor (MPR) 89, 323, 330–332
– cation-dependent 323
– cation-independent 323
– macrophage 332
– pathway 323
mannose binding protein 171, 331
α-mannosidase II (MNSII) 157
mass spectrometry (MS) 40
– analysis 37, 86
– ionization 41
– oligosaccharide mass determination 40
matrix-assisted laser desorption/ionization mass spectrometry (MALDI-MS) 41
matrix-assisted laser desorption/ionization time-of-flight mass spectrometry (MALDI-TOF-MS) 119
megakaryocyte growth and development factor (MGDF) 349
membrane-associated lysosomal protein 326

membrane bound *N*-glycoprotein precursors 167
monoclonal antibody 156, 157
– anti-collagen type II antibodies 94
– 9G8a binds 304
– technology 2
monosaccharides analysis 36
– monosaccharide linkage data 37
– monosaccharide sequence 37
– sequential exoglycosidase digestions 37
mucin-type O-linked glycoprotein glycosylation 28–29
– no consensus sequence 29
muscle eye-brain disease (MEB) 154
mycobacteria cell walls 201

n

N-acetylglucosamine (GlcNAc) 56, 330
– β-*N*-acetylglucosaminidase 171
– residues 115
– sulfate terminal sequence 117
– transferase 115, 116, 117, 118, 127
N-acetyl glucosaminyltransferase (GnT) enzymes 55, 157, 173, 174, 333
– core 2 GlcNAc transferase (C2GnT) 53
– β-1,2 GNT I activity 156
– GnT-I activity 173, 175, 333
– GnT-III gene 94
– transfearse I 333
– transferase II 157
N-acetylmannosamine (ManNAc) 59
N-acetyl-neuraminic acid (NANA) 60
– CMP transport 59
N-glycan processing 174
– functions 174
– genes 174
– machinery 174
N-glycolyl-neuraminic acid (NGNA) 60, 98
N-glycosylation pathway 28, 56, 96, 150, 152, 154, 170, 179, 183
– hybrid/complex *N*-glycans 170
– in fungi 150
– profile 156
– reaction network 56
– reducing deleterious activities 183
– structure 150
N-glycosylation system 194
– acceptor protein 195–197
– bioconjugate vaccines 204–205
– *C.jejuni* 194
– characteristics of 203
– conformational requirements for 196
– eukaryotic system 196
– functions 94
– glycoengineering 205–206
– introduction 194–195
– lipid-linked oligosaccharide (LLO) substrate 197
– N-OTase 197, 198
– PglB 197, 198
– therapeutic (human) proteins 203–204
– undecaprenol-pyrophosphate (Und-PP) lipid 199
N-linked glycans 118–120, 124, 152
– chains 150
N-linked glycoproteins 24–27, 117, 150
– biosynthetic pathway 117
– classes 25
– *N*-glycosylation step 24
– pool 155
natural antibody response 100
negative ion nanospray mass spectrometry 41
Neisseria gonorrhoea 201
non-fucosylated biantennary structures 120
non-Hodgkin's B-cell lymphomas 168
non-human glycans immunogenicity 61
non-mannose receptor proteins 334
– MBL 334
novavax technology 167
NS0 cells 61, 65
nuclear magnetic resonance (NMR) 40
– oligosaccharide analysis 40
– study 93
nuclear overhauser effects (NOEs) 40
nuclear proteins 28
nucleic acid-based drugs 2

o

O-glycosylation 152, 199
– acceptor protein for 200
– bioconjugate vaccines 204–205
– characteristics of 203
– glycoengineering 205–206
– in fungi 152
– in mammals 152
– in neisseria 201–203
– in *Pseudomonas* aeruginosa 200–201
– *Neisseria meningitides* 202
– prokaryotic organisms 199
– therapeutic (human) proteins 203–204
O-linked glycan(s) 31, 153
– core structures 31
– synthesis 32
– regulation 32
O-linked glycoproteins 27–32
O-linked glycosylation 28
– pathway 53

– types 28
O-linked mucin-type glycan core structures synthesis 29
O/*N*-linked glycan chain extension 32
oligosaccharide structures 199
– bacterial oligo/polysaccharide biosynthesis eukaryotes 199
– transferred to proteins 199
oligosaccharyl transferase (OST) enzyme 25, 54
one enzyme-one linkage rule 22
optimal follicle maturation 131

p
P. aeruginosa 201
– glycan structures in 201
– *O*-glycosylation 201
Paget's disease 12
parathyroid hormone (PTH) 269
pauci-mannose structures 333
PEGylation protein 341, 350
– covalent coupling of 341
– importance of 341, 349
– schematic drawing of 342
peptidyl-α-hydroxyglycine α-amidating lyase (PAL) 253
peptidyl-proline *cis-trans* isomerases (PPIases) 277
– ubiquitous proteins 277
peptidylamidoglycolate lyase (PGL) 256
peptidylglycine α-amidating monooxygenase (PAM) 253
– anion exchange (AEX) chromatography 267
– assays for measurement of 254
– bifunctional enzyme 256
– bifunctional frog 264
– Chinese hamster ovary (CHO) cells 267
– cloning/expression of 263
– co-expression of 265
– genomic structure 257
– glycine-extended peptide 256, 265
– human bifunctional 264
– mechanism of action 255
– monooxygenase activity 254
– optimization of 261
– peptide α-amidation reaction 255
– purification of recombinant 267
– recombinant 267, 268
– species distribution of 257
– structure-activity relationships (SAR) for rat 258–260
– substrate specificity 253
– tissue-specific forms of 258
peptidylglycine α-hydroxylating monooxygenase (PHM) 253
– bifunctional enzyme 256
– frog skin 264
– redox cycling 256
periodic acid-schiff (PAS) reaction 34
periplasmic protein disulfide 282
– DsbC 282
piggyBac system 181
pituitary biosynthetic mechanisms 127
pituitary glycoprotein hormones 120
placental enzyme 328
plant cell glycoproteins 334
plant/yeast expression systems 334
plasma cells 81
– derived thrombin forms 223
– glycoproteins 28, 168
– non-proliferating 81
platelet-derived growth factor 112
poly(ethylene glycol) (PEG) 13, 348
– activatation towards amino/thiol residues 345, 346
– general properties 343
– granulocyte colony-stimulating factor 349
– growth hormone releasing hormone (GHRH) 351
– hemoglobin (Hb) 352
– immunogenic proteins 348
– interleukin-2 235
– plasma kinetics 344
– polymerization process 343
– potential effects 347–348
– practical importance 349
– process 13
– protein conjugates 342, 344, 347, 349, 350
– ring-opening polymerization 343
polyacrylamide gel electrophoresis of fluorophore labeled saccharides (PAGEFS) 39
polyhedrin promoter 174–177
polypeptide chain 27
– *N*-glycosylation potential site 27
Pompe's disease 320
post-translational hydroxylation 229
– biological function of 229
– biosynthesis of hydroxylated amino acids 231–233
– hydroxylated biopharmaceuticals 233–234
– β-hydroxylated proteins 235
– partial amino acid sequences of 214
– properties 220
– schematic of the Ca^{2+}-binding site 230
post-translational modification (PTM) 1, 7, 8, 13, 79

– acetylation 7
– ADP ribosylation 7
– amidation/sulfation 12
– γ-carboxylation 11
– disulfide bond formation 10–11
– engineering 12–13
– extending 12–13
– glycosylation 8–10
– β-hydroxylation 11
– importanace 7
– machinery 8, 80
– phosphorylation 7
– profile(s) 7, 12, 14
– proteolytic cleavage 10–11
– structure 14
post-translational processing 118
pregnancy cytotrophoblast cells 126
process analytical technology (PAT) 99
professional antigen presenting cells 94
proinsulin proteolytic processing 10
– schematic representation 10
protein disulfide bond 288
– formation 279, 288
– oxidative pathway 279
protein disulfide isomerase (PDI) 278, 288
– DsbG 285
– proteins folding in ER 289
protein folding mechanism 25
– catalysts classes 277
protein glycosylation (pgl) 17, 18, 23, 33, 52, 194
– biological importance 18
– characterization 33
– detection 33
– glycosylation enzymes 52
– *N*/*O*-linked glycoproteins relationship 23
– *N*-glycosylation pathway 169, 181
– profile 51
protein post-translational modification 2–7
proteins glycosylation 205
protein synthesis 24
proteolytic cleavage 10
– role 10
Pseudomonas strains 201
pulsed amperometric detection (PAD) 35

r

rare genetic disorder, *see* Gaucher's disease
reactive oxygen species (ROS) 255
reagent array analysis method (RAAM) 37
recombinant baculoviral vector 165
recombinant DNA technology 12
recombinant human erythropoietin (rHuEpo) 297
– amino acid 298, 305
– carbohydrate glycoforms 302
– crystal structure of Epo bound 308
– darbepoetin alfa 303
– endogenous Epo (eEpo) 298, 299
– ESAs 310
– glycopeptide mapping of 299
– glycosylation analogs 301, 306
– mammalian cells 297
– Ser/Thr residues 301
– serum Epo (sEpo) 299
– sialic acid-containing carbohydrate 303
– three-dimensional model of 308
– urinary Epo (uEpo) 299
recombinant human glucocerobrosidase 156
recombinant human granulocyte macrophage-colony stimulating factor (rhGM-CSF) 153
recombinant monoclonal antibodies 98–100
recombinant protein 51, 59, 68, 80, 165, 166
– expression 149
– glycosylation 68
– therapeutic glycoproteins 166
– TIMP 1 59
recombinantly-expressed enzyme 327
– carbohydrate-remodeled 327
– cerezyme 327
red blood cell (RBC) 297
– Epo 297
remodeling enzymes 330
– α-mannosidase I 330
rheumatoid factor (RF) 90
rough endoplasmic reticulum (RER) 22, 23, 24
– membrane 24, 25
– secretory pathway 23, 24

s

Saccharomyces cerevisiae 150, 264
Schizosaccharomyces pombe 152
second-generation ERT development 327
– CHO cell-expressed recombinant human GCase-cerezyme production 327
– enzymes biochemical comparison 327–330
– recombinant technology 327
serum acid phosphatase activity 326
serum protein mannan binding lectin (MBL) 94
Sf9 cells 169, 175, 179, 184
– baculovirus-infected 169
– endogenous processing GlcNAcase 184
sialic acid-9-phosphate synthase (SAS) 177
sialyl-Lewis X glycan structures 54
sialyl transferase (ST) enzyme 52, 54, 60, 183
signal recognition particle (SRP) 24

sodium dodecyl sulfate-polyacrylamide gel electrophoresis (SDS-PAGE) 36, 119
soluble periplasmic protein 282
– DsbC 282
spinner cell culture system 327
steroid hormone production 109
Streptococcus pyogenes 89
Streptomyces lividans 265
structural microheterogeneity 114–121
sugar processing enzymes 177
surface plasmon resonance (SPR) 331
systems biology 100

t

tangential flow ultrafiltration (TFF) 267
targeting mechanism 172
therapeutic antibody 97
– cetuximab 97
– glycoform profile 97
therapeutic protein(s) 1, 6, 8, 11, 12, 13, 149, 173
– biopharmaceutical sector 1
– examples 11
– glycosylation characteristics 8–9
– levemir 13
– post-translational modifications 1, 6
– production 154
– remodeling yeast glycosylation 154
thioredoxin reductase (trxB) 286, 287
– gene coding 287
Thomsen–Friedenreich (T/TF) antigen 29
thrombin-thrombomodulin complex 222
– activated form of PC (APC) 222
– zymogen 222
thrombolytic therapy 223
thrombopoietin 349
thyroid-stimulating hormone (TSH) 111
– glycoforms 129
thyrotropin-releasing hormone (TRH) 128
TIM barrel catalytic core 321
time of flight (TOF) analysis 41
tissue plasminogen activator (tPA) 277, 287
total correlation spectroscopy (TOCSY) 40
trans-Golgi 58
– glycosylation process 58
– network 12, 24
transgenic glycosyltransferases 179
– galactosyltransferase 179
– *N*-acetylglucosaminyltransferase II 180
– sialyltransferase 179
transgenic insect cell line 178, 179, 181
– proof of concept 178
– transgenic glycosyltransferases 179
– transgenic sugar processing genes 180
– transposon-based systems use 181
transmembrane protein 91
trimming enzymes 22
– glycosidases 22
2,4,6-trinitrobenzene-1-sulfonate (TNBS) 39
Trypanosoma cruzi 176

u

UDP-*N*-acetylglucosamine (UDP-GlcNAc) 176, 177
– epimerase levels 177
undecaprenol-pyrophosphate (Und-PP) lipid 199
ureidoglycolate lyase (UGL) 256
US food and drug administration (FDA) 326

v

vitamin K-dependent proteins 212, 218
– coumarin-based antagonists 218
– dependent proteins 212
– factor IX (FIX) 229
– factor X (FX) 229
– hemostasis function 223
– liver 219
– polypeptides 217
– redox cycle 218

w

Walker–Warburg syndrome (WWS) 154
weak anion exchange high-performance liquid chromatography (WAX-HPLC) 38
western equine encephalitis virus 167
Wollinella succinogenes 194

x

Xenopus laevis 264
Xigiris, *see* activated protein C

y

yeast engineering 149
yeast expression systems 334
yeast glycosylation 149
– pathway 155